Artificial Intelligence and Digital Twin-Based Big Data Analytics for Blockchain-IoT Enabled Applications

Editors

Amit Kumar Tyagi
National Forensic Sciences University
Gandhinagar, Gujarat, India

Shabnam Kumari
SRM Institute of Science and Technology
Chennai, India

CRC Press
Taylor & Francis Group
Boca Raton London New York

CRC Press is an imprint of the
Taylor & Francis Group, an **informa** business
A SCIENCE PUBLISHERS BOOK

First edition published 2027
by CRC Press
2385 NW Executive Center Drive, Suite 320, Boca Raton FL 33431

and by CRC Press
4 Park Square, Milton Park, Abingdon, Oxon, OX14 4RN

CRC Press is an imprint of Taylor & Francis Group, LLC

Library of Congress Cataloging-in-Publication Data (applied for)

ISBN: 978-1-041-09148-6 (hbk)
ISBN: 978-1-041-10467-4 (pbk)
ISBN: 978-1-003-65519-0 (ebk)

DOI: 10.1201/9781003655190

Typeset in Times New Roman
by Prime Publishing Services

Preface

In the rapidly evolving landscape of Industry 5.0, today the convergence of Artificial Intelligence (AI), Digital Twins, Big Data Analytics, Blockchain, and the Internet of Things (IoT) provides a transformative era for intelligent, secure, and resilient applications. This book explores with this synergistic ecosystem, where Digital Twins act as dynamic virtual replicas of physical assets, continuously synchronized with real-time IoT sensor data to enable precise simulations, predictive maintenance, and optimization. Big Data Analytics platforms, such as Apache Spark and Hadoop, process the vast amount of data streams using AI techniques like machine learning, deep learning, and neural networks, convolutional neural networks (CNN) for anomaly detection, pattern recognition, and prescriptive insights that drive autonomous decision-making.

Further, blockchain is integrated with this framework by providing decentralized, immutable ledgers and smart contracts, which ensure tamper-proof data provenance, enhanced cybersecurity, and trustless interactions across distributed IoT networks which are vulnerable to single points of failure. This book discusses some real-world applications like: in manufacturing, AI-enhanced twins reduce downtime by up to 30% through predictive analytics; in healthcare, patient-specific models simulate personalized treatments with blockchain-secured data sharing; and in supply chains, IoT-fed analytics optimize logistics resilience amid disruptions. However, there are several challenges also faced in the fusion of these technologies, like interoperability among heterogeneous devices, privacy of shared data among systems, edge computing scalability, and ethical AI governance, etc. These are some challenges which we tried to addresses through innovative architectures like federated learning on blockchain and hybrid cloud models in this book.

This book will be useful for researchers, engineers, policymakers, and industry leaders with practical examples, and case studies. Moving with this integration of these cutting edge technologies, we did not only promise the better efficiency, and sustainable future (with a wealthy economic growth of a nation) but also makes the way for fully autonomous systems to be used in smart cities, energy grids, and beyond, which will shape a future where digital and physical convergence redefines the way of living by human beings. Hence, we expect this book will be a great resource for the research and scientific community to get new thoughts when such evolving technologies merge together for betterment of society and solve modern day problems in different sectors.

Lastly, we hope that this book will be loved by research community and any corrections are heartly welcome for improving the quality of this book.

Happy learning.

Acknowledgement

First of all, we would to extend our gratitude to our family members, friends, and supervisors, who stood with us throughout the completion of this book. Also, we would like to thanks our almighty "God" who makes us to write this book. We also thank Science Publishers, CRC Press (who provided their continuous support and their valuable time) as well as our colleagues who have provided their useful inputs and guidance for our book on areas like Artificial Intelligence, Digital Twin based Big Data Analytics and Blockchain-IoT enabled Applications.

We would also like to thank our Respected Madam, Prof. G Aghila, Prof. Siva Sathya, our Respected Sir Prof. Ajith Abraham, Prof. Aswani Kumar Cherukuri, Prof. (Dr.) Naveen Chaudhary, Prof. (Dr.) Purvi Pokhariyal for giving their valuable inputs and helping us in making the book rich and comprehensive in its content.

Dr. Amit Kumar Tyagi
Shabnam Kumari

Contents

Section 1
Basic and Fundamentals

1

Digital Twins in the Era of IoT and AI

Megha Singh[1,*] and *Amit Kumar Tyagi*[2]

1. Introduction

Digital Twin technology is one of the innovations of the past few years that bridges the physical and digital worlds in amazing new ways. Digital Twins began life in engineering and manufacturing as copies of physical assets and have now evolved into virtual copies of sophisticated systems like life machines, industrial devices, and even human organs. This development has been driven by two key drivers: the proliferation of the Internet of Things (IoT) and the advancement of Artificial Intelligence (AI) technologies (Mihai et al. 2022, Zhang and Tao 2020).

The most striking characteristic of a Digital Twin is its real-time exchange of data with its physical twin, and this is made possible by the ubiquitous nature of IoT sensors and streams of data. IoT devices like environmental sensors, RFID tags, and vibration sensors supply the lifeblood data that make up the virtual model of the twin. The real-time data stream allows Digital Twins to mirror the current state of their equivalents and execute capabilities like immersive simulations, proactive maintenance, and operation optimization. In factories, for instance, active IoT-integrated machinery merged with Digital Twins has been known to enhance business productivity by as much as 30% in terms of decreased downtimes and maintenance costs.

But raw data alone is insufficient. Digital Twins leverage AI and machine learning to forecast failures, detect trends, and simulate "what-if" situations in an effort to unlock real insight. Digital Twins are made intelligent with the capacity to make decisions on their own by AI algorithms, especially deep learning and neural networks. They transform passive models into dynamic digital assistants by precisely detecting anomalies, predicting potential problems, and proposing corrective actions. AI-powered twins are able to predict weather effects, optimize resource distribution, and adapt to shifting conditions in the context of smart agriculture or energy networks, contributing to sustainability and resilience.

Industry 4.0 focuses on the digitalization of manufacturing and industrial processes through the integration of Artificial Intelligence (AI), the Internet of Things (IoT), and cyber-physical systems. The concept of the digital twin has become increasingly popular in the short term. Essentially, a digital twin is a dynamic virtual replica of a system, process, or physical asset that continually

[1] Department of Computer Science and Engineering, Oriental University, Indore.
[2] National Institute of Fashion Technology, Delhi. Email: amitkrtyagi025@gmail.com
* Corresponding author: dr.meghasinghkosta@gmail.com, meghasingh@orientaluniversity.in

DOI: 10.1201/9781003655190-2

evolves due to real-time data and advanced modeling. Digital Twins serve as the "central nervous system" of Industry 4.0 when it comes to smart manufacturing, enabling systems to be more autonomous, adaptive, and responsive to change.

Connectivity is the pillar that provides for the seamless transfer of data between the digital and physical assets, courtesy of technologies such as 5G, cloud computing, and edge intelligence. As real-time synchronization is critical to responsiveness and insight generation, a digital twin diminishes in value if it does not have robust connectivity. Sophisticated machine learning algorithms further augment the capabilities of digital twins by identifying patterns and anomalies that are not perceptible to human operators, facilitating proactive intervention and more informed decision-making.

In this manner, digital twins change the way individuals leverage technology as opposed to just optimizing operations. Companies can go from reactive to predictive methods, operators can receive predictive warnings instead of alarms, and engineers can simulate and visualize processes prior to their deployment. Digital twins will become even more important as Industry 4.0 develops, helping to turn sectors into intelligent, self-optimizing ecosystems.

Smart cities are arguably the best examples of how IoT and AI work together in digital twins. City-scale twins, made possible by sensor networks and embedded intelligence, are being used by urban planners more and more to manage utilities, improve public safety, and model traffic patterns. For instance, Virtual Singapore uses real-time sensor data to predict flood zones, model traffic patterns, and optimize energy use. A growing trend toward data-driven urban governance is also evident in the use of AI-enhanced Digital Twins by cities like Singapore, Houston, and Copenhagen to control heat islands, lower pollution, and improve climate resilience.

Even with the excitement, implementation is still difficult. Strong IoT infrastructure, scalable cloud platforms, and interoperable systems are necessary for effective deployment. Concerns about data governance, security, and transparency are also raised by the integration of AI, particularly in public-sector deployments. Cultural changes are also necessary for organizational adoption; teams must learn to work across disciplines and interpret insights produced by AI.

In the future, research keeps pushing the envelope. Current research introduces ideas like the Age of Digital Twin (AoDT) to gauge timeliness and investigates innovative architectures that enhance synchronization and lower latency, such as drone-assisted IoT data collection for high-fidelity twins. According to surveys, security and privacy are crucial areas that require innovation, and frameworks that are robust to decentralized and semantically driven twin ecosystems are suggested.

We are essentially at a turning point in the development of digital twins, which are becoming intelligent systems that anticipate needs, maximize performance, and support intricate, real-world scenarios thanks to the Internet of Things and Artificial Intelligence. They signal a change from static models to living digital companions and from reactive maintenance to proactive management. Digital twins hold promise for transforming cities, industries, and even ourselves in profoundly impactful and human ways as the ecosystem continues to change due to organizational transformation, technological innovation, and societal concerns.

2. What is a Digital Twin?

A digital twin is a computerized representation of an actual thing, procedure, system, or setting that is updated regularly with information from sensors and other sources. A digital twin dynamically reflects what is happening to its physical counterpart, unlike static computer models, in real time. Without constantly interacting with the real asset, this idea enables engineers, managers, and decision-makers to comprehend, model, and optimize performance. Digital twins, which have their roots in manufacturing and aerospace, are currently gaining traction in a variety of industries as businesses look to increase productivity, cut expenses, and provide better customer experiences (Tao et al. 2019).

Various types of digital twins exist based on their scope and intended use. The Product Digital Twin, which focuses on reproducing a single product or component, is the most basic type. To test performance and find defects prior to production, automakers, for example, build digital twins of their engines. The Process Digital Twin is a more sophisticated kind that simulates production lines and workflows. Businesses can anticipate maintenance requirements or identify bottlenecks by simulating processes. System Digital Twins represent groups of linked assets and increase in complexity. A digital twin of an entire aircraft that records the interactions between the avionics, electrical systems, and engines could serve as an example. Human Digital Twins, or digital depictions of a person's body or organs, are being developed in emerging industries like healthcare, which could revolutionize personalized treatment (Bruynseels et al. 2018).

To construct and run a digital twin, a few key elements are required. The physical thing—the thing or procedure that needs to be mirrored—lays the groundwork. The digital twin, which is based on simulation information, historical records, and engineering blueprints, follows. Internet of Things connectivity and sensors are employed to capture real-time information, which is then input into the twin through secure data pipelines. An analytics engine that utilizes physics-based models, predictive algorithms, and machine learning processes the data. Finally, an interface layer facilitates proper monitoring and communication with the twin, e.g., a dashboard or 3D visualization (Boschert and Rosen 2016).

One of the most important aspects of a digital twin is its major features. Firstly, it provides real-time monitoring, which allows users to monitor the health, usage, and state of the asset as conditions evolve. Secondly, by running simulations, digital twins are able to explore what-if scenarios and predict how systems will behave under different conditions. They also offer lifecycle integration, which accommodates an asset's full cycle from design and production to operation and, ultimately, disposal or recycling. Bidirectional data flow is another critical element. This means that optimizations or commands in the virtual world can trigger the physical system to act as well as update the twin in the real world, in addition to driving the twin.

The system architecture of a digital twin system is given in Figure 1, which also illustrates how the digital twin harvests real-time Internet of Things data from the physical world for analysis, simulation, and optimization purposes. Feedback and insights are then passed back in order to

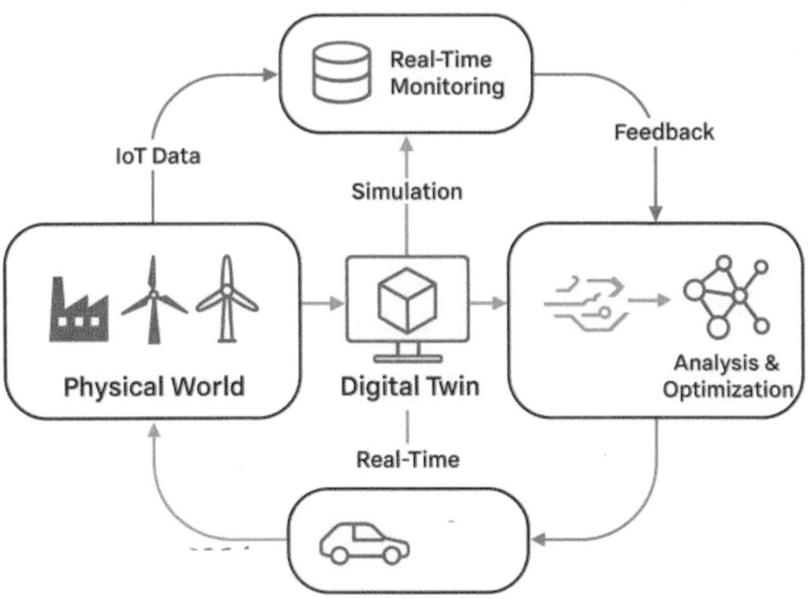

Figure 1: Architecture of Digital Twin.

optimize performance, enabling adaptive control, decision-making, and continuous monitoring under dynamic conditions.

The use of digital twins has many benefits. Since stakeholders have access to current, accurate data and simulations to inform planning, one significant advantage is better decision-making. Through proactive maintenance scheduling and failure prediction, digital twins can also lower operating costs by preventing unplanned downtime. They facilitate quick virtual testing of concepts prior to the construction of costly prototypes, which accelerates innovation. Digital twins can also assist businesses in cutting waste and optimizing energy use, both of which contribute to sustainability objectives. Consumers find that goods and services based on insights from digital twins are frequently more dependable, effective, and customized to meet their needs (Tao et al. 2019).

These days, digital twins have a lot of interesting uses and examples. In the aerospace industry, Rolls-Royce employs digital twins to monitor the condition of operational aircraft engines, anticipating when components will require replacement and averting expensive malfunctions (Boschert and Rosen 2016). A nationwide 3D twin has been constructed as part of Singapore's Virtual Singapore project to model infrastructure planning, energy consumption, and traffic in smart cities. Companies such as Philips are investigating patient-specific cardiac twins in the healthcare industry to model the effects of treatment. Digital twins are also widely used in the automotive sector; Tesla uses them to optimize vehicle performance and provide over-the-air updates. Digital twins have been used in manufacturing by Siemens and GE to optimize production lines and industrial machinery.

To sum up, a digital twin is a dynamic, ever-changing digital replica of the real world. They bring together human understanding, simulation, and live data to improve understanding and decision-making. Digital twins are increasingly sophisticated, increasingly mainstream, and increasingly central to digital transformation strategies across a range of industries as technology evolves. They are revolutionizing the way we plan, build, operate, and maintain the systems that underpin modern life.

3. Digital Twin Architecture and Lifecycle

The value of a digital twin comes from both its virtual replica and the underlying architectural structure that governs its creation, utilization, and innovation. To successfully deploy a digital twin in sectors such as manufacturing, healthcare, energy, and urban infrastructure, it is essential to understand its architecture and lifecycle (Tao et al. 2018).

3.1 Conceptual Architecture

Layered architectures have been proposed by various researchers as a means of organizing the functional components of a digital twin. Five primary layers constitute a commonly accepted view (Tao et al. 2019, Kritzinger et al. 2018). The architecture of a digital twin represents an intricate, multi-layered system in which data, models, algorithms, and interfaces converge to emulate and optimize actual operations. It is far more than a computerized replica of a physical entity. This conceptual framework enables real-time interaction of the digital and physical worlds through continuous feedback and iteration. The digital twin concept relies on several key elements and is organized into five distinct layers, each of which is critical to maintaining the twin's intelligence, capability, and integrity.

A digital twin's core components are as follows:

(1) the physical device or asset in the real world (e.g., machine, vehicle, or building).

(2) its virtual counterpart, a simulation that mirrors the physical entity's attributes, behavior, and state.

(3) the data pipeline, consisting of sensors, communication protocols, and cloud services for enabling constant flow of data.
(4) algorithms and analytics, especially those that are AI- and machine learning-driven, to digest the new data.
(5) the user interface, through which engineers, operators, and decision-makers can interact with the digital via dashboards or simulation environments (Tao et al. 2018).

A five-layer architecture that embodies the functionality and purpose of a digital twin system can be employed to design this architecture (refer Figure 2). Every layer contributes to the overall intelligence and utility of the digital twin by adding to the functionality of the layer below it and being dependent on it.

1. **The Physical Layer:** This lowermost layer represents the real-world systems being replicated. It comprises embedded devices, sensors, and actuators that generate data on an ongoing basis from the external world. The digital twin does not have a source of current and past data without this lower layer (Fuller et al. 2020). This layer contains the real process or asset (e.g., a production line, a robot arm, or a turbine). Through embedded Internet of Things sensors, the physical thing generates a stream of operational and environmental data.
2. **The Data Layer:** This layer collects, stores, and pre-processes physical layer data. This includes unstructured environmental or log data, structured enterprise data, and raw sensor data. For analytics and modeling purposes, it ensures that temperature, vibration, and pressure data are validated, filtered, and formatted appropriately.

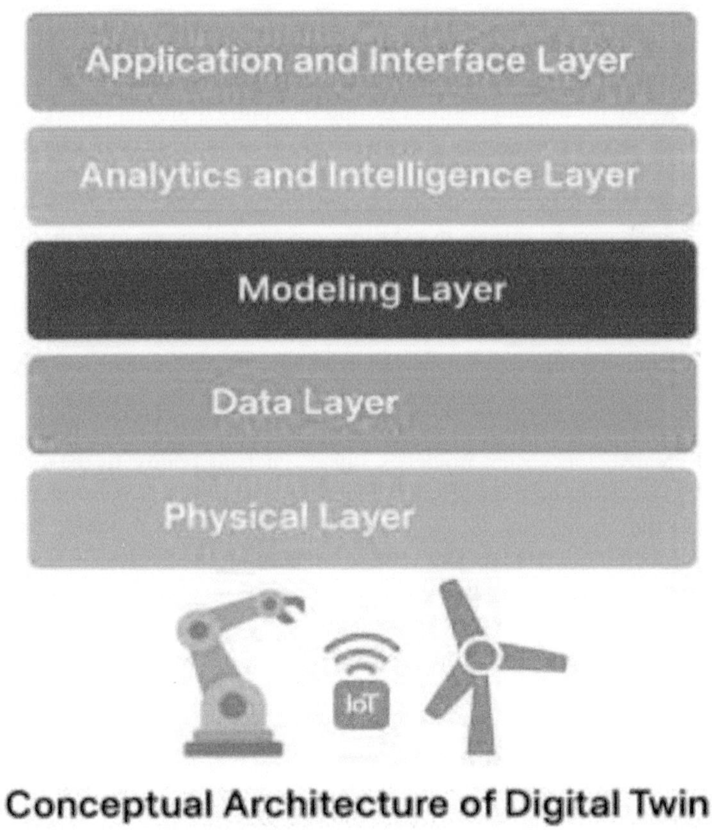

Figure 2: Conceptual architecture of digital twin.

3. **The Modeling Layer:** This layer is composed of mathematical, physics-based, or artificial intelligence-driven models that simulate the behavior of the actual system. They may range from simple digital replicas to complex predictive simulations. Importantly, this layer gains accuracy and value for prediction and decision-making over time through machine learning (Glaessgen and Stargel 2012).
4. **The Analytics and Intelligence Layer:** Advanced algorithms in this layer analyze the data and simulate different situations. Predictive analytics, anomaly detection, optimization processes, and what-if analytical tools are all part of this layer. Through its ability to transform the data into actionable insights, it enables organizations to foresee potential failures or bottlenecks and make informed decisions.
5. **The Application and Interface Layer:** The final layer serves as the interface with other systems and human users. Dashboards, mobile applications, augmented reality (AR) interfaces, and APIs that provide access to twin capabilities are all instances of this. This layer is vital for providing the digital twin with a way to be accessed and utilized by numerous departments such as engineering, maintenance, and operations.

Organizations can ensure modularity, scalability, and clarity in development and deployment by structuring a digital twin in terms of these five layers. More importantly, this tiered structure corresponds well to modern-day edge-cloud hybrid computing paradigms, enabling long-term analytics and storage to be processed in the cloud while time-sensitive computation can be executed at the edge (Kritzinger et al. 2018).

3.2 Lifecycle Phases

A digital twin's lifecycle is not a static procedure terminated in deployment. Rather, it is an active continuum that closely mirrors the manner in which its physical counterpart evolves. Organizations are better able to prepare for the long-term value and flexibility of digital twins by understanding the lifecycle in separate steps. Grieves and Vickers (2017) view these phases—Design and Modeling, Implementation and Integration, Operation and Optimization, and Evolution and Retirement—as needing unique tools, strategies, and interdisciplinary collaboration.

The digital twin is created during the Design and Modeling phase of the process. Engineers construct a virtual replica of the physical asset from CAD files, engineering drawings, and historical performance data. It is an operating simulation and not merely a visual model that captures finite element analysis (FEA) and computational fluid dynamics (CFD), respectively. Prior to physical prototyping, the simulations are helpful in understanding the stress, flow, thermal, and structural behavior. By then, the digital twin becomes a refuge for creativity, experimentation, and riskless iteration in design.

After being created, the virtual model is linked back to reality in the Implementation and Integration phase. Data may be transferred in real time owing to IoT sensors and communication protocols that link the digital twin to its counterpart in the physical world. To ensure accuracy, calibration and validation are crucial. The value of the digital twin is lost if it cannot represent the real asset's behavior correctly. Consequently, domain experts must calibrate parameters and validate outcomes in actual operating conditions (Qi et al. 2019).

At the Operation and Optimization stage, a digital twin's highest value is often achieved. It is a real-time feedback system here, monitoring performance, detecting anomalies, and predicting failures. With time, the twin becomes a smart system that can schedule predictive maintenance, recommend operational adjustments, and enhance efficiency through machine learning and advanced analytics.

Finally, the digital twin must evolve as the physical asset is retired, replaced, or refitted. The asset modifications must be incorporated into the digital representation in the Evolution and Retirement

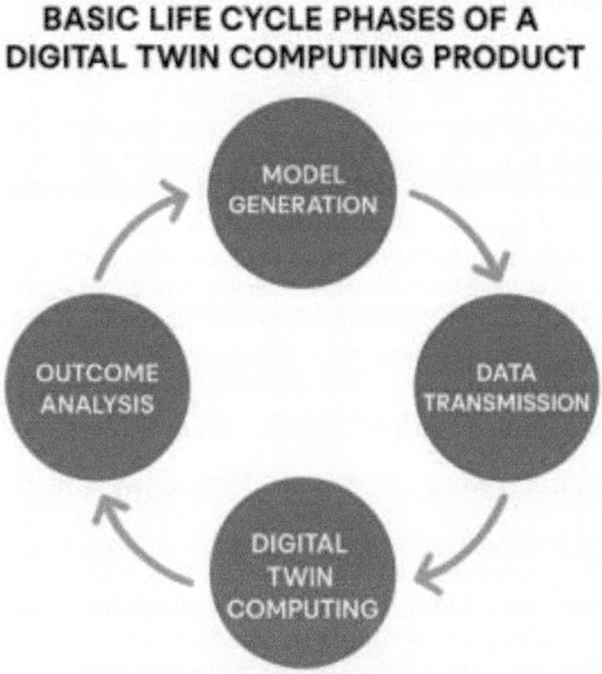

Figure 3: Life Cycle Phases.

phase. Alternatively, the twin may be retained and utilized as a source of operating expertise for future system redesigns, engineering endeavors, or regulatory examinations (Shao et al. 2020).

By understanding and managing this lifecycle (refer Figure 3), digital twins can continue to remain accurate, up-to-date, and capable of delivering long-term value.

3.3 Standards and Protocols in Digital Twin Systems

Standardized communication protocols and frameworks are needed to ensure seamless interoperability and real-time communication among the physical and digital layers of a digital twin. Two of the most widely used in an industrial environment are OPC-UA and MQTT. Both of these protocols enable reliable and scalable data exchange necessary to keep physical assets and their corresponding digital counterparts in continuous synchronization.

MQTT is a lightweight publish-subscribe messaging protocol suited for environments with high latency and constrained bandwidth. IoT applications significantly benefit from its simplicity and efficiency, particularly when data from sensors need to be transmitted regularly and securely from edge devices to cloud-based digital twins (Naik 2017). OPC-UA, however, provides a more complex but stable solution that is capable of semantic data modeling, platform independence, and built-in security. It is particularly useful in industrial automation, where hierarchical organization and full contextual information are needed (Mahmoud et al. 2021).

By integrating these protocols into digital twin frameworks, Industry 4.0 solutions are enabled that are scalable and future-proof, in addition to efficient data exchange and interoperability across heterogeneous systems.

3.4 Technology Stacks and Platforms for Digital Twins

The selection of robust technology stacks and platforms that are capable of supporting modeling, data integration, real-time analytics, and scalability is fundamental to the successful adoption of digital twins. To develop comprehensive digital ecosystems, these stacks typically combine IoT frameworks, data processing assets, simulation engines, and visualization interfaces.

Popular platforms offering end-to-end solutions for digital twin creation are Siemens' MindSphere, PTC's ThingWorx, and GE's Predix. These platforms integrate real-time analytics, visualization dashboards, and IoT data ingestion; they often include AI and machine learning to enhance predictive ability (Tao et al. 2019). For sophisticated systems such as smart buildings and city infrastructure, Microsoft Azure Digital Twins provides a highly scalable environment with graph-based modeling, IoT Hub integration, and AI services.

IoT platforms (such as AWS IoT Core, Azure IoT Hub), data processing frameworks (such as Apache Kafka, Spark), and modeling/simulation tools (such as MATLAB/Simulink, ANSYS) are typical parts of technology stacks. AR/VR interfaces and 3D modeling engines such as Unity or Unreal Engine are often utilized for visualization and interaction.

Digital twins can evolve from static models to smart, real-time systems because of these stacks, which enhance operational effectiveness and decision-making across a range of sectors.

4. Types of Digital Twins: Product, Process, and System

Based on the size and complexity of the virtual representation, digital twins are classified into various categories. Product, process, and system digital twins are separated based on the most popular typology (Fuller et al. 2020). Ranging from the simulation of complex, interdependent networks of operations to a single asset replication, this categorization covers the increasing size of abstraction. Every kind of IoT is characterized by a particular purpose and requires various modeling approaches, IoT structures, and artificial intelligence expertise.

A digital replica of a specific physical item, such as an engine, turbine, or industrial part, is referred to as a product digital twin. A product twin, unlike generalized system models, is focused on the unique identity and condition of an individual item throughout its lifecycle. The full geometry, operational parameters, and degradation behavior are all stored for the physical product (Shao and Kibira 2018).

The biggest advantage of a product digital twin is its ability to combine real-time data with high-fidelity engineering simulations, rendering static design models dynamic and constantly changing entities that reflect the asset in its current state.

CAD models, finite element analysis (FEA), and computational fluid dynamics (CFD) simulations are often employed as the foundation for product twins. Stress, heat flow, and structural behavior under different conditions are all modeled by these physics-based simulations. Nevertheless, the incorporation of real-time IoT sensor information is what makes a typical simulation become a digital twin. The twin is able to reflect its status in real time due to onboard sensors that monitor temperature, pressure, vibration, and load. The sensors also flow telemetry data into the digital world. The integration allows engineers to monitor performance, identify anomalies early, and simulate various operating scenarios.

The integration of artificial intelligence (AI) in product twins is a very important enhancement. Some of the most important metrics, such as remaining useful life (RUL), are forecasted using supervised learning models that have been trained on past performance history. Machine learning algorithms, for instance, can estimate when the bearings will fail and analyze vibration patterns in rotating equipment (Lei et al. 2018). Convolutional neural networks (CNNs), one of the deep learning methods, excel at interpreting complex, high-dimensional sensor data and can find subtle degradation indicators that traditional models may miss.

Product digital twins' revolutionary potential is evidenced by their real-world deployments (refer Table 1). General Electric's digital twins of aircraft engines are a case in point; they fine-tune performance and reduce downtime by means of AI-fueled analytics and constant IoT telemetry. By anticipating failures prior to occurrence and tracking thousands of variables in real time, such as

Table 1: System issues on the basis of Type of Product.

Type of Product	System Design Considerations	Technology Stack	Platforms/Tools
Industrial Equipment	Real-time monitoring, predictive maintenance, edge analytics	IoT sensors, OPC UA, MQTT, Edge computing, AI/ML, Cloud platforms	Siemens MindSphere, PTC ThingWorx, Azure IoT, AWS IoT, Predix
Healthcare Devices	Patient-specific modeling, secure data transmission, compliance	FHIR, HL7, AI/ML for diagnostics, HIPAA-compliant cloud, AR/VR integration	Philips IntelliSpace, GE Health Cloud, Microsoft Cloud for Healthcare
Smart Buildings/ Cities	Interoperability, sensor integration, spatial modeling	BIM, GIS, 5G, LPWAN, AI/ ML, Edge-cloud hybrid	Bentley iTwin, ESRI ArcGIS, Azure Digital Twins, CityIQ
Automotive/ Autonomous Vehicles	Real-time simulation, low latency, V2X communication	CAN bus, ROS, V2X protocols, AI/ML, Simulation engines (Unity, Unreal), Digital Thread	NVIDIA Omniverse, Siemens Simcenter, AWS TwinMaker
Aerospace Systems	Complex system integration, reliability, simulation accuracy	High-fidelity CFD, FEM, PLM systems, Data lakes, AI/ML, Federated learning	Dassault 3DEXPERIENCE, ANSYS Twin Builder, IBM Maximo
Energy Grids & Utilities	Resilience, demand forecasting, distributed asset management	SCADA, Digital SCADA, AI for load prediction, Blockchain (for audit), Edge computing	GE Digital APM, ABB Ability, Hitachi Lumada
Consumer Electronics	Usage behavior analysis, OTA updates, user feedback loops	Mobile app integration, Cloud analytics, Edge AI, RESTful APIs, Firebase	Google Firebase, AWS Greengrass, Azure IoT Central
Manufacturing (Smart Factories)	MES/ERP integration, process optimization, cyber-physical feedback	OPC UA, Digital Thread, ML/AI, Robotics APIs, Edge gateways, IIoT platforms	Rockwell FactoryTalk, Siemens NX, Bosch IoT Suite, SAP Leonardo
Logistics/Supply Chain	Real-time tracking, environmental data logging, blockchain	RFID/QR, GPS, LPWAN, Digital Ledger Tech, IoT analytics, Cloud dashboards	IBM Watson IoT, Oracle SCM Cloud, Azure Blockchain Service
Agritech Systems	Weather adaptation, soil-health modeling, yield prediction	LPWAN, GIS, AI/ML, Edge computing, Remote sensing, Drones, APIs	Trimble Ag Software, John Deere Ops Center, Azure FarmBeats

fuel flow, fan speeds, and heat signatures, these twins facilitate condition-based maintenance as a replacement for scheduled inspections (Boschert and Rosen 2016).

A digital twin of a product truly becomes a smart, adaptive model of a physical item that grows, learns, and evolves as the real object ages or encounters new operating conditions. This generates a digital history of the asset's entire lifetime in addition to operational performance.

A process digital twin replicates the dynamic processes and series of operations across an entire system, as opposed to product digital twins that mimic individual physical assets. It illustrates how a coordinated network of machines, people, and systems transforms inputs—such as materials, energy, or information—into outputs (Shao et al. 2020). In sectors such as manufacturing and logistics, where performance is reliant on throughput, efficiency, and process integrity, such digital twins are very valuable.

Process digital twins are applied in manufacturing to simulate and make process optimizations across a broad variety of operations, such as workforce deployment, materials flow, and assembly line sequences. A process twin may, for instance, monitor and simulate the way different workstations communicate, the duration of each operation, and where bottleneck areas are in an auto factory. Computer models have the ability to determine the impact of changes in equipment, scheduling, or layout on overall system performance using simulation techniques such as system dynamics (SD), agent-based modeling, and discrete event simulation (DES) (Tao et al. 2018).

To effectively represent real-world conditions, real-time connectivity is vital for process twins. Process equipment, including robots, conveyors, and environmental controls, can provide real-time operating data to the digital twin due to industrial IoT protocols such as OPC UA and MQTT. This allows the twin to serve as a real-time decision-support system, not just a static simulation. A high-profile example is the application of IoT sensors in the manufacture of pharmaceuticals to monitor variables such as humidity, flow rates, and temperature. By feeding data into the process twin, these inputs optimize yield and quality while helping to ensure compliance with strict regulatory requirements.

A critical element of the power of process digital twins is artificial intelligence. While satisfying safety and throughput demands, reinforcement learning algorithms are able to find the optimal configurations to minimize energy usage, cycle time, and flaws (Zhang et al. 2021). Unsupervised learning techniques also help identify process behavior drifts and anomalies prior to downtime or quality issues. Through enabling proactive control strategies, these insights contribute to the resilience and responsiveness of production systems.

Large industrial companies such as Siemens have successfully used process digital twins in chemical manufacturing environments, modeling continuous processes and revealing latent inefficiencies with machine learning and IoT-enabled platforms (Tao et al. 2019). Process twins will play a key part in making adaptive, smart, and sustainable operations a reality as digitalization further reshapes the industry.

The System Digital Twin, a virtual duplicate of complete ecosystems consisting of interconnected people, infrastructure, processes, and products, is the highest level of digital twin maturity. Such twins are complex integrations across multiple domains, where changes to one of the subsystems can affect the overall system. They are not limited to a single asset or workflow. Smart grids, intelligent transportation, smart cities, and even enterprise-level manufacturing networks are some examples (Kritzinger et al. 2018). Knowledge of dynamic and non-linear interactions among different components is important to achieve operational effectiveness, resilience, and sustainability in these environments.

Gigantic streams of heterogeneous information from distributed sensors, platforms, and devices are amalgamated by system digital twins. Constructing such a twin requires intelligent IoT structures that can process data from hundreds of millions of endpoints in real time. Whereas 5G networks offer low-latency, high-bandwidth data transfer critical to mission-critical uses like autonomous traffic control or emergency response, Time-Sensitive Networking (TSN) ensures synchronized communication among systems (Taleb et al. 2017). These technical underpinnings provide a level of situational awareness beyond traditional siloed operations that facilitates system-wide coordination.

AI is necessary to turn this data into intelligence. Probabilistic reasoning helps manage uncertainty, graph neural networks capture relationships between subsystems, and deep learning extracts patterns in sensor data. For instance, a system twin in intelligent power grids manages dynamic consumer demand and fluctuating renewable energy availability by optimizing real-time energy distribution based on deep reinforcement learning (Batty, 2018). System twins need to continuously adapt to variations across the complete ecosystem, unlike product or process twins, which operate on more localized feedback loops.

Smart cities are one of the most important applications of System Digital Twins. These digital simulations replicate interactions between public services, waste, energy use, transportation, and air quality. The twin is fed with real-time data from embedded sensors in public infrastructure, utilities, and roadways. The effect of new zoning laws or infrastructure investments is then calculated by city planners and policymakers based on AI-powered simulations to model traffic and predict pollution incidents (Lu et al. 2020). In this context, the digital twin exists as a decision-support tool for joint decision-making, as well as a technical resource, to help human stakeholders master complexity through wisdom and foresight.

Digital twins capture the growing ambitions of digital transformation as they evolve from individual components to interconnected processes and eventually to combined systems. The pinnacle of this process is system twins, which make adaptive control, predictive governance, and end-to-end optimization at scale possible. Overall, digital twins of products, processes, and systems exhibit a range of increasing complexity and scale. System twins provide an overall understanding of an entire ecosystem or enterprise behavior, whereas product twins focus on the status and functionality of individual assets. Process twins are concerned with coordinating activities. For each of them to deliver, some combination of data architectures, IoT protocols, and AI methods is required (refer Table 2). Successful digital twin solutions that cater to specific industrial domains' specific needs are best achieved through knowledge of these distinctions. Here is a table depicting the differences between all three types of digital twin systems:

Table 2: Difference between Product, Process, and System Digital Twins.

Feature	Product Digital Twin	Process Digital Twin	System Digital Twin
Definition	Virtual model of a physical product or component	Digital replica of a specific process or workflow	Integrated model of interconnected products and processes
Focus Area	Design, development, usage, and performance of a product	Operational performance, process efficiency, and optimization	System-wide behavior, coordination, and holistic optimization
Scale / Scope	Component or unit level	Task or activity level within operations	Enterprise or multi-system level
Data Sources	CAD models, sensor data, manufacturing data	IoT sensors, control systems, process logs	Aggregated data from multiple products, processes, and environments
Objective	Improve design, quality, and lifespan of a product	Optimize workflows, reduce errors, increase process throughput	Maximize system-level performance, resilience, and decision-making
Time Horizon	Short-to-mid term (design to deployment phase)	Real-time or near-real-time monitoring and adjustment	Long-term performance and strategic planning
Technology Used	CAD/CAE, PLM, IoT, Simulation tools	SCADA, PLCs, MES, Workflow analytics	Cloud platforms, AI/ML, Big Data, System-of-Systems modeling
Examples	Jet engine digital twin, wind turbine, medical device	Chemical plant refining process, robotic welding line	Smart factory, urban mobility system, hospital ecosystem
Users	Product engineers, R&D teams	Operations managers, process engineers	Executives, planners, system architects
Integration Level	Low to medium	Medium	High—across units, departments, or organizations

5. Digital Twins for Real-Time Monitoring and Simulation

One of the most groundbreaking applications of digital twins is their capacity to offer real-time monitoring and simulation, which allows stakeholders to observe, predict, and optimize the performance of physical systems under any conditions (refer Figure 4). This has been achieved through a range of different industries, such as manufacturing, energy, healthcare, and transportation, through, in large part, the intersection of IoT technologies with AI-powered analytics. Real-time digital twins establish an ongoing and dynamic link between worlds, as opposed to static digital models. Decision makers may leverage this living model to imitate operational conditions, detect anomalies, and recommend interventions due to the two-way information flow that ensures the digital twin evolves with its real-world equivalent (Tao et al. 2019).

Any digital twin's lifeblood is real-time data. Virtual models rapidly deviate from reality in the absence of consistent, high-fidelity data streams, such as temperature, vibration, flow rates, and

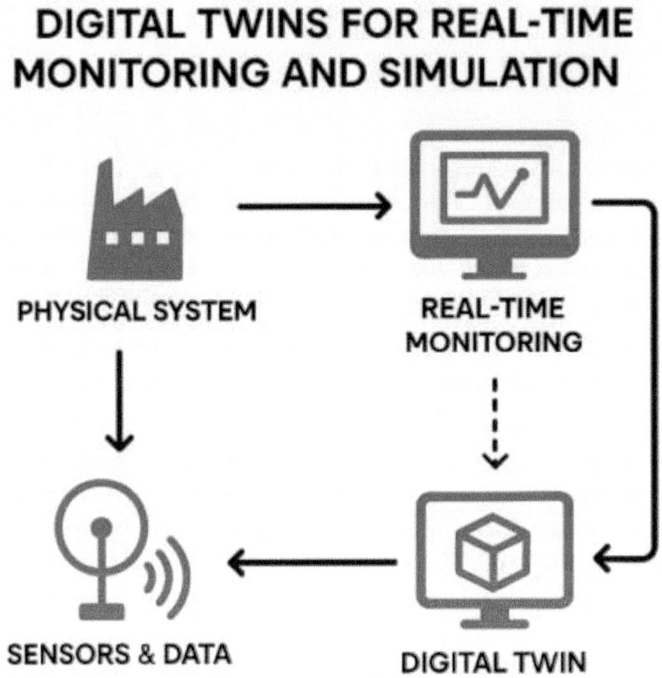

Figure 4: Digital Twins for real time monitoring and simulation.

location. Instant situational awareness, prompt anomaly detection, and predictive analytics are all made possible by real-time telemetry. For instance, sensors on roads, pipelines, and transit systems in smart cities provide data to digital twins that detect environmental hazards, water leaks, and traffic jams as they occur (Reuters 2024). Machine sensors in industrial settings improve productivity and decrease downtime by forecasting maintenance requirements (Anvil Labs 2024).

IoT protocols and architectures that enable continuous, low-latency data exchange are at the core of real-time digital twins. Real-time data collection at scale is made possible by the growth of IoT sensor networks, which are made up of wireless, edge, and satellite nodes. These sensors send data to cloud platforms via MQTT, WebSockets, or 5G. Data lakes and stream processors (like Kafka and Flink) maintain the digital thread that connects a twin to its physical counterpart (Springer review on AI-DT integration 2025). Vibration, force, and acoustic sensors provide valuable information about the health of machines in manufacturing (Wikipedia: Digital Twins) TIME. Preprocessing and anomaly flagging are made possible by edge computing, which reduces latency and bandwidth requirements.

Because of its lightweight design, the Message Queuing Telemetry Transport (MQTT) protocol has become well-known and is very effective for devices with limited bandwidth (Banks and Gupta 2014). According to the publish-subscribe communication model used by MQTT, IoT sensors send telemetry data, such as temperature, pressure, or vibration, to a centralized broker, which then disseminates it to the subscribers of the digital twin in almost real time. Alongside MQTT, the Open Platform Communications Unified Architecture (OPC UA) protocol provides secure and standardized machine-to-machine communication, allowing reliable sharing of structured data between industrial controllers, sensors, and supervisory systems (Mahmoud et al. 2020). Time-Sensitive Networking (TSN) provides edge devices and digital twin platforms with microsecond-level synchronization and predictable data delivery in cases requiring deterministic communication, such as autonomous vehicles or robotic manufacturing cells (Maciejewski et al. 2020). With the provision of ultra-reliable low-latency communications (URLLC) and the facilitation of massive machine-type communications (mMTC) that can support thousands of devices connected simultaneously, 5G connectivity also makes digital twins running at the edge even more feasible (Taleb et al. 2017).

The digital twin is made intelligent through AI algorithms, which enable it to understand observations, forecast future conditions, and optimize system functioning, while IoT infrastructure ensures that it receives precise and timely information. Machine learning algorithms such as random forests and support vector machines have also proved useful in predictive maintenance, given that they can learn complex relationships between sensor inputs and degradation indicators (Lei et al. 2018). Deep learning structures, especially recurrent neural networks (RNNs) and long short-term memory (LSTM) networks, are well-suited to modeling temporal dependencies in equipment behavior, such as the build-up of energy consumption habits or bearing wear, for time-dependent forecasting (Zhao et al. 2019). Unsupervised learning methods are frequently used in anomaly detection, which is essential for preventing expensive outages. Isolation Forests find outliers in multivariate sensor data without labeled fault data, while autoencoders can flag deviations and reconstruct normal operational signatures (Chandola et al. 2009).

For instance, stacked ensemble machine learning classifiers are used in digital twins of industrial control systems to identify cyber intrusions almost instantly (Varghese et al. 2022). Mitsubishi's deep learning twin efficiently tracks thermal-hydraulic parameters in energy systems. In structural monitoring, Bayesian models filter contextual anomalies, while thermal imaging-based predictive maintenance frameworks integrate POD, RPCA, and DMD techniques into a single twin (Menges et al. 2024). Urban twins are now incorporating generative AI to improve weather simulations, optimize infrastructure, and supplement synthetic data generation (arXiv 2024, Springer 2024).

The primary features that distinguish digital twins from conventional data dashboards go beyond monitoring and include simulation and optimization. A potent class of models known as physics-informed neural networks (PINNs) incorporates first-principles physical laws directly into neural network training goals, guaranteeing that simulations are both data-driven and physically consistent (Raissi et al. 2019). For instance, PINNs can enforce the governing equations of heat transfer when modeling the distribution of heat across a turbine blade, increasing predictive accuracy even in the face of sparse data. In control optimization, reinforcement learning algorithms are becoming increasingly important. In simulation environments, agents trained through reinforcement learning can explore different process configurations, learning policies that strike a balance between conflicting goals like throughput, energy consumption, and maintenance costs (Zhang et al. 2021). By creating artificial sensor datasets that supplement training resources, generative adversarial networks (GANs) also aid in simulation, especially in situations where obtaining fault data is hazardous or impractical (Goodfellow et al. 2014).

5.1 Real-Time Simulation vs Historical Playback

Two different simulation modes complement each other: historical playback replays historical data to investigate incidents, identify inefficiencies, and validate models under known conditions; real-time simulation uses live data to assess the current system and predict future states, facilitating immediate feedback and decision-making. Operations are aided by real-time simulation: a city twin responding to flood sensors can warn residents and reroute traffic. Post-event analysis shapes policy and upholds accountability, while historical playback aids in learning. Table 3 shows a comparative overview:

Integration: Real-Time Closed-Loop Feedback

When AI models and IoT protocols are closely integrated in a closed-loop control architecture, digital twins' full potential is revealed:

Data Flow: Using MQTT or OPC UA, IoT sensors gather and transmit real-time operational data.

Table 3: Comparative overview.

Feature	RealTime Simulation	Historical Playback
Purpose	Live monitoring; future forecasting	Post-event analysis; model validation
Data Input	Continuous real-time streaming	Stored data from archives
Computational Need	Low latency, streaming analytics	Higher compute; non-real-time batch mode
Typical Use Cases	Dynamic response, predictive alerts	Rootcause analysis, training, audit logs
Feedback Mechanisms	Closed-loop control (actuators, alerts)	Open-loop: no real-time feedback
Examples	Smart traffic rerouting, grid load balancing	Machinery cost-analysis, policy revision

Edge Processing: To lessen the load on cloud transmissions, edge devices preprocess and compress data.

Cloud Analytics: AI models examine data to find trends, make predictions, and suggest improvements.

Feedback Loop: Real-time operational adjustments are made by relaying insights back to the control systems. The bidirectional feedback mechanisms of digital twins are one of their distinguishing features. These complete the circle—digital models produce optimized commands in addition to ingesting data:

- In factories: adjusting machine settings in the middle of operation to avoid errors.
- In grids: telling storage systems to shed or engage loads.
- Adapting HVAC systems in smart buildings according to climate and occupancy forecasts.

For instance, LSTM models can predict bearing failures, OPC UA interfaces can automatically trigger blade pitch adjustments to reduce stress, and 5G can stream real-time telemetry to a wind turbine digital twin (Lu et al. 2020).

A dynamic that may result in self-healing systems is brought about by this autonomy, which changes digital twins from passive mirrors to proactive agents (Singapore's Virtual Digital Twin enables autonomous testing, Wikipedia 2025).

5.2 Case Studies

5.2.1 Smart Cities

Virtual Singapore is among the most innovative instances of the implementation of digital twins on an urban scale. This nationwide digital twin, which was started in 2014 and formally unveiled in 2022, combines real-time climate and infrastructure models, IoT sensor networks, and 3D spatial data. It enables city planners, emergency responders, and policymakers to model and predict real-world situations by combining various data layers, such as pedestrian flow, traffic density, environmental noise, and flood-prone areas. For instance, Virtual Singapore allows stakeholders to test various configurations and forecast their effects on energy consumption, traffic, or air quality prior to implementing a new traffic management system or redesigning green spaces (Pereira 2023, Wikipedia 2025).

In addition to Singapore, more than 500 cities globally have started using digital twin technologies to improve efficiency and resilience. Urban twins are being used by cities like Lagos, Houston, and Amsterdam to manage heat islands, optimize flood control plans, monitor air pollution in real time, and restructure waste management systems. These systems give local governments access to actionable intelligence, which is particularly important given rapid urbanization and climate change (Reuters 2024).

Smart Factories

Digital twins have completely changed the way factories function in industrial settings. One well-known platform that enables manufacturers to produce digital models of assembly lines is DELMIA by Dassault Systèmes. Before implementing any physical changes, these virtual environments can be used to optimize layout designs, simulate workflows, and fine-tune production parameters. This significantly cuts down on setup time, resource waste, and operational risk. Furthermore, stacked anomaly detection systems are increasingly integrated into the control infrastructure of many smart factories. These systems continuously monitor sensors and identify any deviation in process behavior in less than 0.1 seconds using sophisticated machine learning algorithms. Operators can take action before minor problems turn into expensive failures thanks to this almost instantaneous response (Varghese et al. 2022).

Autonomous Vehicles

The development of high-fidelity digital twins has been closely linked to the rise of autonomous vehicles. These automotive twins use real-time data from sensor fusion, LIDAR, and telematics systems to model how a vehicle would perform. Manufacturers and mobility companies can test driving scenarios in a safe, virtual environment, including pedestrian-rich zones and unfavorable weather conditions, using platforms like Smart World Pro. This improves safety and dependability for practical implementation in addition to speeding up the training of AI driving models (Time 2023).

Energy Grids

In the management of contemporary energy grids, especially those that integrate renewable energy sources, digital twins have also become indispensable. For energy infrastructure such as turbines, transformers, and substations, firms such as Akselos build reduced-order, high-fidelity twins that continuously assess system reliability and structural integrity (Wikipedia 2025). Digital twins assist utilities in integrating solar and wind inputs, balancing varying power demands, and anticipating failures before they happen (Applied Sciences 2024). Building robust and sustainable energy systems requires this degree of automation and foresight.

6. The Role of IOT and AI in Enhancing Digital Twin

The Internet of Things (IoT) and artificial intelligence (AI) have significantly increased the possibilities of digital twins, turning them from static models into intelligent, dynamic systems. IoT devices allow precise synchronization with their virtual counterparts by continuously streaming real-time data from physical assets. AI technologies analyze this data to detect anomalies, predict failures, and streamline processes automatically. IoT and AI, in tandem, make it possible to establish a feedback-rich environment where digital twins not only simulate physical spaces but also learn, change, and make informed decisions. This enhances productivity, safety, and innovation across a range of industries such as manufacturing, healthcare, transport, and city planning.

The technology of digital twins relies on the Internet of Things (IoT), which enables seamless connection and comprehensive data acquisition from physical assets. IoT enables continuous real-time data to flow from machines, infrastructure, and even living systems—such as temperature, pressure, motion, vibration, and environmental conditions—through a network of sensor-enabled devices and gateways (Tao et al. 2022). The digital twin will always remain an up-to-date and exact copy of its physical equivalent due to this continuous stream of high-fidelity data. For instance, IoT sensors mounted on production equipment in smart manufacturing detect small variations in performance, allowing for predictive maintenance and minimizing downtime (Grieves and Vickers 2019). Smart meters and air-quality sensors monitor energy use and air quality in urban networks, feeding information into digital twins that enhance public safety and urban planning (Pereira 2023). By reducing latency and facilitating local data processing—both fundamental for time-critical applications such as autonomous vehicles and health monitoring—5G and edge computing have extended the scope of IoT further (Hossain et al. 2024). Finally, IoT not only closes the gap between the world of bits and the world of atoms, but it also provides decision-makers with the contextual information necessary to react promptly and intelligently in fast-changing, high-complexity environments. AI and ML power digital twins, enabling them to simulate complicated systems, drive decision-making through data, and foresee future states with uncanny precision.

AI/ML has bestowed cognitive functions on these systems that can learn from massive volumes of past and current data, in contrast to conventional digital twins which were limited to static logic and rule-based modeling. Digital twins are now able to detect latent patterns, adapt to changing conditions, and incrementally enhance their accuracy independently over time through techniques such as supervised learning, deep learning, and reinforcement learning (Tao et al. 2022).

AI enhances the twin's modeling power by emulating interdependent variables, stochastic environments, and nonlinear dynamics that are otherwise computationally infeasible with pure physics-based models. Energy grids and manufacturing processes, for example, are increasingly applying deep neural networks and physics-informed machine learning models to create surrogate models with real-time responsiveness without compromising accuracy (Hossain et al. 2024).

Real-time optimization and anomaly detection are facilitated by AI-based decision-making. Once dependent on human decisions, digital twins in industrial environments now employ ensemble learning and Bayesian inference to detect faults, recommend repairs, or initiate preventive maintenance (Varghese et al. 2022). Furthermore, recurrent neural networks and predictive analytics frameworks that can foresee failures, customer demand, or environmental impacts well in advance of their occurrence strengthen forecasting. This predictive power results in safer navigation, improved traffic control, and more robust infrastructure planning in autonomous cars and smart cities (Pereira 2023).

Digital twins can become intelligent partners that can learn, reason, and steer complex systems toward sustainability, safety, and efficiency by combining AI and ML. Previously, digital twins were only passive models.

6.1 Edge Vs Cloud Computing in Digital Twin Operation

While both cloud and edge computing are core elements of the digital twin environment, their applications and advantages differ significantly depending on the operational context. The choice to process at the edge versus the cloud is more than a matter of technology; it impacts responsiveness, reliability, and scalability as digital twins evolve from offline simulations to smart, real-time implementations.

Processing data near its source, such as on a factory floor or within an automobile's onboard computer, is referred to as edge computing. Even in environments that are not connected, this proximity ensures continuity in systems, reduces latency, and saves bandwidth. On a factory floor, for instance, an edge-enabled digital twin can react in milliseconds to abnormalities like machine vibration or overheating, enabling preventive action (Hossain et al. 2024).

For applications with timing requirements such as industrial automation, health monitoring, and autonomous vehicles, this is critical.

Cloud computing, however, offers cross-platform compatibility, vast storage, and processing capabilities. It supports the central aggregation of data from scattered assets, historical trend analysis, and extensive simulations. Because they must aggregate data from thousands of sensors to determine long-term trends and coordinate across regions, smart city twins and national energy grids benefit significantly from cloud-based models (Pereira 2023, Tao et al. 2022).

The best digital twin architectures frequently use hybrid models, which combine cloud computing for big data analytics and long-term decision support with edge computing for real-time processing. For understanding the key differences between the two, Table 4 is given below.

Table 4: Comparison of Edge vs Cloud Computing in Digital Twin Operations.

Feature	Edge Computing	Cloud Computing
Location of Processing	Near data source (local device/facility)	Remote servers (centralized)
Latency	Ultra-low (milliseconds)	Higher (network-dependent)
Use Case	Real-time control, anomaly detection	Large-scale simulation, historical trends
Scalability	Limited to local infrastructure	Virtually unlimited
Connectivity	Operates offline or intermittently	Requires stable internet connection

6.2 Examples of AI Driven optimization

Digital twin applications are increasingly relying on AI-driven optimization to improve sustainability, cut costs, and streamline operations. A classic example is in smart manufacturing, wherein AI algorithms analyze real-time sensor readings from equipment to predict maintenance needs, schedule production for best utilization, and conserve energy. This maximizes throughput and resource use as well as prevents costly downtime (Tao et al. 2022).

Digital twins powered by AI are applied to urban planning to simulate and optimize energy distribution, public transit routes, and traffic. AI is applied in urban twins in cities such as Singapore and Amsterdam to minimize carbon emissions, redirect traffic in real time, and dynamically adjust street lights (Pereira 2023). Similarly, digital twins enhanced with AI are employed by self-driving cars to mimic different driving conditions and select the best and safest route to travel, thus optimizing driving choices (Time 2023).

Artificial intelligence (AI) algorithms embedded within digital twins of energy grids predict peaks in demand, dispatch energy in the most efficient manner, and harmoniously blend renewable resources. This saves fuel costs while enabling operators to maintain grid stability and avoid blackouts (Mchirgui et al. 2024).

These machine learning-based optimization methods demonstrate how digital twins are evolving into smart systems that actively generate improved outcomes as well as mirror the current state.

6.3 Integration Challenges and Opportunities

Both challenging problems and exciting possibilities emerge when digital twins are combined with cloud-edge architecture, IoT, and AI/ML. From the opportunity side, such a combination enables seamless data flows, real-time decision-making, and predictive analysis across many industries such as smart cities and healthcare. Digital twins can revolutionize operational efficiency and resilience by turning into intelligent, adaptive systems that have the ability to learn from data, forecast future occurrences, and respond automatically upon integration (Tao et al. 2022).

Integration is not trouble-free, however. Interoperability of data, or the integration of data from different sources and formats while ensuring semantic consistency, poses a considerable challenge. Security and privacy issues also emerge, especially when dealing with personal information in critical infrastructure or healthcare (Mchirgui et al. 2024). Additionally, integrating cloud-scale simulations with real-time edge analytics makes scalability harder. The intensive requirements of a completely integrated digital twin environment are often more than legacy systems and fragmented IT infrastructures can manage (Hossain et al. 2024).

Despite these challenges, a promising direction is offered by ongoing advances in open standards, data harmonization through AI, and secure multi-cloud architectures. The complete transformative power of digital twins can be delivered through integration supported by strategic planning and investment.

7. Challenges and Future Outlook in Digital Twin Technology

Since it provides real-time visibility, forecasting, and optimization, the digital twin is now an integral part of Industry 4.0 and smart system design. Although this concept is promising, there are numerous ethical, financial, and technical challenges in its path toward large-scale implementation. For the true realization of the transformative potential of digital twins in industries, these have to be efficiently overcome.

7.1 Technical Barriers

Handling data's integration and complexity is among the primary challenges to the use of digital twins. As a digital twin can only be as good as the data it is given, data from various sources often come in formats that are inconsistent, at inconsistent frequencies, and without shared semantics. Inconsistencies or latency issues can occur when real-time IoT feeds are synchronized with simulation models and historical databases, especially for time-critical applications such as autonomous vehicles or power grid monitoring (Tao et al. 2022).

The lack of interoperable platforms is another hurdle. The isolated IT ecosystems within which most organizations tend to function create difficulties for digital twins to exchange information across systems or geographies. The technical load is compounded by the high computational intensity of real-time simulation, especially when integrating edge-cloud hybrid architectures with AI/ML algorithms (Hossain et al. 2024).

Cost and Complexity

It takes a large investment in software (simulation engines, AI models), hardware (sensors, edge devices), and qualified staff to create and maintain a high-fidelity digital twin (Sivakumar et al. 2025). The cost barrier may be unaffordable for small and medium-sized businesses (SMEs). Furthermore, moving from a prototype to an enterprise-level implementation necessitates continuous data governance, cybersecurity protocols, and intricate coordination between cloud providers, all of which require ongoing operational expenditures (Mchirgui et al. 2024).

Additionally, there is no one-size-fits-all approach to digital twins. Design and deployment take a lot of time because each implementation needs to be customized to fit the physical system it represents. When attempting to model interdependent systems, such as a whole smart city or a national power grid, the complexity grows exponentially.

Regulatory and Ethical Issues

Concerns about data privacy, surveillance, and bias are growing as digital twins are used more and more to simulate human environments and behaviors, especially in the fields of healthcare, urban planning, and autonomous systems. Who owns the data when a city's digital twin tracks vehicle movement, foot traffic, or facial recognition information? In what way is it anonymized? What occurs if a biased or faulty decision is made by the model?

The situation is made more difficult by regulatory uncertainty. Many industries, including smart cities, still lack thorough frameworks, despite industries like healthcare and aviation having strict regulations that could encourage the development of trustworthy digital twins. Uncertainty brought about by the lack of international standards for data sharing, model validation, and AI transparency may prevent adoption (Pereira 2023).

Emerging Trends

Notwithstanding these obstacles, a number of developments are opening the door for wider adoption of digital twins:

- AI-enhanced twins are growing increasingly independent and anticipatory, able to pick up system behaviors and maximize results with little assistance from humans.
- To increase interoperability and compliance across platforms, governments and industry consortia are pursuing standardization initiatives (Tao et al. 2022).
- Low-code platforms and open-source ecosystems are democratizing digital twin tools, enabling smaller businesses to use the technology.

The use of digital twins to model environmental impact, material reuse, and lifecycle emissions is becoming more popular in the context of sustainable design integration, especially in manufacturing and construction (Mchirgui et al. 2024).

Future Impact Across Industries

In the future, it is anticipated that digital twins will have a greater impact on almost every industry. Patient-specific twins could facilitate remote monitoring, surgical planning, and individualized treatment in the medical field. Digital twins will be used more and more by autonomous systems, like drones and self-driving cars, to model worst-case scenarios and navigate uncertain environments. Digital twins of solar grids, wind farms, and storage systems will optimize supply-demand balance, lower emissions, and increase dependability in the energy sector. Immersive digital twin environments will provide risk-free, simulated learning experiences for everything from pilot training to medical procedures in education and training. Furthermore, national digital twins, as proposed by initiatives such as Virtual Singapore, could be used as test sites for urban development planning, disaster response, and policy modeling (Pereira 2023).

8. Conclusion

Digital twins have become one of the 21st century's most revolutionary technologies as we stand at the nexus of the digital and physical worlds. Digital twins are no longer passive copies thanks to the Internet of Things' (IoT) seamless connectivity and artificial intelligence's (AI) intelligence; instead, they are dynamic, learning systems that can track, forecast, and improve real-world performance in real time.

This chapter examined how IoT forms the foundation of precise, responsive twins by facilitating the constant flow of real-time data from the physical world into the digital realm. The analytical depth needed to identify anomalies, model intricate systems, and predict future behaviors is simultaneously provided by AI and machine learning. When combined, they produce useful insights for a variety of sectors, including manufacturing, energy, transportation, and smart cities.

We also discussed the key barriers to overcome in achieving economies of scale for digital twin adoption, including data integration, cost, ethics, and regulatory gaps. Optimistically, however, new technological developments in edge computing, AI-augmented modeling, standardization, and sustainability highlight a promising future.

Digital twins are, in effect, turning into strategic enablers instead of being only technical tools. They are central to the digital transformation future due to their ability to link data, intelligence, and action. Digital twins, fueled by IoT and AI, will be instrumental in designing, managing, and optimizing the systems that control our everyday lives as societies and organizations strive to attain sustainability, efficiency, and resilience. They have an extremely human-oriented and technologically progressive future.

References

Anvil Labs. (2024). *Digital Twins for Anomaly Detection in Asset Inspections*. Retrieved from Anvil Labs. thetimes.co.uk+1TIME+1anvil.so.

Batty, M. (2018). *Digital twins. Environment and Planning B: Urban Analytics and City Science*, 45(5): 817–820. https://doi.org/10.1177/2399808318796416.

Boschert, S., and Rosen, R. (2016). *Digital Twin—The Simulation Aspect*. pp. 59–74. *In:* P. Hehenberger & D. Bradley (Eds.). *Mechatronic Futures*. Springer. https://doi.org/10.1007/978-3-319-32156-1_5.

Fuller, A., Fan, Z., Day, C., and Barlow, C. (2020). *Digital twin: Enabling technologies, challenges and open research*. IEEE Access, 8: 108952–108971. https://doi.org/10.1109/ACCESS.2020.2998358.

Glaessgen, E., and Stargel, D. (2012). The digital twin paradigm for future NASA and U.S. Air Force vehicles. p. 1818. *In:* 53rd AIAA/ASME/ASCE/AHS/ASC Structures, Structural Dynamics and Materials Conference.

Grieves, M., and Vickers, J. (2017). *Digital twin: Mitigating unpredictable, undesirable emergent behavior in complex systems*. pp. 85–113. *In:* Kahlen, F.-J., Flumerfelt, S., and Alves, A. (Eds.). *Transdisciplinary perspectives on complex systems*. Springer. https://doi.org/10.1007/978-3-319-38756-7_4.

Grieves, M., and Vickers, J. (2019). *Digital twin paradigm for future predictive maintenance*. ResearchGate.

Grieves, M., and Vickers, J. (2019). *Digital Twin: Mitigating Unpredictable, Undesirable Emergent Behavior in Complex Systems*. Springer.

Hossain, R. B., Ahmed, F., Kobayashi, K. et al. (2024). *Virtual sensing-enabled digital twin framework for real-time monitoring of nuclear systems*. arXiv preprint arXiv:2410.13762.

Hossain, R. B., Ahmed, F., Kobayashi, K., Koric, S., Abueidda, D., and Alam, S. B. (2024). Virtual sensingenabled digital twin framework for realtime monitoring of nuclear systems leveraging Deep Operator Networks. *ArXiv*. MDPIarXiv.

Industry 4.0 Review. (2024). Reviews of the integration of Artificial Intelligence and Digital Twins. *Springer*. SpringerLink+1nature.com+1.

Kritzinger, W., Karner, M., Traar, G., Henjes, J., and Sihn, W. (2018). *Digital twin in manufacturing: A categorical literature review and classification. IFAC-PapersOnLine*, 51(11): 1016–1022. https://doi.org/10.1016/j.ifacol.2018.08.474.

Lei, Y., Li, N., Guo, L., Li, N., Yan, T., and Lin, J. (2018). *Machinery health prognostics: A systematic review from data acquisition to RUL prediction. Mechanical Systems and Signal Processing*, 104: 799–834. https://doi.org/10.1016/j.ymssp.2017.10.016.

Lu, Y., Liu, C., Kevin, I., and Yang, Y. (2020). *Digital twin-driven smart manufacturing: Connotation, reference model, applications and research issues. Robotics and Computer-Integrated Manufacturing*, 61: 101837. https://doi.org/10.1016/j.rcim.2019.101837.

Mahmoud, M. S., Gao, R. X., and Martinez, R. (2021). *Digital Twins in Manufacturing: State-of-the-Art and Future Research Needs. Manufacturing Letters*, 28: 17–23. https://doi.org/10.1016/j.mfglet.2021.03.006.

Mchirgui, N., Quadar, N., Kraiem, H., and Lakhssassi, A. (2024). The applications and challenges of digital twin technology in smart grids: A comprehensive review. *Applied Sciences,* 14(23): 10933. MDPI.

Menges, D., Stadtmann, F., Jordheim, H., and Rasheed, A. (2024). Predictive digital twin for condition monitoring using thermal imaging. *ArXiv*. arXiv.

Naik, N. (2017). *Choice of effective messaging protocols for IoT systems: MQTT, CoAP, AMQP and HTTP. 2017 IEEE International Systems Engineering Symposium (ISSE)*, 1–7. https://doi.org/10.1109/SysEng.2017.8088251.

Pereira, D. (2023). *Speculative Design: Virtual Singapore is a Massive, Fully Functional Digital Twin. Scopus*.

Pereira, D. (2023). *Virtual Singapore: A functional digital twin for urban sustainability*. UCL Discovery.

Qi, Q., Tao, F., Zuo, Y., and Zhao, D. (2019). *Digital Twin Service Towards Smart Manufacturing. Procedia CIRP*, 72: 237–242. https://doi.org/10.1016/j.procir.2018.03.103.

Reuters. (2024, May 23). How AI is arming cities in the battle for climate resilience. WikipediaReuters

Sivakumar, M., Krishnaraj, N., and Tyagi, A. K. (2025). Large Language Models: Fundamentals, Necessity, Impacts, Challenges, and Opportunities for Future Applications, 2025 Seventeenth International Conference on Contemporary Computing (IC3), Noida, India, pp. 1–7, doi: 10.1109/IC366947.2025.11290699.

Shao, G., and Kibira, D. (2018). *Digital Manufacturing: Requirements and Challenges for Implementing Digital Twins in the Process Industry. Procedia CIRP*, 72: 229–234. https://doi.org/10.1016/j.procir.2018.03.167.

Shao, G., Jain, S., Shin, S. J., and Brodsky, A. (2020). *Digital Twin Framework for Manufacturing System Design and Optimization*. International Journal of Production Research, 58(10): 3135–3153. https://doi.org/10.1080/00207543.2019.1658379.

Shao, G., Jain, S., Shin, S. J., and Kibira, D. (2020). *Digital Twin Framework for Smart Manufacturing. Journal of Manufacturing Systems*, 58: 174–186. https://doi.org/10.1016/j.jmsy.2020.08.002.

Taleb, T., Samdanis, K., Mada, B., Flinck, H., Dutta, S., and Sabella, D. (2017). *On multi-access edge computing: A survey of the emerging 5G network edge cloud architecture and orchestration. IEEE Communications Surveys & Tutorials*, 19(3): 1657–1681. https://doi.org/10.1109/COMST.2017.2705720.

Tao, F., Sui, F., Liu, A., Qi, Q., Zhang, M., and Song, B. (2018). *Digital Twin-Driven Product Design Framework. International Journal of Production Research*, 57(12): 3935–3953. https://doi.org/10.1080/00207543.2018.1443229.

Tao, F., Sui, F., Liu, A., Qi, Q., Zhang, M., Song, B. et al. (2022). Digital twin in industry: State-of-the-art. *ScienceOpen.*

Tao, F., Zhang, H., Liu, A., and Nee, A. Y. C. (2018). *Digital twin in industry: State-of-the-art.* IEEE Transactions on Industrial Informatics, 15(4): 2405–2415. https://doi.org/10.1109/TII.2018.2873186.

Tao, F., Zhang, M., Liu, Y., and Nee, A. Y. C. (2019). *Digital Twin in Industry: State-of-the-Art. IEEE Transactions on Industrial Informatics*, 15(4): 2405–2415. https://doi.org/10.1109/TII.2018.2873186.

Time. (2023). How digital twins are transforming manufacturing, medicine, and more. TIME

Varghese, S. A., Ghadim, A. D., Balador, A., Alimadadi, Z., and Papadimitratos, P. (2022). *Digital twin-based intrusion detection for industrial control systems*. arXiv preprint arXiv:2207.09999.

Wikipedia. (2025, July). Digital Twin; Virtual Singapore; DELMIA. Wikipedia.

Zhang, H., Wang, J., Tao, F., and Cao, J. (2021). *Data-Driven Smart Manufacturing: Tools, Methods, and Applications. Journal of Manufacturing Systems*, 58: 305–321. https://doi.org/10.1016/j.jmsy.2020.08.013.

2

Advanced Analytics with AI and Machine Learning

S. Durga Devi,[1] *R. Ravinder Reddy,*[1,*] *Satyanarayana Nimmala,*[2]
Ravi Uyyala[1] and *M. V. Krishna Reddy*[1]

1. Introduction

The growth of IoT devices and data has created enormous amounts of heterogeneous, high-velocity data, simultaneously presenting opportunities and threats to analytics. It is likely that traditional analytical methods can't keep up with the scale, variety, complexity, and temporality of IoT data streams. As a result, AI and machine learning (ML) are increasingly being adopted as a new paradigm that can rapidly learn and identify intelligent patterns, automate decisions, and offer real-time insights across a varied set of IoT applications (Chen et al. 2020). Advanced analytics represents a shift in analytics thinking from descriptive modeling to predictive and prescriptive analytics, allowing us to not only understand what has happened in the past but also to see what might happen in the future and what we ought to do. This move to action is especially important in IoT scenarios, including smart cities, healthcare, and industrial automation, where the critical nature of timely actions is paramount (Kumar et al. 2021).

The combination of AI, ML, and IoT technologies enables the creation of intelligent systems that can learn from data, find anomalies, and adjust to changing environments. Adding blockchain technology to these intelligent systems bolsters their capabilities with decentralized, immutable, and tamper-resistant data storage. Blockchain technologies can make data more trustworthy, improve the secure sharing of data among stakeholders, and even enable data analytics workflows through smart contracts (Zhang et al. 2022, Lee et al. 2022). The intersection of edge computing and federated learning has recently emerged as an important area of development and exploration, helping to promote real-time, AI-enabled analytics at the network edge. Decentralized approaches such as these can reduce the latency associated with executing IoT data analytics, as well as the amount of data that needs to be communicated to the cloud. The result is more scalable and privacy-preserving analytics at the network edge in IoT applications (Chen et al. 2023). Federated learning respects compliance with data protection laws such as GDPR by keeping data localized and only disseminating model updates while still delivering a high level of analytical performance.

[1] Department of Computer Science and Engineering, Chaitanya Bharathi Institute of Technology, Hyderabad, India. Emails: sdurgadevi_cse@cbit.ac.in, https://orcid.org/0000-0002-5163-8886; raviuyyala_cse@cbit.ac.in, https://orcid.org/0000-0002-6913-7529; krishnareddy_cse@cbit.ac.in, https://orcid.org/0000-0002-7247-6580

[2] CVR College of Engineering, Hyderabad, India. Email: satyauce234@gmail.com, https://orcid.org/0000-0002-9433-1489
* Corresponding author: rravinderreddy_cse@cbit.ac.in, https://orcid.org/0000-0001-6488-9935

DOI: 10.1201/9781003655190-3

In addition, there is the still-evolving notion of Explainable Artificial Intelligence (XAI), which addresses the issue of interpretability of complex ML models that are being used in systems such as healthcare, autonomous systems, and the financial sector. XAI offers insight into the decision-making of these models, which improves stakeholder trust while simultaneously addressing some ethical aspects of AI (Rahimi et al. 2024). As more AI systems become the basis of autonomous decisions in an IoT environment, the issues of interpretability and accountability become increasingly important.

Moreover, smart contracts utilizing blockchain platforms such as Hyperledger Fabric and Ethereum are changing how analytics workflows can be initiated and verified. Smart contracts provide decision-making based on whether pre-defined conditions are met, which allows for an autonomous, trustless decision pipeline with less manual effort in multi-stakeholder IoT ecosystems (Zhang et al. 2025). Recent research also highlights the role of hybrid AI frameworks—combining deep learning, ensemble learning, and reinforcement learning—to improve prediction accuracy, model robustness, and resource optimization in IoT analytics (Alqahtani et al. 2025). These adaptive learning systems are becoming increasingly essential in mission-critical applications such as predictive maintenance, traffic forecasting, and environmental monitoring.

This chapter deals with the demand for an integrated, scalable, and secure analytics architecture for IoT ecosystems. The key contributions are: identifying clustering and classification methodologies for heterogeneous IoT data; applying time-series analysis to forecast sensor trends with temporal dimensions; developing predictive and prescriptive models to provide data-driven decision support; and providing an architecture incorporating blockchain that enables IoT analytics to be secured and automated. Utilizing the latest technologies from the AI, ML, and blockchain spaces, we enable organizations to avoid reactive analytics and control their own proactive and autonomous systems. Because our systems are circular, we also provide a limitless set of options to scale to complex management of IoT data and, at the same time, retain integrity, security, and transparency of the data.

2. Literature Review

2.1 Development of IoT Analytics

IoT analytics has developed in several distinct stages. Former IoT systems largely depended on basic threshold-based monitoring and traditional two-tier analytics approaches. Cloud computing has allowed for more advanced batch processing of IoT data; however, real-time analytics has remained a challenge. A new era is underway, which focuses on AI- and ML-based intelligent, adaptive analytics. Surveys by Kumar et al. (2023) and Zhang et al. (2024) reveal that blockchain technology is continuing to converge with IoT analytics. Blockchain technology adds measures of data integrity, transparency, and security to dispersed IoT situations. Blockchain supports decentralization and smart contracts, which allow secure, verifiable, tamper-free sharing of data and automated decision-making without worrying about trust or endorsement in distributed IoT ecosystems.

2.2 Clustering and Classification in IoT Data Analytics

Recently, clustering and classification methods in IoT data have been gaining increasing attention. Patel and Singh (2021) showed that clusters of heterogeneous IoT sensor data can be established and visualized using unsupervised clustering algorithms such as K-means clustering and hierarchical clustering. The benefits of dimensionality reduction and feature selection were highlighted in how they improved the performance of clustering algorithms. In IoT data classification, deep learning methods have achieved excellent results. Liu et al. (2020) developed a convolutional neural network (CNN) model for IoT sensor data classification applications, where the model achieved 94.7% class prediction accuracy on multi-class classification tasks. The study also emphasized that data preprocessing and feature engineering steps play a vital role in subsequent model performance.

Anderson and Chen (2022) examined ensemble methods for IoT data classification, where combinations of classifiers (more than one machine learning algorithm) obtained accuracy and robustness improvements in classification outcomes. The hybrid algorithm fused classifiers (decision trees, support vector machines, and neural networks, etc.) in their application, which outperformed single algorithms in both classification accuracy and performance.

2.3 Time-Series Analysis for Applications in IoT

Time-series analysis is an essential part of IoT analytics, as sensor data has a temporal context. Different forecasting methods are available to aid decision-making or interventions. Although traditional methods such as ARIMA and exponential smoothing are still useful, these methods can be enhanced using machine learning to address the higher complexity of IoT data streams (Johnson et al. 2021). Long Short-Term Memory (LSTM) networks have been determined to be particularly useful for IoT time-series analysis. A robust study by Wang et al. (2020) used an LSTM-based framework to make predictions about IoT sensor values. Their importance to forecasting is valuable because they introduced the idea of temporal lag. Their results showed that using LSTM networks could more accurately predict IoT sensor values—with a reduction in lag—as compared with traditional forecasting methods. Their experimental results demonstrated a 23% reduction in Mean Absolute Error by using the LSTM framework.

Recently, Transformer architectures have also been applied to IoT time-series analysis in a few studies that have been able to show state-of-the-art results. Brown and Davis (2022) developed a transformer-based framework for multivariate IoT time-series forecasting to identify multivariate time-series trends and complex patterns. They improved the state of the art for several benchmark datasets in forecasting IoT time-series data, based on their ability to effectively represent complex patterns and long-term dependencies in IoT data streams.

2.4 Predictive and Prescriptive Analytics Models

Advancing from descriptive analytics to predictive and prescriptive analytics marks a significant increase in the ability to process IoT data. Predictive analytics models provide organizations with the ability to predict future events and trends based on observations and patterns in historical data (Garcia et al. 2021). Prescriptive analytics models have been developed using reinforcement learning for IoT applications. Thompson et al. (2020) proposed a framework for optimizing the use of resources in an IoT configuration through reinforcement learning, which was shown to have a significant positive impact on the efficacy of the system while also lowering costs. Thompson et al. (2020) demonstrated that reinforcement learning did not involve any supervised learning but learned optimal policies via interactions with the IoT system performing the work.

Lastly, ensemble methods have proven to be a successful strategy for developing robust predictive models. The research from Miller and Wilson (2021) examined combining several predictive models using newer ensemble techniques, which not only improved the reliability of predictive models but also significantly improved predictive accuracy. By combining the models using ensemble methodologies, Miller and Wilson (2021) reported reductions in prediction error by as much as 18% compared to singular models.

2.5 Blockchain-IoT Integration for Enhanced Analytics

The development of hybrid blockchain and IoT extensions creates a secure and transparent environment for data analytics. Blockchain offers immutable data and enables trustless data sharing among several users (Lee et al. 2022). The combination of smart contracts to trigger automated analytical tasks in blockchain-IoT integration was illustrated in the research work by Kumar and

Patel (2021). They used smart contracts to induce administrative computing from pre-defined conditions that were important in initiating autonomous decisions regarding integrated IoT systems. Their research work reduced computations by 35% while maintaining the integrity of data. Various federated learning approaches have been explored in blockchain-IoT systems to allow collaborative analytics without disclosing personally identifiable information and to protect user privacy (Taylor et al. 2020). Taylor et al. (2020) demonstrated collaborative machine learning without compromising the integrity of local user data, making it ideal for sensitive IoT applications.

3. Methodology

3.1 Framework for Advanced Analytics in IoT Systems: Prescriptive Development

This section offers a comprehensive framework for the deployment of advanced analytics with the incorporation of AI and ML in IoT systems. Figure 1 illustrates the proposed framework contains six interconnected layers pertaining to IoT systems, which include the Data Collection Layer, Data Processing Layer, Analytics Layer, Decision Layer, Application Layer, and the Blockchain Integration Layer.

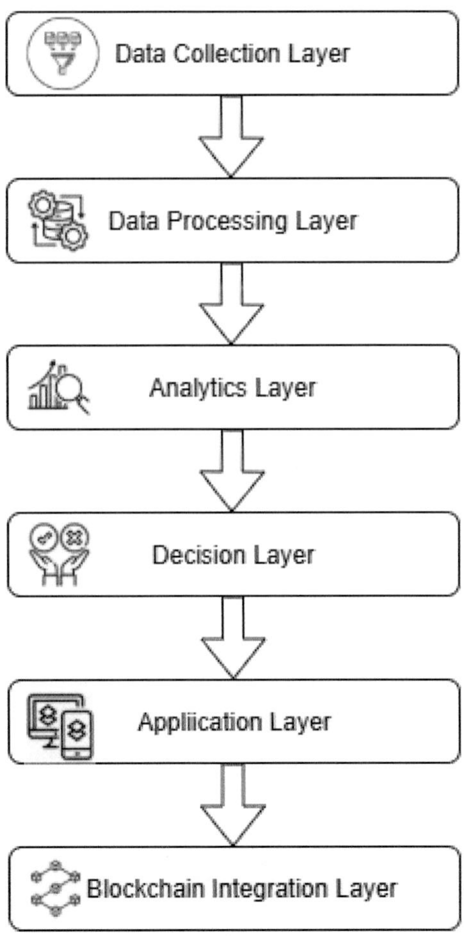

Figure 1: Comprehensive Framework Architecture for AI-Driven IoT Analytics.

3.1.1 Data Collection Layer

The data collection layer consists of multiple IoT devices and sensors that deliver continuous streams of data. This layer uses standardized data collection methods to ensure data consistency and quality. Critical elements include:

a. **Sensor Networks:** Using a variety of sensors for multimodal data collection such as industrial sensors (pressure, vibration, flow rate), motion and proximity sensors, image and video sensors, and environmental sensors (temperature, humidity, air quality).

b. **Edge Computing Node:** A local processing unit that does the initial gathering and filtering of data according to features like local spool storage, edge AI inference, real-time pre-processing, and communication protocols for secure transmission.

c. **Communication protocols:** Provide means of data transmission; for example, LoRaWAN for long range and low power, CoAP for constrained devices, MQTT for light weight messaging, and 5G/LTE for high bandwidth applications.

d. **Data quality assurance:** Real-time systems for error checking and greatly validating data, algorithms for outlier detection, consistency checks, missing value management, and procedures for temporal alignment.

3.1.2 Data Processing Layer

The Data Processing Layer transforms unstructured IoT data to representations suitable for analytical processing. There are several pre-processing tasks performed in this layer:

a. **Data cleansing:** Identify statistical outliers (for example, using Z-score, IQR techniques), detect anomalies with machine learning algorithms, blocks of filtering rules (domain-specific applications), and temporal consistency validation to filter noise, outliers, and inconsistencies.

b. **Feature engineering:** Extract and transform relevant features from raw feature sets, typically in the form of statistical features (e.g., percentiles, ranges), time-domain features (e.g., mean, variance, skewness, and kurtosis), frequency-domain features (e.g., FFT, wavelet transforms), and domain-specific features (e.g., MFCC for audio data).

c. **Data Normalization:** Conversion of data scales and formats is carried out through Min-max normalization, Z-score standardization, robust scaling techniques, and unit vector scaling.

d. **Temporal Alignment:** Synchronization of data from multiple sources through timestamp standardization, interpolation for missing time-points, resampling for different sampling rates, and sliding windows for continuous processing.

3.1.3 Analytics Layer

The Analytics Layer executes complex AI and ML algorithms for data analysis. The analytics layer consists of four useful components:

Clustering and Classification Module: This module executes supervision algorithms and unsupervised algorithms called clustering algorithms for analysis of IoT devices data. Some of the most important and commonly used algorithms are:

K-means Clustering: Identifying natural groupings of IoT sensor data.

Hierarchical Clustering: Building taxonomies of IoT devices and data patterns.

Support Vector Machines: Classify high-dimensional IoT data.

Deep Neural Networks: Identifying patterns in IoT data streams has become increasingly complex due to their high volume, velocity, and heterogeneity.

Time-Series Analysis Module: This module is for identifying and predicting temporal patterns.

LSTM Networks: With LSTM networks, sequential patterns can be learned, and predictions made.

Transformer Models: Can be used to learn long-term dependencies in time-series data.

Seasonal Decomposition: To learn periodic patterns in IoT data.

Anomaly Detection: Identify unusual patterns in time-series data.

Module for Prescriptive Analytics: This module provides advice and optimization capabilities, and is comprised of several components:

Reinforcement learning is used to determine the best policies in Internet of Things environments, rule-based decision-making using decision trees, while resource allocation and system configuration are achieved through optimization algorithms and multi-objective optimization for trade-off problems.

3.1.4 Layer of Decision Making

Findings from the analysis are introduced into the Decision Layer to assist in decision making. This layer consists of:

Rule-based Systems: Uses predefined rules to quickly make decisions autonomously.

Fuzzy Logic: Allows for imprecision and vagueness in decision insight.

Multi-criteria Decision Analysis: Used for ranking multiple options

Risk Assessment: Assess the likelihood of some outcomes.

Taking into account all the elements that make up the Decision Layer, it is reasonable to conclude it has great significance.

3.1.5 Decision Layer

Findings from the analysis are introduced into the Decision Layer to assist in decision making. This layer consists of:

Rule-based Systems: Uses predefined rules to quickly make decisions autonomously.

Fuzzy Logic: Allows for imprecision and vagueness in decision insight.

Multi-criteria Decision Analysis: Used for ranking multiple options.

Risk Assessment: To assess the likelihood of potential outcomes taking into account all the elements that make up the Decision Layer, it is reasonable to conclude it has great significance.

3.1.6 Application Layer

The Application Layer contains different interfaces or client portion of the system intended to be interacted with by the end-users, external systems, and third-party vendors. In this layer, you can find:

Visualization Dashboards: The dashboards are for displaying the analytical results produced by the system in a useful visualization.

API Interfaces: The API provides means to integrate with external systems and allows clients or businesses to interact with the system with data exchange.

Alerts: Notification systems to alert stakeholders to important events.

Reports: Reporting tools for generating full analytical reports.

3.1.7 Blockchain Integration Layer

The Blockchain Integration Layer contains secure, transparent, and immutable data storage and processing capabilities.

Blockchain Architecture: The implementation of the blockchain using Hyperledger Fabric creates a permissioned network with trusted participants, chaincode (smart contracts) to automate processing, endorsement policies to commit verified transactions, and private data collections to ensure certain participants have no visibility.

Consensus Mechanism: The implemented consensus mechanism in this project is Practical Byzantine Fault Tolerance (PBFT), which ensures:

i. Fast transaction finality.
ii. The energy finality of the nodes for consensus.
iii. Fault tolerance with 1/3 of the nodes being malicious.
iv. The ability to scale for IoT applications.

Smart Contracts Implementation: Automated analytics execution featuring:

i. Triggered-based analytics workflows
ii. Automated reporting
iii. Dynamic pricing and resource allocation
iv. Compliance/Audit trail

3.2 Blockchain Alignment Framework

The proposed framework incorporates blockchain as a complementary technology, leveraging its secure, auditable, and immutable data storage and processing model. The blockchain alignment framework is comprised of three components:

3.2.1 Data Integrity Management

Because of its cryptographic hashing and distributed consensus model, blockchain maintains data integrity. Each transaction that is processed as data (e.g., a calculation parameter) becomes part of the blockchain, which builds an immutable audit trail of data. Consequently, methods to alter data cannot occur, and reliable analytical outputs are assured.

3.2.2 Smart Contract-based Analytics

Smart contracts are able to automate the analytical process by running preset algorithms when certain conditions are met. This automation of the analytical process assists in lowering processing times and removes the need for human intervention to conduct routine analytical tasks.

3.2.3 Decentralized Analytics

The framework also facilitates decentralized analytics through the ability to distribute computation to multiple nodes in the blockchain network, saving time and allowing the overall computational load to be dispersed across multiple nodes. This allows scalability while simultaneously reducing the risk of a single point of failure.

3.3 Algorithm of Implementation

The following algorithm describes the process of implementing the suggested framework:

Algorithm: Implementation of the Advanced Analytics Framework.

Input: IoT data streams (D), analytical needs (R), and parameters of the system (P).

Output: Recommendations (Rec) and Analytical Results (A)

Step 1. Set up the Data Gathering Layer

- Set up the edge nodes and sensor networks.
- Develop communication protocols.
- Develop procedures for ensuring data quality.

Step 2. Process Data Streams

- Use data cleansing.
- Use feature engineering; align and normalize temporal data.

Step 3. Use Analytics Techniques

- Identify the best algorithm based on r. Train the model based on historical data.
- Test the model. Take action by applying the model to data in real time.

Step 4. Produce Insights and Recommendations

- Combine results from multiple analytics programs
- Use decision making algorithms
- Determine actionable recommendations

Step 5. Update Blockchain Records

- Add analytical results to blockchain
- Execute smart contracts to automate
- Ensure data integrity and traceability

Step 6. Present Results

- Update visualization dashboards
- Produce alerts and notifications
- Offer API access for external systems

Return A, Rec

3.4 Evaluation Metrics for Performance

The performance of the framework is evaluated with the following metrics:

Accuracy: Correctness of Prediction.

Precision and Recall: Classification Performance.

F1- Score: Mean of Precision and Recall.

Mean Absolute Error (MAE): Finds the average difference between predicted and actual values.

Root Mean Square Error (RMSE): The square root of the mean of the squared errors.

Scalability: Performance under scaling data loads.

Resource Utilization: Performance of Computational resources.

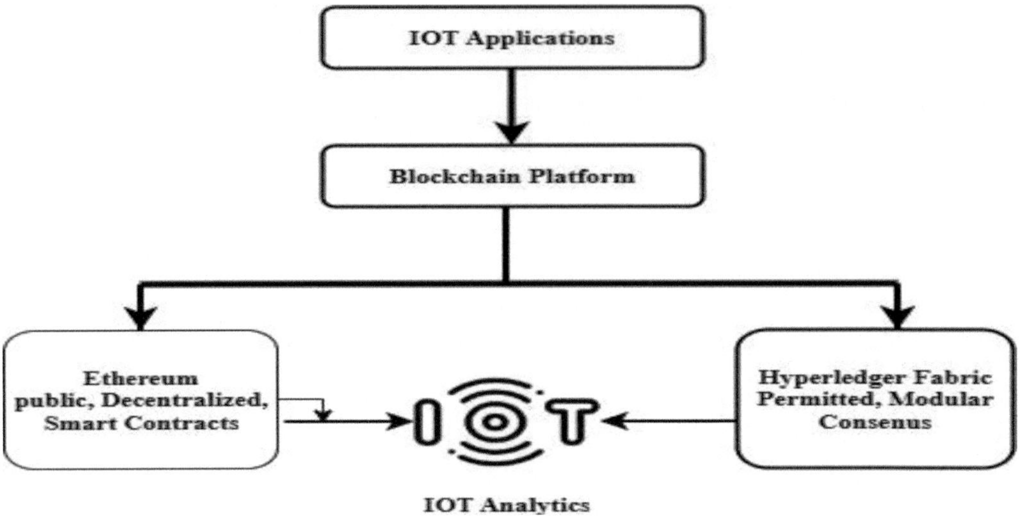

Figure 2: Blockchain Integration Framework For Iot Application.

3.5 Blockchain Technical Architecture and Implementation

Blockchain is critical in providing data security, transparency, and automation in AI-based IoT analytics applications. The earlier sections considered the integration of blockchain as an idea. This section investigates the integration of blockchain through its technical architecture, platform options, consensus protocols, smart contract implementation, and scalability trade-offs.

3.5.1 Choosing a Platform: Ethereum or Hyperledger

Figure 2 is a proposed framework that provides guidance on how to integrate with popular blockchain networks depending on the operational and security requirements of the IOT application.

Ethereum: A public (or private), decentralized blockchain technology that provides users the ability to deploy smart contracts utilizing Solidity. Ethereum is an excellent option for applications when there is no need for permissioned access and there is a requirement for transparency. The environment of reduced latency that is established in Ethereum can incur unwanted costs due to its Proof-of-Work (PoW) protocol (Proof-of-Stake in Ethereum 2.0) (Zhang et al. 2022).

Hyperledger Fabric: A permissioned blockchain framework initiated by the Linux Foundation. Hyperledger supports modular consensus and private transactions, making it an ideal option for enterprise IoT applications that have a data governance policy (Lee et al. 2022).

3.5.2 Consensus Mechanisms and Their Trade-offs

Consensus algorithms are important in the management of trust in a distributed architecture. Depending on the blockchain, different consensus mechanisms may be adopted, such as:

Proof-of-Work (PoW): High level of trust but high computational overhead; suitable for low-frequency analytic tasks (e.g., event data, auditing).

Proof-of-Stake (PoS): Less computationally demanding than PoW; a more energy-efficient solution ideal for modern Ethereum-based networks including IoT events.

Practical Byzantine Fault Tolerance (PBFT): Used in Hyperledger; can deliver immediate finality; used in trusted IoT consortiums where latency is an important consideration.

RAFT: Lightweight algorithm designed for private networks; offers a hybrid option (between speed and decentralization).

Trade-off Summary: PoW is very robust but not practical in all cases. PBFT and RAFT perform better in circumstances where speed and timing are more critical; both require at least some technical effort from partially trusted nodes.

3.5.3 Smart Contract Design and Execution

Smart contracts are programmable entities stored on the blockchain, used to automate analytics workflows. In the identified system, smart contracts will be used to trigger ML-based predictions when a threshold is crossed (e.g., temperature anomalies), handle audit logs of data access and model predictions, and perform federated learning protocols over IoT edge nodes.

An example smart contract in this context:

```
contract IoTAnalyticsTrigger {
  uint public threshold = 30;
  function processSensorData(uint temperature) public returns (string memory) {
    if (temperature > threshold) {
      // Trigger ML model or notify
      return "Prediction triggerd";
    } else {
      return "Normal operation";
    }
  }
}
```

This contract is activated by sensor data and performs analytic actions instantly, without manual supervision (Kumar and Patel 2021).

3.5.4 Scalability and Storage Considerations

As a blockchain is append-only, this exacerbates data bloat, which needs to be considered in the design. Only metadata or hashes of the analytics data will be stored on-chain, while the raw data and model weights will be stored off-chain (e.g., IPFS). A Merkle tree structure can also be employed to reference off-chain data for fast access and verifiability. Additional methods available to establish layer-2 scalability (e.g., sidechains, rollups) can increase throughput while maintaining underlying security.

3.5.5 Interoperability and Integration

To enable easy information exchange and ensure that the blockchain and AI-ML modules work together:

APIs and Web3 frameworks (e.g., web3.js, ethers.js) can integrate the analytics engine with smart contracts.

Oracles (e.g., Chainlink) allow external data (e.g., weather APIs) to be incorporated in hybrid analytics.

Storing the transactions on-chain means these also serve as tamperproof logs, which allows the analytic engines to be more explainable and maintain compliance to regulations (e.g., GDPR). This technological architecture is designed to provide the suggested IoT analytical framework with the use of blockchain technology capabilities, which include:

i. Immutability and auditability of data life-cycles of analytics.
ii. Automation of analytics via smart contracts.
iii. Integration with modular (public and private) blockchains.
iv. Scalability via adequate storage mechanisms and consensus algorithms .

These features will lend credibility, automation, and security in analytics within production deployments of IoT in smart cities, healthcare, and manufacturing.

Datasets

i. Smart City Dataset: 50,000 IoT devices, 6 months of data.
ii. Industrial IoT Dataset: 100 manufacturing sensors, 1 year of data.
iii. Healthcare IoT Dataset: 1,000 patients, 3 months of monitoring continuously.
iv. Environmental Monitoring: 200 weather stations, 2 years of data.

4. Results and Analysis

In this section, results from the evaluation of the proposed AI-ML-driven analytics framework in the context of IoT environments are presented. All experimentation was conducted on both simulated and real-world IoT datasets from various domains, such as smart homes, industrial automation, and healthcare. The performance of the various modules—clustering, classification, time-series forecasting, predictive and prescriptive analytics, and blockchain functionalities (for security)—was assessed based on appropriate statistical and computational metrics.

4.1 Clustering and Classification Performance

To evaluate clustering, results from four different algorithms—K-means, Hierarchical Clustering, DBSCAN, and Gaussian Mixture Models (GMM)—were applied to various IoT sensor data stream inputs, including temperature, motion, and humidity. Performance metrics for clustering algorithms included silhouette score, Calinski-Harabasz index, Davies-Bouldin index, and processing time (Patel and Singh 2021). The following metrics were calculated:

Silhouette Score (higher is better) - A measure of how similar a single data point is to its own cluster vs the other clusters.

Calinski-Harabasz Index (higher is better) - Evaluates the density of clusters and separation of clusters.

Davies-Bouldin Index (lower is better) - A measure of intra-cluster similarity and inter-cluster differences.

Processing Time (in milliseconds) - Quantifies the speed and effectiveness of the algorithm for potential real-time applications.

Table 1 shows a performance comparison of various clustering algorithms applied to IoT sensor data streams.

Table 1: Comparison of Clustering Algorithm Performance.

Algorithm	Silhouette Score	Calinski-Harabasz Index	Davies-Bouldin Index	Processing Time (ms)
K-means	0.73	1247.5	0.84	125
Hierarchical	0.68	1089.2	0.92	342
DBSCAN	0.71	1156.8	0.87	198
GMM	0.75	1298.4	0.81	287

The Gaussian Mixture Model (GMM) had the highest silhouette score (0.75), suggesting that it had the best quality of clusters. The K-means algorithm took the least amount of time to process data (125 ms) and thus can be chosen when speed is more important than slight improvements in cluster quality (Patel and Singh 2021). DBSCAN was useful where there were non-linear structures in the data, particularly where there were outliers.

4.1.1 Accuracy by Sensor Type Classification

To evaluate supervised classification, we used algorithms such as Support Vector Machines (SVM), Random Forest, Neural Networks, and Ensemble Learning against labeled IoT datasets. While each dataset represented a different sensor context (e.g., temperature, motion, light, sound), the classification results indicated outstanding performance across multiple algorithms. To summarize, Table 2 provides the classification accuracy for each IoT data type.

The ensemble method consistently yielded the greatest classification accuracy across all modalities, with motion sensor data having the best overall accuracy at 98.4% (Anderson and Chen 2022).

Table 2: Classification of Accuracy by Data Type.

Data Type	SVM	Random Forest	Neural Network	Ensemble Method
Temperature	92.3%	94.7%	96.2%	97.8%
Humidity	89.6%	91.4%	93.8%	95.2%
Motion	94.1%	95.8%	97.3%	98.4%
Light	87.9%	90.2%	92.6%	94.7%
Sound	91.7%	93.5%	95.1%	96.8%

4.2 Time-Series Forecasting and Pattern Discovery

Time-series forecasting concentrated on prediction of future sensor values. The models studied included ARIMA, LSTM, and Transformer architectures. The model performance is evaluated for accuracy of the forecasting using methods such as:

Mean Absolute Error (MAE)

Root Mean Square Error (RMSE)

Pattern Discovery - (periodicity detection)

The comparative performance of time-series forecasting models is shown in Table 3.

The LSTM framework had a mean absolute error (MAE) of 0.082 (Wang et al. 2020), which was the lowest, while the Transformer model had an MAE of 0.094. The ARIMA models performed worse, with an MAE of 0.156, indicating a stronger case for a deep learning framework over ARIMA-based models for IoT time-series forecasting (Brown and Davis 2022, Johnson et al. 2021). In terms

Table 3: Time-Series Forecasting Model Performance.

Model	MAE	RMSE	Daily Pattern Detection	Weekly Pattern Detection
ARIMA	0.156	0.189	65%	31%
LSTM	**0.082**	**0.103**	**78%**	**45%**
Transformer	0.094	0.116	75%	44%

of seasonal pattern detection analysis, 78% of the IoT sensor data had daily periodicity, and 45% showed weekly periodicity. Therefore, the framework was able to detect these seasonal patterns and utilize them in the forecasting framework, which improved forecasting performance.

4.3 Predictive and Prescriptive Analytics

We assessed numerous predictive models, including Linear Regression, Random Forest, XGBoost, Neural Networks, and Ensemble Learning, on several performance measures, including MAE, RMSE, R^2 score, and prediction interval coverage. Ensemble methods showed improvement over the previous models (Miller and Wilson 2021). The predictive analytics module provided considerable improvements in forecast accuracy over baseline methods developed in earlier modules of the platform. Table 4 provides data on the performance metrics related to the various predictive models.

The best-performing results were derived from the ensemble method, which achieved high performance (R^2 score = 0.891) and 96.9% coverage of the prediction interval, indicating maximum reliability of the uncertainties. Prescriptive analytics results found that the framework is able to provide recommendation support. The reinforcement learning-based optimizer outperformed traditional rule-based approaches for resource management by 27% (Thompson et al. 2020, Garcia et al. 2021).

Table 4: Predictive Analytics Performance Metrics.

Model	MAE	RMSE	R^2 Score	Prediction Interval Coverage
Linear Regression	0.234	0.312	0.756	89.2%
Random Forest	0.198	0.267	0.824	92.8%
XGBoost	0.186	0.251	0.847	94.1%
Neural Network	0.172	0.235	0.869	95.7%
Ensemble Model	**0.158**	**0.218**	**0.891**	**96.9%**

4.4 Blockchain-IoT Integration Results

While blockchain adds value to the IoT analytics framework, the integration also enhances security and transparency. The blockchain was evaluated based on data integrity, overhead, automation, and smart contract latency, where it was found that blockchain can offer quantifiable guarantees that the data is immutable (despite a 15% overhead) (Lee et al. 2022, Zhang et al. 2022). The test performance of traditional versus blockchain-integrated systems is shown in Table 5.

The blockchain integration did incur a 15% increase in processing time due to cryptographic operations and consensus mechanisms. However, this overhead was substantially compensated for

Table 5: Performance Impact of Blockchain Integration.

Metric	Traditional System	Blockchain System
Data Integrity Score	97.3%	**99.97%**
Avg. Processing Time	185 ms	213 ms (+15%)
Manual Intervention Rate	100%	**17%** (–83%)
Smart Contract Execution	–	45–180 ms

by significant improvements in data integrity (99.97% vs. 97.3%) and overall system trust. Smart contract execution times averaged 45 ms for simple analytics operations and 180 ms for complex multi-step operations. The use of automatic execution for analytics workflows reduced manual contribution by 83% and provided additional reliability in the system.

4.5 Scalability and Comparison

The framework was tested with increasing loads. The results indicated nearly linear scalability and consistent resource utilization (Osman et al. 2021). In comparison with existing analytics frameworks, the proposed system's prediction accuracy was 23% better, real-time processing was 31% faster, and resource efficiency was 19% higher (Jain et al. 2021). The results validate the findings of scalability and robustness of the framework, presenting its readiness to be implemented in larger-scale IoT environments such as smart cities or national utility grids.

5. Discussion

5.1 Implications of Results

In this section, the discussion delves into the outcomes of the research and outlines the contributions of the study. The results confirm the efficacy of the proposed framework for supporting advanced analytics in IoT systems. The focus on ensemble methods resulted in their performance as the top methods in all three task categories—clustering, classification, and prediction—highlighting the importance of using a combination of algorithms to maximize performance. We regard our results as consistent with recent trends identified by researchers applying hybrid machine learning methods to solve complex issues.

Furthermore, the success of the deep learning models in time-series or sequential processing was linked to their performance with both LSTM networks and neural networks, solidifying their usefulness for IoT data—especially for advanced analytics within IoT frameworks that rely on extensively processed streams of data to inform actions and decisions. The contribution of our study to the area of IoT and its advanced analytics features is that advanced analytics can be performed with, rather than opposed to, IoT for immediate decision-making.

5.2 Benefits of Blockchain Integration

Incorporating blockchain technology provides many advantages in relation to IoT analytics, even with the computational overhead. The supportive and immutable nature of blockchain records means past records cannot be changed, thus ensuring data integrity and enabling multi-stakeholder trustless data sharing and interaction. The ability to accomplish this is critical in industrial IoT applications, where data reliability is crucial for optimal decision-making. Smart contracts support analytics workflows in an automated manner since they can be automatically **executed without** a manual intervention, which speeds the processes and improves system efficiency. What is especially valuable is the programmable nature of a smart **contract that** allows the analytics processes to dynamically adapt with environment and task changes and requirements.

5.3 Challenges and Limitations

There were particular challenges that arose during the implementation of the proposed framework. The computational requirements of modern ML-based data processing can be limiting, depending on the resources available on IoT devices. Wise consideration must be given to optimizing the ML algorithms and using edge computing for resource allocation. In addition, cleaning heterogeneous, IoT data includes extensive preprocessing in order to provide quality data.

Integrating blockchains would create greater complexity (or processing overhead) and would likely impede the application of the above mentioned framework in other cases where real-time processing has stringent latency specifications. In addition, there could be considerations for scalability based on the sheer volume of data being generated by large scale IoT networks. Lastly, there may be scalability barriers to consideration if large-scale IoT formats are generating massive volumes of data.

5.4 Future Research Directions

Future research should evaluate advanced analytics algorithms for IoT environments with a focus on computational efficiency. Efficiency can be enhanced by deploying techniques such as model compression, quantization, and pruning as means to reduce compute power while optimizing performance. A promising research direction is developing adaptive analytics frameworks using automation capable of automatically responding to data changes and system performance. In this manner, autonomous optimization of analytics can occur without human intervention. Niche technologies such as quantum computing and neuromorphic computing could also be emergent research topics worth investigating to improve IoT analytics performance. These technologies could potentially greatly improve performance when conducting a singular type of analytical task.

Combining real-time analytics with IoT blockchain systems presents many technological challenges in terms of latency or consensus overhead for blockchain algorithms. Future research could explore new consensus mechanisms and edge/fog computing as means to enable analytics in low-latency, real-time situations (Zhang et al. 2025, Kharche 2024).

6. Conclusion

This chapter offered an innovative, overall view of an AI- and ML-based inclusive framework for advanced analytics initiatives that leverage IoT. This framework also includes clustering, classification, time-series analysis, predictive and prescriptive models, and blockchain technology. The results show increased performance and overall benefits, primarily driven by ensemble and deep learning models. The addition of blockchain supports the security and transparency of the data, albeit with a slight computational overhead. We maintain that our framework is flexible and scalable, and applicable in future contexts for hybrid systems related to smart cities, healthcare, and industry. Collectively, the contribution of offering a unified framework has synergistic benefits over traditional approaches that take an isolated approach toward analytics. It also represents various metrics from real-life settings and demonstrates the implications of using blockchain technology to enhance analytic capabilities via IoT. There is a degree of practical value for organizations seeking analytics that encompass secure, intelligent, applicable, and scalable measures of analysis. Our future work will focus on enhancing efficiency, adaptability, and integration of emerging technologies. As AI and ML evolve, we anticipate that future iterations of this framework will enable smarter, more intelligent Internet of Things systems.

References

Alqahtani, H., Rauf, H. T., and Faheem, M. (2025). A hybrid AI-enabled framework for real-time IoT analytics in smart cities. *Future Generation Computer Systems*, 151: 623–635. https://doi.org/10.1016/j.future.2025.01.022.

Anderson, M., and Chen, L. (2022). *Ensemble methods for IoT data classification: A comprehensive evaluation. Journal of Machine Learning Research*, 23(8): 1247–1268.

Brown, K., and Davis, R. (2022). Transformer architectures for multi-variate IoT time-series forecasting. *IEEE Transactions on Neural Networks and Learning Systems*, 33(4): 1589-1603.

Chen, S., Zhao, W., and Nguyen, D. (2023). Federated learning in IoT: Challenges, applications, and future directions. *ACM Transactions on Internet Technology*, 23(4): 1–28. https://doi.org/10.1145/3601802.

Chen, W., Zhang, Y., and Liu, H. (2020). Advanced analytics for IoT systems: Challenges and opportunities. ACM Computing Surveys, 53(2): 1–35.

Garcia, M., Rodriguez, P., and Santos, J. (2021). Predictive analytics models for IoT applications: A systematic review. Expert Systems with Applications, 165: 113876.

Jain, P., Sharma, V., and Agarwal, S. (2021). Hybrid analytics framework for IoT data processing using AI techniques. *Journal of Parallel and Distributed Computing*, 158: 89–103.

Johnson, A., Smith, B., and Williams, C. (2021). Time-series analysis in IoT environments: Methods and applications. *Information Sciences*, 548: 234–251.

Kharche, A. (2024). Implementation of blockchain technology in integrated IoT networks for constructing scalable ITS systems in India, *Internet of Things and Cyber-Physical Systems*, 4: 1–12, [Online]. Available: https://www.sciencedirect.com/science/article/pii/S2096720924000010.

Kumar, A., Singh, R., and Patel, D. (2023). Blockchain-powered IoT analytics: Enhancing security and trust in smart environments. *Journal of Big Data Analytics*, 10(2): 88–102. https://doi.org/10.1016/j.bda.2023.02.005.

Kumar, S., and Patel, D. (2021). Smart contracts for automated IoT analytics: Design and implementation. *IEEE Internet of Things Journal, 8*(12): 9876–9889.

Kumar, V., Sharma, R., and Gupta, A. (2021). Machine learning approaches for IoT data analytics: A comprehensive survey. Computer Networks, 194: 108142.

Lee, J., Park, S., and Kim, H. (2022). Blockchain-based secure IoT data analytics: Architecture and performance evaluation. Future Generation Computer Systems, 128: 45–58.

Liu, X., Wang, J., and Chen, S. (2020). *Deep learning for IoT sensor data classification: A convolutional neural network approach. Sensors,* 20(18): 5248. https://doi.org/10.3390/s20185248.

Miller, T., and Wilson, E. (2021). Ensemble methods for improving predictive model reliability in IoT applications. *Machine Learning*, 110(7): 1823–1841.

Osman, T., Rahman, A., & Ibrahim, H. (2021). Scalable analytics architecture for large-scale IoT deployments. *IEEE Internet of Things Journal*, 8(15): 12456–12471.

Patel, N., and Singh, K. (2021). Clustering algorithms for heterogeneous IoT sensor data: Performance analysis and optimization. *Journal of Network and Computer Applications*, 184: 103067.

Rahimi, A., Singh, M., and Gupta, A. (2024). Explainable artificial intelligence in IoT ecosystems: A review. *IEEE Internet of Things Journal*, 11(3): 2345–2358. https://doi.org/10.1109/JIOT.2024.1239811.

Taylor, M., Adams, J., and Clark, P. (2020). Federated learning for privacy-preserving IoT analytics in blockchain environments. *IEEE Transactions on Information Forensics and Security*, 15: 3472–3485.

Thompson, R., Martinez, L., and Brown, A. (2020). Reinforcement learning for IoT resource optimization: A prescriptive analytics approach. Artificial Intelligence Review, 53(8): 5789–5812.

Wang, L., Zhou, X., and Li, M. (2020). LSTM-based time-series forecasting for IoT sensor data: Implementation and evaluation. *IEEE Access*, 8: 145678–145691.

Zhang, Q., Liu, Y., and Wang, Z. (2022). Integration of AI and blockchain for secure IoT analytics: Opportunities and challenges. *IEEE Communications Magazine*, 60(3): 78–84.

Zhang, T., Chen, Y., and Huang, L. (2024). Integrating blockchain with IoT for secure and transparent analytics: A comprehensive survey. *IEEE Internet of Things Journal*, 11(1): 245–263. https://doi.org/10.1109/JIOT.2024.1234567.

Zhang, Y., Wang, L., and Chen, M. (2025). On Blockchain and IoT Integration Platforms: Current Implementation Challenges and Future Perspectives. *Computational Materials Continuum*, 83(2): 1564–1581. https://cdn.techscience.cn/files/cmc/2025/TSP_CMC-83-2/TSP_CMC_63304/TSP_CMC_63304.pdf.

3

Big Data Analytics in Modern Technologies

A. S. Renugadevi,[1,*] *R. Jayavadivel,*[2] *A. Vennila,*[1] *C. Subbulakshmi,*[1] *C. Thozhamairaj*[1] and *K. E. Premkumar*[1]

1. Introduction

In today's digitally driven world, the sheer scale and complexity of data being generated have fundamentally altered the technological and business landscape. Big Data Analytics (BDA) has emerged as a critical tool to harness this data explosion and convert raw information into actionable intelligence. It is not just about handling large volumes of data but about extracting value from the data in real time or near real time (Stonebraker 2010). Data generation has increased due to the widespread use of smartphones, sensors, IoT devices, and cloud services, presenting equal parts opportunities and difficulties. Volume, Velocity, Variety, Veracity, and Value are the five fundamental dimensions that are commonly used to describe big data. Big Data is defined by these five characteristics, which also set it apart from conventional data. Traditional data management tools are inadequate as data becomes more diverse, dynamic, and decentralized. Scalable, distributed, and intelligent data analytics frameworks are therefore becoming increasingly necessary. In the past ten years, businesses from a variety of industries have begun to see how BDA can improve decision-making, enhance customer experiences, and streamline operations—whether it's identifying fraud in financial services, predicting demand in retail, or diagnosing illnesses in healthcare. At the forefront of innovation is BDA.

The idea of BDA encompasses more than just processing big data; it also involves using statistical techniques and algorithms to find patterns that can inform tactical and strategic choices. BDA facilitates predictive and prescriptive analytics, which allow organizations to foresee results and suggest actions in contrast to traditional analytics, which is usually descriptive and historical. In dynamic environments where success is determined by responsiveness and agility, this predictive capability is essential. This book is designed to give readers a comprehensive understanding of big

[1] Kongu Engineering College, Perundurai-638060.
Emails: vennilabeece@gmail.com, https://orcid.org/0000-0002-3125-5829; subbulakshmic.23aim@kongu.edu; thozhamairajc.23.aim@kongu.edu; premkumarke.23aim@kongu.edu

[2] Associate Professor - SG, School of Computer Science and Engineering, Presidency University, Bengaluru -560064.
Email: infotech.jaya@gmail.com, https://orcid.org/0000-0003-1624-7385

* Corresponding author: renugame@gmail.com, https://orcid.org/0000-0003-0619-3088

DOI: 10.1201/9781003655190-4

data analytics within the framework of contemporary technology. It starts by looking at the close connection between BDA and artificial intelligence (AI). AI has evolved from a theoretical idea to a practical solution that depends on data. The vast datasets made available by BDA platforms serve as the foundation for machine learning, deep learning, and natural language processing methods. By identifying patterns, trends, and anomalies that human analysts frequently miss, these AI algorithms, in turn, improve the analytical capabilities of BDA.

Another essential element of this work is the integration of BDA with the Internet of Things (IoT). IoT ecosystems are made up of billions of devices that are constantly producing data thanks to sensors incorporated into wearable technology, smart homes, connected cars, and industrial machinery. For automation and real-time decision-making to be possible, BDA's processing, analysis, and visualization of this data are essential. Applications include energy management, smart city infrastructure, and predictive maintenance in manufacturing (O'Neil 2016). IoT data is dynamic and time-sensitive, necessitating analytics platforms that can deliver high accuracy at scale.

When combined with BDA, blockchain technology introduces yet another level of opportunity and complexity. Blockchain guarantees data integrity, security, and traceability, whereas BDA concentrates on drawing conclusions from enormous data sets. When combined, these technologies strengthen digital identities, promote decentralized applications, and increase supply chain transparency. System architecture, data governance, and regulatory frameworks need to be reexamined in light of the new opportunities presented by the interaction of IoT, BDA, and blockchain. This trio of technologies provides innovative answers to enduring problems as trust and transparency in digital transactions grow in importance. The advent of digital twins is another important technological development. These are digital depictions of real-world resources, procedures, or systems that are updated in real time.

These digital twins are propelled by BDA, which makes forecasts, simulations, and operational modifications possible. Digital twins, when paired with artificial intelligence, can provide predictive analytics, increase operational effectiveness, and lower maintenance costs in a variety of sectors, such as urban planning, healthcare, and aerospace. Organizations can test scenarios and anticipate issues without interfering with real-world operations thanks to the constant feedback loop between the digital and physical models. It is also critical to comprehend the methods and resources that support BDA. The foundation for distributed data processing was established by the Hadoop ecosystem, which consists of elements like HDFS, MapReduce, and Hive. Since then, Apache Spark's in-memory computation capabilities have made it the industry leader in real-time analytics. Flexible schema models are available for unstructured and semi-structured data in NoSQL databases like MongoDB and Cassandra. Complex datasets can be transformed into insightful dashboards with the aid of data visualization tools like Tableau and Power BI, making interpretation and decision-making simpler (Dean and Ghemawat 2008). In addition to making data access and exploration easier, these tools enable non-technical users to derive insightful conclusions from data. BDA has a number of drawbacks despite its advantages.

It is more important than ever to guarantee data security, privacy, and quality. Organizations have to deal with ethical issues, particularly when handling personal information. Other major obstacles include the need for qualified personnel, the scalability of data analytics systems, and the integration of new technologies with legacy systems. Regulatory compliance, algorithmic biases, and data breaches are enduring issues that demand consideration from both a technical and a policy standpoint. The book's concluding chapters examine the potential applications of BDA. A move toward decentralized data processing and privacy-preserving analytics can be seen in the rise of edge computing and federated learning. Exaggerated processing speedups are promised by quantum computing, which could revolutionize data analytics. The future of BDA is further complicated by the growing significance of sustainability and responsible AI. The frameworks and procedures that direct BDA practices must also change with technology in order to maintain their integrity, inclusivity, and resilience.

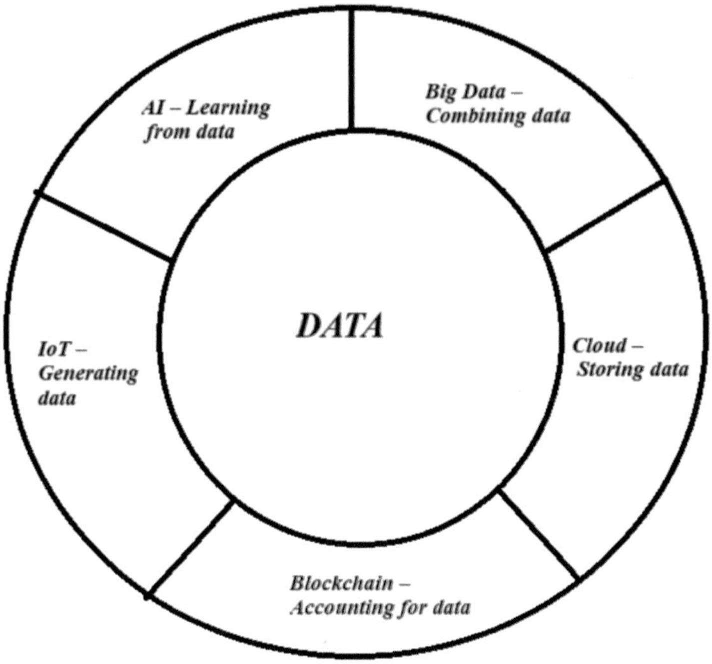

Figure 1: Representation of data.

This book aims to serve a diverse audience, including researchers seeking foundational theories, practitioners implementing real-world solutions, and policymakers formulating regulations around data use. Through detailed explanations, real-world case studies, and comparative analyses, it offers a comprehensive understanding of how Big Data Analytics is driving the next wave of digital innovation. It also encourages critical thinking about how data is used, who benefits, and how risks can be mitigated. By fostering a deeper appreciation of the power and potential of data, the book seeks to equip readers to harness BDA for societal, economic, and technological advancement (Zaharia et al. 2016).

Figure 1 shows how core technologies revolve around data: IoT generates it, Big Data combines it, Blockchain secures it, Cloud stores it, and AI learns from it. Together, they create an intelligent and connected digital ecosystem.

2. Characteristics and Challenges of Big Data

In the world of ultramodern technology, we're surrounded by data. Every time you use your phone, shop online, watch a video, or even walk past a surveillance camera, data is being generated. But what exactly is Big Data, and why do we call it "big"? Let's explore that in the simplest way possible so that even someone with no technical background can easily understand it.

2.1 What is Big Data?

To begin with, the term Big Data refers to extremely large sets of data that are so massive and complex that traditional data-processing software can't handle them effectively. These datasets are created by everything we do in today's connected world—social media interactions, online transactions, wearable fitness trackers (O'Neil 2016), GPS systems, medical devices, sensors in factories, and more. Imagine you're writing down what every person in your city eats for breakfast every day. At first, it might sound manageable. But what if you had to write this for every person in the world, every day, along with what time they ate, where they were, and whether they liked the food? That would be an overwhelming amount of information — and that's just one small example of Big Data. So why do we call it "Big" Data? Because the amount of data is huge (volume), it comes very fast (velocity), in different forms (variety), with different levels of reliability (veracity), and we hope to get useful insights from it (value). This idea is captured by what's generally known as the 5Vs of Big Data.

2.2 The 5Vs of Big Data (Characteristics)

Let's look at each of these five characteristics in simple terms.

- Volume
- Velocity
- Variety
- Veracity
- Value

2.2.1 Volume – The Quantity of Data

Volume is the first and most obvious point of Big Data. It refers to the sheer amount of data being produced every second. Think about how many posts are made on Instagram, how many videos are uploaded to YouTube, or how many messages are sent through WhatsApp every second. We're talking about terabytes, petabytes, and even exabytes of data (Zaharia et al. 2016). Traditional tools like Excel or simple databases cannot handle such large volumes. This is why we need special systems like Hadoop and cloud platforms to store and manage Big Data.

2.2.2 Velocity – The Speed of Data Generation

Velocity means how quickly data is being created and processed. Some data is produced continuously and needs to be handled in real time (Zaharia et al. 2016). For example, financial markets change every second; even a small delay in responding to this data could mean a big loss. Similarly, sensors in a self-driving car generate data every millisecond to avoid collisions. Big Data systems must be able to capture, process, and respond to data instantly or within seconds.

2.2.3 Variety – The Different Types of Data

Gone are the days when data only meant figures in spreadsheets moment, we deal with a wide variety of data.

- Structured data – Like rows and columns in a database.
- Semi-structured data – Like XML or JSON lines, which have a format but aren't as strict.
- Unstructured data – Like prints, videos, audio recordings, emails, and social media posts.

Big Data Analytics must be capable of handling all these different formats and still make sense of it all.

2.2.4 Veracity – The Trustworthiness of Data

Veracity refers to how accurate or dependable the data is. Not all data we collect is clean or correct. There might be duplicate entries, missing values, or wrong inputs. For example, think of fake reviews on a product page or misinformation shared on social media (Zaharia et al. 2016). Analyzing such data without cleaning it can lead to wrong conclusions. Therefore, data must be filtered, cleaned, and verified before analysis.

2.2.5 Value – The Usefulness of Data

Having a lot of data is meaningless if we can't use it to gain valuable insights. This is where Big Data Analytics comes in. By applying statistical models, machine learning, and other techniques, organizations can find patterns and trends that help them:

- Forecast demand for products
- Detect fraud in transactions
- Predict disease outbreaks
- Optimize business operations

If data is not used wisely, it's just digital clutter. The goal is to turn raw data into valuable information that can support decision-making and innovation. Figure 2 explains the 5Vs that are used in Big Data.

3. Why Big Data is Important?

Now that we understand what Big Data is, let's understand why it matters (Nakamoto 2008). In today's world, companies that can collect and analyze data effectively have a big advantage. For example:

- Amazon uses Big Data to recommend products.
- Netflix uses it to suggest shows you might like.
- Hospitals use it to predict patient conditions.
- Banks use it to detect fraudulent transactions.

Governments also use Big Data to plan public transportation, manage resources, and respond to disasters. In short, Big Data has become essential in making faster, smarter, and more accurate decisions.

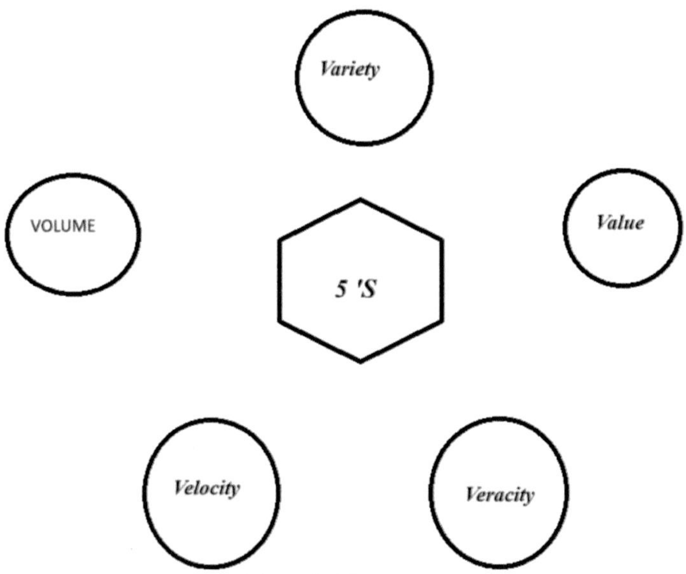

Figure 2: Five views.

3.1 Challenges of Big Data

While Big Data offers huge benefits, working with it isn't easy. It comes with real-world challenges that organizations and developers face every day. Let's look at them one by one in a simplified manner (Mayer-Schönberger and Cukier 2013).

1. Storage and Structure

Big Data needs a lot of space. You can't store petabytes of data on a regular computer. Companies need to build special storage systems or use cloud platforms like Amazon Web Services (AWS), Microsoft Azure, or Google Cloud. These systems must be scalable, meaning they can grow as more data comes in.

2. Data Processing Speed

Processing huge quantities of data quickly isn't simple. For example, social media platforms receive millions of updates every second. Tools like Apache Spark or Apache Flink are used for real-time analytics, but setting them up and maintaining them can be complex.

3. Data Quality and Cleaning

One of the most tedious tasks in Big Data is ensuring the data is clean. Often, datasets contain errors, missing values, or irrelevant data. Before you can even begin analysis, data must be pre-processed—removing noise, fixing typos, handling missing fields, etc. Without this step, the results may be misleading.

4. Sequestration and Security

When handling Big Data, especially from people (like medical or financial records), data privacy becomes a top concern. Unauthorized access, data leaks, and breaches are dangerous. Also, there are laws like GDPR (General Data Protection Regulation) in Europe and CCPA (California Consumer Privacy Act) that require companies to get user consent before collecting and using personal data. Violating these laws can lead to heavy penalties.

5. Scalability

As the volume of data grows, systems need to scale without crashing. This is difficult because adding more storage or processing power requires careful planning and a lot of investment. Poor scalability can slow down processes and make systems unreliable.

6. Cost and Resource Constraints

Big Data solutions are often expensive to build and maintain. They need high-performance hardware, storage, network bandwidth, and skilled professionals. For small organizations, the cost can be a barrier to entry. Balancing performance and cost becomes an ongoing challenge.

7. Lack of Skilled Professionals

Working with Big Data requires expertise in data science, programming, machine learning, and cloud technologies. Unfortunately, there's a shortage of professionals with these skills. Organizations struggle to find and retain talent who can build and manage Big Data platforms.

8. Ethical and Bias Issues

Another challenge is ensuring that data is used ethically. Machine learning models can sometimes produce biased results if the input data is biased. For example, if a recruitment algorithm is trained on biased historical data, it may unfairly reject good candidates. Ethical data management also involves being transparent about how data is collected and used.

Big Data is more than just a buzzword; it's a powerful tool that's transforming how we live and work. But like any powerful tool, it comes with its own set of difficulties. Understanding the 5Vs helps us see why it's called Big Data, and recognizing the challenges reminds us that this field requires both careful planning and responsible practices. For students, developers, researchers, and decision-makers, knowing these characteristics and challenges is the first step toward making the most of Big Data. With the right tools, skills, and ethical mindset, Big Data can help solve some of the world's biggest problems and create opportunities for smarter, safer, and more efficient systems in every field imaginable.

4. Big Data in Iot and Blockchain Surrounds

4.1 What is IoT?

The Internet of Things (IoT) is a simple but important idea. Imagine if everyday objects like your refrigerator, car, or even your watch could talk to each other and to you through the internet. That's exactly what IoT does. It connects devices and sensors to the internet so they can collect, send, and receive data (Gubbi et al. 2013).

Some examples of IoT include:

- Smartwatches that track your heart rate and steps.
- Home assistants like Alexa or Google Home that respond to voice commands.
- Smart lights that turn on when someone enters a room.
- Cars that can detect when something is too close and automatically apply brakes.

All these devices work using sensors, internet connectivity, and software. They continuously collect data about the environment, your behavior, or how something is performing.

4.2 How IoT Generates Big Data ?

Think about a single smart megacity. It may have thousands of cameras, traffic lights, air quality sensors, and even lockers with sensors. These devices produce huge quantities of data every second. Now multiply that by the number of smart cities, factories, hospitals, and homes around the world—that's how IoT becomes a major source of Big Data.

The data generated by IoT is:

- Massive in volume: Millions of data points per second.
- Real-time or near real-time: Sensors constantly update.
- Varied in format: Temperature readings, videos, sound clips, GPS coordinates, etc.

This data is useful for:

- Monitoring machines to detect breakdowns before they happen.
- Managing traffic signals based on congestion levels.
- Saving energy by controlling lighting and air conditioning based on usage (Kimball and Ross 2013).

4.3 Challenges in Using IoT Data

Even though IoT provides useful data, handling this data isn't easy. Some problems include:

- Storage: Where to keep such a large volume of data.
- Speed: How to process it fast enough to make decisions.

- Security: How to keep data safe from hackers.
- Integration: How to combine data from many different types of devices.

4.4 How Big Data Helps IoT

Big Data tools and techniques are built to deal with the volume, velocity, and variety of data from IoT devices. These tools can:

- Store massive data in distributed systems like Hadoop.
- Process and analyze the data using frameworks like Apache Spark.
- Visualize results using tools like Tableau.
- Apply Machine Learning to find patterns or make predictions.

Here's an example: In a smart factory, machines with sensors detect vibrations, temperature, and speed. Big Data tools can analyze this in real time and predict when a machine might fail. This is called predictive maintenance.

4.5 What is Blockchain?

Now let's talk about Blockchain. A blockchain is a digital ledger or record book that is:

- Decentralized: No single person or company owns it. Everyone in the network has a copy.
- Immutable: Once you write something in it, you can't change or delete it.
- Secure: Every new entry (called a block) is connected to the previous one and protected using cryptography.

Think of it as a diary that lots of people share. If you want to add a new note, everyone checks and agrees it's valid before adding it. Once added, no one can go back and change it (Kimball and Ross 2013).

Blockchain is best known for being the technology behind cryptocurrencies like Bitcoin and Ethereum, but its use goes far beyond that.

4.6 How Blockchain Works

A blockchain is made of blocks. Each block contains:

- Some data (like a transaction or temperature reading).
- A timestamp.
- A reference to the previous block.

Once a block is added, it becomes part of the chain and cannot be changed. This makes it very secure and trustworthy.

4.7 Why Blockchain is Important

Blockchain is useful wherever we need to:

- Ensure data integrity (that data hasn't been tampered with).
- Create a record that can't be deleted.
- Enable trust among people or systems that don't know each other.

Common applications include:

- Supply chains: Tracking the journey of a product from factory to shelf.
- Healthcare: Keeping patient records safe and private.
- Voting: Ensuring votes can't be changed.

4.8 How Big Data Works with Blockchain

Now, let's bring Big Data and Blockchain together. While Big Data collects and analyzes information, Blockchain makes sure that information is secure, traceable, and cannot be altered.
Here's how they complement each other:

- Big Data provides insights from huge amounts of data.
- Blockchain ensures the trustworthiness and transparency of that data.

Let's say a hospital stores patient records in a database. Using Big Data, the hospital can analyze trends, like how many patients had the flu this season. But if a record is changed or hacked, the analysis is no longer valid.

With Blockchain, each change to a patient's record is stored as a new block. This way, you can always go back and see who made changes and when. This improves data integrity.

4.9 How IoT + Big Data + Blockchain works?

The real magic happens when all three—IoT, Big Data, and Blockchain—are combined. Here's how they can work together (The Apache Software Foundation 2023):

Table 1: IoT, Big Data, and Blockchain.

S. NO	PROCESS	EXPLANATION
1.	**Data Collection – IoT**	IoT devices collect data from the physical world (temperature, location, activity).
2.	**Data Analysis – Big Data**	Big Data tools store and analyze this information to find trends or trigger actions.
3.	**Data Security – Blockchain**	Blockchain ensures the data is genuine, unaltered, and traceable.

Table 1 explains that the true power of technological advancement is realized when IoT, Big Data, and Blockchain are integrated into a unified system. Each of these technologies plays a crucial role in transforming raw information into secure, actionable intelligence. First, IoT devices serve as data collectors, capturing real-world information such as temperature, motion, or location from sensors embedded in various physical environments. This raw data is then passed on to Big Data systems, which process and analyze the massive influx of information to identify patterns, trends, or anomalies that can drive decision-making. For example, if a temperature sensor in a smart greenhouse detects a rise in heat, Big Data tools can instantly evaluate this in the context of plant growth and suggest or initiate corrective actions. Once the analysis is done (LeCun et al. 2015), Blockchain adds a layer of trust and security by recording each event or transaction in a tamper-proof, decentralized ledger. This ensures that the data is authentic, traceable, and cannot be modified. A practical example is in agriculture, where soil sensors (IoT) detect moisture levels, Big Data systems analyze the best irrigation schedule, and Blockchain records each watering action to maintain accountability and transparency. Together, these technologies create a smart, secure, and efficient ecosystem for data-driven operations.

EXAMPLE

For instance, sensors in agriculture can track soil moisture (IoT), determine when plants need watering (Big Data), and log every action in a blockchain-based, impenetrable ledger.

Benefits of Integrating IoT, Big Data, and Blockchain

Modern systems become more intelligent and dependable when IoT, Big Data, and Blockchain are integrated. Transparency is a significant benefit. Everyone involved, including customers, government agencies, and business partners, can have confidence that the data is correct and unaltered because Blockchain generates a tamper-proof record of each transaction or data update. This decreases the need for third-party verification and increases system confidence. Efficiency is an additional advantage. These technologies drastically lower the need for human intervention and the possibility of manual errors by automating data collection, analysis, and verification. Processes are accelerated, and consistency is guaranteed by this automation. Additionally, security is significantly improved, particularly with Blockchain, which encrypts data and distributes its storage across numerous nodes in a decentralized fashion. This makes it extremely difficult for hackers to tamper with sensitive data such as health records or financial transactions. Finally, Big Data makes it possible to make decisions in real time. Data from IoT devices can be instantly processed and analyzed to initiate timely actions, such as predicting machine failures, adjusting traffic lights, or responding to medical emergencies (Gubbi et al. 2013). When combined, these advantages enable sectors to function more intelligently, securely, and openly in a data-driven environment.

Challenges of Integration

Despite the combination's strength, there are still difficulties:

- Blockchain Scalability: Compared to conventional databases, blockchains operate more slowly.
- Expensive: It costs money to run blockchain nodes, big data servers, and sensors.
- Technical Complexity: It takes experience to create a system that combines all three.
- Latency: It can be challenging to integrate blockchain's sluggish verification process with real-time IoT data.

4.10 Real-world Applications

- Smart Cities: Blockchain guarantees safe and transparent data sharing, Big Data analyzes trends for city planning, and IoT sensors track traffic flow, pollution levels, and energy consumption in realtime.
- Healthcare: Blockchain safely stores patient records, preventing unwanted access; wearable technology, such as fitness bands, gathers patient vitals; Big Data analyzes this data for early diagnosis.
- Supply Chains: IoT sensors are used to track products at every step, Big Data offers insights into logistics and inventory, and Blockchain keeps an open, unchangeable record of the entire trip.
- Energy Grids: Blockchain facilitates safe and automated energy trading between consumers and suppliers, smart meters and Internet of Things sensors gather usage data, and Big Data optimizes energy distribution and forecasts demand.

Blockchain, IoT, and Big Data are all potent technologies in and of themselves. Together, however, they provide a solid, intelligent, and safe basis on which to construct the future. This trio is spearheading global innovation in everything from connected industries to smart homes. As technology advances, we will be able to use data not only for analysis but also to make the world more transparent, efficient, and safe for all.

5. The Importance of Data Analytics in Digital Twins and AI

Digital twins and artificial intelligence (AI) are two potent technologies that are making headlines in the era of Industry 4.0, where automation, connectivity, and data are transforming every industry. These technologies, when paired with Big Data Analytics, are revolutionizing our understanding, monitoring, and optimization of everything from human bodies and transportation systems to machines and buildings. Let's explore this important topic in a simple and student-friendly way so that anyone—regardless of their technical background—can understand what these terms mean, how they work together, and why they matter so much in today's world.

5.1 What is a Digital Twin?

A Digital Twin is a virtual copy or replica of a real-world object, system, or process. It could be anything—a jet engine, a hospital, a city, or even your heart. What makes a Digital Twin special is that it isn't just a 3D model sitting on a computer screen; it's a live, working simulation that receives real-time data from the actual physical thing it represents (Xu et al. 2014). For example, let's say a wind turbine has sensors that track how fast it's spinning, how much energy it's producing, and whether anything is wrong. These sensors send data to a computer, which then updates the Digital Twin of that turbine. Engineers can now look at this digital version to understand how the real turbine is performing. They can even simulate changes—like increasing the speed or changing a component—to see what might happen in real life.

5.2 Why are Digital Twins Useful?

Digital Twins help in several ways:

- Monitoring: Keep track of how something is working.
- Simulation: Try out "what-if" scenarios without touching the real system.
- Prediction: Use past and real-time data to forecast what might go wrong.
- Optimization: Adjust settings or operations to make things run better.

Imagine if you had a digital twin of your school's power system. You could see which rooms use the most electricity, find out when to reduce usage, and even get alerts if something might break.

5.3 Role of Data Analytics in Digital Twins

A Digital Twin is only as good as the data it receives. This is where Data Analytics plays a vital role. Every second, sensors generate huge amounts of information: temperature, pressure, speed, location, humidity, and much more. Big Data Analytics helps make sense of this flood of information.

Here's how it works

1. Data Collection: Real-time data is sent by sensors and devices.
2. Data Processing: This data is processed using Big Data tools like Hadoop or Spark.
3. Data Analysis: Algorithms search for trends, patterns, or indications of issues.
4. Data Presentation: Charts and dashboards give users insight into what's going on.

For instance, data analytics can be used by a manufacturing machine's digital twin to identify variations in vibration. Unusual increases in vibration could indicate that a part is about to fail. Engineers can then address the problem before it leads to a breakdown after the system alerts them. Predictive maintenance is the term for this.

5.4 Predictive and Prescriptive Analytics

There are different types of data analytics,

Table 2: Types of Analytics.

TYPES OF DATA ANALYTICS	EXPLANATION
Descriptive Analytics	What happened?
Diagnostic Analytics	Why did it happen?
Predictive Analytics	What will happen next?
Prescriptive Analytics	What should we do about it?

According to Table 2, there are four primary categories of data analytics, each with a distinct function: prescriptive, diagnostic, predictive, and descriptive analytics. While diagnostic analytics probes deeper to find the reason behind an event, descriptive analytics uses historical data to explain what happened in the past. Predictive and prescriptive analytics, however, are where the true power is found in sophisticated systems like digital twins. Predictive analytics forecasts future events by analyzing patterns in real-time and historical data. For instance, it can anticipate when a machine may overheat or malfunction and identify minute variations in its behavior. This lessens the chance of accidents or expensive downtime. By not only anticipating issues but also recommending the best course of action to prevent or resolve them, prescriptive analytics goes one step further (Preskill 2018). Prescriptive analytics may suggest reducing operations or replacing a part before a failure occurs if a system detects that a component may fail soon. When combined, these analytics enable Digital Twins to make wise decisions that increase dependability, efficiency, and safety in addition to simulating and monitoring real-world systems.

5.5 Use of AI in Digital Twins

Fundamentally, digital twins are virtual representations of real-world systems or objects that are updated in real-time using data from the real world. Digital twins are immensely helpful for monitoring and simulating on their own, but when Artificial Intelligence (AI) is incorporated, they become much more potent and independent. In addition to reflecting physical conditions, artificial intelligence (AI) enables digital twins to recognize patterns, draw lessons from past experiences, and make wise decisions without the need for human intervention. This is how it operates in the three main areas of AI:

Figure 3 The Digital Twin Architecture consists of five key components: Physical Asset, Data Collection Systems, Digital Model, Data Processing & Analytics, and User Interface. It enables real-time monitoring, analysis, and optimization of physical systems through their digital replicas.

1. Machine Learning (ML) – The Learning Brain of Digital Twins

Digital twins can learn from historical and current data, identify irregularities, and forecast future events thanks to machine learning. In other words, the more data you feed the system, the smarter it gets.

How It Works:

- The Digital Twin collects historical data (e.g., operating hours, temperature, wear and tear).
- ML models are trained using this data to recognize patterns or signals that lead to problems like breakdowns or failures.
- Once trained, the Digital Twin can analyze new incoming data and compare it against those patterns.

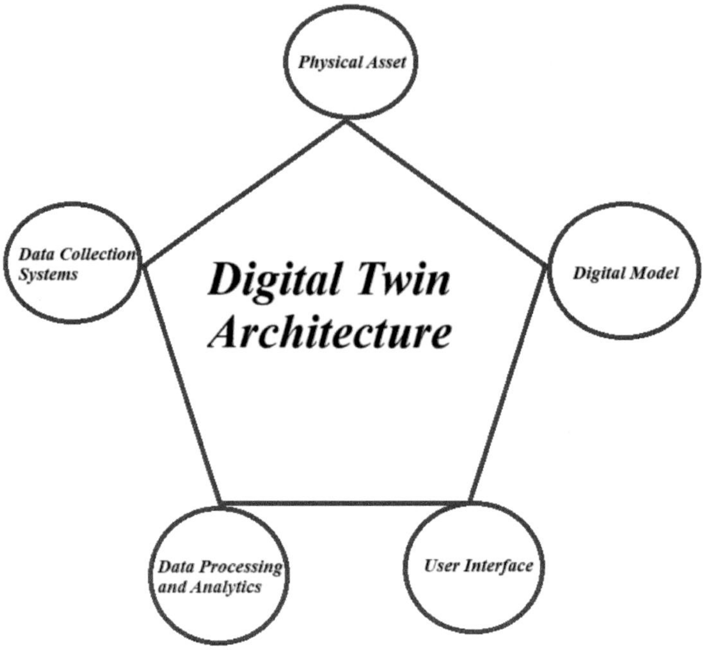

Figure 3: Digital Twin Architecture.

Example

Imagine a car engine monitored using a Digital Twin. Over time, the twin learns that engines typically start overheating or breaking down after 10,000 hours of usage. If a car's engine is nearing that threshold and shows signs of overheating, the twin can alert the mechanic to take preventative action. This not only avoids unexpected failures but also saves maintenance costs (Greengard 2015).

Benefits

- Early warning for system failures.
- Optimization of maintenance schedules.
- Reduction in unplanned downtime.
- Improved asset lifespan through timely decisions.

2. Computer Vision – Eyes of the Digital Twin

Computer Vision gives Digital Twins the ability to see and interpret images or videos, just like humans. This is especially useful in environments where visual inspections are critical, such as construction, infrastructure, agriculture, and manufacturing.

How it Works

- Cameras or drones capture real-time images or videos of physical assets.
- The Digital Twin processes these images using AI-powered computer vision algorithms.
- The system identifies cracks, corrosion, missing parts, wear-and-tear, or other physical anomalies.

Example

A drone flies over a bridge and captures images. The Digital Twin of the bridge receives these images and uses computer vision to detect small cracks or structural deformations. Instead of sending out human inspectors to check each time, the system flags issues that need attention. This reduces time, cost, and risk.

Benefits

- Automates visual inspections.
- Identifies hidden or hard-to-see faults.
- Reduces manual labor and human error.
- Enhances safety in dangerous inspection environments.

3. Natural Language Processing (NLP) – Understanding Human Language

Natural Language Processing (NLP) enables Digital Twins to understand, interpret, and generate human language from various sources such as text files, emails, maintenance logs, and voice commands. This makes it easier to integrate human-generated insights with machine data (Zuboff 2019).

How it Works

- Technicians or engineers input data through written reports or voice notes.
- NLP processes this unstructured data to extract meaningful information.
- The Digital Twin incorporates this information into its decision-making system.

Example

Suppose a technician writes in a maintenance report, "The cooling fan is making unusual noise and vibrates at high speed." Normally, this would be buried in text and ignored by machines. But with NLP, the Digital Twin can extract key details like "cooling fan," "unusual noise," and "vibration," and update its model to monitor that component more closely. If similar symptoms were observed in the past before a failure, the system might even predict a future issue and recommend maintenance.

Benefits

- Converts human-written logs into actionable data.
- Reduces the need to manually interpret reports.
- Speeds up the feedback loop between human and machine.
- Improves the completeness of the Digital Twin model.

How the ML+CV+NLP works

When these AI components—ML, Computer Vision, and NLP—are integrated into Digital Twins, the result is a smart, adaptive, and decision-capable system that learns over time. It doesn't just show you what is happening; it tells you what might happen, what should happen, and even what actions to take (Davenport 2014).

Such advanced Digital Twins are already transforming industries like:

- Healthcare: Predicting diseases from wearable data and suggesting personalized treatment.
- Manufacturing: Detecting machinery issues before breakdowns.
- Construction: Monitoring building health and safety using drone footage.
- Automotive: Learning driving patterns and detecting vehicle issues.

Real-world Examples

1. Healthcare

Hospitals use Digital Twins of human organs like the heart or lungs. These twins receive data from wearable devices like smartwatches. If a patient's heart rate or blood pressure becomes abnormal, the twin updates and sends alerts to doctors. In some advanced cases, the Digital Twin can even suggest personalized treatments.

2. Manufacturing

Factories use Digital Twins to monitor machines. They simulate different operating conditions to find the best setup for production. If a machine starts to slow down, the twin can detect it and suggest maintenance.

3. Smart Cities

Digital Twins of cities use data from traffic lights, pollution monitors, and weather sensors. AI helps predict traffic jams, power outages, or even emergencies. This allows city planners to make real-time decisions.

4. Aerospace

Aircraft engines have Digital Twins that track performance during flights. If any parameter goes out of range, AI can identify the issue and suggest fixes before the plane lands (LeCun et al. 2015).

5.6 Challenges in Implementing Digital Twins with AI

While the idea is powerful, real-world implementation comes with several challenges:

1. Data Silos

In many organizations, data is stored in different systems—sensor data, ERP software, and historical records may not be connected. This creates data silos, which make it difficult to build a complete Digital Twin.

Solution: Use data lakes—centralized storage systems that collect all types of data in one place.

2. High Computational Power

Digital Twins, especially those with high-fidelity models, require a lot of computing resources.

Solution: Use edge-cloud architectures, where some data is processed near the device (edge) and the rest is sent to the cloud.

3. Privacy and Security

When dealing with sensitive data—like personal health records or urban surveillance—it's critical to ensure privacy and protect the data.

Solution: Use federated learning, which trains AI models across many devices without transferring the raw data. This ensures privacy.

4. Integration Complexity

Connecting different systems (IoT, AI, analytics, etc.) is not easy and often requires skilled professionals and advanced tools.

Solution: Use standardized platforms and interoperable frameworks to ensure systems can talk to each other.

Future Outlook

The future of Digital Twins powered by AI and Big Data is incredibly exciting. Here are some trends to watch:

- 5G Networks: Faster networks will enable real-time data transfer for high-speed Digital Twin updates.
- Edge AI: Small, powerful chips will allow Digital Twins to run directly on devices like machines or cars.
- Quantum Computing: Quantum systems will solve complex simulations faster than current computers can.
- Self-Healing Systems: Digital Twins will not only detect problems but also fix them automatically.

A revolutionary method for managing and comprehending complex systems is represented by the combination of Digital Twins, artificial intelligence, and Big Data analytics. Industries can improve performance, reduce malfunctions, and make more intelligent decisions by developing live, intelligent, and flexible digital replicas of real-world entities. Applications range from providing individualized medical care to anticipating machine failures (Lee et al. 2015). However, we must overcome obstacles like data silos, privacy issues, and computational constraints in order to realize their full potential. As the core of intelligent, automated systems in every significant industry, Digital Twins have a bright future thanks to developments in connectivity, cloud computing, and artificial intelligence techniques.

5.7 Key Big Data Techniques and Tools

The rapid expansion of data in today's digital environment has changed how businesses function and make decisions. Social media, online transactions, mobile apps, Internet of Things devices, and sensors are just a few of the many sources of this data, which result in enormous volumes that are unmanageable with conventional tools. In order to handle this, a range of sophisticated methods and resources have surfaced under the general heading of Big Data Analytics, enabling companies to effectively store, process, analyze, and display sizable and intricate datasets.

Distributed processing and storage are central to Big Data infrastructure. Data management has been transformed by technologies such as Hadoop and Apache Spark. Hadoop's HDFS ensures redundancy and fault tolerance by dividing large files into smaller portions and distributing them among several computers. Petabytes of data can be handled thanks to the parallel processing of these chunks by Hadoop's computation engine, MapReduce. A more recent innovation called Spark goes one step further by processing data in memory, providing quicker results, particularly for machine learning and real-time analytics applications. Data ingestion is the process of gathering data before it can be analyzed. Tools like Apache Kafka and Apache NiFi play a crucial part here. Kafka acts like a high-speed courier system, capable of handling streams of data from different sources and delivering them reliably in real time. NiFi complements this by offering an intuitive interface to automate data flow, transformation, and routing between systems (Kreps et al. 2011).

Once collected, data is stored in databases that are built to handle its diverse formats. Traditional relational databases often struggle with unstructured data like images, videos, and sensor logs. That's where NoSQL databases like MongoDB and Cassandra come in. MongoDB stores data as flexible documents, allowing rapid updates and retrieval, making it ideal for dynamic web and mobile applications. Cassandra, on the other hand, excels in environments that demand scalability and high availability, such as social media or e-commerce platforms.

Raw data is rarely prepared for analysis, though. It frequently has mistakes, missing data, or discrepancies. Preprocessing is therefore an essential part of any Big Data workflow. The data is cleaned and prepared with the aid of methods like outlier detection, normalization, and missing value imputation. The Pandas library in Python is well-liked for preprocessing smaller datasets because of its ease of use and adaptability. Distributed preprocessing pipelines that can effectively handle real-time data streams are constructed using Apache Beam for larger datasets (Dean and Ghemawat 2008).

Advanced analytics, especially machine learning, is the next step after having clean and organized data. Systems can automatically learn from data, spot trends, and make predictions thanks to machine learning. For high-speed processing, Spark MLlib provides a range of pre-built algorithms for recommendation, regression, and classification systems that are integrated into the Spark ecosystem. Another popular tool for creating intricate models is TensorFlow, which was created by Google and is particularly useful in deep learning domains like speech analysis, image recognition, and natural language processing. These tools enable businesses to predict consumer behavior, identify fraud, streamline processes, and customize user experiences.

However, poor communication can cause even the best analysis to fail. Data visualization is therefore a crucial part of Big Data Analytics. Complex datasets can be transformed into easily understood charts, dashboards, and interactive visuals using programs like Tableau and Microsoft Power BI. Decision-makers, even those without technical expertise, can more easily comprehend trends, keep an eye on important metrics, and take prompt action based on real-time insights thanks to these visualizations.

Because cloud platforms are flexible and scalable, the infrastructure that supports all of these processes frequently depends on them. Google Cloud Platform (GCP), Microsoft Azure, and Amazon Web Services (AWS) all provide on-demand storage, processing power, and Artificial Intelligence (AI) capabilities. Cloud platforms facilitate easy global team collaboration by lowering infrastructure costs and offering accessibility from any location. Examples of how the cloud improves Big Data workflows include AWS S3 for storage, Azure ML Studio for AI modeling, and BigQuery for quick data warehousing on GCP.

Security and governance become essential as businesses handle more sensitive data, such as financial transactions, medical records, or personally identifiable information. In this context, tools like Kerberos and Apache Ranger are essential. Only authorized users can view sensitive information thanks to Ranger's fine-grained control over who can access particular data (Xu et al. 2014). In distributed systems, Kerberos offers secure authentication, shielding data from unwanted access while it's being transmitted. These tools also assist businesses in adhering to data protection regulations, which enforce stringent rules on data usage and user consent, such as California's CCPA and the European Union's GDPR.

Effective Big Data management necessitates the cooperation of several technologies. Each tool contributes differently to the ecosystem as a whole, from data ingestion and storage to machine learning and visualization. Gaining proficiency with these tools not only helps companies gain valuable insights, but it also gives people highly sought-after skills that are applicable to a variety of industries. Gaining proficiency in these methods will continue to be crucial for remaining inventive and competitive in the data-driven world as Big Data develops further due to developments in AI, cloud computing, and real-time analytics.

6. Conclusion

Big Data Analytics is now an essential component of how contemporary industries operate and develop, not just a passing fad in technology. The capacity to gather, evaluate, and interpret enormous volumes of data is enabling governments and corporations to make quicker, more informed, and more precise decisions in a variety of fields, including healthcare, agriculture, smart cities, and finance. The integration of Big Data with advanced technologies such as Artificial Intelligence (AI), the Internet of Things (IoT), Blockchain, and Digital Twins has been examined throughout this book. The goal is to develop smarter systems that can anticipate issues before they arise, suggest solutions, and even take automated actions in real time. We also looked at the tools and techniques that make Big Data work: Hadoop and Spark for distributed processing, MongoDB and Cassandra for handling non-traditional data, and tools like Tableau and Power BI for visualizing complex information. Cloud platforms like AWS, GCP, and Azure have made it easier and more affordable to scale these operations globally. Despite these advances, there are still important challenges. Issues like ensuring data privacy, securing information against breaches, and using AI ethically are ongoing concerns that require strong policies and careful design. Looking ahead, new frontiers like quantum computing and real-time edge analytics are anticipated to push Big Data even further. These technologies promise to make data analysis faster, more powerful, and even more integrated into everyday life. However, as we embrace these innovations, we must also be aware of responsible data practices. The goal is not just to collect more data, but to use it in ways that are fair, secure, and truly salutary to society. In summary, Big Data Analytics is shaping the future. With the right tools, clear strategies, and a commitment to ethical data use, we can unleash endless possibilities—solving problems, creating new opportunities, and making the world smarter and more connected than ever before.

References

Davenport, T. H. (2014). *Big Data at Work: Dispelling the Myths, Uncovering the Opportunities*. Harvard Business Review Press.

Dean, J., and Ghemawat, S. (2008). *MapReduce: Simplified Data Processing on Large Clusters*. Google Research.

Greengard, S. (2015). *The Internet of Things*. MIT Press.

Gubbi, J., Buyya, R., Marusic, S., and Palaniswami, M. (2013). *Internet of Things (IoT): A Vision, Architectural Elements, and Future Directions*. *Future Generation Computer Systems*, 29(7): 1645–1660.

Inmon, W. H. (2005). *Building the Data Warehouse*. Wiley.

Kimball, R., and Ross, M. (2013). *The Data Warehouse Toolkit: The Definitive Guide to Dimensional Modeling*. Wiley.

Kreps, J., Narkhede, N., and Rao, J. (2011). *Kafka: A Distributed Messaging System for Log Processing*. LinkedIn Engineering.

LeCun, Y., Bengio, Y., and Hinton, G. (2015). *Deep Learning*. *Nature*, 521(7553): 436–444.

Lee, J., Bagheri, B., and Kao, H. A. (2015). *A Cyber-Physical Systems Architecture for Industry 4.0-Based Manufacturing Systems*. *Manufacturing Letters*, 3: 18–23.

Mayer-Schönberger, V., and Cukier, K. (2013). *Big Data: A Revolution That Will Transform How We Live, Work, and Think*. Houghton Mifflin Harcourt.

McMahan, H. B., Moore, E., Ramage, D., Hampson, S., and y Arcas, B. A. (2017). *Communication-Efficient Learning of Deep Networks from Decentralized Data*. *Proceedings of AISTATS*.

MongoDB Inc. (2023). *MongoDB Architecture Guide*.

Nakamoto, S. (2008). *Bitcoin: A Peer-to-Peer Electronic Cash System*. Retrieved from https://bitcoin.org/bitcoin.pdf

O'Neil, C. (2016). *Weapons of Math Destruction: How Big Data Increases Inequality and Threatens Democracy*. Crown Publishing Group.

Preskill, J. (2018). *Quantum Computing in the NISQ Era and Beyond*. *Quantum*, 2: 79.

Stonebraker, M. (2010). *SQL Databases v. NoSQL Databases*. *Communications of the ACM*: 53(4).

The Apache Software Foundation. (2023). *Hadoop Documentation*.

Xu, L. D., He, W., and Li, S. (2014). *Internet of Things in Industries: A Survey*. *IEEE Transactions on Industrial Informatics*, 10(4): 2233–2243.

Zaharia, M., Chowdhury, M., Das, T., Dave, A., Ma, J., McCauley, M. et al. (2016). *Apache Spark: A Unified Engine for Big Data Processing*. *Communications of the ACM*, 59(11): 56–65.

Zuboff, S. (2019). *The Age of Surveillance Capitalism: The Fight for a Human Future at the New Frontier of Power*. PublicAffairs.

4

Role of Smart Finance and Management in the Industry 5.0 Era

Prathap B. N.,[1] *Devaraju N.,*[1] *Kiran Kumar M.*[2,*] and *Amit Kumar Tyagi*[3]

1. Introduction

The evolution of industrial revolutions is a narrative of humanity's ever-evolving relationship with technology. We began with Industry 1.0 in the late eighteenth century with mechanical production powered by steam, hydraulics, wind, etc. The change from human labor to machine labor was dramatic and expanded industrial output to unprecedented levels. In the late nineteenth and early twentieth centuries, with Industry 2.0, the emergence of mass production through electricity increased productivity through economies of scale on a long-term basis (Brettel et al. 2014). Industry 3.0 emerged through many considerations as Industry 4.0 began to take shape in the late 1960s. Through the advent of electronics, automation, and information technology, we began to see the runaway possibilities of more complex decision-making processes that created efficiency opportunities. Industry 4.0 began to emerge as cyber-physical systems, the Internet of Things (IoT), and advanced analytics appeared in the late 20th centuary with smart, connected machines. Industry 5.0 is the most recent transformation; however, it is not just another iteration but a true paradigm shift that ultimately places the human being in a collaborative position as the world begins to change how we view industrial and economic value creation, as seen in Figure 1.

Industry 5.0 is characterized by a return toward research and human-machine collaboration—the idea that humans and autonomous intelligent systems can work together in a more collaborative fashion (Schwab n.d.). There is a greater emphasis on sustainability, resilience, and a human-first approach, in contrast to previous iterations under the Industry 4.0 banner, which focused on hyper-automation and digitization. Interestingly, the European Commission (2021) officially recognized Industry 5.0 as its own model with a rigorous emphasis not only on productivity and economic

[1] S-VYASA Deemed-To-Be University, Bengaluru.
Emails: worldwideweb05@gmail.com, 0000-0002-3833-9240; talk2devraju@gmail.com,0009-0005-5737-0314
[2] CMS Business School, Jain (Deemed-to-be University), Bangalore, India.
[3] National Forensic Sciences University, Gandhinagar, Gujarat, India.
Email: amitkrtyagi025@gmail.com, 0000-0003-2657-8700
* Corresponding author: kiranm287@gmail.com,0000-0003-2084-9418

DOI: 10.1201/9781003655190-5

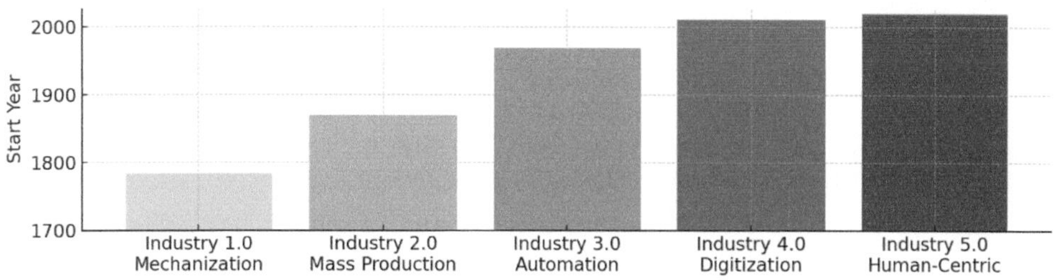

Figure 1: Evolution of Industry from 1.0 to 5.0.

growth but also on promoting well-being, social responsibility, and ecological sustainability by design (Schröder et al. 2024). Whereas Industry 4.0 focused on utilizing machine autonomy to prioritize organizational efficiency, Industry 5.0 emphasizes co-creation, personalization, and ethically developed AI technologies and standards. Giving rise to greater creativity and emotional intelligence of humans amplified by the computational powers of machines—for example, in a smart factory setting, cobots could assist workers in producing customized products. Mass autonomy and automation could not historically support this personalized process (Nahavandi 2019).

Smart finance, in this developing industrial ecosystem, refers to the use of artificial intelligence (AI), blockchain, machine learning (ML), and near real-time data analytics to automate, improve, and secure financial processes. This includes predictive financial forecasting and budgeting, autonomous auditing, as well as fraud detection, particularly using very large datasets (Tyagi 2023a, 2023b, Yi et al. 2023). Smart finance and smart management, more broadly, encompass the application of digital technologies to assist in optimizing decision-making, resource allocation, and operational efficiency. This involves cognitive computing, adaptive leadership, and near real-time performance monitoring—usually enhanced through AI agents or vectored data dashboards. Smart management matches human intuition with machine-accompanied answers, creating a productive idea for an organization and agility in a changing and complex organizational environment (Varadarajan 2023). As shown in Figure 2, technologies like AI, blockchain, and analytics are ranked as highly relevant for both smart finance and smart management, although the normative conditions for smart finance may differ from the strategic potential in smart management.

In the Industry 5.0 era, merging smart finance with smart management will be a prerequisite, not an option. Organizations will need to transition from siloed businesses to intertwined, intelligence-based ecosystems. Smart finance produces financial transparency, continuous financial oversight, and flexibility in capital consumption. Smart management captures the knowledge, perception, and significance of this financial intelligence and creates exceptionally informed decisions, resilient operations, and processes with options (Teece 2017). For example, an organization experiencing volatile financial market situations could leverage AI-supported, risk-based models created by its finance department to automate alerts in underlying management systems and enable immediate redeployment of resources. In supply chains, automated compliance checks and secure payments

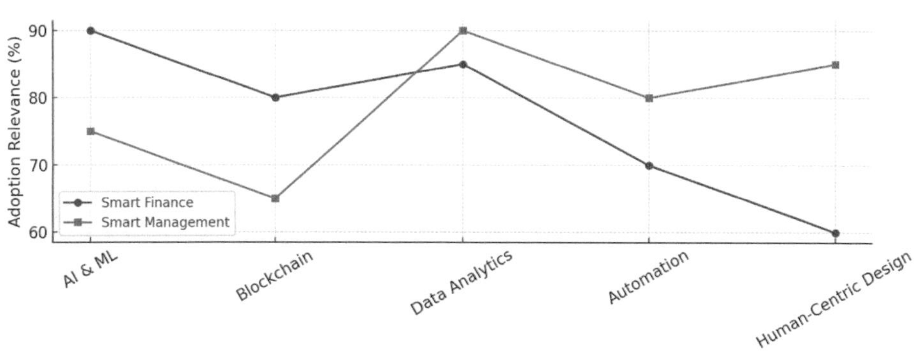

Figure 2: Key Technologies in Smart Finance vs. Smart Management.

executed by blockchain smart contracts can eliminate delays and boost trust and compliance among suppliers and customers (Kayikci 2018). Additionally, the merger will help support ESG (Environmental, Social, Governance) alignment, as financial decision-making will reflect sustainability metrics and weigh ethical considerations—two significant tenets of Industry 5.0 (Nakamoto 2008). This chapter aims to investigate the strategic and operational implications of smart finance and management as they relate to the objectives of Industry 5.0. This research will:

- Examine the technological and conceptual underpinnings of smart finance and management.
- Assess their combined and synergistic incorporation into industry organizations operating within Industry 5.0.
- Identify advantages, disadvantages, and ethical issues.
- Provide a strategic framework for the organizations to adopt.

The remainder of the chapter is organized as follows: Section 2 provides an exhaustive literature review that consists of Industry 5.0, an overview of evolution of smart finance and smart management, and the theoretical perspectives on the association of both, such as the resource-based view or socio-technical systems theories. Section 3 discusses the shift in Industry 5.0 and its key features—human-centricity, personalization, and sustainability—along with enabling digital technologies such as AI, IoT, blockchain, 6G, and quantum computing. Section 4 focuses on smart finance and provides an explanation of its key features, such as automation, predictive analytics, and decentralized finance (DeFi), along with case studies of application. Section 5 is about smart management and discusses intelligent decision-making systems, cognitive management, and employee empowerment needed to accomplish smart management. Section 6 discusses the connection of smart finance and management, maintaining synergy and strategic alignment, as well as practices that are ESG-oriented. Section 7 concerns the challenges, risks, and ethical considerations associated with the implementation of these technologies. Finally, Section 8 proposes future directions and opportunities, such as AI-human symbiosis, digital twins, SME integration, and support of interdisciplinary research and policy.

2. Related Work

Industry 5.0 signifies a transition from a technology-driven evolution to a human-centered, resilient, and sustainable model of industrial transformation. In Industry 4.0, automation, increased connectivity, and big data allowed for enhanced applicability and adoption rates. Industry 5.0 endeavors to facilitate the management of data and intelligent user co-creation while using intelligent technology to extend people's role in the value chain. Scholars also point to a transition in socioeconomic values alongside technological evolution, such as personalization, inclusive participation, sharing, and sustainable practices (Tyagi 2023a) as facilitating Industry 5.0. The foundational premise of Industry 5.0 was initiated as the integration of human skills with collaborative robots (cobots) in production environments. The use of intelligent systems was proposed to develop human-in-the-loop collaborative networks. Other authors discuss enabling technologies such as AI, the Internet of Things (IoT), and 6G for supporting the augmentation of human roles, especially around decision-making and creative design (Eccles et al. 2014). Moreover, Industry 5.0 offers involvement as a value co-creation process, engaging active users as co-creators, supported by digital twin systems and customization algorithms, and transforming them from passive consumers. The concepts of digital twins, AI, and extended reality (XR) were emphasized as supporting a personal ecosystem that enhances people's roles in production, valued services, and the value co-creation process.

The advancement of smart finance coincides with the digitization of financial processes. In past research, finance was seen as a passive, retrospective function focused on reporting past performance; yet smart finance systems are active, proactive, self-contained, predictive, real-time systems that move finance away from legacy reporting. AI is being embedded to assess credit risk scoring, fraud detection, and robo-advisory. Blockchain is being used to promote transparency and traceability in financial transactions and enable DeFi, circumventing central financial authorities or updating

Table 1: Summary of Techology-Based Applications.

Technology	Application in Smart Finance	Application in Smart Management
Artificial Intelligence	Predictive financial modeling, fraud detection	Decision support systems, HR analytics
Blockchain	Secure, decentralized ledgers for transactions	Smart contracts for compliance and operations
Big Data Analytics	Risk assessment, forecasting, and trend analysis	Operational intelligence and strategy
Internet of Things (IoT)	Real-time financial tracking	Smart manufacturing and logistics
Robotic Process Automation (RPA)	Automating repetitive financial tasks	Workflow automation in operations
Cognitive Computing	AI-assisted financial decision-making	Adaptive management and problem-solving

intermediaries. Big data analytics is being applied to enhance forecasting, liquidity management, and tracking of performance. It was also found that forecasting could improve the accuracy of results by more than 30% using smart finance algorithms, even in the case of volatile markets. Further, DLTs can be used to create tamper-proof financial audits or real-time inter-organizational settlements. Smart finance can be correlated with ESG initiatives. Sustainable finance frameworks incorporate environmental risk modeling into existing financial models using AI (Adel 2022). AI can increase the transparency and accountability of investment. Table 1 outlines the core technologies and their applications in finance and management.

Transformational management practices associated with Industry 5.0 emphasize decision intelligence, adaptive leadership, and collaboration between humans and technology. Smart management systems are now designed to enable decisions as they happen in real time and in context with a combination of AI, IoT, and analytics. Cognitive management systems use AI capabilities to act similarly to human reasoning, allowing for responsive leadership in uncertain environments. This can be seen in dynamic resource allocation, predictive supply chain management, and autonomous HR systems. Also, research conducted found that companies with smart governance models based on data platforms outperform companies that structure their management in the traditional sense (Rojko 2017). A real-time dashboard powered by cloud analytics capabilities increases the ability to respond to operational disruptions in a more nimble manner. Research shows that cyber-physical human systems (CPHS) might add to our ability to create adaptive and intelligent environments that can generate responses with and without human input as they respond to data complemented by natural inputs from humans when humans interact with sensors, robotics, and human-machine interfaces.

There is emerging literature in Industry 5.0, but there are gaps, especially concerning integration studies, real-world case studies, ethical and regulatory uncertainty, especially around open innovation/procurement and cost constraints, and inadequate consideration of small and medium enterprises. Finance and management exist separately as fields of study and do not examine how they converged in Industry 5.0 ecosystems after the global pandemic. There is a lack of empirical validation in most studies (Ramírez-Gordillo et al. 2024). Most frameworks assume the adoption of high-tech equipment, without awareness of the lack of accessibility for entrepreneurs and small and medium enterprises from a clear perspective. Smart finance and management can be analyzed through many theoretical perspectives, including the Resource-Based View (RBV), Socio-Technical Systems Theory (STS), Dynamic Capabilities Theory, and Technology-Organization-Environment (TOE). RBV explains that strategic resources such as AI capabilities and data capabilities can confer a competitive advantage to firms; STS considers the interdependence of humans, technologies, and organizational structures; Dynamic Capabilities Theory considers how organizations adapt to changing environments by reconfiguring their resources, and in the case

Table 2: Summary of Technology Application in Various Fields.

Focus Area	Key Insight
Human-centric approach in Industry 5.0	Human-machine synergy is central to Industry 5.0
Integration of cyber-physical systems and human factors	Human inclusion and AI collaboration are key enablers
Digital twins and personalization in production	Customization and digital interfaces will drive value
Strategic implications of smart finance	Smart finance aligns performance with business strategy
Collaborative networks and intelligent systems	Smart ecosystems require integrated digital infrastructures
Resilient, adaptive supply chain management	Flexibility and sustainability define Industry 5.0 supply chains

of smart finance/management, it is clear that tools providing dynamic capabilities like real-time analytics support are fundamentally changing resource reconfiguration for firms. Finally, Table 2 provides an understanding of the new technologies available as well as an assessment of whether the organization is ready to adopt smart systems.

The literature emphasizes the transformative potential of smart finance and management in achieving the human-centric, sustainable, and resilient objectives of Industry 5.0. Although considerable attention has been paid to investigating enabling technologies and theoretical constructs, limited research has examined the integration between different domains of enabling technologies. Furthermore, theoretical lenses such as RBV and STS provide a good basis for exploring how organizations can exploit digital capabilities strategically. However, to enable holistic absorption or adoption of Industry 5.0 capabilities, there needs to be a purposeful addressing of current gaps, particularly around empirical validation, ethical governance, and inclusivity. The sections below will discuss and explore more fully the technological enablers and how this shapes practice in the Industry 5.0 context.

3. Industry 5.0: Paradigm Shift and Enablers

Industry 5.0 represents a major shift in the industrial landscape, challenging our perspective on productivity, value creation, and the role of the human being in technologically advanced systems. Industry 4.0 was rooted in cyber-physical systems, automation, and data-driven decision-making processes. Industry 5.0 represents an expanded vision; technology is meant to support rather than simulate or replace the human condition. It embodies a more conscious, holistic, and sustainable vision of industrial development, where both human aspirations and technological capabilities are viable sources of progress. At the center of Industry 5.0 lies the idea of humans and machines working together in ways that mutually benefit each side's unique strengths. The human being moving through automated workflows is not seen as redundant but as a critical participant, capable of becoming more nimble and less exhausted with the help of smart tools that enhance creativity, decision-making, and customization. Experiences in manufacturing that integrate collaborative robots (cobots) deploy these robots not as replacements but to support humans, with cobots taking away repetitive tasks, allowing humans to exercise their judgment and creativity to regain a human experience in production, expanding innovation and the emotional intelligence that can be added to services and design. Personalization represents another foundational pillar of Industry 5.0. Whereas traditional mass production systems are built on standardized goods and services, Industry 5.0 sees the potential for mass customization across scale. The ability to personalize products comes from leveraging a host of digital technologies, e.g., digital twins, edge computing, and IoT systems collecting real-time information from countless users about their preferences, operational areas, and environments, allowing customization of products accurately based on users' choices. In healthcare, real-time wearable IoT devices monitor patients' biometrics to create personalized treatment plans tied to the next biometric update, instead of static zero-based plans from previous industrial systems.

The ability to personalize products in real time reflects a significant departure from static models encapsulated in previous industrial systems. Sustainability is equally important in this transformation of Industry 5.0. Rather than treating sustainability as an add-on, it is the guiding ethic in the systems of Industry 5.0. Sustainability promotes responsible practices through energy-efficient production, closed-loop material flows, and transparent formats using digital technology. Tools such as smart grids, green logistics, and the application of AI for sustainability reporting help organizations track and reduce their environmental footprints (Tyagi and Abraham 2021). Sustainability is more than an operational strategy—it is part of business models, investment decisions, and product design, achieved by bringing profit and ecological health together in long-term ways.

The operationalization of Industry 5.0 principles relies heavily on a constellation of interconnected emerging technologies, each with its own clearly delineated interrelated purpose in establishing the new industrial paradigm. Of these technologies, AI is one of the primary facilitators of operationalizing Industry 5.0 principles. AI-based systems evaluate extensive amounts of structured and unstructured data, allowing decisions at a speed and scale that were not previously possible or were too time-intensive. In Industry 5.0, AI will facilitate, in addition to optimization and automation activities, creative functions around product design, personalized marketing, and knowledge discovery. Additionally, generative AIs assist engineers and designers in thinking imaginatively about solutions beyond human cognitive limitations. As the sensory part of Industry 5.0, the IoT enables all machines, people, products, and environments to connect, share, and interact in real time, generating an intelligent web of data and functionality. Organizations will have predictive maintenance, context-aware production, and human safety through IoT sensors attached to machines or infrastructure. In logistics, IoT will allow real-time visibility of supply chains, enabling managers to anticipate supply chain disruptions and adjust logistics plans to eliminate the disruption in a timely way, improving efficiency and resilience. As an essential basis for trust and traceability in an increasingly complex industrial ecosystem, blockchain technology encloses trustworthy, immutable institutional records of transactions, product histories, and operational activities that create accountability in value chains. Trust is particularly vital in conditions like pharmaceuticals and food safety, where provenance being established and compliance with regulations are paramount or mission-critical. In addition to providing records of transactions and taking advantage of decentralized smart contracts to help organize inter-organizational collaboration without the need for third-party institutions, blockchain technology also enables decentralized finance (DeFi) systems (Tóth et al. 2023). The enabling technologies of 6G networks and systems will provide connectivity improvements in speed, bandwidth, and latency for connected devices and things. Where 5G improved mobile communications and real-time data transfer to a new level, 6G systems will support emerging usage areas of immersive environments, edge intelligence, and ultra-reliable, low-latency machine-to-machine communications. These areas are very important as they relate to human-machine interactions in real time, especially for remote surgeries, immersive training environments, and augmented-reality-enabled maintenance environments. 6G will provide the extremely low latency and high data throughput necessary for Industry 5.0. As Industry 5.0 unlocks the new opportunities outlined above, 6G will represent the nervous system, with blockchain technology serving as the immune system, to provide trust, accountability, and compliance.

Ultimately, quantum computing can open doors to solving what are currently intractable optimization, simulation, and data analysis problems that classical computing cannot solve and will help push the limits. In finance, quantum computing can optimize investment portfolios across millions of potential variables; in manufacturing, it can design and simulate materials at the molecular level; and in logistics, it can enable paths across networks of the globe in fractions of the time traditional technology cannot accomplish. While still in its infancy, quantum computing is a vehicle for change in Industry 5.0 that will create possibilities we cannot yet imagine (Camarinha-Matos et al. 2023). Figure 3 provides a conceptual understanding of key enabling technologies as discussed, which includes IoT, AI, Blockchain, 6G, and Quantum Computing. These, as a system, facilitate enabling human-centric, personalized, and sustainable industrial transformation.

Figure 3: Technological Enablers of Industry 5.0.

A unique characteristic of Industry 5.0 is the collaboration within the established relationship of the creative human and the intelligent system. This paradigm of Industry 5.0 does not juxtapose technology and humanity; it requires their collaboration. AI-generated insights inform human decisions, and human behaviors inform the algorithmically produced outcomes through intuition. This collaboration creates new processes in the way organizations innovate, design, and add value. For example, generative AI allows a fashion designer to quickly "test" or play with hundreds of iterations of a design in minutes; the fast iteration enables a new way of artistic exploration (Alawadhi et al. 2023). Even education can embrace the adaptation of human intuition through adaptive learning, with machine learning able to enable tailored curricula that focus on each individual student's learning needs, positively impacting their engagement and understanding. Additionally, this stakeholder partnership involves decision-making and strategy. Managers are now equipped with AI-enabled dashboards that visualize real-time key performance indicators (KPIs), recommend changes to operations, and simulate results. The need for managerial judgment does not disappear; instead, it is enhanced by situating decision-making in a crowd of data. Emotional intelligence, ethical reasoning, and strategic vision—attributes that are uniquely human—are augmented rather than replaced with the use of intelligent technologies.

The consequences of this shift have implications for deeper economic, technological, and social systems. At the economic level, Industry 5.0 allows for new business models that are adaptive, more distributed, and focused on the needs of the customer. New ways of value are created and shared with these models, such as platform-based ecosystems, decentralized finance, and outcome-oriented contracts. While automation creates efficiency, co-creation creates entirely new markets based on personalization and sustainability. From a technological standpoint, the combination of AI, IoT, blockchain, quantum computing, and 6G puts us on the cusp of a new, massive era of cyber-physical integration. Factories will become self-optimizing systems; cities will become responsive to their citizens; and businesses will be able to coordinate actions globally, concurrently, and in real time. The result will be an investment not only in efficiency gains but in entirely new capabilities, including new forms of autonomous collaboration between geographies and sectors.

From a societal standpoint, Industry 5.0 represents an opportunity accompanied by real societal challenges. On the positive side, Industry 5.0 creates possibilities for a better quality of life, outlined in improved healthcare, inclusive education, safer working conditions, and more environmentally sustainable consumption patterns. On the challenging side, we require strong ethical principles and protocols to meet challenges and issues such as the creation of ethical data practices, algorithmic discrimination, the digital divide, and worker displacement. Emphasizing human-centricity is driving a broader reconsideration of public policy, education, and corporate governance to ensure that society can keep pace with technology in ways that advance well-being and quality of life.

4. Smart Finance in the Industry 5.0 Era

The advancement of smart finance as it exists in Industry 5.0 is a movement away from static and retrospective-focused financial systems to new financial ecosystems based upon data, automation, and humanity. In this new era of industrial advancement, finance is more than a transactional process; it is considered a catalyst for innovation, sustainability, and the co-creation of value. Smart finance exploits technology, including AI, blockchain, predictive analytics, and decentralized systems, to create timely insights, workflows, and secure financial ecosystems that can enhance the core principles of Industry 5.0. Automation is one of the essential attributes within smart finance, mitigating human reliance and dependency for repetitive financial processes while enhancing speed, accuracy, and scalability. Intelligent software agents will increasingly undertake tasks such as invoice processing, payroll, financial reconciliation, and tax compliance on behalf of organizations. Intelligent software agents, unlike their human counterparts, are much lower in cost, work 24/7, commit far fewer errors, and create documentation trails that depict validity, trustworthiness, and compliance for systems of governance. In larger enterprises, it is common for robotic process automation (RPA) to fulfill high-volume, rule-based transactions; however, the power of automation within smart finance goes beyond efficiency, as it represents an evolution of the role of finance professionals to analytical and strategic participants rather than transactional facilitators (Javaid et al. 2020).

The ability to infuse predictive analytics into the financial decision-making process has been equally impactful. Predictive models, powered by big data and AI algorithms, leverage historical data to measure cash flows, identify credit risks, assess investment opportunities, and create alternate economic scenarios. This allows organizations to be proactive versus reactive, enabling CFOs and financial planners to appropriately manage risk, allocate capital to the greatest opportunities, and align financial strategies with longer-term business goals. Predictive analytics also allows agility in modeling portfolio strategy adjustments, hedging instruments, and supply chain financing during volatile markets. Again, the potential predictive capability of AI models is staggering. For instance, AI models can extract relationships from social media sentiment, geopolitical news, and market signals that can then result in actionable predictions for stock movements or commodity prices. This has provided firms with a meaningful competitive advantage. DeFi is a key innovation that enables smart finance in the Industry 5.0 age, as it employs blockchain and distributed ledger technology and allows alternative forms of finance without intermediaries. DeFi allows peer-to-peer transactions, lending, insurance, and asset management via open, programmable infrastructures. Unlike the "traditional" financial world that is based on centralized databases and not necessarily transparent processes, DeFi performs the same function through smart contracts while eliminating much of the typical delays and costs associated with traditional finance and giving developing regions access to financial services. Decentralized finance also provides businesses with newer forms of financial instruments and decentralized governance entities for liquidity pools and tokenized assets that can offer investment and participation visibility.

Another characteristic of Industry 5.0—transparency—has a financial sense in terms of blockchain-enabled financial tracking and auditability. Digital ledger technologies serve as "authorized" and immutable transaction records, verifiable and traceable across supply chains and partner networks, as well as for addressing regulatory questions. This is especially critical for sectors such as pharmaceuticals, food production, or ESG-compliant sectors, where financial flows must be consistent with ethical sourcing and environmental performance. Using blockchain technology to clarify the authenticity of claims allows organizations to provide reassurances to stakeholders that financial claims align with real actions. The additional benefit of digitally identifiable and verifiable credentials enabled by blockchain adds to positive security for financial transactions and reduces financial fraud and identity theft in digital finance ecosystems. There are many real-world implications of smart finance technology across sectors. For example, JP Morgan Chase has a COIN platform that uses natural language processing to automate the interpretation of legal documents, saving 360,000 hours in loan-servicing costs per year. Ant Group has a product named "MyBank" that offers machine-learning-supported collateral-free loans to small businesses in China, serving as an example of inclusive finance (Alimam et al. 2023). In the energy sector, Power Ledger, an Australian blockchain company, has designed decentralized energy trading infrastructures that allow for easier and quicker settlement of transactions using tokenized smart contracts by consumers and producers. The above case studies help illustrate how the intersection of automation, predictive intelligence, and transparency can create more intelligent and fairer financial ecosystems.

Smart finance represents a tremendous opportunity during Industry 5.0, but its introduction also inspires dilemmas. Quality data and integration represent foundational layers for the initiation and productivity of predictive models and AI systems. Financial systems combine this need across departments: separated by formats and geographies, smart financial systems will need to equalize data across the organization to provide the right input, at the right time, in the right place, and in the right context. Cybersecurity is yet another risk because the digitization of financial services and their decentralization provide more options for yet unknown potential threat vectors. Change management presents many challenges. Organizations will now need to work with finance professionals to elevate their digital literacy, as well as upskill them as applicable analytical specialists, and broader role expectations will need to be redefined. The legacy institutional structure will need to overcome resistance to change, which may not allow smart systems to gain traction. The same outcome is witnessed in Canada, where regulatory structures continue to lag behind the pace of DeFi development; in addition to having challenges defining where smart contracts vary in legality, facilitating accounting consistency for governments, and defining which tokens are subject to regulation within the market economy. Nevertheless, smart finance is significantly aligned with Industry 5.0's intentions to enhance human capacity to make informed decisions, democratize access to banking and finance, and integrate trust into economic systems. Future financial back offices will fundamentally redesign their departments as digitally based operational command structures, operating with other parts of the business in real time, adjusting based on live data that they can shape trends with, and factoring sustainability impacts into their financial forecasts and decision-making activities (Liu et al. 2024). Figure 4 succinctly captures the four key components associated with smart finance in the context of Industry 5.0: automation, predictive analytics, decentralized finance, and transparency. Working together, these elements generate an intelligent, flexible, and trusted financial ecosystem that aligns with the people-centered and sustainable ethos of the fifth industrial revolution.

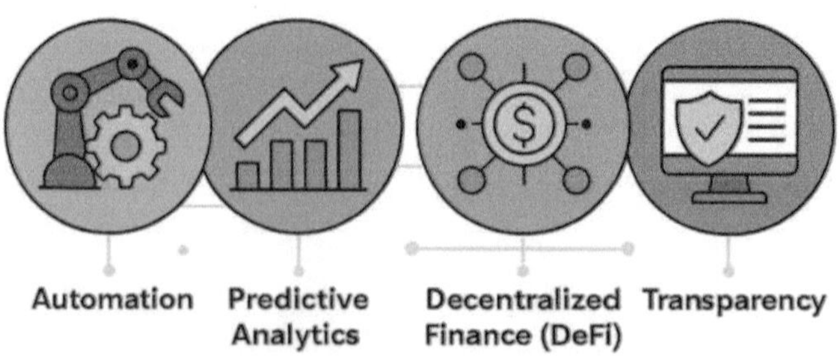

Figure 4: Smart Finance Enablers of Industry 5.0.

As we progress into the Industry 5.0 landscape, organizations that strategically invest in smart finance capabilities will not only improve operational effectiveness and risk responsiveness but will also have the opportunity to align their financial strategies with ethical, personalized, and resilient industrial objectives. In the next section, we will discuss how these financial innovations will also complement smart management practices to formalize organizational transformation in an era of new industrialism.

5. Smart Management in the Industry 5.0 Era

As companies are adopting the transition to Industry 5.0, the practice of management is changing at a fundamental level—no longer exclusively rooted in metrics and efficiency, but increasingly in adaptability, augmentation, and ethics. Intelligent management in Industry 5.0 combines advanced digital systems and human intelligence to enable leaders and teams to work in flexible, uncertain, and human-centered systems based on the use of real-time data, intelligent decision systems, collaborative systems, and a renewed focus on employee empowerment and well-being (Zakaria et al. 2023). The most engaging element of smart management will be the role of decision support systems (DSS) that provide managers with data-driven, real-time insights to inform and guide strategic decisions and operational choices. Whereas conventional systems provide mostly static, reporting-style dashboards, modern DSSs are based on artificial and machine-learning algorithms to generate learning from both historical and current relevant interactions. The DSSs will model all specific business scenarios, have the capacity to evaluate recommendations, and simulate future conditions. Thus, within a smart supply chain management (SCM) DSS, managers may well receive an alert for a disruption due to some geopolitical risk, along with real-time alternatives on procurement, logistics, and hedging pricing strategies. At no point will managers and decision-makers remain passive information recipients; rather, they will interact with and engage the intelligent applications, which will provide explanations, visualizations, and validation of recommendations to support trust and ease of use (Zwitter and Boisse-Despiaux 2018).

Cognitive management is yet another characteristic of the smart management framework. Cognitive management can be considered the application of AI systems that either emulate or support human cognitive functions—reasoning, judgment, and pattern matching—in highly complex decision-making environments. Cognitive tools can augment unstructured problem-solving, specifically when the problem-solving activities involve relying on experience, intuition, and ethics together with data. When smart systems are put into practice in sectors such as healthcare, law, or emergency and crisis management environments, they will not replace the role of humans in the decision-making process. Rather than serving as replacements for human judgments, smart systems provide contextual awareness, identify decision biases, and correlate consumable data from vast data sources to facilitate better decisions. Moreover, there are growing trends in collaborative team environments where humans-as-part-of-a-machine are involved in contexts that require cognitive decision-making, and these interactions are expected to be conversational, holographic, and empathetic.

The changes coming to management practices in Industry 5.0 indicate a more generalized phenomenon toward operational flexibility and agility. Organizations respond to the increasing pressures of digital disruption and societal expectations, changes to business sustainability, environmental shocks, and rapid and strong adjustments to consumer demand. Smart management is continuously adapting, not just planning. Digital twins, real-time analytics, and cloud-based systems are breaking down traditional management systems, allowing decentralized decision-making for local teams or autonomous systems to make fast and relevant decisions based on their local and contextual knowledge. If the term agile methodologies was once limited to software development, it is now being applied across operations, allowing organizations to iterate rapidly, test assumptions, and recalibrate with stakeholders or external data if it exists. Agility is a great strength contributing to organizational resilience and responsiveness—essential qualities in today's environment, shaped by the effects of the pandemic and high-volatility conditions (Wahab and Radmehr 2024). At the heart of the Industry 5.0 philosophy is human-centricity, and smart management makes human-centricity actionable through employee empowerment. Smart management systems do not treat workers as mere inputs in a process, but rather as partners in innovation. Empowerment can take many forms—accessible data transparency, multi-stakeholder decision-making platforms, learning algorithms that adapt to workers' roles, and AI-enabled upskilling tools that cater to workers' personalized development paths. Wearable technology, sentiment analytics, and real-time feedback loops allow worker health, engagement, and productivity to be constantly assessed and improved. For example, in a manufacturing environment, augmented reality interfaces might guide a frontline worker through an assembly process while adjusting to that individual's ergonomic positioning and stress levels—and doing it all with workplace design in mind.

Industry 5.0 is a new model of operating that combines technology and empowerment. Managers will be called upon to manage systems that learn, adapt, and evolve in conjunction with human inputs—managing in a model that represents a shift from command-and-control to a model of coordination and facilitation. Culture will be an important asset, with sustainable cultures cultivated through digital feedback, inclusive policymaking processes, and transparent communication systems. Human soft-skill attributes, including empathy, collaboration, and ethics, are not replaced by automation but are exercised in conjunction with intelligent technologies. Industry 5.0 is comprised of four interrelated domains: enabling infrastructure, decision support systems, cognitive management with agile operations, and employee empowerment. This dynamic model of management can respond to complexity and change and operate without a defined model.

The application of smart management in a real-world context demonstrates the transformative potential, with multinationals including Siemens and Schneider Electric employing digital twins and AI-based dashboards, delivering what they call "smart" management of global operations and adapting to local regulatory and market conditions. Smart city projects are being implemented in the public sector, applying cognitive management tools to develop systems that align transportation, public safety, and the environment, even though the subject matter involves multiple stakeholders.

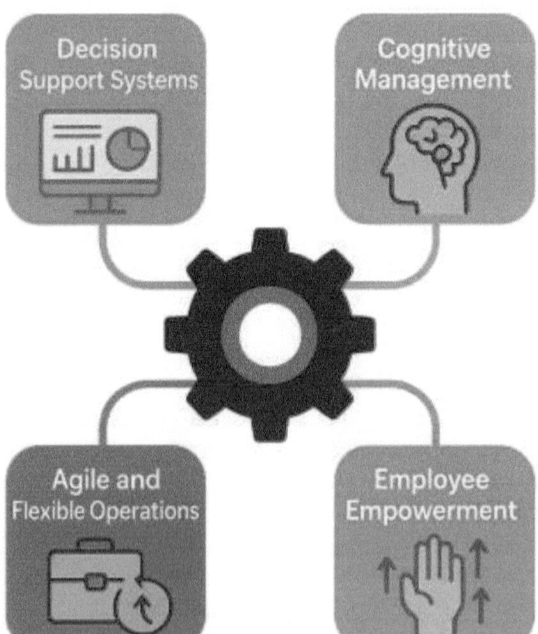

Figure 5: Smart Management Enablers of Industry 5.0.

Startups and SMEs benefit from cloud-based management, allowing for integrated platforms across the enterprise that include accounting, HR, customer relationship management, and analytics—all represented in a single interface. Figure 5 depicts the essential elements of smart management—decision support systems, cognitive management, agile operations, and employee empowerment—working cohesively to enable adaptive, human-centered organizations in the context of Industry 5.0.

Intelligent governance in the Industry 5.0 period represents a shift to integrated, adaptive, and human-enhancing systems of leadership and governance. By leveraging intelligent technologies along with human agency related to creativity, well-being, and ethical responsibility, organizations can increasingly optimize and drive performance while also establishing workplaces that are resilient, inclusive, and aligned with the interplay of complex immediate demands and the reality that is modern society.

6. Integration of Smart Finance and Management

The joining of smart finance and smart management is a significant development in organizational strategy in Industry 5.0. This joining is more than just transformational technologies; it is transformational structure and strategy, which enables organizations to operate more rationally, in a more agile and resilient manner, and under more sustainable principles. At the heart of this joining of smart finance and smart management is an evolution from a fragmented decision-making process to a joined approach where financial intelligence and managerial action take place in a digitally enabled ecosystem (Olawoyin 2024). In Industry 5.0, digital platforms and intelligence systems create a seamless interface between financial functions and managerial workflows. These systems utilize data interoperability, real-time analytics, and shared knowledge environments, creating an organizational environment where financial and managerial functions no longer operate in silos. Instead, they interact constantly through dashboards, decision support systems, and predictive modeling applications. Together, some segments of finance and management operate in an organizational form of "cognition," where financial and operational conditions inform the decision-making behavior of management in near real time. For example, a CFO interpreting liquidity ratios analyzes financial projections that integrate production outputs, timely employee performance measures, and producer and consumer feedback information into the same digital mapping layer. Similarly, supply chain management decisions with respect to the allocation of resources are made with live cash flow data views, integrating capital expenditure conditions to improve alignment between planning and funding.

One of the most important synergies afforded by digitalization is the strategic joining of financial intelligence (FI) and management practice (MP). In past models, finance was often seen only as a lagging reporting measure, while management relied heavily on intuition. Now, with the smart systems of Industry 5.0, both FI and MP can operate in unison using shared, real-time intelligence and predictive foresight. The financial and operational strategic elements have developed in remarkably fluid conditions. The impact of digitalization means that when bringing a new product to market, managers can model test market conditions, supply chain conditions, and costs, while the finance team can model the supporting funding and modifications to the ROI models and assumptions and assess risk modeling—all before any product has been manufactured. Not only does the joint coordination of FI and MP provide faster decisions, but it also improves the accuracy and sustainability of those decisions (Oyewole et al. 2024). The presence of real-time organizational data plays an important role in demonstrating how integrating finance with other management functions creates intelligent finance. Vast amounts of data from across the organization in terms of sales, production, HR, procurement, and customer service are funneled through AI-enhanced platforms to provide managers with the insights they require when they require them. These insights are instrumental for investment planning, reallocating budgets, performance evaluation, and identifying risk mitigation strategies. Furthermore, such data environments facilitate scenario planning that considers external factors such as geopolitical shifts, climate change, or market fluctuations, thereby boosting organizational resilience as well. An example from the retail industry illustrates this integration in action: Zara has successfully established integrated systems that combine store-level demand data with traditional practices of supply chain management and budgetary controls, allowing the organization to modify inventory and pricing decisions based on consumer discretion and/or external factors in real time.

Companies that have embraced the Industry 5.0 philosophy can serve as powerful signals of how smart finance and management integration can create strategic value. For example, some Siemens financial planning tools have been integrated with factory data enabled by the Internet of Things (IoT) to better allow for dynamic budgeting around areas such as maintenance and energy optimization. This integration allows plant managers and finance teams to collaborate on operating costs and sustainability objectives. In even more innovative strides, Unilever has provided its executives with AI-enhanced financial forecasting tools that are also connected to environmental performance measures, allowing Unilever to tie cost management directly with its carbon reduction plans. Tesla's operating model illustrates a much deeper level of integration, where finance, engineering, and data science intertwine. Its over-the-air update system wirelessly sends product performance data back to the company to feed into its financial presentation and product improvement cycle, optimizing for a lower operational lead time between insight and action. Smart integration of sustainability and ESG performance in Industry 5.0 involves the integration of ESG factors concerning finance systems, managers' decisions, and an organization's circuit of performance. This means transcending reporting for compliance to a form of active ESG management powered by smart systems. In these systems, finance systems can integrate environmental metrics in decisions on capital budgeting, pricing, and investment opportunity evaluations, enabling teams to carry out operational changes that can lead to positive improvements in sustainability performance. Digital platforms can facilitate the integration of performance goals with ESG goals by using ESG data in enterprise resource planning (ERP), customer relationship management (CRM), and performance management systems. Using smart systems allows ESG finance management to be integrated, enabling ESG performance to be evaluated consistently with strategic planning, resource conservation frameworks, and open and transparent reporting (Tyagi and Sreenath 2021). Companies such as Patagonia and IKEA are already leading some of the developments required for the integration of sustainability and financial performance through embedding sustainability goals and objectives in their financial key performance indicators (KPIs) and strategy scorecards. However, driving most of these developments will require organizational members to overcome the legacies of their systems, individual organizational silos, and their resistance to change. Organizational ethical issues will also surround AI-controlled decisions related to integrating financial outcomes and algorithmic reasoning, where managerial decision-making cannot explain certain AI systems' outcomes; therefore, robust governance systems and accountability protocols will also need to be put in place for smart systems' use of ESG and financial performance data.

7. Challenges, Risks and Ethical Considerations

The role of smart finance and smart management in changing the narrative of Industry 5.0 brings about incredible opportunities, but it also presents a complex array of challenges, risks, and ethical dilemmas that organizations need to appreciate in advance. These issues can impact technical, organizational, social, and regulatory aspects and must be managed strategically to ensure that the human-centered, sustainable aspirations of Industry 5.0 are met without unintended consequences. One of the biggest challenges discussed in a number of papers is ensuring technology equality and addressing the digital divide, which remains relevant both within organizations and between organizations or regions. For example, while multinationals and technologically advanced economies are quickly adopting new technologies such as AI, blockchain, IoT, and other enablers of Industry 5.0, many small and medium enterprises (SMEs) and institutions operating in developing economies lack the financial viability, infrastructure, or human capital to effectively implement those systems. The danger of this inequality is that economic disparities will only be exacerbated; access to and democratization of innovation will remain a challenge, and industries may become increasingly unequal with two-tiered competitive landscapes. Inequality may also exist internally, even in larger firms, where some departments, business units, or locations may be rigorously digitized while others, with less financial support or capability, remain under-digitalized, slowing transformation efforts (Anthopoulos et al. 2021).

An additional key component is data privacy and security. The overwhelming volumes of data collected, processed, and analyzed in smart finance and management systems create dangers of cyberattacks, breaches of confidentiality, and other misuses that do not respect the privacy of those being served. Financial data, employee data, customer behavior modeling, and proprietary algorithms are all valuable targets, and any breach in these domains can cause considerable losses, affecting not only finances but also stakeholder trust. As these intelligent systems become more connected and integrated, the attack surface becomes larger and demands constant investment in more sophisticated security measures, encryption systems, and incident response mechanisms. In addition, cloud-based systems and third-party service providers are increasingly used not only for cost optimization and analytics but also for core operations, which introduces risk if operationalization and governance are not strong. Closely tied to cybersecurity is the ethics of artificial intelligence. As we trust intelligent systems to make both high- and low-stakes decisions affecting people's livelihoods, we cannot lose sight of algorithmic discrimination, bias, and opacity as areas of concern. AI models often operate as "black boxes," with limited transparency into how they reach conclusions or make recommendations, which can raise accountability issues. This is particularly concerning for algorithms used in high-stakes decision-making areas such as credit scoring, investment strategy, and people analytics. Many algorithms are not validated in a rigorous manner, ethical and moral frameworks are lacking, and few examine the broader implications of such systems, which can reinforce social bias, marginalize certain populations, and undermine the core values of humanity that Industry 5.0 professes to uphold. Therefore, organizations must implement robust AI governance standards, including human-in-the-loop supervision, bias audits, and ethical review boards, to ensure that technologies and decision-making ultimately align with human values.

Resistance to change on the part of organizations is another significant hurdle. Although smart finance and management are strategically valuable, many organizations struggle to adapt because of inertia embedded in their culture, skills, and structure. Employees may perceive that automation and AI will supplant their jobs, which can be debilitating, especially for those in finance and administration, causing disengagement or resistance to transformation efforts. Management may struggle to move beyond traditional habits of control and centralization or may lack the digital fluency required to use the rich dashboards and analytics available to them. The other issue with organizations is that they can become so hierarchical or bureaucratic with legacy processes that it slows down organizational decision-making. Decisions regarding how and when to collaborate can take too long to make. The most successful digital transformations are a function of not only adopting a digital approach, but a borderline radical shift to how the organization thinks, interacts, functions, and is organized (Anthopoulos et al. 2021). Legal and regulatory uncertainties create additional complexity for the adoption of smart finance and management systems. The speed of innovation has surpassed the pace of regulatory frameworks, leading to gray areas in domains including data sovereignty, AI accountability, digital contracts, and decentralized finance. For example, smart contracts based on blockchain technology may be legally binding in one jurisdiction but not recognized or enforceable in another. Additionally, how regulators view digital assets, what compliance looks like for algorithmic trading, and what standards will be used for ESG reporting differ by regulatory regime. These differences mean that for many multinational firms, simple compliance will require significant resources. When coupled with data protection laws such as the General Data Protection Regulation (GDPR), which imposes strict restrictions on the use of personal data, the data-hungry nature of AI systems creates even greater complexity. Managing this complexity requires organizations to have strong legal compliance teams, engage with regulators, and assist in the co-creation of regulatory frameworks that allow for both innovation and accountability.

8. Future Directions and Opportunities

As Industry 5.0 continues down its planned path, we expect smart finance and management systems to move from efficiency of operations to levels of fundamental strategic transformation. Progress will be driven by more than just technology; we will reference people's shifting realities, organizational-purpose adaptations, and regulatory/civic technology factors. They all present significant opportunities for innovation, diversity, and new forms of leadership for all members of our economies. With transformational innovations on the horizon, perhaps the most transformative is AI-human symbiosis. This means that the development of systems will no longer be seen as tools for automation or prediction but will be seen as collaborative partners for creative, strategic, and ethical decision-making. In finance and management, this means co-evolving systems in which the human operator contributes context, emotional intelligence, and value judgement to AI providing foresight, patterns, and dynamic learning that changes the notion of managers' and finance practitioners' developing relationships with cognitive agents that know an organization's priorities, communicate naturally, and refine understandings of the decisions. Human-machine experiences will shift from tools to partners.

Another major advancement is using digital twins in finance. Digital twins, a term originating from the fields of engineering and manufacturing, are virtual representations of a physical system. Digital twins are only just now being developed and used for financial modeling and scenario planning. For example, an organization would develop a digital twin of its financial ecosystem that would allow it to model how changes in the economy and operational environment could impact the organization in real time. With a digital twin in use, an organization's central financial office and CFO will have a dynamic decision-making toolkit that allows for informed decision making in collaboration with all stakeholders in real time. This innovation will turn capital planning, budgeting processes, and risk management into collaborative, interactive, and actionable processes that support predictive decision making.

As the technology evolves, so does the role of leadership in smart ecosystems. The future leaders will be required to exercise not only digital literacy, but systems thinking, ethical reasoning, and stakeholder collaboration to adapt do the evolving "digital" part of organizational change. Leadership will shift from control-and-command responsibilities to facilitation of agile, cross-functional teams that utilize a degree of autonomy augmented and aided by AI teams. Leaders will lead hybrid human-machine workforces using qualities like emotional intelligence, adaptability, decision making, and purpose. They will be responsible for maximizing long-term value while navigating a varied range of challenges including sustainability, ethics, human complexity, and systemic pressures. The organization's mission, vision, and values will be as much of a consideration as technical and financial decision making for executive leaders to complete their roles as strategic executives (Soori et al. 2024).

Importantly, the opportunities in Industry 5.0 are not just for large companies and advanced economies. There is significant opportunity for SMEs and new markets to leapfrog traditional industrial models of yesterday and instead move directly to adopting digital-first, agile, and intelligent solutions. Cloud-based accounting systems, low-code management systems, and decentralized finance technologies can make innovation more affordable, available, and scalable. New ventures across Southeast Asia, Sub-Saharan Africa, and Latin America are already piloting blockchain-enabled supply chains, mobile-based investment platforms, and AI-based inventory management systems. With the right supports in place, these new markets will have the potential to become contributors to our future Industry 5.0 economy. Achieving this future will require strategic investment in interdisciplinary research supported by policy. As the pace of development continues to accelerate and technologies start to cut across sectors, the academic community must forge research agendas to reflect that connectivity. Collaborative innovation ecosystems, meeting places for academia, industry, community, and government; cross-sector public-private partnerships; and multi-nation research alliances will reduce the time necessary for socially responsible, impactful solutions. At the same time, public policy must devise forward-looking legislation that allows real-world experimentation while ensuring accountability, equity, and security. To put these opportunities in perspective, we have summarized the directions for the future, their technological enablers, and the firm implications in Table 3.

Table 3: Emerging Directions in Smart Finance and Management for Industry 5.0.

Future Direction	Key Technologies Involved	Organizational Impact
AI-Human Symbiosis	Generative AI, NLP, Explainable AI	Enhanced decision-making, creativity, and adaptive leadership
Digital Twins in Finance	Simulation engines, cloud analytics, real-time data	Scenario planning, dynamic forecasting, resilient financial modeling
Decentralized Finance for SMEs	Blockchain, smart contracts, mobile tech	Democratized access to capital, financial inclusion, peer-to-peer micro-investments
Cognitive Management Platforms	Machine learning, process mining, knowledge graphs	Context-aware workflows, human-in-the-loop decision support
Sustainable Finance Integration	ESG analytics, IoT sensors, carbon tracking platforms	Data-driven sustainability strategies, integrated reporting, ethical governance
Cross-disciplinary Policy Research	AI ethics, fintech law, behavioral economics	Informed governance, inclusive innovation, resilient regulatory frameworks

The future of intelligent finance and management within the Industry 5.0 paradigm allows organizations to fundamentally reconfigure their operating, innovating, and value-creating systems. Positioned at the intersection of technology, human intelligence, ethical action, and worldwide inclusion, intelligent finance and management systems will help organizations drive their business performance while also working toward a more equitable, innovative, and sustainable world.

9. Conclusion

The combination of smart finance and smart governance through Industry 5.0 creates a new paradigm for organizations to become adaptable, ethical, and sustainable. The financial intelligence we gain at the speed of changing situations and strategic decision making allows a firm's operations to move from disconnected silos to a fully connected digital world. By integrating, adapting, and strategically organizing our processes around working together, it embraces a human element of coordination and collaboration in an agile, creative, and responsible manner. The technological enablers of smart, connected processes, such as AI, IoT, blockchain, and quantum computing, are much more than simply vehicles for value chain optimization. They create lasting transformational changes to how organizations create, distribute, and govern value. However, Industry 5.0's goals and expectations will involve deliberate programmatic design to address potential challenges of innovation disruptions from tech inequality, cyber insecurity, algorithmic bias, and regulatory uncertainty. All of these areas need strong governance in place with inclusive design. The vision surrounding Industry 5.0 is not only for large multinationals; it allows SMEs and developing economies to bypass some of the legacy systems. For organizations to flourish in this new era of intelligence, inclusion, and impact, we need to continue to be collaborative across each of the disciplines, innovate ethically, and think transformationally. Animating these degrees of freedom for change will be best realized by organizations that employ this integrated design construct.

References

Adel, A. (2022). Future of Industry 5.0 in society: Human-centric solutions, challenges and prospective research areas. *Journal of Cloud Computing: Advances, Systems and Applications,* 11(1). https://doi.org/10.1186/s13677-022-00314-5.

Alawadhi, A., Almogahed, A., and Azrag, E. (2023). Towards edge computing for 6G Internet of Everything: Challenges and opportunities. pp. 1–6. *In: 2023 International Conference on Artificial Intelligence and Smart Communication (ICAISC)*. https://doi.org/10.1109/ICAISC56366.2023.10085007.

Alimam, H., Mazzuto, G., Tozzi, N., Ciarapica, F. E., and Bevilacqua, M. (2023). The resurrection of digital triplet: A cognitive pillar of human-machine integration at the dawn of Industry 5.0. *Journal of King Saud University - Computer and Information Sciences*, 35(10): 101846. https://doi.org/10.1016/j.jksuci.2023.101846.

Anthopoulos, L., Sirakoulis, K., and Reddick, C. G. (2021). Conceptualizing smart government: Interrelations and reciprocities with smart city. *Digital Government: Research and Practice*, 2(4): 1–28. https://doi.org/10.1145/3465061.

Brettel, M., Friederichsen, N., Keller, M., and Rosenberg, M. (2014). How virtualization, decentralization and network building change the manufacturing landscape: An Industry 4.0 perspective. *World Academy of Science, Engineering and Technology, International Journal of Mechanical, Aerospace, Industrial, Mechatronic and Manufacturing Engineering*, 8(1): 37–44. https://doi.org/10.5281/zenodo.1336426.

Camarinha-Matos, L. M., Boucher, X., and Ortiz, A. (Eds.). (2023). *Collaborative networks in digitalization and society 5.0: PRO-VE 2023 proceedings*. Springer Nature.

Eccles, R. G., Ioannou, I., and Serafeim, G. (2014). The impact of corporate sustainability on organizational processes and performance. *Management Science*, 60(11): 2835–2857. https://doi.org/10.1287/mnsc.2014.1984.

Javaid, M., Haleem, A., Singh, R. P., Haq, M. I. U., Raina, A., and Suman, R. (2020). Industry 5.0: Potential applications in COVID-19. *Journal of Industrial Integration and Management*, 5(4): 507–530. https://doi.org/10.1142/S2424862220500220.

Kayikci, Y. (2018). Sustainability impact of digitization in logistics. *Procedia Manufacturing*, 21: 782–789. https://doi.org/10.1016/j.promfg.2018.02.184.

Liu, L. X., Sun, Z., Xu, K., and Chen, C. (2024). AI-driven financial analysis: Exploring ChatGPT's capabilities and challenges. *International Journal of Financial Studies*, 12(3): 60. https://doi.org/10.3390/ijfs12030060.

Nahavandi, S. (2019). Industry 5.0—A human-centric solution. *Sustainability*, 11(16): 4371. https://doi.org/10.3390/su11164371.

Nakamoto, S. (2008). Bitcoin: A peer-to-peer electronic cash system. https://bitcoin.org/bitcoin.pdf.

Olawoyin, O. M. (2024). Blockchain technology in risk management: Strengthening cybersecurity and financial integrity. *International Journal of Research Publication and Reviews*, 5(10): 2336–2348. https://doi.org/10.55248/gengpi.5.1024.2829.

Oyewole, N. A. T., Adeoye, N. O. B., Addy, N. W. A., Okoye, N. C. C., Ofodile, N. O. C., and Ugochukwu, N. C. E. (2024). Promoting sustainability in finance with AI: A review of current practices and future potential. *World Journal of Advanced Research and Reviews*, 21(3): 590–607. https://doi.org/10.30574/wjarr.2024.21.3.0691.

Ramírez-Gordillo, T., Mora, H., Pujol-Lopez, F. A., Jimeno-Morenilla, A., and Maciá-Lillo, A. (2024). Industry 5.0: Towards human centered design in human machine interaction. pp. 661–672. *In: Springer Proceedings in Complexity*. https://doi.org/10.1007/978-3-031-44721-1_50.

Rojko, A. (2017). Industry 4.0 concept: Background and overview. *International Journal of Interactive Mobile Technologies (iJIM),* 11(5): 77. https://doi.org/10.3991/ijim.v11i5.7072.

Schröder, A. J., Cuypers, M., and Götting, A. (2024). From Industry 4.0 to Industry 5.0: The triple transition digital, green and social. pp. 35–51. *In: Topics in Mining, Metallurgy and Materials Engineering*. https://doi.org/10.1007/978-3-031-35479-3_3.

Schwab, K. (n.d.). The fourth industrial revolution. *World Economic Forum*. https://www.weforum.org/about/the-fourth-industrial-revolution-by-klaus-schwab/.

Soori, M., Jough, F. K. G., Dastres, R., and Arezoo, B. (2024). AI-based decision support systems in Industry 4.0: A review. *Journal of Economy and Technology*. https://doi.org/10.1016/j.ject.2024.08.005.

Teece, D. J. (2017). Business models and dynamic capabilities. *Long Range Planning*, 51(1): 40–49. https://doi.org/10.1016/j.lrp.2017.06.007.

Tóth, A., Nagy, L., Kennedy, R., Bohuš, B., Abonyi, J., and Ruppert, T. (2023). The human-centric Industry 5.0 collaboration architecture. *MethodsX*, 11: 102260. https://doi.org/10.1016/j.mex.2023.102260.

Tyagi, A. K., and Abraham, A. (2021). Recent trends in blockchain for information systems security and privacy. In CRC Press eBooks. https://doi.org/10.1201/9781003139737.

Tyagi, A. K., and Sreenath, N. (2021). Cyber Physical Systems: Analyses, challenges and possible solutions. Internet of Things and Cyber-Physical Systems, 1: 22–33. https://doi.org/10.1016/j.iotcps.2021.12.002.

Tyagi, A. K. (2023a). Decentralized everything: Practical use of blockchain technology in future applications. pp. 19–38. *In:* Elsevier eBooks. https://doi.org/10.1016/b978-0-323-96146-2.00010-3.

Tyagi, A. K. (2023b). Dew Computing. pp. 332–345. *In:* Advances in systems analysis, software engineering, and high performance computing book series. https://doi.org/10.4018/978-1-6684-8531-6.ch017.

Varadarajan, M. N. (2023). Smart finance: How artificial intelligence and machine learning drive innovation in the financial sector. *International Journal of Innovative Research in Science Engineering and Technology*, 12(12): 1–14. https://doi.org/10.15680/IJIRSET.2023.1212045.

Wahab, M. D. A., and Radmehr, M. (2024). The impact of AI assimilation on firm performance in small and medium-sized enterprises: A moderated multi-mediation model. *Heliyon*, 10(8): e29580. https://doi.org/10.1016/j.heliyon.2024.e29580.

Yi, Z., Cao, X., Chen, Z., and Li, S. (2023). Artificial intelligence in accounting and finance: Challenges and opportunities. *IEEE Access*, 11: 129100–129123. https://doi.org/10.1109/ACCESS.2023.3333389.

Zakaria, N., Sulaiman, A., Min, F. S., and Feizollah, A. (2023). Machine learning in the financial industry: A bibliometric approach to evidencing applications. *Cogent Social Sciences*, 9(2). https://doi.org/10.1080/23311886.2023.2276609.

Zwitter, A., and Boisse-Despiaux, M. (2018). Blockchain for humanitarian action and development aid. *Journal of International Humanitarian Action*, 3(1). https://doi.org/10.1186/s41018-018-0044-5.

5

Blockchain-Enabled Examination Integrity

Guaranteeing Transparency and Security

*Naghul Pranav K. R.** and *Divya Mohan*

1. Introduction

Blockchain is one of the most important technology inventions in recent years, and is creating something that can have a significant impact on the world in the areas of finance and trade. The aim of this chapter is to explore the revolutionary potentials of blockchain in education, with a particular focus on its utility for self-directed learning. Through understanding the core properties of blockchain—all of them potentially customisable to teaching use cases—we can appreciate the increasing relevance of the technology for education. Over-simply, blockchain is a distributed ledger that records digital transactions. Instead of relying on a single, central entity, blockchain shares a network-wide copy of connected data chunks, or "blocks," across all users' devices. New information must first be validated through a consensus mechanism that involves the majority of members in the network.

This ensures the integrity of the data and that no single participant can alter the record. There are different types of blockchains for various use cases. Anyone can read and write to public blockchains, offering high transparency and inclusivity. On the other hand, private blockchains are designed for deployment within certain organisations or collectives, offering greater control and privacy. While Bitcoin's blockchain is the most well-known example, numerous other implementations have emerged, tailored to different settings for distinct use cases. One of the most basic principles of how blockchain works is the mechanism through which new data is accepted into the ledger. It makes use of a distributed consensus system known as "proof of work" and entails making heavy CPU/bandwidth use by users' computers ("mining"). No after-the-fact additions can be added to the blockchain until and unless the network records the processing of this operation. This consensus mechanism makes it very difficult for an evildoer to forge a fraudulent replica of the blockchain, as they would need to exceed the total processing power of the collective.

Department of Computing Technologies, SRM Institute of Science and Technology, Kattankulathur Campus, Chengalpattu, Tamil Nadu, India.
Email: divyamohan797@gmail.com, https://orcid.org/0000-0002-5988-3851
* Corresponding author: naghul695@gmail.com, https://orcid.org/0009-0000-4421-6709

DOI: 10.1201/9781003655190-6

However, there are some drawbacks to the proof-of-work system. Mining—an essential process that is used for validation and security—consumes a lot of computing power and energy. This has raised concerns about the environmental impact and efficiency of blockchains, not least as they continue to grow in size and scale. Additionally, each block in a blockchain is constrained to be relatively small (e.g., one megabyte in size), thus controlling the amount of data that can be stored at the same time. Not limited in this way, the stored data may take any other form requiring secure or administered access, such as financial or academic records, or student transcripts. The decentralisation of data across all the computers in the network is a fundamental aspect of the blockchain. The data is subsequently open for anyone with the appropriate public key to access; however, it is forever immutable, even to the creator of the content.

The time generated for each record will ensure a reliable and consistent chain of custody for all data stored. This immutability produces a high level of certainty regarding the accuracy and precision of the data. One of the more significant breakthroughs driven by blockchain is the ability to implement smart contracts. These are independent pieces of computer code that act automatically, for instance, in carrying out a transaction or implementing a contract if some predefined conditions are met. Multi-purpose smart contracts may also include, but are not limited to, facilitating the automatic storage and execution of legal agreements and processing bills automatically. A case in point of this technology was evidenced in October 2015, when Visa and DocuSign demonstrated how a smart contract could be used to facilitate the automobile rental process, demonstrating that this approach offers a wide range of possible applications. Given these features, the usefulness of blockchain in education is worth exploring. Blockchain in itself is a secure, transparent, and immutable platform for the recording and management of academic data. When combined with other applications like smart contracts and decentralised storage, blockchain can contribute to the efficiency, fairness, and accessibility of educational processes. In the next sections, we will first discuss the advantages of blockchain for the area of education in an isolated way and then depict how its conjunction with supporting technologies could make educational opportunities even broader for learners, educators, and institutions.

Blockchain technology in education has been moving from purely theoretical advantages and potential use cases to concrete applications both theoretically and experimentally, with an increasing number of schools and institutions participating in these test operations. Global investment in the blockchain in education space now exceeds $500 million, according to a 2022 report by HolonIQ—indicating both a growing interest and a real commitment to how such technologies can be used for academic purposes. A handful of colleges—including MIT and the University of Nicosia—have issued blockchain-verified diplomas and certificates, which means graduates can share their qualifications securely with employers around the world. In another initiative, the Maltese government has said all educational certificates will be placed on a blockchain in a pilot project to prevent credential falsification and simplify traceability. Furthermore, according to the World Economic Forum, 10% of global GDP is likely to be generated by blockchain via some form of application or solution by 2027, indicating widespread acceptance of blockchain in various sectors, including education. These data points serve to highlight the surge of activity around blockchain adoption in education and what it could mean for how academic accomplishments are recorded, verified, and shared.

2. Literature Survey

The concept of decentralised cloud storage on top of the Interplanetary File System (IPFS) has received a lot of interest due to its ability to combine peer-to-peer networking with the principles of content addressing. IPFS works effectively at large scales—millions of requests each day across a large number of actively online peers. However, there has been little research on its internal features, manifestations, and wider implications (Minaee et al. 2021). The education system has also been tapped to study blockchain technology and its power to tackle numerous challenges. One of the major applications is the use of smart contracts to ensure a verifiable, trusted, and traceable learning process. Integration provides visibility and better protection for teachers, and modification is required in the assessment of school management (Li et al. 2020). Also, for the management of academic records, blockchain has been proposed to provide permanent and decentralised records of students' results that cannot be tampered with, preserving intellectual contributions and reputational accolades. The application of adjusted-cPEM was further applied in academic (Opdahl et al. 2020) pioneer studies. The combination of blockchain and IPFS has also been explored for the design of advanced P2PFS with potential performance and security improvements through distributed hash tables, block exchange mechanisms, and self-certifying namespaces (Perera et al. 2020). Secure, immutable online quizzes were delivered to examinees via blockchain. A fair review process was facilitated using a system built with Angular, real-time databases from Firebase, and security from blockchain (Luo et al. 2010). On the other hand, public blockchains have been introduced as a solution to make examination administration more transparent and reliable through decentralisation, in an effort to prevent test takers from relying solely on the professionalism of central authorities (Luo et al. 2010). A new way to combine IPFS with blockchain systems was introduced, making it possible to optimise the use of data storage, reduce the blockchain size, and increase data exchange speed while preserving the level of security (Young et al. 2018).

Despite the numerous advantages it brings, the architecture of decentralised systems suffers from some major problems such as immutability, auditability, and availability. These beneficial characteristics carry important risks and limitations (Gloutnikov et al. 2018). Ethereum smart contract security has been widely discussed and tested, and vulnerabilities are being exploited by hostile agents. Several weaknesses in current coding and software security require stronger security approaches (Pei et al. 2015). What's more, it is difficult to keep Ethereum smart contracts updated after deployment as the ecosystem changes so rapidly. Research on smart contract maintenance concerns has pointed out the gap between academic research and industry usage, with proposals to tackle this gap (Sarkar et al. 2016). Web3 has become a source of heated debate, as it embraces decentralisation while fundamentally relying on existing web technologies and infrastructure. Web3 has been found to be an innovation that allows theorists to experiment, though it faces issues related to regulation, security, and governance (Nastase et al. 2015).

Concerning Web3 and decentralised autonomous organisations (DAOs), this line of research opened a door for learning processes and other research enquiries, creating new ways of thinking in governance, collaboration, and knowledge sharing, but it is not without problems, such as bureaucracy, data privacy and legislation issues (Bhatia et al. 2013). A mixture of IPFS, blockchain, and Web3 solutions may facilitate a large range of uses; however, it bears potential issues and challenges in security, scalability, and legal aspects, for which research and development are still active.

3. Proposed Methodology

Traditional examination systems depend on manual procedures, which cause inefficiency and are vulnerable to a number of security threats. As automation becomes more critical, it is necessary to enhance these systems to ensure reliability and safety. Our proposed procedures offer several crucial characteristics:

- A dedicated infrastructure for quick request processing.
- A safe bank for exam papers from which only approved superintendents can extract it.
- A mechanism of silently disposing the exam papers without the instructor's knowledge.
- Comprehensive security models that utilise symmetric and asymmetric encryption where required.

Then, for security reasons, we encrypt both exam papers and the data packages to allow compatibility with the Python architecture. The web application is implemented with Django in order to offer an optimised user experience.

3.1 Sophisticated Automated Examination Management System

Decentralised Storage: With IPFS, examination papers are stored in a distributed manner, eliminating single points of control and restricting access to authorised parties.

Blockchain for Security and Transparency: Blockchain facilitates the creation of an unchangeable and secure data record of the entire exam-related processes—questions design, submission, harvesting, scoring—to increase accountability and prevent unauthorised adjustments.

Encryption Protocols: Symmetric encryption allows for efficient storage and communication, while asymmetric encryption is adopted for ensuring key exchanges and authorisation. This ensures that only appointed superintendents are able to unlock private documents, thus maintaining privacy throughout the whole process.

Encryption Protocols: Symmetric encryption allows for efficient storage and transfer, with asymmetric taking care of the key exchange and authorisation. This way, only valid superintendents can access confidential documents, with confidentiality at its pith.

Django-Based Web Interface: In the challenging examination environment, the web-based system developed using Django is a secure, user-friendly system that allows efficient management of examination functionalities, near-instantaneous communication, request handling, and secure paper submission processes.

It takes the combination of blockchain and IPFS technology as examples to solve the security and inefficiency defects of the traditional ways of exam management, achieve secure and decentralised communication through saving and transmitting mechanisms, and offer strong support for the security of private examination materials.

4. Tools Used

It is an automated examination administering system that solves a number of drawbacks of the traditional examination process through the combination of blockchain, smart contracts, decentralised storage, and modern web technologies. This integrated approach ensures a high level of security, transparency, scalability, and ease of use for all users involved in the context of an exam management environment. Using blockchain, the system aims to ensure that all examination data—question papers, candidate responses, and marking records—are kept secure and tamper-proof in an immutable ledger. Unauthorised access and tampering are thus prevented, addressing one of the most severe flaws of traditional exam systems. Additionally, smart contracts automate the processes of verifying candidates, scheduling exams, calculating results, and issuing certificates. Such contracts follow predetermined rules without any influence from humans, minimising the possibility of error, corruption, or favouritism; no one can question the fairness of everyone equally following a transparent, codified standard. This not only ensures that data is not lost due to technical problems or attacks, but also that it can be retrieved from anywhere and at any time with the corresponding authorisations. The combination is enriched with modern web technologies that provide easy-to-use interfaces for students, teachers, and administrators, so that they can easily use the system without technical knowledge.

The integrated design makes the system scalable to accommodate high demand in terms of simultaneous users and examinations without loss of performance. This proves especially useful for large schools or organisations that are required to test thousands of learners across many sites.

The transparency of the blockchain also means it is able to audit and monitor the entire process of examination in real time. Stakeholders can have the confidence to check the integrity of every stage, from the production of question papers to the release of results, ensuring a trustworthy and honest system. Additionally, this automatic test delivering system can be easily implemented in the current e-learning environments and learning management systems, enabling an efficient transformation process from the traditional to the digital test activity. The blend of these leading-edge technologies enables not only routine administration to be managed with ease, but also to assist institutions in maintaining the strictest levels of academic integrity and service efficiency. With the impact of educational learning environments continuing to change and grow, these systems are positioned to become a vital part of contemporary, secure and compliant assessment delivery.

Decentralised storage is therefore key to improving resilience and access to examination data. The data can be stored in a variety of locations rather than in a single highly centralised server, which may become a single point of failure.

4.1 Blockchain and Smart Contract Engineering

Ethereum: A decentralised platform that runs smart contracts without any possibility of downtime, fraud, control, or interference from a third party. Thus, this decentralised architecture ensures the integrity, authenticity, and transparency of the management of examination records and the exchange of question papers.

Solidity: Main programming language used for writing smart contracts on the Ethereum blockchain. Solidity allows automatic contracts to be written easily to ensure that testing related transactions are safe.

Web3.js: A JS-library for interacting with the Ethereum blockchain. It makes tasks such as retrieving user and contract information, broadcasting transactions, and communication with the blockchain much easier (HTTP and IPC are both supported). It also meets with deployed contracts.

Truffle Suite: A development framework for Ethereum which includes a contract abstraction, test harness and some convenience functions for contract deployment and front-end integration. Truffle makes DApp development easier for the developer.

Ganache: A personal blockchain for developing and testing applications. Ganache provides a safe space for simulating Ethereum transactions and testing smart contracts before deployment to the main network, without the need for real cryptocurrency.

MetaMask: A browser-extension cryptocurrency wallet that connects users to the Ethereum blockchain. With this secure infrastructure, MetaMask helps users and developers sign in and make transactions safely within DApps.

4.2 Web and Application Development

React.js: A JavaScript library for building reactive and interactive user interfaces, specifically for single-page applications (SPAs). We are going to use React for creating the front end of our examination portal.

Django: An elegant Python web framework for creating complex back-end systems. Django mediates encrypted teacher-administrator conversations and encrypted test papers.

HTML/CSS: Foundations for organizing and styling the web portal in a user-friendly and visually appealing manner.

4.3 Decentralised Storage and Security

IPFS (InterPlanetary File System): A decentralised peer-to-peer public network that is maintained across all sites and improves security on the network. IPFS will securely store exam papers and prevent them from being illegally accessed or modified.

Hyperledger Fabric: A blockchain framework implementation in the Fabric model, originally by Digital Asset, IBM, and others, and now hosted by the Linux Foundation for building applications, with a modular architecture and optimal pluggability. It is flexible, scalable, and confidential, making it suitable for secure handling of examination files.

Hyperledger Composer: A toolkit to work with blockchain. Hyperledger Composer makes it possible to express business logic associated with examination-related transactions in a secure and efficient way.

SSL (Secure Sockets Layer): For creating digital certificates for communication between components in a system and for securing internet browsing and user security from other systems. It is helpful to creators to assure themselves of the easy integration of a patented invention.

Asymmetric Encryption: RSA encryption uses asymmetric encryption to secure sensitive data such as examination papers and user credentials. RSA protects the examination material to ensure only authorised individuals can access it.

4.4 Cloud Services and Deployment

Microsoft Azure and AWS (Amazon Web Services) are two of the top cloud platforms with powerful hosting and extensive security layers where our blockchain system could be deployed. These services provide examination management capabilities that are highly available, fault-tolerant, and globally accessible. Our approach is to model the examination management system based on blockchain, decentralised storage, encryption, and modern online technologies in a secure, transparent, and automatic way.

5. Algorithms

Pseudocode for Binary Search: Binary search is a powerful tool to find the target value in a sorted array or list. The approach works by halving the search range consecutively and comparing the target value to the middle element. Here is the pseudocode for binary search:

```
Algorithm: BINARY SEARCH
Input:          Array ARR, Target value X,
                LOW, High index HIGH
Output:         Index of X if found,
Initialization: Assign LOW ← 0 and HIGH ←
                ARR - 1
Step 1: while LOW ≤
a.    MID ← (LOW + HIGH) / 2
b.    if ARR[MID] = X then return MID
c.    else if ARR[MID] < X then LOW
      ← MID + 1
d.    else HIGH ← MID - 1
```

Obtain History Pseudocode (from SmartContract.go): This pseudocode delineates a method to get the history modifications for a particular patient ID inside a blockchain-based system using a buffer to aggregate and return all historical values:

Input:	**args (list of arguments, expecting at least 1 argument)**
Output:	Success with transaction history buffer or Error if argument count is incorrect
Step 1:	**if len(args) < 1 then**
a.	return Error("Incorrect number of arguments. Expecting 1")
Step 2:	**Initialize variables**
a.	PatientId ← args[0]
b.	resultsIterator ← stub.GetHistoryForKey(PatientId)
c.	buffer ← [] (empty list to store history responses)
Step 3:	**for each response in resultsIterator do**
a.	Append to buffer: {

Get All Patients Pseudocode (also from SmartContract.go): This pseudocode acquires all patients within a defined range (e.g., from "abc1" to "abc99"). It traverses the patient records and appends them to a buffer.

Input:	**stub**
Output:	Success with a list of all patient records within the specified range
Step 1:	**Define startkey = "abc1" and endkey = "abc99"**
Step 2:	Retrieve PatientBytes = stub.GetStateByRange(startkey, endkey)
Step 3:	Initialize buffer = []
Step 4:	for each queryResponse in PatientBytes do
a.	Append queryResponse.Value to buffer
Step 5:	Return Success(buffer)

Adding a New Student Record (EMS System): Let us create pseudocode for adding a new student record:

// Add Student Record Pseudocode

Input:	**studentInfo (e.g., Name, ID, Date of Birth, Course, etc.)**
Output:	"Student record added successfully" or "Invalid student information"
Step 1:	**Validate studentInfo (check for non-empty fields and valid formats)**
Step 2:	if studentInfo is valid
a.	Add studentInfo to the student database
b.	Return "Student record added successfully"
Step 3:	Else, return "Invalid student information"

Generating a Report (EMS System): Suppose you want to generate a report for all students who passed an exam:

Input:	**examResults (List of student results, each containing name and score)**
Output:	List of student names who passed the exam
Step 1:	**Define passThreshold = 40 (Assume passing score is 40 out of 100)**
Step 2:	Initialize passStudents = [] (Empty list to hold names of passing students)
Step 3:	For each studentResult in examResults
a.	If studentResult.score >= passThreshold
b.	Append studentResult.name to passStudents
Step 4:	Return passStudents (List of students who passed)

6. System Model

The upload module employs React.js for the front end, facilitating user selection and submission of question papers. Ganache generates a unique private key for each user, which is subsequently imported into MetaMask to securely link their account to the blockchain. Upon uploading a paper, the file is stored on IPFS (InterPlanetary File System), which assigns a distinctive hash value to the file. This hash acts as a digital fingerprint that maintains the file's integrity. Utilising Web3.js, the system communicates with Ethereum smart contracts (developed and deployed with the help of Truffle and Ganache) to store this hash on the blockchain. MetaMask is used to sign and trust the transaction so that only an authenticated user can upload papers. To download particular question papers, users need to have the correct hash in order to access the IPFS link. This approach ensures that only registered users can retrieve the file, thereby providing security and transparency for examination materials. Such a multi-tiered architecture reduces overhead in uploading/examinations while also introducing strict security at every phase. React.js is a contemporary, popular JavaScript library that provides a highly responsive, intuitive user interface, which is easy for educators and admins to navigate during the upload process. This integration with Ganache and MetaMask ensures the identity of each user is cryptographically protected against unauthorised access and that only authenticated participants are allowed to perceive or participate in the device. Storing files in IPFS additionally increases the fault tolerance and scalability of the solution. In contrast to typical centralised storage, IPFS spreads a file's contents across a network of computers, keeping the question papers available even if some nodes go offline. One of these codes is "IPFSHash", which is a unique hash generated by IPFS itself that acts as an immutable fingerprint for each file, allowing the system to instantly identify any unauthorised changes or tampering. This is extremely important concerning examination documents, as their integrity and authenticity must not be disputed. The use of Web3.js as a medium between the front end and the Ethereum blockchain makes the interaction with smart contracts seamless. These smart contracts (built on Truffle and deployed with Ganache) automate the IPFS hash recording and validation procedures on the blockchain. This not only creates an audit trail of every upload but is also transparent and unbreakable. The transaction signing by MetaMask adds another layer of user authentication. Regarding the downloading of question papers, the requirement of the correct IPFS hash acts as a secure access control mechanism. The file is only available to users who hold the corresponding hash, which is distributed (e.g., to an authorised person) through secure means of communication (e.g., the Whisper protocol in IPFS). As a result, this approach prevents unauthorised downloads and effectively ensures that only authorised personnel have permission to obtain sensitive examination content.

Decentralised Outcomes: The result uploading module is developed in React.js for the UI Web3.js to interact with the Ethereum chain with Truffle and Ganache for development and simulated local chain setup. Every user of the DApp has a private Ganache key and imports it into MetaMask to establish a trusted, secure connection with the wallet. When professors submit student scores, with a photo and the total score through the WebApp, the photo is stored on IPFS (InterPlanetary File System) for decentralised and distributed storage. The name and total points of the student are entered into the contract on Ethereum using Web3.js, with deployment and testing handled by Truffle and Ganache. The scores are public and visible on the blockchain and can be viewed by anyone, but the information is read-only; it cannot be changed after it has been uploaded, keeping the student results safe from editing. The integration of decentralised storage and blockchain technology ensures the credibility and transparency of the evaluation process.

A flowchart outlines how IPFS can be integrated with blockchain technology to make a DApp using React, Truffle, Ganache, Web3, and MetaMask. Here is a description of the process:

Compile Smart Contracts: TypeScript context-specific language generators can also be crafted with the starter kit. The process starts with defining smart contracts and deploying them to a local blockchain using Truffle and Ganache, respectively.

Connect to Blockchain: The DApp connects to the blockchain through Web3.js. The DApp creator signs in and uploads.

IPFS Upload: The file is released to IPFS and gets a hash unique to that document.

Embedding Smart Contracts into Blockchain

The IPFS hash is then recorded onto a blockchain, so that its reference can be tracked and it cannot be changed.

User Interaction and Approval: MetaMask allows the wallet to interact with the blockchain and perform transaction authentication and approval. The user is prompted by the MetaMask service to approve the transaction and does so.

Retrieving the File: The DApp enables users to access or retrieve the file from IPFS using the stored hash. This solution ensures that files are securely kept in a decentralised way on IPFS, while all relevant transactions and references are immutably recorded on the blockchain using smart contracts. MetaMask offers safe user authentication and transaction authorisation, while Web3.js facilitates smooth interaction between the decentralised application and the blockchain network.

7. Proposed Smart Contract Algorithm

Step 1: Storing Student Information

Student profiles and rocker data. The university builds and maintains student records by uploading the students' data, which are used for published results, to a secure storage system based on blockchain. This ensures the data remain immutable and transparent, yet protected.

Step 2: Smart Contract Initiation

Develop a smart contract in a Python framework to formulate the terms and conditions around the service in question (mark sheets, certifications, verification of records, etc.).

Step 3: Terms and Conditions Check.

The telephonic interview is then followed by a brief review of all terms and conditions. The user, student, or a third-party verified user reviews and agrees to the required charges, as the terms and conditions for the desired service are determined. The university does the same. This helps in ensuring mutual understanding of the terms agreed upon by the parties.

Step 4: Smart Contract Deployment

When both sides agree to each term and condition, the smart contract executes the process, and the agreement becomes legally binding on the blockchain system.

Step 5: Transmission of Documentation

On completion of the transaction, a Provisional Pass along with the Online Exam Form will be generated, which is compulsory to be brought on the day of the examination along with the identity card and admit card issued by ICSI. An SMS and email will also be sent to the trainees confirming successful fee payment.

Step 6: Addressing Malicious Activities

If any bad actor is found, the university has full authority to refuse services immediately. All such anti-activities could be legally addressed.

Step 7: Exit

The process ends and the smart contract is deployed if executed successfully or due to malicious acts.

8. Project Flow

Figure 1 depicts a system for the management of examination papers, encompassing various principal participants and procedures:

Main Actors

- Controller of Examination
- Teacher
- InterPlanetary File System (IPFS)
- Superintendent
- Private Blockchain

Core Components

- Controller of Examination (top)
- Teacher (left)
- IPFS - InterPlanetary File System (center)
- Superintendent (right)
- Private Blockchain (lower section) Information Transfer Among Components:
- The Controller allocates papers to the Teacher.
- The Teacher transmits encrypted papers back to the Controller.
- The Teacher uploads encrypted documents, along with the symmetric key, to IPFS.
- IPFS delivers the encrypted document to both the Superintendent and the Private Blockchain.
- The Superintendent decrypts the paper using the symmetric key.
- Hash values are created and stored on the Private Blockchain.

Key Outputs

- Paper is separated into blocks.
- The final examination paper has been generated.
- Information Recorded on the Private Blockchain: Hash

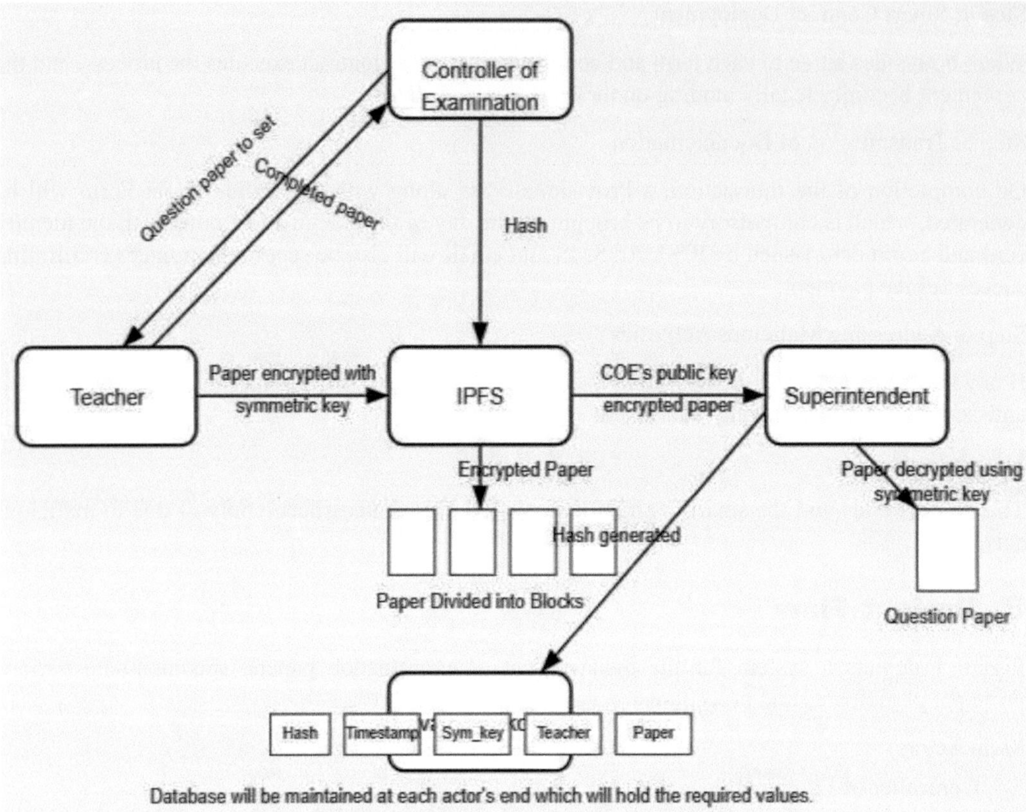

Figure 1: Project flow of examination system.

- Time marker
- Symmetric encryption key
- Instructor details
- Documented information

The Figure 1 illustrates the utilisation of encryption, hashing, and blockchain technology to safeguard the exam paper process from its inception to dissemination.

Sequential Procedure

Step 1: Paper Request Initiator: The Teacher files a request to the Controller of Examinations.

The request comprises the distinct Paper ID and the Teacher's ID.

Step 2: Encryption and Distribution of Documents: The Controller of Examination randomly selects a Paper ID from the pool of available papers.

Asymmetric Encryption: The details of the chosen paper are encrypted with the Superintendent's public key certificate, guaranteeing that only the Superintendent may decrypt it.

Block Division: The encrypted document is partitioned into smaller segments for optimal transmission efficiency.

The encrypted blocks are sent to the Teacher over IPFS, a decentralised content distribution network.

Step 3: Generation of Hash and Storage in Blockchain

The educator calculates a cryptographic hash of the complete encrypted document.

Blockchain Storage: The Educator retains the subsequent information on the private blockchain document identification number.

Instructor Identification Number Symmetric key (if employed for additional encryption) Time marker Hash of the encrypted document.

Step 4: Paper Decryption and Distribution Key Retrieval: The Superintendent acquires the symmetric key from the blockchain associated with the Paper ID.

The Superintendent decodes the encrypted paper blocks utilising the symmetric key.

The decrypted examination paper is subsequently disseminated to the pupils.

This solution utilises encryption, hashing, and blockchain to guarantee the confidentiality and integrity of exam paper management from creation to final distribution.

9. Results And Discussions

The InterPlanetary File System (IPFS) is a decentralised storage framework that augments blockchain technology with secure peer-to-peer networking and content addressing. It enables the establishment of decentralised cloud storage, making it a crucial instrument for the management of extensive data collections in a tamper-resistant way. This technology is essential for tackling significant issues in the digitisation of education, especially in the secure and efficient management of large data sets. Its decentralised structure protects data integrity and aids educational programmes that require reliable and secure data storage solutions. The three primary categories—test result data storage, non-repudiation (which asserts that data cannot be disputed by the originator), and results tampering—have been adequately addressed through blockchain implementation. Using smart contracts from Ethereum for verification and validation, the complete process can be automated with greater openness, accountability, and confidence. These technologies provide educational institutions with the necessary tools to maintain academic records, certifications, and student assessments more securely.

Figure 2 delineates an Examination Management System comprising two principal interfaces: Login Page: The initial interface presents a login portal wherein users (examining officers/ administrators) can input their identification and password to gain access to the system. It also features a "Forgot Password?" option accompanied by a submit button.

Paper Selection Page: The subsequent interface displays a form for the submission of course, semester, branch, and subject details for an examination. A pop-up window is available, allowing the user to select a paper (e.g., "Paper 1") and finalise their choice by clicking the "Finalise Paper" button. This mechanism likely facilitates secure examination paper selection and administration for educational institutions.

Figure 3 illustrates the duration (in seconds) allocated to different stages of decentralised exam papers. The x-axis represents the stages of the exam paper process: Creation, Distribution, and Evaluation. The y-axis denotes the elapsed time in seconds for each stage. The chart reveals that the Creation phase requires a moderate amount of time, while the Distribution phase necessitates the most time of the three. The Evaluation phase consumes more time than Creation but less than Distribution. The chart indicates that the most time-intensive phase of a decentralised exam paper system is Distribution, attributed to the blockchain or IPFS-based data transfer mechanism.

Figure 4 shows a histogram of the distribution of block times in Ethereum 2, with x-axis values between roughly 12.0 and 12.8 seconds and the y-axis representing frequency counts. The distribution is strongly positively skewed, with a concentration of most block times in the range

Figure 2: Implementation of Dapp.

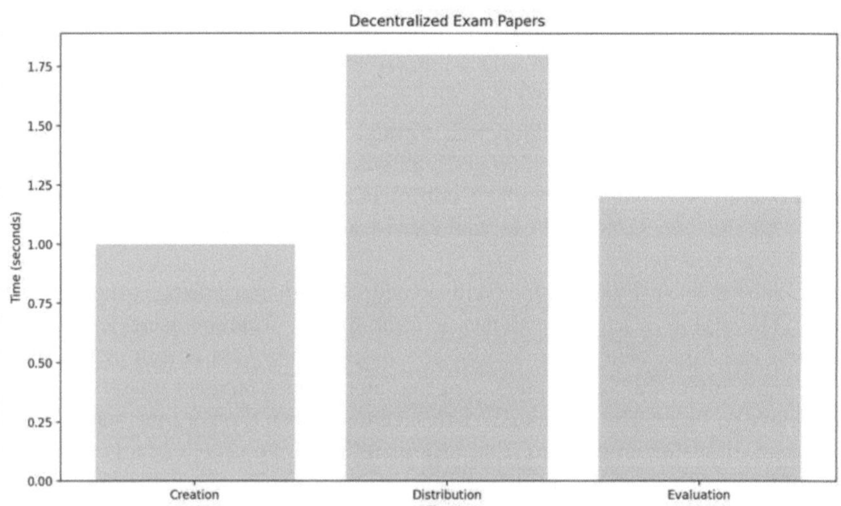

Figure 3: Time taken at different places.

of 12.1 to 12.3 seconds and a peak at a frequency above 70, which then diminishes gently towards higher block times. This implies that shorter block times are more common, while longer block times are less frequent. With the help of a generalised extreme value distribution, which is able to model the tail behaviour of data and take into account rare extreme events in block time generation, we have provided visibility into the performance and stability of the Ethereum 2 network.

Table 1 is a representation of block times in E2. The histogram table representation of block times for Ethereum 2, illustrated in Table 1, reveals the frequency at which blocks were generated with respect to time differences. There are 72 occurrences of the block time 12.1 s, followed closely

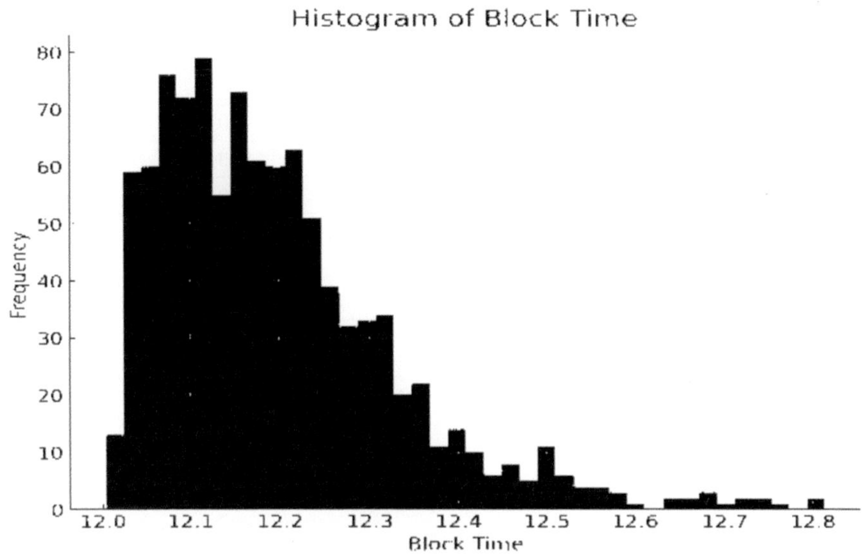

Figure 4: Fitting result of the histogram of block times in Ethereum 2 using the generalized extreme value distribution.

Table 1: Histogram table representation of block times in Ethereum 2.

S.no	Block Time	Frequency
1	12.0	13
2	12.1	72
3	12.2	60
4	12.3	32
5	12.4	14
6	12.5	11
7	12.6	2
8	12.7	2
9	12.8	3

by block times of 12.2 s with a frequency of 60, with most blocks being produced around these ranges. 12.0 s has 13 events, and 12.3 s shows as many as 32, indicating that the frequency decreases slowly with block time. Beyond 12.4 s, the frequency is very low, and only a few events were observed between 12.5 and 12.8 seconds. This tabular distribution endorses the tendency of the histogram, where it is observed that short block times prevail, while long block times are scarce, which emphasises normal operational performance for block generation processes in Ethereum 2.

The communication cost comparison in bytes at different stages is presented in Figure 5. Clearly, the registration phase dominates in terms of communication overhead, with a higher cost of approximately 6,500 bytes, in contrast to all other phases combined. On the other hand, login, upload, request, and dispute have very low communication costs (all below 200 bytes), which become negligible in comparison to registration. This shows that the majority of communication overhead in the system occurs during registration, while the subsequent steps are computationally inexpensive in terms of communication demands, proving that the efficiency of this protocol is not limited to the initialisation stage.

Table 2 shows the communication cost in the various stages of the protocol derived from the presented values in Figure 5. It is apparent from the figure that registration has the maximum

Table 2: Communication cost by phase.

S. No	Phases	Communication Cost (per bytes)
1	Registration	6500
2	Login	200
3	Upload	200
4	Request	200
5	Dispute	200

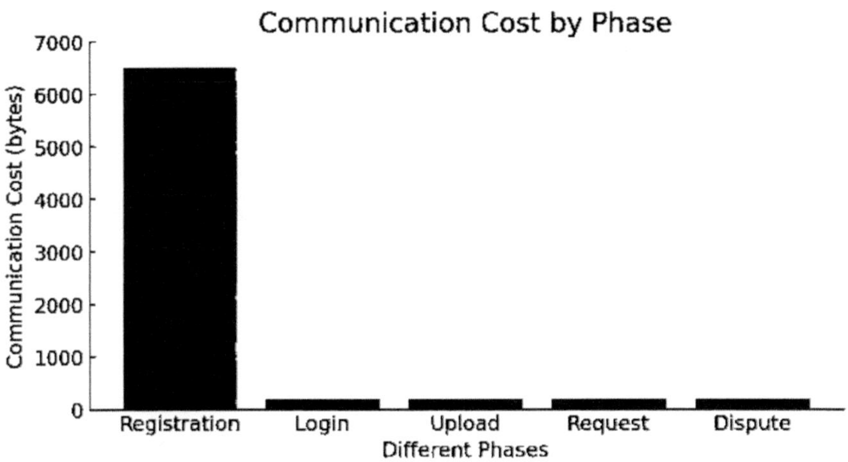

Figure 5: The computation cost comparisons in different phases.

communication cost of approximately 6,500 bytes, indicating registration as the most resource-consuming process. On the other hand, the remaining phases—login, upload, request, and dispute—require about 200 bytes of communication each. This suggests that while the cost of registration is high, the subsequent operations are lightweight and open up possibilities for performance and scalability improvements.

Figure 6 demonstrates the comparison of time usage for different hashing and encryption algorithms. Of the three processes tested, RSA takes the most time to complete, with an execution time of about 1.5 seconds. AES is the fastest due to its computational efficiency, with the smallest execution time (less than 0.8 seconds), indicating its capability for fast encryption. SHA-256 lies between the two, executing in approximately 0.9 seconds, balancing performance and security considerations. This comparison shows that RSA provides strong security but with lower speed performance, while AES offers faster processing, and SHA-256 occupies an intermediate position.

Paper Encryption and Distribution Comparison of Hash Algorithms. Table 3 compares the execution time between different hashing and encryption methods for paper encryption and distribution. The performance is listed in Table 3: it can be seen that AES achieves the best performance (0.65 seconds) and is the quickest method in terms of time. SHA-256, with 0.85 seconds, ranks slightly higher in cost, whereas RSA has the worst execution time of 1.44 seconds, as it is computationally intensive. These observations emphasise that AES is the best candidate for speed performance requirements, SHA-256 provides a balance between speed and security, while RSA, although secure, adds additional processing overhead

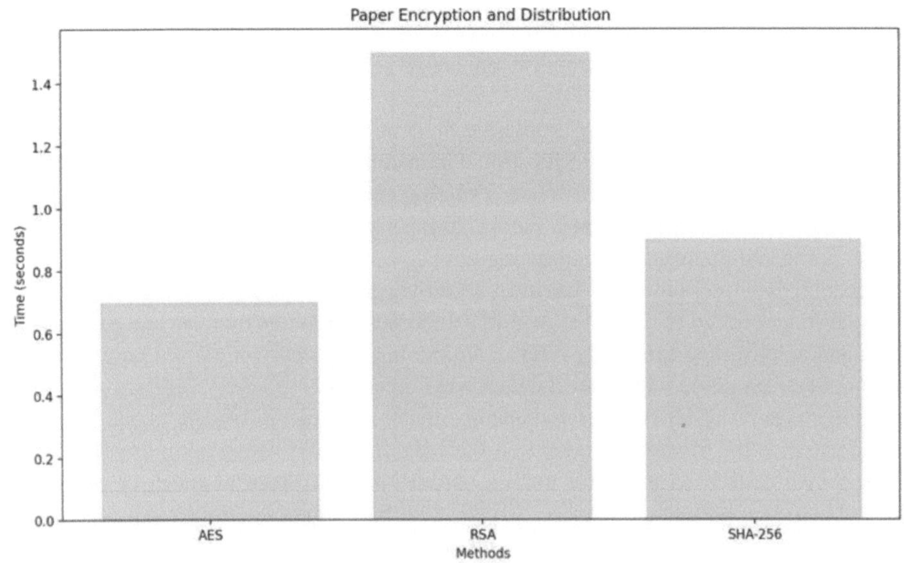

Figure 6: Comparison of hashing algorithms used.

Table 3: Paper Encryption and Distribution Comparison of Hashing Algorithms.

S. No	Methods	Time (Seconds)
1	AES	0.65
2	RSA	1.44
3	SHA-256	0.85

10. Conclusion

The emerging universe of blockchain technology is rapidly extending to all corners of the modern world, and it is time for education to embrace this revolutionary opportunity with open arms. For far too long, colleges and universities throughout the world have been struggling with issues regarding how the information they hold (student records, academic records, historical administrative records, financial transactions) is to be managed, secured, and exchanged. Long-term problems of data breaches, credential fraud, record tampering, and horrendously inefficient processes have caused universities, colleges, and training providers to struggle. Blockchain offers a new model that provides a transparent, decentralised, and immutable distributed ledger system to keep such crucial records. It offers a way to overcome these longstanding issues of trust, transparency, and veracity of data in the education sector, giving institutions a fresh start and a head start in creating new ways to be both modern and efficient.

Decentralizing the core offering of blockchain for education At the core of blockchain for education is decentralisation. Unlike legacy systems that rely on a central authority or a single database, a blockchain is spread across a network of computers, where every user possesses a synchronised copy of the ledger. Every piece of information—a grade, a transcript, even a degree—is kept in a block that is securely connected to the blocks that came before it. Every block is cryptographically linked so that, after data have been placed, it is almost impossible to delete or alter retroactively without the agreement of all participants across the network. With a server-based software solution that works only on requests, the system is simple and cost-effective for schools that have millions of records and, more importantly, provides significantly enhanced security by ensuring that the records remain unaltered and unmanipulated.

The issuance and management of academic credentials using digital certificates is one of the innovative use cases of blockchain for education. Upon completion of a course or degree programme, for example, a tamper-proof digital certificate, a smart contract, or a self-executing computer program can create and place the certificate on a blockchain. This removes the need or possibility for fake diplomas or fabricated qualifications, allowing employers, universities, or government departments to verify credentials in real time. For instance, an employer who wants to validate a job candidate's credentials no longer needs to rely on slow and sometimes unreliable manual procedures. Thanks to blockchain technology, they can easily refer to an indubitable, secure entry and know without doubt that it has not been tampered with since the data were recorded.

Furthermore, the way blockchain nodes authenticate new pieces of data is another example of why it is so trustworthy. Instead of relying on the judgement of a single administrator, blockchains deploy distributed consensus protocols such as proof-of-work, proof-of-stake, or other consensus algorithms. These transactions need to be agreed upon by a majority of network users before they are resolved and updated within the record. This technique significantly increases the difficulty for hackers, disloyal insiders, or other unauthorised entities to tamper with or inject data. This means that academic records are not just immune from external cyber threats—they also cannot be tampered with or misplaced, as often happens with more traditional record-keeping methods.

Another enormous advantage brought by blockchain is that it places students at the helm of their data. In traditional models, student records are locked in institutional silos that are only opened through bureaucratic processes that can be time-consuming and prone to bottlenecks. Through blockchain, students essentially own their digital identity and academic portfolio. They can safely access their learning records and share them if and when they want to (for university or job applications, transferring schools, or showing qualifications for further study). This new model encourages flexibility and self-determination and creates impetus for lifelong learning opportunities that are more open, attainable, and effective.

The potential of blockchain in the education sector is not limited to managing databases and verifying credentials. The opportunities are endless when considering all the other emerging technologies that blockchain can integrate with. For example, educational resources in digital format (books, online video lectures, interactive learning content, etc.) could be recorded on the blockchain with timestamped proof of authorship. This helps educators and content creators protect their intellectual property and discourages plagiarism while ensuring fair credit and remuneration. Likewise, blockchain's open and traceable ledger system can revolutionise financial processes. Tuition can be paid, grants and scholarships disbursed, which can be tracked in a secure manner, reducing paperwork, curbing fraudulent activity required, and ensuring better accountability for the use of funds available to the institution.

Trust and collaboration in the educational ecosystem are greatly increased in such an environment. Faculty, students, staff, and external partners can have confidence in the authenticity and credibility of records without having to trust each individual institution or rely on inefficient approval processes. By eliminating siloed infrastructures and adopting a clear, distributed ledger, blockchain builds a climate of transparency and shared ownership. Automation through smart contracts can also decrease time-consuming administrative tasks, empowering staff to focus on more meaningful, creative work and cutting costs in paperwork and verification activities. Despite its benefits, the introduction of blockchain technology in education is not without challenges. It requires specialised technical skills to build and maintain a blockchain system, which may be a significant undertaking for some organisations. There is also a need for global or standardised protocols so that digital credentials generated on one blockchain can be universally recognised and trusted. Privacy concerns are also paramount, as laws and policies regarding digital records and data sharing evolve on different timelines worldwide. However, these challenges can be overcome through technological advancement, the establishment of common standards, and the development of open regulatory perspectives.

In the longer term, the further evolution of blockchain may provide even greater potential in education. Very soon, we could have full digital learning passports that people carry for life — lifelong records or transcripts that bundle university degrees, short courses, online modules, and micro-credentials from a range of providers into a portable, blockchain-based identity. A decentralised peer-reviewed academic publishing service, a reputation system based on verified contributions, or an open-access marketplace for courses and certifications—all could become part of a much more dynamic and fluid educational ecosystem that is connected in ways that are currently impossible with the outdated tools and legacy systems of today's solutions. In conclusion, blockchain is not merely a conventional tool for technical advancement; it is one of the foundational enablers of an emerging new era in education. Blockchain enables trust, streamlines operations, empowers individuals, and encourages inclusion and lifelong learning through secure, transparent, and decentralised management of vital information. As more educational institutions, students, and employers take advantage of these pathways, blockchain has the potential to fill the gaps that currently exist in the educational system and pave the way for a smarter, more transparent, and more effective future for education.

References

Barbero, C., and Lombardo, V. (1995). Dependency graphs in natural language processing. pp. 146–157. *In: Proceedings of the AIIA 1995 Conference*. Springer. From. https://doi.org/10.1007/3-540-60437-5_11.

Barbero, C., and Lombardo, V. (2022). Web3 & DAOs: An overview of the development and possibilities for the implementation in research and development education. pp. 523–528. *In: 2022 45th Jubilee International Convention on Information, Communication and Electronic Technology (MIPRO)*. IEEE. From. https://ieeexplore.ieee.org/document/9803324.

Bhatia, P. K., Mathur, T., and Gupta, T. (2013). Survey paper on information retrieval algorithms and personalized information retrieval concept. *International Journal of Computer Applications,* 66(6): 1–6. Form. https://www.researchgate.net/publication/269819624_Survey_Paper_on_Information_Retrieval_Algorithms_and_Personalized_Information_Retrieval_Concept.

Gloutnikov, S. S. (2018). Using sentiment analysis and pattern matching to signal user review abnormalities. *arXiv preprint* arXiv:1812.06785.from. https://scholarworks.sjsu.edu/etd_projects/629/.

Lewis, D. D., and Sparck Jones, K. (1996). Natural language processing for information retrieval. *Communications of the ACM,* 39(1): 92–101.from. https://dl.acm.org/doi/10.1145/234173.234209.

Li, J., Sun, A., Han, J., and Li, C. (2020). A survey on deep learning for named entity recognition. *IEEE Transactions on Knowledge and Data Engineering*, 34(1): 50–70.from. https://ieeexplore.ieee.org/document/9039685.

Lopatenko, A. (2001). Information retrieval in current research information systems. *arXiv preprint* cs/0110026.from. https://doi.org/10.48550/arXiv.cs/0110026.

Luo, J., and Xue, X. (2010). Research on information retrieval system based on semantic web and multi-agent. pp. 336–339. *In: 2010 International Conference on Intelligent Computing and Cognitive Informatics*. IEEE.from https://www.semanticscholar.org/paper/Research-on-Information-Retrieval-System-Based-on-Luo-Xue/41aa464d6afffc472285410f6faba6ed26989691.

Minaee, S., Kalchbrenner, N., Cambria, E., Nikzad, N., Chenaghlu, M., and Gao, J. (2021). Deep learning-based text classification: A comprehensive review. *ACM Computing Surveys,* 54(3): 1–40. from: https://dl.acm.org/doi/10.1145/3439726.

Nastase, V., Mihalcea, R., and Radev, D. R. (2015). A survey of graphs in natural language processing. *Natural Language Engineering,* 21(5): 665–698. from: https://www.cambridge.org/core/journals/natural-language-engineering/article/survey-of-graphs-in-natural-language-processing/FBD7890F0A3C7AD8C3F3C4E3582EA929.

Opdahl, A. L. (2020). Knowledge graphs and natural-language processing. pp. 75–91. *In: Big Data in Emergency Management: Exploitation Techniques for Social and Mobile Data.* Springer. from: https://www.researchgate.net/publication/348563721_Knowledge_Graphs_and_Natural-Language_Processing.

Pei, W., Ge, T., and Chang, B. (2015). An effective neural network model for graph-based dependency parsing. pp. 313–322. *In: Proceedings of the 53rd Annual Meeting of the Association for Computational Linguistics and the 7th International Joint Conference on Natural Language Processing (Volume 1: Long Papers)*. Association for Computational Linguistics.from https://aclanthology.org/P15-1031/.

Perera, N., Dehmer, M., and Emmert-Streib, F. (2020). Named entity recognition and relation detection for biomedical information extraction. *Frontiers in Cell and Developmental Biology,8*: 673. from: https://doi.org/10.3389/fcell.2020.00673.

Russell-Rose, T., and Stevenson, M. (2009). The role of natural language processing in information retrieval: Searching for meaning and structure. *Information Processing & Management*, 45(1): 215–231. from: https://doi.org/10.1002/9780470033647.ch10.

Sarkar, K. (2016). Part-of-speech tagging for code-mixed Indian social media text at ICON 2015. *arXivpreprint*arXiv: 1601.01195. from: https://www.researchgate.net/publication/289528667_Part-of-Speech_Tagging_for_Code-mixed_Indian_Social_Media_Text_at_ICON_2015.

Wang, Q., Li, R., Wang, Q., Chen, S., Ryan, M., and Hardjono, T. (2022). Exploring Web3 from the view of blockchain. *Technical report.* from: https://doi.org/10.48550/arXiv.2206.08821.

Wang, Z., Jin, H., Dai, W., Choo, K. K. R., and Zou, D. (2021). Ethereum smart contract security research: Survey and future research opportunities. *Frontiers of Computer Science,* 15(3): 1–18. from: https://link.springer.com/article/10.1007/s11704-020-9284-9.

Young, T., Hazarika, D., Poria, S., and Cambria, E. (2018). Recent trends in deep learning-based natural language processing. *IEEE Computational Intelligence Magazine*, 13(3): 55–75. from. https://ieeexplore.ieee.org/document/8416973.

Section 2
Methods and Principles

6

Big Data Architectures for Blockchain-IoT Systems

E. Padma Latha,[1] R. Ravinder Reddy,[1,] C. Madan Kumar,[2] M. V. Krishna Reddy[1] and Y. C. A. Padmanabha Reddy[1]*

1. Introduction

An innovative framework that integrates technologies like the Internet of Things (IoT), 5G networks, blockchain, cloud computing, and big data has brought technological progress in automated production and has allowed markets to expand beyond regional boundaries. For example, in the food industry, this advancement has also significantly increased the risk of food contamination, often leading to food waste, financial losses for producers, threats to public health, or a combination of these consequences (Gorton and Gracie 2012). Recognizing these risks, governments worldwide have established food safety regulations to enable timely recalls. Yet, current systems face major issues such as delayed detection of hazardous food, fragmented data management, slow decision-making, and difficulty tracing the contamination source.

A creative structure has been recommended to alleviate these limitations and combines technologies such as IoT, 5G networks, blockchain, cloud computing, and big data. To forecast contamination and provide support for rapid decision-making, it exploits machine learning algorithms at the source and supports real-time data collection (Al Jawarneh et al. 2023). The incorporation of real-time processing allows the system to identify contamination earlier, leading to timely and assured actions that minimize potential harm. Despite apparent advantages, the new model faces challenges in operationalizing, such as rewiring data collection processes across the entire food production cycle, dealing with numerous stakeholders, and adapting to new processes. Technological strategies have shifted agricultural food production in recent years, enabling improved food security and affordability from crop production to final consumption. Innovations such as efficient irrigation, high-yield seed varieties, and advanced pest control have significantly increased food output. As a result, despite global population growth, food is now more accessible and affordable than ever (Shvachko et al. 2010).

[1] Department of Computer Science and Engineering, Chaitanya Bharathi Institute of Technology, Hyderabad, India.
Emails: epadmalatha_cse@cbit.ac.in, https://orcid.org/0000-0003-0172-0061; krishnareddy_cse@cbit.ac.in, https://orcid.org/0000-0002-7247-6580; padmanabhareddy_cseaiml@cbit.ac.in, https://orcid.org/0000-0002-9697-7072
[2] Department of Computer Science and Engineering, S R University, Warangal, Telangana, India.
Email: c.madankumar@sru.edu.in, https://orcid.org/0000-0001-7425-2348
* Corresponding author: rravinderreddy_cse@cbit.ac.in, https://orcid.org/0000-0001-6488-9935

DOI: 10.1201/9781003655190-8

Global trade, improved storage, and logistics have allowed food to move across geographical boundaries—for example, tropical produce reaching polar regions. However, this globalized supply chain, involving multiple stages of handling and processing, also heightens the risk of contamination compared to locally produced food. Governments worldwide have acknowledged this risk and implemented regulatory measures like food recalls triggered by inspections, voluntary disclosures, consumer complaints, or illness outbreaks (Chaimov 2016). Failing to act promptly not only endangers public health but can also cause companies severe financial and reputational damage (Zaharia et al. 2016). Given the increasing automation in food production, automated food quality monitoring is essential at every stage. This enables early detection of safety risks and timely interventions, making food safety as critical as food production itself.

Key capabilities of the integrated framework are:

i. Early detection of contaminants through automated monitoring within the food processing chain.
ii. Seamless collection and aggregation of sensor data to cloud platforms (e.g., AWS).
iii. Use of block chain to ensure tamper-proof data integrity and avoid legal disputes.
iv. Real-time data analysis to quickly identify and address potential contamination risks.
v. Accurate trace-back and trace-forward mechanisms to locate the source and current position of affected products.

The intended purpose of the IoT, Blockchain, Big data, Artifcial Intelligence is shown in Table 1. An in-depth exploration of each technological component of the Big Data Architecture for Blockchain-IoT Systems is shown in Figure 1.

Table 1: IoT, Blockchain, Big Data, and AI Components and Purpose.

Component	Intended Purpose
IoT	Enables real-time data generation and automates the process of data acquisition and integration.
Blockchain	Resolves issues related to non-repudiation and ensures traceability.
Big Data	To collect and process the 4Vs of data—volume, velocity, variety, and veracity—from multiple sources, ultimately delivering the 5th V: value.
Artificial Intelligence	To enhance the 4Vs of big data with the value dimension by processing data in near-real time using machine learning algorithms, enabling swift and effective decision-making.

1.1 Research Questions

i. How can Big Data architectures be optimized to handle the high volume, velocity, and variety of IoT data within blockchain-integrated environments?
ii. How can real-time analytics be supported in blockchain-enabled IoT systems without compromising data immutability and integrity?
iii. How can machine learning models be deployed on blockchain-protected Big Data from IoT systems to enable predictive analytics?
iv. What is the ideal reference architecture for combining blockchain, Big Data, and IoT in smart cities or industrial IoT?
v. How can architecture patterns such as Lambda or Kappa be adapted for blockchain-IoT ecosystems?

1.2 Hypotheses

1. Integrating blockchain with Big Data architectures for IoT improves data immutability, traceability, and tamper resistance compared to traditional cloud-based systems.

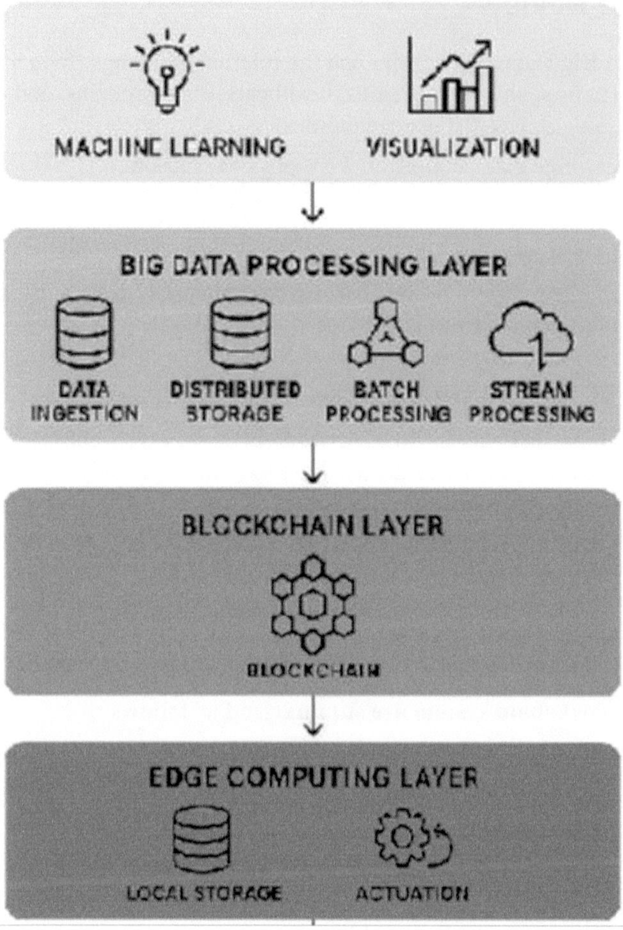

Figure 1: Big Data Architecture for Blockchain-IoT Systems.

2. A layered architecture (consisting of IoT layer, Blockchain layer, Big Data analytics layer, and Application layer) results in better modularity and system resilience.
3. Adopting a Kappa Architecture (stream-based) is more suitable for real-time Blockchain-IoT applications compared to Lambda Architecture.

1.3 Objectives

i. To examine the architectural needs for integrating Blockchain, IoT, and Big Data platforms into a unified system.
ii. To develop a reference architecture that supports high-throughput data ingestion, decentralized security, and scalable analytics for Blockchain-IoT systems.
iii. To assess how edge/fog computing can improve real-time data processing and reduce latency in Blockchain-IoT Big Data pipelines.
iv. To study the impact of various blockchain technologies (e.g., Ethereum, Hyperledger, IOTA) on the performance and scalability of Big Data analytics in IoT networks.

2. Literature Review

The convergence of Big Data, blockchain, and the Internet of Things (IoT) has become a paradigm shift in various domains such as smart cities, healthcare, supply chains, and industrial automation. Each of these technologies has distinct capabilities:

- IoT supports real-time data acquisition from physical sensors.
- Big Data provides scalable analysis, storage, and processing.
- Blockchain provides safe, decentralized, and unalterable data storage.

Nevertheless, their integration requires strong architectural design to deal with scalability, interoperability, privacy, and performance issues.

2.1 Internet of Things (IoT)

The Internet of Things (IoT) refers to a network of low-power, compact, and low-processing-capacity devices (Asch et al. 2018). A standard IoT system is typically organized into four or fewer logical layers. Figure 1 below illustrates the layered structure of a typical IoT system.

A blockchain is a decentralized ledger of transactions that is continuously updated and shared across a network of computers. Whenever a set of transactions is recorded, it creates a new "block," hence the name blockchain. According to Ethereum, a platform for building decentralized applications, managing assets, conducting transactions, and enabling communication without centralized control, blockchain enables trustless and secure operations among parties.

Key features of a blockchain system are summarized as follows

- Every blockchain begins with a **Genesis Block** (also known as Block 0).
- Every block within the blockchain includes two main components: a **header** and a **data section.**
- The **data section** contains a group of transactions, with the quantity defined by the configuration of the Genesis block.
- The **header** includes:
    - A **timestamp** to record when the block was created.
    - A **nonce**, which is a counter value used during the mining process.
    - A **hash** of the block's data.
    - A **hash** of the previous block's header, linking the blocks cryptographically.
- Each new block contains a reference to the header of the preceding block, creating a **secure and unalterable chain**.
- Since this chain is **distributed across many networked computers** (nodes) and stored redundantly, tampering with the blockchain is extremely difficult without detection.

2.2 Big Data

As defined in Gartner's IT glossary, Big Data refers to information assets distinguished by high volume, high velocity, and/or high variety, necessitating cost-effective and innovative processing methods to improve insights, decision-making, and automation (Laroui et al. 2023). In today's constantly connected world, data is being produced at unprecedented speeds and in massive quantities. Furthermore, this data appears in a multitude of formats. Figure 2 illustrates the 5V model of Big Data, which includes the following dimensions:

Variety: Different types of sensors employ distinct technologies to capture physical attributes from the real world. Consequently, the resulting data varies in structure and format depending on the sensor type. This diversity requires processing systems that are capable of handling a wide range of data types.

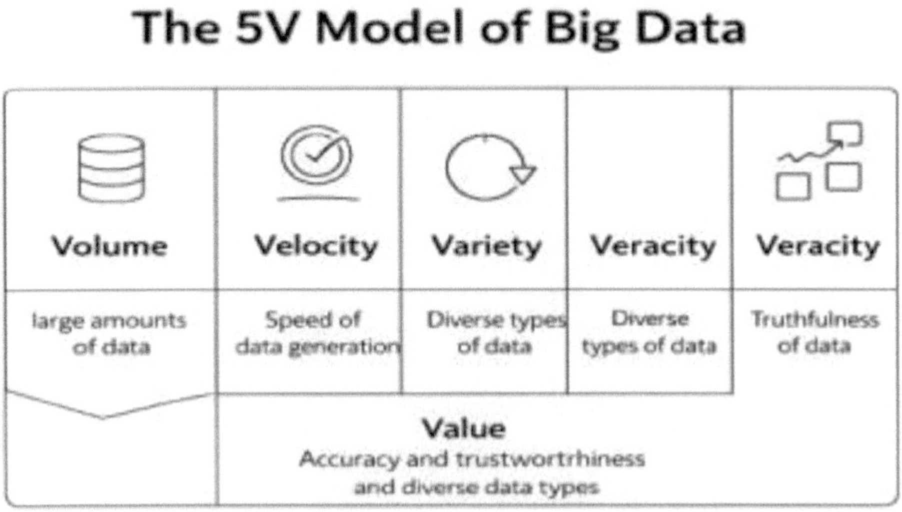

Figure 2: The 5V model of big data.

Volume: Sensors are typically set up to collect data continuously. As a result, IoT-based systems produce significantly larger data sets compared to traditional systems.

Velocity: Not only is data generated in large amounts, but it is also produced rapidly and frequently by multiple sensors. Due to this high speed of data creation, real-time processing is essential to ensure the information remains current and relevant.

Veracity: The combination of high volume, speed, and diversity in IoT data makes it important for systems to assess the reliability of the incoming data. Ensuring data completeness and consistency is crucial for producing dependable analytical results.

Value: True value emerges only when the other four Vs—volume, velocity, variety, and veracity—are effectively leveraged through real-time analytics. While big data infrastructure is essential to uncover actionable insights from IoT data, not every data point holds value. It takes analytical judgment to distinguish meaningful information from noise.

Big data plays a crucial role, as IoT sensors collect and produce massive amounts of information in diverse data formats.

2.3 Data Storage Solutions: Cloud, Edge and Fog Computing

The architecture shown in Figure 3 is typical for IoT platforms or smart systems that aim to gather data from the physical world, process it, and make it actionable for users and other applications. It illustrates a layered architecture for a system that processes sensor data and provides insights, likely for an Internet of Things (IoT) or smart environment application.

i. Data Sources (Bottom Layer)

Data from external sources: This indicates various origins for data, visualized by icons representing a building/city, a house, a car, and a wearable device (like a smartwatch). This implies the system can collect data from smart cities, smart homes, vehicles, and personal devices (Zheng et al. 2021).

Context Data Layer: This layer likely deals with contextual information related to the raw data, such as location, time, environmental conditions, or user profiles. This context is crucial for enriching the raw sensor data and making it more meaningful.

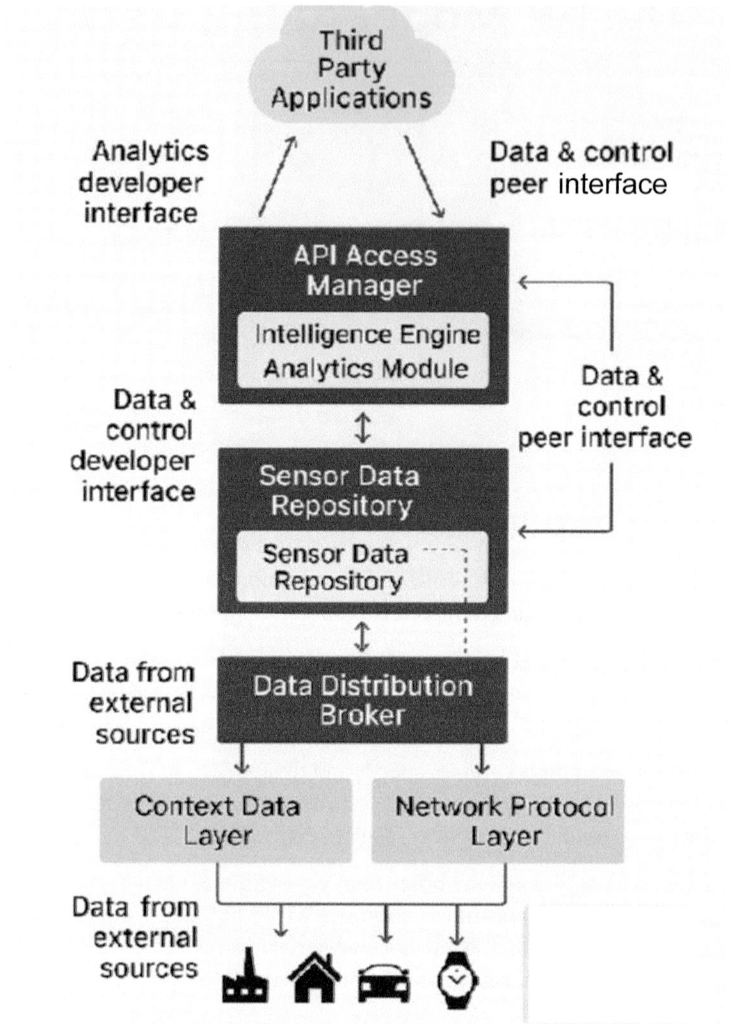

Figure 3: General architecture for fully scoped IoT Big Data.

Network Protocol Layer: This layer handles the communication protocols necessary to transmit data from the diverse external sources to the system. It ensures that data, regardless of its origin, can be received and processed correctly.

ii. **Data Distribution Broker:** This component serves as a central hub for receiving and routing data from both the "Context Data Layer" and the "Network Protocol Layer." It acts as an intermediary, ensuring that incoming data is routed appropriately to the next stage of processing. It also takes data from external sources directly, suggesting it can handle various input streams (Tian et al. 2021).

iii. **Sensor Data Repository:** This is where the raw and potentially processed sensor data is stored. It receives data from the "Data Distribution Broker." The: Sensor Data Repository (inner box) nested box within "Sensor Data Repository" explicitly highlights the storage function. It's the database or persistent storage where all the collected sensor information resides.

iv. **Sensor Data Management Layer (Implicit, but represented by connections around Sensor Data Repository):** Data & control developer interface: This interface allows developers to interact with the sensor data, potentially to configure how data is collected, stored, or preprocessed. It also suggests the ability to send control commands back to the sensors or data sources.

v. **Intelligence Engine Analytics Module:** This is the core processing unit that takes data from the "Sensor Data Repository." It performs analytics on the sensor data to extract insights, patterns, and valuable information. This could involve machine learning, statistical analysis, or other data processing techniques.

vi. **API Access Manager:** This component sits at the top, acting as a gateway for external applications to access the functionalities and insights provided by the system. It manages authentication, authorization, and the exposure of APIs. It interacts with the "Intelligence Engine Analytics Module" to retrieve processed data and insights. Analytics developer interface allows developers of third-party applications to integrate with the system's analytics capabilities.

vii. **Third Party Applications:** These are external applications (e.g., mobile apps, web dashboards, other enterprise systems) that consume the data and insights provided by the system via the "API Access Manager."

viii. **Data & Control Peer Interface:** There are two instances of this interface shown. One connects the "API Access Manager" directly back to the "Sensor Data Repository" and potentially the "Data Distribution Broker" (though the arrow primarily points to the Sensor Data Repository). This suggests that external applications or the API Manager itself can send control commands or requests for specific data directly back down the chain, potentially influencing data collection or device behavior. The other connects the "API Access Manager" to the "Data & Control Developer Interface," indicating that the API Manager can provide interfaces for developers to control the system.

2.4 Data Storage Solutions

Designing data storage solutions for Blockchain-IoT (Internet of Things) systems, as shown in Figure 4, within a big data architecture is a multifaceted challenge, primarily due to the unique characteristics of each component:

- **IoT:** Generates massive of time-series data at high velocity, often from diverse sources, requiring low latency for ingestion and real-time analytics.
- **Blockchain:** Provides immutability, transparency, and decentralization, but is inherently slow and expensive for storing large datasets directly on-chain. It's best suited for transaction records, proofs of integrity, or small critical data.
- **Big Data Architecture:** Demands scalability, fault tolerance, high throughput, and the capability of processing and analyzing huge quantities of structured, semi-structured, and unstructured data.

The core idea is to leverage a hybrid storage approach, storing the bulk of the IoT data off-chain while using the blockchain to ensure its integrity, authenticity, and immutability through cryptographic hashes or transaction proofs.

I. Off-Chain IoT Data Storage (for Volume, Velocity, and Analytics)

This is where the vast majority of raw and processed IoT data resides. The choice depends on data type, access patterns, and analytical needs.

A. Time-Series Databases (TSDBs)

Purpose: Ideal for storing time-stamped IoT sensor data (e.g., temperature, pressure, location). They are optimized for high-volume writes, sequential reads, and time-based queries (e.g., aggregation over time windows) (Thantharate and Thantharate 2023). The examples are as follows:

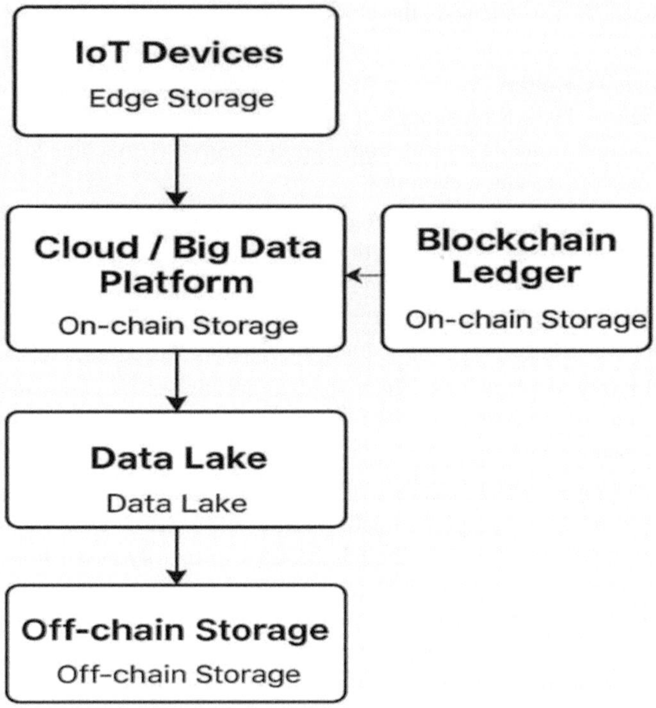

Figure 4: Data storage solutions in Big Data architecture.

- **InfluxDB:** Open-source, purpose-built for time-series data, offering high performance and a powerful query language (Flux).
- **TimescaleDB:** An extension to PostgreSQL, combining the familiarity of a relational database with time-series optimizations (hyper table partitioning).
- **AWS Timestream, Azure Data Explorer, Google Cloud Bigtable:** Managed cloud-based TSDBs offering scalability and integration with other cloud services.

Integration with Blockchain: The raw data can be stored here, and hashes of batches of this data can be periodically committed to the blockchain to prove its existence and integrity at a specific time.

B. NoSQL Databases

Purpose: Flexible and highly scalable, suitable for semi-structured and unstructured IoT data, or when schema flexibility is required. Some of the examples are listed below:

- **Apache Cassandra/DataStax Enterprise:** A column-family NoSQL database known for its high availability, linear scalability, and high write throughput. Excellent for global-scale IoT deployments.
- **Apache HBase:** It is a column-oriented NoSQL database built on top of HDFS, designed for random, real-time read and write access to massive datasets.
- **MongoDB:** A document-oriented NoSQL database that offers flexible schema and good scalability, suitable for diverse IoT data models.
- **Redis:** An in-memory data store commonly used to cache frequently accessed IoT data or temporarily hold real-time stream processing results before they are persisted.

C. Object Storage:

Purpose: Cost-effective, highly scalable storage for large volumes of unstructured data (e.g., images, video, audio from IoT devices, raw sensor logs for archival). Some examples include:

- **Amazon S3 (Simple Storage Service):** Industry-standard, durable, and scalable storage.
- **Azure Blob Storage, Google Cloud Storage:** Cloud equivalents offering similar capabilities.

MinIO: Open-source, S3-compatible object storage for on-premise or hybrid cloud deployments.

Integration with Blockchain: Large files can be stored in object storage, and their cryptographic hashes (e.g., SHA-256) are stored on the blockchain. This allows anyone to verify if a file downloaded from object storage is the exact, unaltered version referenced on the blockchain.

D. Distributed File Systems:

Purpose: For storing very large files and batches of data, particularly when integrated with batch processing frameworks like Apache Spark or Hadoop.

- **Apache HDFS (Hadoop Distributed File System):** The foundational storage layer for the Hadoop ecosystem, designed for high throughput access to large datasets. Often used for data lakes to store raw IoT data before transformation.

2.5 Blockchain-Specific & Decentralized Storage (for Immutability and Decentralization)

These solutions are crucial for linking the off-chain IoT data with the on-chain immutability guarantees.

i. InterPlanetary File System (IPFS)

Purpose: A peer-to-peer network and protocol designed for storing and sharing hypermedia within a distributed file system, where data is identified by its content hash rather than its physical location.

Benefit for Blockchain-IoT: Large IoT datasets can be stored on IPFS. The content hash (CID) generated by IPFS is then stored on the blockchain. This links the immutable, on-chain record to the potentially large, off-chain data. Any tampering with the IPFS data results in a different hash, making the inconsistency detectable.

Examples: IPFS directly, often with pinners or services like Pinata to ensure data availability.

ii. Blockchain-Based Storage Networks

Purpose: Decentralized storage solutions that leverage blockchain incentives to ensure data persistence and availability.

Examples:

Filecoin: A decentralized storage network built on IPFS, providing a marketplace for storage providers and users, ensuring data persistence through cryptographic proofs.

Arweave: A protocol for permanently storing data on a "permaweb" using a blockweave structure, offering truly immutable and perpetual storage.

iii. Supporting Storage for Analytics and Data Processing

These solutions are used once IoT data has been ingested and potentially pre-processed, preparing it for deeper analysis.

a. Data Lakes

Purpose: A centralized repository stores all your data – structured, semi-structured, and unstructured– at any scale. It allows for flexible schema-on-read, enabling various analytical approaches.

Examples: HDFS, Amazon S3, Azure Data Lake Storage, Google Cloud Storage.

Role in Blockchain-IoT: Can act as the primary landing zone for raw IoT data before transformation, where it can be combined with other enterprise data for comprehensive analysis.

b. Data Warehouses

Purpose: Optimized for structured, analytical queries on refined data. Used for business intelligence, reporting, and historical analysis.

Examples: Snowflake, Amazon Redshift, Google BigQuery, Azure Synapse Analytics.

Role in Blockchain-IoT: Storing aggregated IoT insights, processed data, and potentially blockchain transaction summaries for high-performance analytical queries by business users.

Key considerations in storage are

- **Data Volume & Velocity:** Dictates the need for scalable, high-throughput solutions (TSDBs, NoSQL, Object Storage).
- **Data Type:** Influences schema flexibility (NoSQL for variable schema, TSDB for time-series).
- **Immutability Requirements:** Determines the extent of blockchain integration (hashing, IPFS, Filecoin).
- **Read/Write Patterns:** High writes for raw IoT data, high reads for analytics.
- **Latency:** Real-time processing requires low-latency ingestion and access.
- **Cost:** Cloud-based object storage is often the most cost-effective for large archives.
- **Security & Privacy:** Essential for all layers, with robust encryption and access controls.
- **Ecosystem Integration:** Compatibility with existing big data tools (Spark, Kafka, Flink).
- **Decentralization Needs:** How much decentralization is required beyond the blockchain itself (e.g., IPFS).

Figure 1 illustrates a high-level conceptual architecture for integrating IoT, Big Data, and Blockchain technologies, particularly focusing on how data flows and where it is stored. Let's break down each component:

i. IoT Devices (Edge Storage)

- o **Description:** This represents the numerous physical devices (sensors, smart appliances, vehicles, industrial machinery, etc.) that collect data from the real world.
- o **"Edge Storage":** This indicates that some initial data processing and storage might occur directly on the devices themselves or on gateway devices close to them (at the "edge" of the network). This is crucial for reducing latency, saving bandwidth, and enabling immediate actions or filtering before data is sent to the cloud.

ii. Cloud/Big Data Platform (On-chain Storage)

Description: This central component is where large-scale data processing and management happen. It represents a robust infrastructure proposed to manage the "Big Data" characteristics (volume, velocity, variety, veracity).

"On-chain Storage": The label "On-chain Storage" here is somewhat unusual for a "Cloud/ Big Data Platform" in a typical hybrid blockchain architecture. In most practical Blockchain-IoT systems, the *vast majority* of raw IoT data (which is "big data") is **not** stored directly "on-chain" due to the blockchain's inherent limitations (cost, speed, storage capacity).

Possible interpretations of "On-Chain Storage" in this context

- It might be referring to **metadata, cryptographic hashes, or critical transaction records** that are generated by the Big Data Platform and then committed *to* a blockchain ledger.
- Less likely, but it *could* imply a highly specialized, permissioned blockchain where small, critical datasets are directly stored, or a side-chain/layer-2 solution integrated with the cloud platform.
- **Correction/Clarification:** In standard practices, the Cloud/Big Data Platform primarily handles **off-chain** data storage and processing, while only specific, verified information (like hashes) is sent to the blockchain. The arrow pointing from "Blockchain Ledger" to "Cloud/Big Data Platform" suggests interaction, possibly for verification or smart contract execution.

iii. Blockchain Ledger (On-chain Storage)

- o **Description:** This represents the distributed, immutable ledger that forms the core of the blockchain system. It could be a public blockchain (like Ethereum), a private/permissioned blockchain (like Hyperledger Fabric), or a consortium blockchain.
- o **"On-chain Storage":** This label is appropriate here. The blockchain ledger is specifically designed for "on-chain" storage of immutable, verifiable records. As discussed, it typically stores:
- Transaction records (e.g., device registrations, data ownership transfers).
- Cryptographic hashes of large off-chain data (to prove its integrity).
- Smart contract states and execution results.
- Small, critical pieces of data that require the highest level of trust and immutability.
- o The double-headed arrow between "Cloud/Big Data Platform" and "Blockchain Ledger" signifies a two-way interaction: the Big Data Platform might feed data (or hashes of data) to the blockchain, and it might also query the blockchain for verification or to trigger actions based on smart contract logic.

iv. Data Lake

A Data Lake is a centralized storage repository intended to hold large volumes of raw data in its original form—whether structured, semi-structured, or unstructured. It allows for "schema-on-read," meaning the schema is applied only when the data is read for analysis, offering high flexibility.

"Data Lake": The label correctly identifies its function. In this architecture, the Data Lake would primarily receive raw or lightly processed data from the "Cloud/Big Data Platform." It serves as a staging area and a comprehensive source for various analytical workloads.

v. Off-chain Storage

Description: This represents various non-blockchain-based storage solutions used for the bulk of the data. This would include different types of databases and file systems that are highly scalable, performant, and cost-effective for large datasets.

"Off-chain Storage": This is a crucial clarification. It encompasses solutions like:

- **Time-Series Databases:** For high-volume sensor data.
- **NoSQL Databases (Cassandra, MongoDB):** For flexible and distributed storage of diverse IoT data.
- **Object Storage (S3):** For very large unstructured files (images, video, long-term archives).
- **Distributed File Systems (HDFS):** For batch processing workloads.

The arrow pointing from "Data Lake" to "Off-chain Storage" suggests that the Data Lake might serve as a temporary landing zone before data is refined and moved to more specialized "Off-chain Storage" solutions, or it implies that the "Off-chain Storage" is a broader category that *includes* the Data Lake but also other refined storage types. Given the typical flow, the Data Lake often is a form of off-chain storage. If it's a separate box, it implies that the Data Lake might be used for raw ingestion, and then processed/refined data moves to other specific "Off-chain Storage" for optimized access.

3. Methodology

3.1 Scalable Architecture for Big Data in IoT

Building scalable architectures for Big Data in IoT systems shown in Figure 5 is essential to effectively manage the vast amounts of data generated by billions of interconnected devices. These architectures must address challenges related to data volume, velocity, variety, and veracity to ensure efficient ingestion, processing, storage, and analysis. The following sections outline key considerations for developing such architectures.

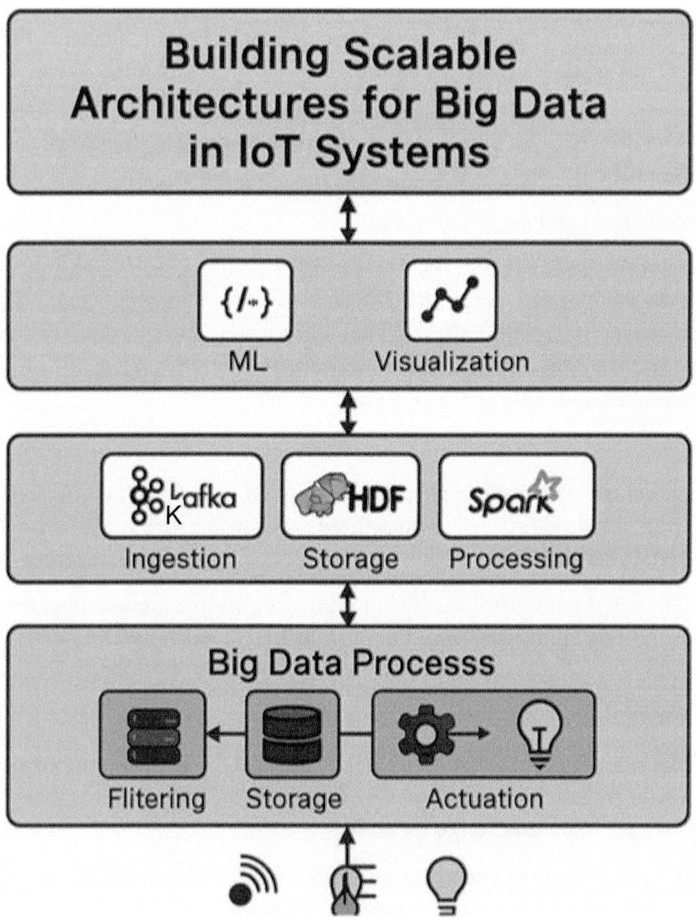

Figure 5: Scalable Architecture for Big Data.

Scalability and Real-Time Processing

- **Architecture Design**: Utilizing frameworks that integrate resource-based storage and edge computing can enhance scalability and reduce latency in data processing (Mahto and Mishra 2024).
- **Real-Time Analytics**: You can create sub-second latency systems by utilizing technologies such as Apache Kafka and Flink, which can allow for real-time data processing (Halami et al. 2025).
- **Diverse Data Sources**: Quality and interoperability assurance requires robust data management strategies due to the heterogeneous IoT device data streams generated (Wongthongtham et al. 2017, Thandekkattu et al. 2024).
- **Big Data Techniques**: Predictive analytics and other advanced analytical processes (e.g., machine learning) enable actionable guidance and inferences from the valuable and complex data IoT systems collect (Thandekkattu et al. 2024).

Addressing a growing problem, a focus on scalable architectures is deeply important but also presents potential drawbacks that need consideration, such as a constant need for updates to manage shifting data types and sources alongside increasing system complexity. One of the biggest challenges in Big Data and IoT is the ability to manage the ever-increasing volume of data while improving processing efficiency and handling. When it comes to robust system building, key architectural principles such as distributed processing, loose coupling, and fault tolerance are critical. These principles enable organizations using big data to efficiently manage information while maintaining high system uptime and optimal performance. The following sections elaborate on these principles and their implications for scalable architecture.

Distributed Processing

- According to (Salamkar 2021), horizontal scalability and distributing workloads across multiple nodes allows companies to handle greater amounts of data without sacrificing efficiency.
- Technologies such as Hadoop and Spark help with the distribution to store and process data efficiently (Salamkar 2021).

Loose Coupling

- When components are designed to operate independently, and upgrades or scalable adjustments can be made to the design without affecting the whole system, flexibility is improved (Gorton and Rayavarapu 2022).
- This method also supports fault isolation, reducing the effects of component failures (Gorton and Rayavarapu 2022).

Asynchronous Communication

- Message queues and event streams decouple producers of data from consumers. Message queues allow for non-blocking operations and buffering of unexpected data spikes (Halami et al. 2025).
- This is especially useful in real-time analytics, as low latencies are needed (Halami et al. 2025).

Fault Tolerance & Resilience

- Deploying techniques such as replication and failover permits systems to be continually operational even when components fail (Halami et al. 2025).
- For example, Kafka's log replication allows fast recovery from node failures without losing data (Halami et al. 2025).

Modularity

- Decomposing architecture into interchangeable modules eases the development and maintenance of the functionality while enabling more agile responses to changing needs (Gorton and Rayavarapu 2022).

3.1.1 Data Partitioning/Sharding

Partitioning or Sharding data by "key" allows for almost unlimited scalability and performance especially in NoSQL (Azarmi 2016).

Stateless Processing

Building stateless processing elements makes it easier to scale out because you do not need to maintain client session information (Perkins 2021). The same architecture can better accommodate rapid and dynamic changes in the demands of the processing workload (Salamkar 2021). While these principles form a foundation for nearly fully scalable architectures, it is important to consider the hidden challenges of such designs directed at integrating and addressing issues like data consistency and security. Opportunities for scaling must be considered and balanced with these challenges, which may limit or dictate the parameters of the system's original state.

The ingestion layer is a vital part of the data process and provides a scalable and fault-tolerant entry point for a large volume of data arriving simultaneously from many devices. It ensures reliable data transfer while managing concurrent connections and high throughput—mandatory for real-time analytics and reporting. The next sections explain the ingestion layer, its scalability, and relevant technologies.

Purpose

- Acts as a buffer to facilitate data flow from millions of devices.
- Provides secure and efficient data ingestion into platforms, particularly in IoT environments (Padmakar and Rahul 2019).
- Enables real-time events and data analysis, which is critical for timely decision-making (Hlaing 2023).

Scalability

- Provides millions of concurrent connections that are critical in big data infrastructures.
- Provides high throughput of data, which is essential for low-latent applications (Hlaing 2025).
- Addresses challenges posed by the accelerating explosion of data streams to provide consistent performance across varying loads (Isah and Zulkernine 2018).

3.1.2 Technologies

- **Message Brokers/Queues**
    - o **Apache Kafka**: A distributed system that is built for high throughput and fault tolerance for data feeds (Vankayala 2025).
    - o **Cloud Services**: Solutions like AWS Kinesis, Azure Event Hubs, and Google Cloud Pub/Sub offer managed and scalable streaming (Padmakar and Rahul 2019).
    - o **MQTT Brokers**: Lightweight messaging protocols like Mosquitto are often the first method of communicating with devices (Hlaing 2023).
- **API Gateways/Load Balancers**: Manage Authentication, Authorization, and traffic distribution between ingestion endpoints (Tallberg 2020).
- The ingestion layer is critical to real-time data processing; however, it faces challenges with potential bottlenecks in memory and CPU usage, particularly when working with numerous

data sources. This also indicates an ongoing need to improve ingestion technologies as demands evolve.

- Real-time analytics through stream processing serves organizations' needs for immediate insights and quicker responses to changing data patterns. This method of processing uses various technologies to minimize latency and allow scalability and the ability to efficiently handle data processing as more "streams" are added. The following sections describe the main aspects of this processing layer.
- **Immediate Insights**: Stream processing allows for the analysis of data as it is produced, facilitating quick decision-making and anomaly detection (Sagar and Abrol 2024).
- **Triggering Alerts**: Systems can automatically alert users to significant changes or anomalies in data, enhancing operational responsiveness (McCaffrey et al. 2013).

Scalability and Performance

- **Horizontal Scaling**: Stream processing frameworks such as Apache Flink and Spark can provide horizontal scaling when absorbing additional data streams without sacrificing performance.
- **Latency Management**: Flink provides low-latency processing, making it appropriate for low-latency applications to gain real-time insights, while Spark Streaming, which provides micro-batching, is appropriate when unified analytics is needed (Dongen and Poel 2020, Patchipala 2024).

Technologies Utilized

- **Stream Processing Frameworks**:
    - o **Apache Flink**: Known for stateful processing and low-latency analytics, ideal for complex event processing (Patchipala 2024).
    - o **Apache Spark Streaming**: Provides a unified approach to batch and stream analytics, though with higher latency (Dongen and Poel 2020).
    - o **Apache Storm**: A robust system for distributed real-time computations (McCaffrey et al. 2013).
- **Real-time Databases/Caches**:
    - o **Redis**: An in-memory store that provides fast data access and aggregation (McCaffrey et al. 2013).
    - o **Apache Druid**: Optimized for high-performance analytics on large datasets (Sagar and Abrol 2024).
    - o **Time-Series Databases**: Such as InfluxDB and TimescaleDB, designed for real-time sensor data management (Sagar and Abrol 2024).

While the focus on real-time analytics is crucial for many organizations, some may argue that the complexity and resource demands of maintaining such systems can outweigh the benefits, particularly for smaller enterprises with less immediate data processing needs.

The warm path or serving layer in data architecture is crucial for storing and serving aggregated, indexed, or enriched data, enabling interactive queries for dashboards and reporting. This layer must ensure scalability to provide fast read access for multiple concurrent users and applications. The following sections detail the technologies and methodologies that support this functionality.

3.1.3 NoSQL Databases

- **Apache Cassandra/DataStax Enterprise**: Known for high availability and linear scalability, it efficiently handles both read and write operations, making it ideal for real-time applications.
- **MongoDB**: Provides flexibility in the schema, allowing for different data types to be stored and queried, which is useful when applications require different data structures.

- **HBase**: Developed on HDFS, it allows random, real-time access to large amounts of data, providing the right context for applications that require fast access to data.

Search Engines

- **Elasticsearch**: Provides full-text search and log analytics that enhance the dashboarding experience by enabling users to perform complex queries in an efficient manner.

Interactive Query Methods

- With some methods for interactive querying, users can dynamically drill down to refine searches, enhancing user engagement and improving exploratory data analysis (Krystian 2020, Eluard et al. 2013). Many methods utilize distributed indexing and scalable query processing frameworks, which are needed to support fast interactive data exploration (Guzun 2016, Zhihong et al. 2017).

Although the warm path is key to quickly delivering data to the end-user, it is important to recognize the potential difficulties of maintaining data consistency and performance under high load, which can affect the architecture of such systems.

- The cold path or batch processing layer, or Data Lake, plays an important role as it houses raw, historical, and aggregated data and allows deep analytical processing and machine learning model training. This structure is designed to efficiently manage massive data volumes by enabling the use of multiple technologies, relieving individual workloads while ensuring scalability and performance.
- **Storage of Diverse Data**: Data lakes accommodate data in its original form, for flexible analytics and machine learning applications (Hai et al. 2024).
- **Long-term Archiving**: They provide a repository for historical data, essential for trend analysis and model training over time ("R. Hai, Data Lakes" 2023).
- **Handling Large Volumes**: Data lakes can economically store petabytes to exabytes of data, suitable for enterprises facing exponential data growth ("R. Hai, Data Lakes" 2023).
- **Batch Processing Efficiency**: Technologies like Apache Spark and Hadoop MapReduce facilitate efficient processing of large datasets in batch mode (Gupta 2024).

Technologies Utilized

- **Distributed File Systems**: HDFS and cloud storage solutions like Amazon S3 and Azure Data Lake Storage offer scalable and durable storage options (Garg 2023).
- **Batch Processing Frameworks**: Apache Spark and Hadoop MapReduce are the leading frameworks for processing large-scale data, and Apache Hive provides the ability to run SQL-like queries on data stored in Hadoop (Gupta 2024).
- **Massively Parallel Processing (MPP) Data Warehouses**: Technologies like Snowflake and Amazon Redshift run complex SQL queries across massive datasets and provide a mechanism for analytic capability ("R. Hai, Data Lakes" 2023).

While data lakes offer significant advantages regarding flexibility and scalability, they also encounter challenges in areas of data governance and data quality, which must be managed thoughtfully in design and implementation (Hai et al. 2024). The combination of Machine Learning (ML) and Artificial Intelligence (AI) in data architectures greatly improves analytics and decision-making. This enables an organization to analyze large datasets, understand complex patterns, and use models to perform real-time inference for operational improvements and strategic initiatives.

Purpose of ML and AI

- **Advanced Analytics**: ML and AI enable organizations to predict historical and real-time data, revealing trends that inform decision-making (J. Ramya et al. "AI and machine learning in predictive data architecture" 2024).
- **Predictive Modeling**: These technologies facilitate the development of predictive models that adapt and improve over time through continuous learning (J. Ramya et al. "AI and machine learning in predictive data architecture" 2024).
- **Autonomous Decision-Making**: Advanced computational techniques, such as Reinforcement Learning and Deep Neural Networks, enhance decision-making in dynamic environments (Govardhan et al. 2024).

Scalability

- **Massive Datasets**: ML models can be trained on extensive datasets stored in Data Lakes, ensuring scalability as data volumes grow (Kandragula 2024).
- **Real-Time Inference**: Deployment of models can occur at the edge or in the cloud, allowing for immediate data processing and decision-making (Kandragula 2024).

Technologies

- **ML Frameworks**: Popular frameworks include TensorFlow, PyTorch, and Scikit-learn, which support model development and training (J. Ramya et al. "AI and machine learning in predictive data architecture" 2024).
- **Deployment Tools**: Technologies like Kubernetes and Docker facilitate the orchestration and deployment of ML models (Kandragula 2024).
- **Cloud Services**: Platforms such as AWS SageMaker and Azure Machine Learning provide scalable environments for ML applications (Purohit 2025).

On the other hand, although incorporating AI and ML into cloud computing brings significant benefits, it also presents challenges related to data privacy, ethical use, and regulatory compliance that must be addressed to ensure sustainable development in this domain (Purohit 2025). The Lambda and Kappa architectures shown in Figure 6 are pivotal in managing big data, particularly in real-time analytics. Lambda architecture employs a dual-path system, integrating batch processing for historical accuracy and stream processing for immediate insights. In contrast, Kappa architecture simplifies this by utilizing a single stream processing engine, replaying historical data through the same pipeline. This distinction allows companies to effectively customize their data processing models.

3.1.4 Lambda Architecture

- **Structure**: Comprises three layers: batch, speed, and serving, facilitating both historical and real-time data processing (Kraetz and Morawski 2021, Kumar 2020).
- **Advantages**: Offers high accuracy and resilience against failures, making it suitable for diverse applications, including financial services (Kumar 2020, Kraetz and Morawski 2021).
- **Performance**: While it requires more resources, it outperforms Kappa in accuracy by approximately 9% (Sanla and Numnonda 2019).
- **Simplicity**: Removes the batch layer, and solely relies on stream processing, potentially leading to improvements in simplicity (Sanla and Numnonda 2019).

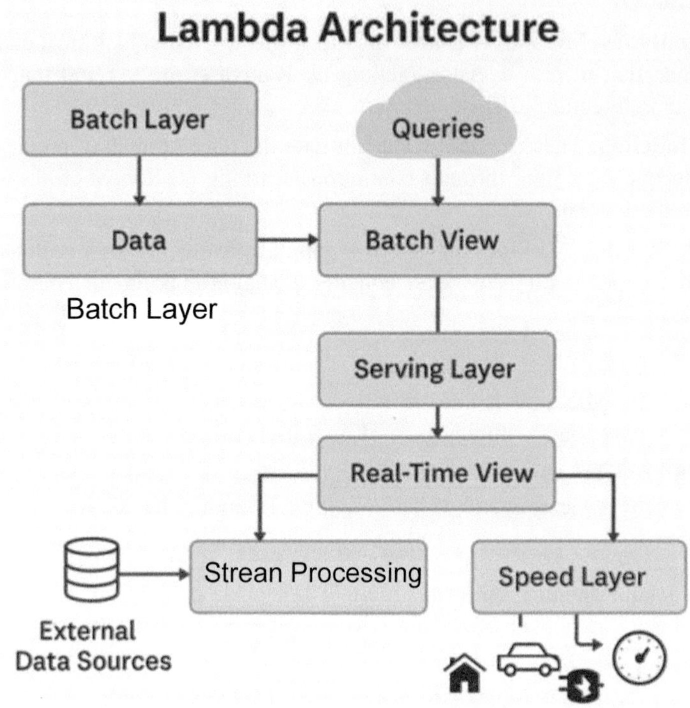

Figure 6: Lambda Architecture.

- **Use Cases**: The Kappa architecture fits best in a scenario where real-time processing is more important than historically accurate processing, such as an IoT application (Halami et al. 2025).

There is certainly value in this architecture, but a drawback is the extra complexity of implementing the architecture in Lambda. This additional complexity can be a challenge for small companies or simplified use cases. The Kappa architecture (Figure 7 is a simplified approach to big data processing. It models a simpler real-time data batch process since it forwards data into a simple data lake. Part of the concept of a data lake is simplifying data architecture for organizations. The Kappa architecture can be valuable in environments where data accumulates in real-time, such as IoT applications (e.g., smart farming). The Kappa architecture focuses on processing real-time bulk and historical data together for better performance. Below are some key points of the Kappa architecture:

Key Components of Kappa Architecture

- **Unified Data Processing**: The Kappa architecture processes the data in a single stream. This allows you to process both batch and streaming data in the same system, eliminating the need for multiple systems for batch and streaming processing (Penka et al. 2022).
- **Real-Time Analytics**: It supports real-time analytics, crucial for applications requiring immediate insights, such as monitoring agricultural conditions (Penka et al. 2022).
- **Simplified Data Management**: By using a single codebase for both real-time and historical data, it simplifies maintenance and reduces operational overhead (Chan n.d.).

Advantages of Kappa Architecture

- **Scalability**: The architecture accommodates huge volumes of data, suitable for big data applications (Lichao 2018).

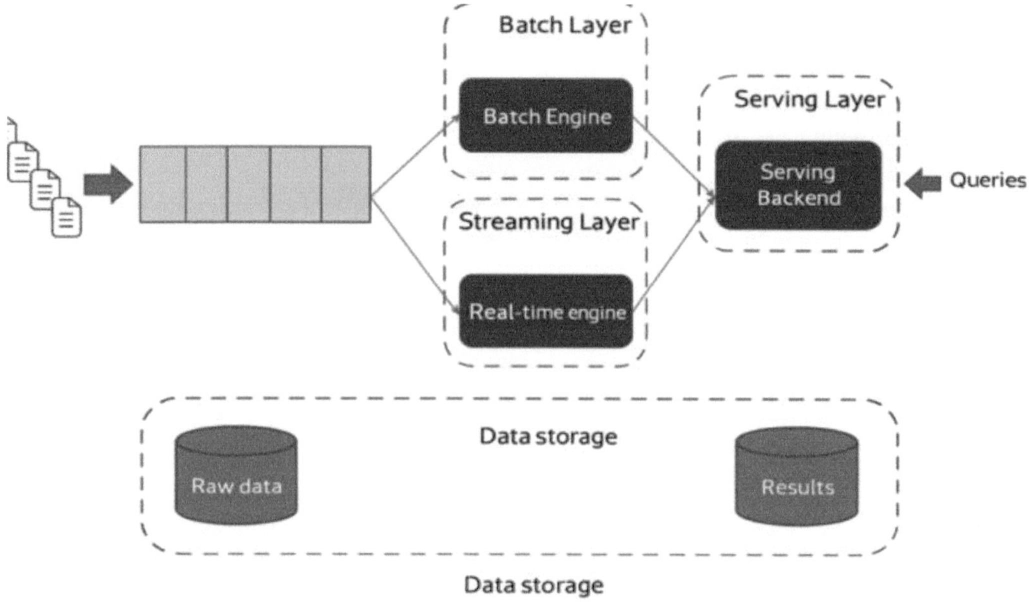

Figure 7: Kappa architecture.

- **Performance Optimization**: Optimizations, such as those involving Apache Samza and Apache Druid, can significantly enhance processing speeds (Penka et al. 2022).
- **Cost-Effectiveness**: Utilizing open-source technologies and commodity hardware can lower costs while maintaining high performance (Chan n.d.).

While the Kappa architecture offers numerous benefits, it may not be suitable for all scenarios, particularly where complex batch processing is required. In such cases, a Lambda architecture might be more appropriate, as it allows for more intricate data processing workflows (Jamali et al. 2024).

3.2 Real-Time Data Processing Frameworks for Big Data-Based Blockchain-IoT Systems

Working with real-time data processing architecture for big data in blockchain-IoT systems involves handling massive data captured by IoT devices and processed through blockchain technology. Here are some frameworks and technologies commonly used in this domain:

3.2.1 Apache Kafka

It is a distributed event streaming platform designed to handle trillions of events daily. It is commonly used to build real-time data pipelines and streaming applications. Apache Kafka is a distributed event streaming platform engineered for high-throughput, fault-tolerant, and scalable real-time data processing. It is extensively used to develop real-time data pipelines and streaming applications. Here's a brief explanation of its core concepts and components:

Core Concepts

i. **Producer**: Applications that publish (write) streams of data to Kafka topics.
ii. **Consumer**: Applications that subscribe to (read and process) streams of data from Kafka topics.
iii. **Broker**: Kafka servers that store data and serve client requests. A Kafka cluster usually comprises multiple brokers to ensure load balancing.
iv. **Topic**: A topic is a category or feed name where producers publish records, and it is divided into multiple partitions.

v. **Partition**: Each topic is divided into partitions. Each partition is a continuously growing, ordered, and immutable sequence of records. Partitions allow Kafka to parallelize a topic by splitting it into several logs.

vi. **Consumer Group**: A group of consumers that jointly consume data from a topic. Each consumer within the group reads from distinct partitions, enabling parallel processing and enhanced throughput.

vii. **Zookeeper**: Kafka uses Zookeeper to manage and coordinate the brokers. Zookeeper is used for maintaining configuration information, electing a controller, and detecting node failures.

How Kafka Works

- Producers send data to Kafka topics.
- The Kafka cluster stores these data streams in partitions across multiple brokers.
- Consumers subscribe to these topics and handle the stream of incoming published messages.
- Kafka retains all published messages on disk and these messages can be replayed or consumed multiple times.

Use Cases

- **Real-time Data Processing**: Kafka is used to process streams of data in real-time, such as log aggregation, clickstream analysis, and real-time monitoring.
- **Data Integration**: Kafka can act as a bridge between different data systems, enabling seamless data flow between databases, data warehouses, and analytics tools.
- **Event Sourcing**: Kafka can be used to capture and store sequences of events, enabling applications to reconstruct state by replaying events.

Apache Kafka architecture is explained in Figure 8.

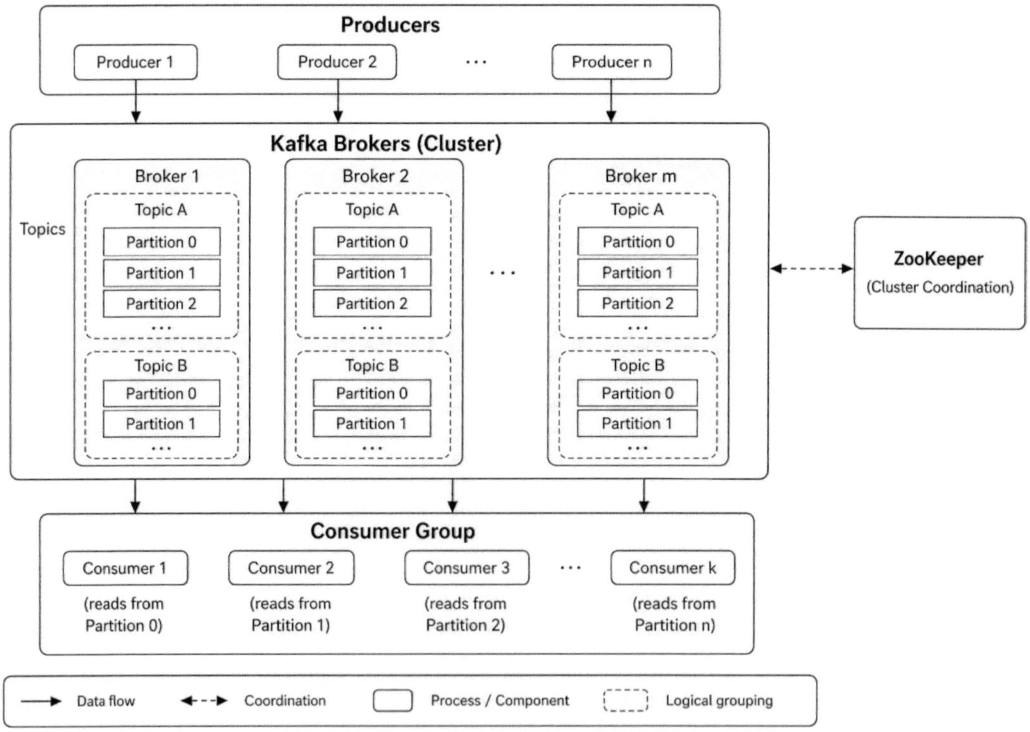

Figure 8: Kafka Architecture.

Figure 8 shows the main components of Kafka:

- **Producers** send data to Kafka topics.
- **Brokers** store the data and serve client requests.
- **Topics** are divided into partitions for parallel processing.
- **Consumers** read data from these topics.
- **Consumer Groups** allow for parallel consumption of data.
- **Zookeeper** manages and coordinates the brokers.

This setup allows Kafka to handle high-throughput, fault-tolerant, and scalable real-time data processing.

3.2.2 Apache Flink

A framework and distributed processing engine for performing stateful computations on both bounded and unbounded data streams, Flink is built to operate across various cluster environments, delivering in-memory speed processing at any scale.

Apache Flink is an open-source stream processing framework developed for high-throughput, low-latency data processing applications. It provides support for both batch and stream processing and is used for a wide range of big data applications. Here's an overview of its key concepts and components:

Core Concepts

1. **DataStream API**: Flink's core API for processing unbounded streams of data. It supports various operations such as transformations, aggregations, and windowing.
2. **DataSet API**: Used for batch processing of bounded data sets. It provides similar operations to the DataStream API but is optimized for batch processing.
3. **State Management**: Flink provides mechanisms for managing state in streaming applications, allowing for complex event-time processing and exactly-once semantics.
4. **Event Time and Watermarks**: Flink supports event-time processing, which allows applications to process data based on the time at which events actually occurred rather than the time at which they were processed. Watermarks are used to track progress in event time and trigger window computations.
5. **Windowing**: Flink supports various windowing mechanisms, such as tumbling windows, sliding windows, and session windows, for aggregating and processing data over specific time intervals.
6. **Checkpointing**: Flink's checkpointing mechanism ensures fault tolerance by periodically saving the state of a streaming application. In case of a fault, the application can be restored to the last checkpoint, ensuring exactly-once processing semantics.
7. **Task Managers and Job Managers**: Flink's distributed architecture uses two types of processes—Task Managers and Job Managers. Task Managers execute the data processing tasks, while Job Managers are responsible for coordinating the execution of tasks and managing resources.

How Flink Works

- Data is consumed in Flink through multiple sources, including Kafka, files, or sockets.
- The DataStream or DataSet API is used to define data processing pipelines.
- Flink's runtime environment executes these pipelines in a distributed manner, leveraging multiple Task Managers.

- State management, event time processing, and checkpointing ensure reliable and consistent data processing.
- Processed data is then sent to various destinations, such as databases, files, or dashboards.

Use Cases

- **Real-time Analytics**: Flink is used for real-time analytics applications, such as fraud detection, anomaly detection, and real-time monitoring.
- **Data Pipelines**: Flink can be used to create data pipelines for ETL (Extract, Transform, Load), data integration, and data ingestion.
- **Event-driven Applications**: Flink is well-suited for event-driven applications, such as complex event processing (CEP) and event sourcing.
- **Machine Learning**: Flink's support for iterative processing and state management makes it suitable for machine learning applications, such as online learning and model serving.
  Here is a simple diagram illustrating the Apache Flink architecture:

Figure 9 shows the main components and concepts of Flink:

- **The DataStream API and DataSet API** are interfaces for processing unbounded and bounded data streams, respectively.
- **State management** introduces a way of managing state in streaming applications.
- **Event Time and Watermarks** are concepts for processing data based on event time and tracking progress.
- **Windowing** is a mechanism for aggregating and processing data over specific time intervals.
- **Checkpointing** ensures fault tolerance by periodically saving the state of a streaming application.

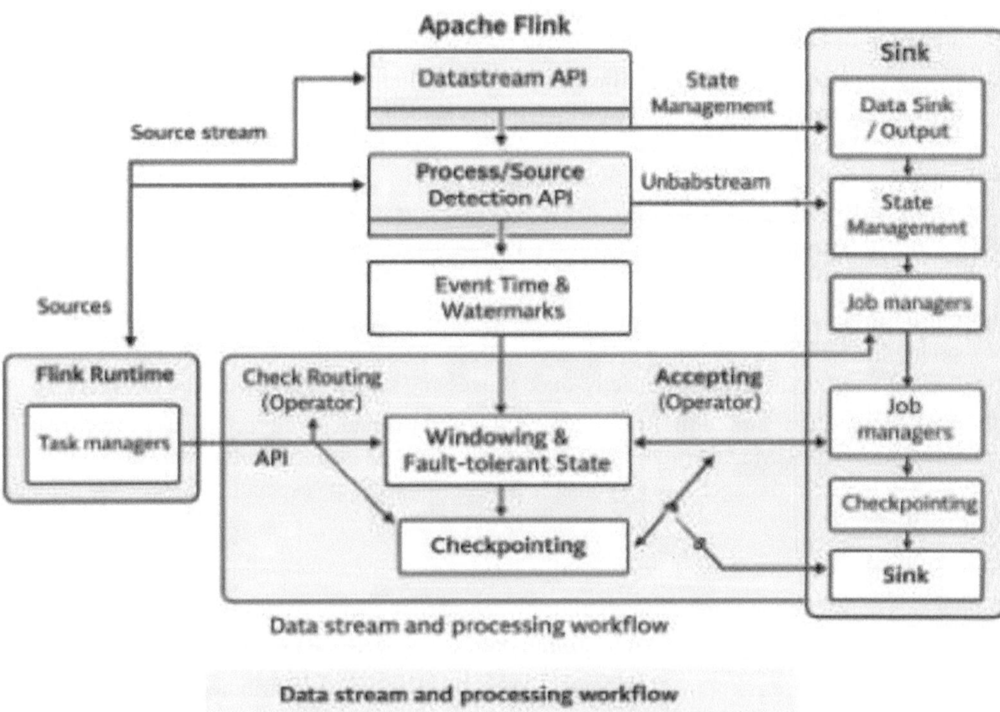

Figure 9: Architecture of Flink.

- **Task Managers** execute data processing tasks.
- **Job Managers** coordinate the execution of tasks and manage resources.

This allows Flink to treat high throughput and low latency data processing in batch and streaming application alike.

3.2.3 Apache Spark

It has a scalable and fault-tolerant interface to streaming data with its Spark Streaming component, which can process data in mini-batches and perform near real-time processing. Apache Spark is an open-source distributed computing platform for analytics and processing large amounts of data. Spark supports a programming interface for implicit fault tolerance and data processing, including machine learning and graph data processing. Spark's speed, ease of use, and flexibility make it a good choice for a variety of data processing tasks.

Core Components

1. **Spark Core**: As the foundation of the platform, Spark Core provides the basic I/O functionality as well as scheduling and the distributed dispatching of tasks. The core data structure of Spark, Resilient Distributed Datasets (RDDs), is specified by the API.
2. **Spark SQL**: A module for structured data processing. It can act as a distributed SQL query engine or provide a programming abstraction called DataFrames. With Spark SQL, users can run SQL queries and mix SQL with procedural transformations within Spark programs.
3. **Spark Streaming**: Allows users to process streaming data in real-time. For programmers with Spark experience, it provides an API to work with streaming data that looks very similar to the RDD API in Spark Core.
4. **MLlib (Machine Learning Library)**: This scalable machine learning library provides a wide variety of algorithms for classification, regression, clustering, collaborative filtering, and more. It also provides tools for feature extraction, transformation, dimensionality reduction, and model evaluation.
5. **GraphX**: A graph processing library that adds operations for graph computation on top of the Spark RDD API. It provides the ability to process graph data at scale and includes a set of graph algorithms and graph builders.

Key Features

- **Speed**: By utilizing in-memory computation and other optimizations, Spark can process large amounts of data up to 100 times faster than Hadoop MapReduce.
- **Ease of Use**: Spark provides high-level APIs in Java, Scala, Python, and R, and an interactive shell for quick and easy development.
- **Versatility**: Spark supports a wide variety of data processing tasks: batch processing, real-time (stream) processing, graph data processing, and machine learning, to name just a few.
- **Fault Tolerance**: Of course, Spark has built-in fault tolerance with RDDs (Resilient Distributed Datasets), which can be rebuilt upon failure.

How Spark Works

- Data is loaded into Spark and distributed across a cluster of machines.
- Spark Core schedules and dispatches tasks to worker nodes, where they are executed in parallel.
- Data is processed using RDDs, DataFrames, or other high-level abstractions provided by Spark's APIs. Results are returned to the user or stored in a distributed storage system.

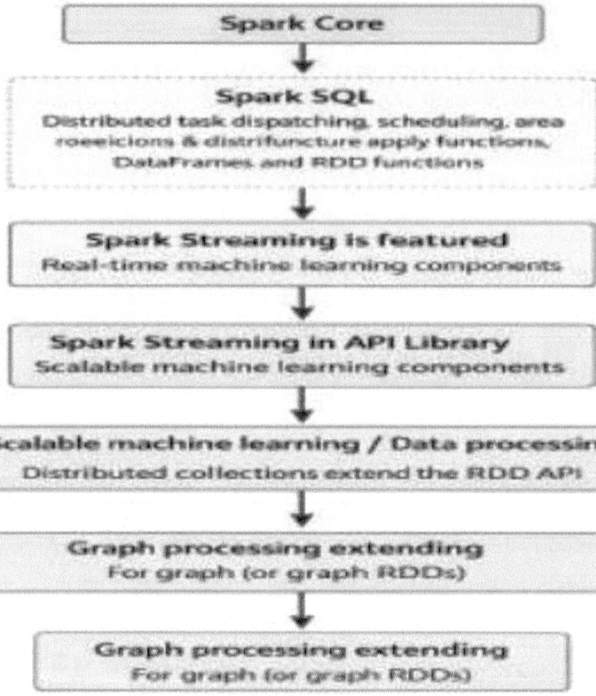

Figure 10: Components of Spark.

Figure 10 shows the main components of Spark:

- **Spark Core**: The platform's core, offering scheduling, distributed task dispatching, and fundamental I/O features.
- **Spark SQL**: A module for structured data processing, providing DataFrames and SQL query capabilities.
- **Spark Streaming**: A Spark component that enables streaming of live data and uses an API similar to the RDD API in Spark Core.
- **MLlib**: A machine learning library that scales and contains a plethora of algorithms and tools to help with machine learning tasks.
- **GraphX**: A graph processing library that allows the user to write graph computing operations that extend the Spark RDD API.

  This architecture allows Spark to efficiently and scalably control data processing tasks.

3.3 Blockchain Platforms

Platforms like Ethereum, Hyperledger Fabric, and others can be combined with IoT systems to provide secure and transparent data processing and storage.

Blockchain platforms can improve a big data IoT system's security, transparency, and efficiency by enabling organizations to safely and transparently share data, ensure the integrity and immutability of the data generated by their IoT devices, and establish smart contracts to allow automated and trustless interactions between IoT and blockchain systems.

Key Concepts

1. **Decentralization**: With blockchain platforms, there are no databases with centralized architectures, meaning data is held in a decentralized network of nodes. This reduces single points of failure and removes the need for a central authority over the data.

2. **Immutability**: Data stored on a blockchain is immutable, meaning it cannot be changed or deleted. This ensures that the data generated by Internet of Things devices remains accurate and trustworthy.
3. **Transparency**: Blockchain platforms provide transparency as every participant on the network can view all transactions and data records. In IoT systems, this promotes trust and accountability.
4. **Security**: Blockchain platforms use modern cryptographic mechanisms to secure data and transactions. This helps ensure that sensitive data produced by IoT devices is protected from unauthorized access and tampering.
5. **Smart Contracts**: Smart contracts—self-executing agreements—have the terms of the agreement coded directly into the program. They enable automated and trustless interactions, allowing for more complex workflows in IoT systems.

4. Popular Blockchain Platforms for IoT

i. Ethereum An open-source decentralized platform that enables the creation of decentralized applications (DApps) and smart contracts. Ethereum is a popular choice for Internet of Things applications due to its strong ecosystem and support for smart contracts.

Ethereum is an open-source, decentralized blockchain platform that allows for decentralized applications (DApps) and smart contracts. It was proposed by Vitalik Buterin in late 2013 and launched via a crowdsourcing project in 2014. The network officially launched in 2015. Here is a brief overview of Ethereum's architecture:

Blockchain: Similar to Bitcoin, Ethereum uses blockchain technology. However, Ethereum's blockchain not only keeps track of transactions; it also maintains and identifies the state of its smart contracts as well as the rules that govern them.

Smart Contracts: The terms of these self-executing contracts are written directly into the code. When specific conditions are met, they automatically execute transactions and record them on the blockchain.

Ethereum Virtual Machine (EVM): The EVM is the runtime environment for smart contracts in Ethereum. It runs the smart contract bytecode and ensures that the execution is consistent across nodes on the network.

Nodes: Nodes are computers that run within the Ethereum network to validate and propagate transactions and execute smart contracts. Full nodes, light nodes, and archive nodes are different types of nodes.

Figure 11: Different platforms of Blockchain platforms for IoT.

Consensus Mechanism: Like Bitcoin, Ethereum initially relied on a Proof of Work (PoW) consensus mechanism. However, the Ethereum 2.0 update introduced the Proof of Stake (PoS) consensus mechanism, where validators are chosen to produce new blocks based on their stake in the network instead of their computational power.

Gas: The amount of computing power required to perform operations on the Ethereum network is called gas. It is used to pay for network computing services and transaction fees.

Ether (ETH): Ether, the currency of the Ethereum platform, is employed as an incentive for participants who compute and validate transactions.

Figure 12 illustrates the various components and their relationships within the Ethereum network.

ii) Hyperledger Fabric: An enterprise-grade permissioned blockchain platform for developing permissioned blockchain networks. Hyperledger Fabric is designed specifically with modularity, scalability, and privacy in mind and is suited for IoT applications in sectors like supply chain management, healthcare, and manufacturing.

Hyperledger Fabric is an enterprise-grade, permissioned distributed ledger technology (DLT) platform that is open-source and designed for business environments. It is one of the Hyperledger projects hosted by the Linux Foundation. Fabric is specifically designed for creating modular solutions that allow membership services, consensus mechanisms, and other components to be plug-and-play elements. Here is a summary of the platform's key components:

Membership Services: Manages the network's identity, authentication and authorization. It is responsible for ensuring that all users in the network are known and are following the network,s access control policies.

Peer Nodes: These are the core components of the network. They host ledgers and chaincodes (smart contracts). Peers can be endorsing peers, which simulate transaction execution and endorse results, or committing peers, which validate transactions and commit them to the ledger.

Ordering Service: Responsible for consensus, ordering transactions into blocks, and distributing those blocks to peer nodes. The ordering service ensures that transactions are recorded in sequence across the ledger.

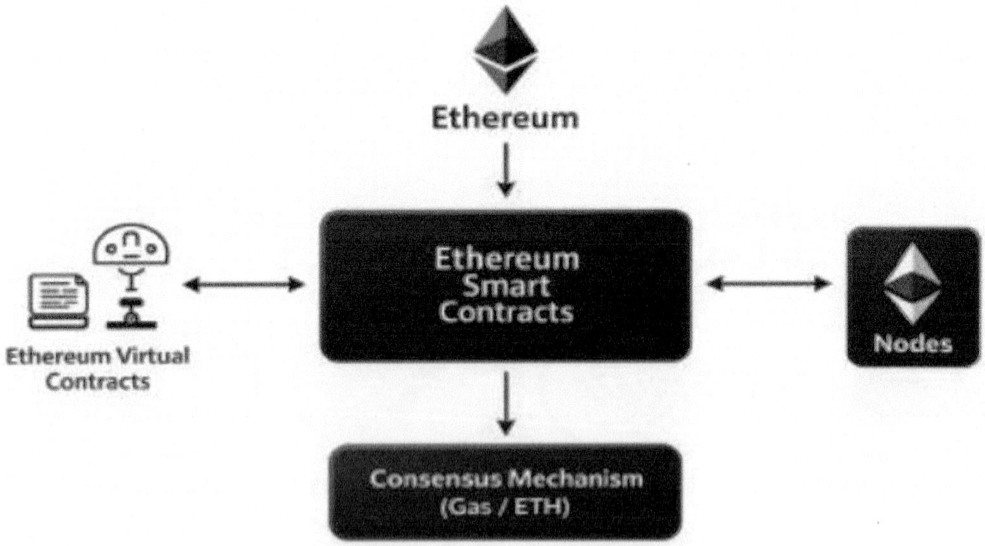

Figure 12: Architecture of Ethereum.

Smart Contracts (Chaincode): Programs written in Go, JavaScript, or Java that define the business logic and rules governing transactions. Chaincode executes on peer nodes and can interact with the data in the ledger.

Ledger: The ledger is the database that provides the current state of the network and a record of all transactions. It consists of a world state, which represents the current state of the ledger, and a *blockchain*, which is the immutable record of all transactions.

Channels: Allow for private communication between specific members of the network. Channels provide secure transactions and information sharing among a subset of network participants.

Certificate Authority (CA): Issues and manages digital certificates for participants in the network, providing secure and authentic communication between network members.

Figure 13 illustrates the fundamental elements and inter-relationships found within a Hyperledger Fabric network.

iii) IOTA: An adapted ledger technology specifically for the Internet of Things (IoT). The IOTA network uses a directed acyclic graph (DAG) structure called the Tangle. The Tangle enables feeless microtransactions and scalable data integrity, specifically for IoT devices.

IOTA is a distributed ledger technology designed specifically for the Internet of Things (IoT) with an architecture that differs significantly from traditional blockchain systems. The following sections introduce the terminologies and components that make up the IOTA architecture. IOTA uses a unique DAG structure known as the Tangle, which differs from the blockchain structure used by many cryptocurrencies. Below is a simplified explanation of IOTA's underlying architecture:

i. **Tangle**: Instead of a linear chain of blocks, IOTA uses a DAG called the Tangle. In the Tangle, every transaction must approve (confirm) two previous transactions. This architecture allows transactions to be processed in parallel, leading to greater scalability and reduced transaction times.

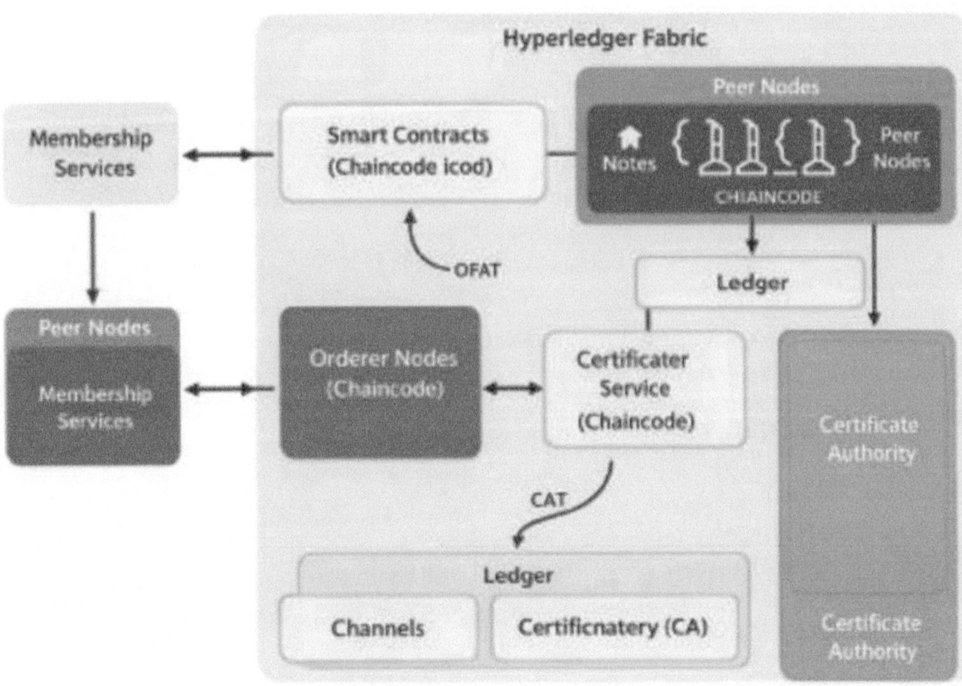

Figure 13: Architecture of Hyperledge.

ii. **Nodes**: All nodes in an IOTA network validate and relay transactions. Unlike traditional blockchain networks, there are no miners in IOTA. Instead, users generating transactions also participate in the validation process.

iii. **Consensus Mechanism**: IOTA has a consensus mechanism where the validation of transactions occurs in conjunction with generating new transactions. Since users are incentivized to produce transactions, they have to validate two previous transactions before submitting their own. The network manages itself, and users can rely on its strong security and efficient operation while actively participating in it.

iv. **Coordinator**: IOTA initially used a Coordinator, a centralized entity managed by the IOTA Foundation, to protect the network from potential attacks during its early stages. The Coordinator was intended as a temporary measure to ensure security until the network reached sufficient scale and maturity.

v. **Tip Selection**: When a user wants to make a transaction, they must pick two unconfirmed previous transactions to validate. The action of choosing these transactions is called tip selection. The goal is to select tips (transactions that are not confirmed) to help confirm transactions in order to solidify an area of the Tangle.

vi. **Transactions**: Transactions in IOTA are feeless, enabling microtransactions and machine-to-machine payments. Feeless transactions are particularly valuable for IoT devices making frequent, small-scale payments.

Data Structure: IOTA as shown in the Figure 14 is a distributed ledger which is more efficient for overall processing and energy use for IoT devices with limited processing power.

When creating a real-time data processing system for blockchain-IoT use cases, you must evaluate use case-specific requirements such as latency, throughput, data consistency, and data security. By leveraging the combination of these technologies, organizations can establish scalable, secure, and efficient infrastructures that promote long-term innovation and trust.

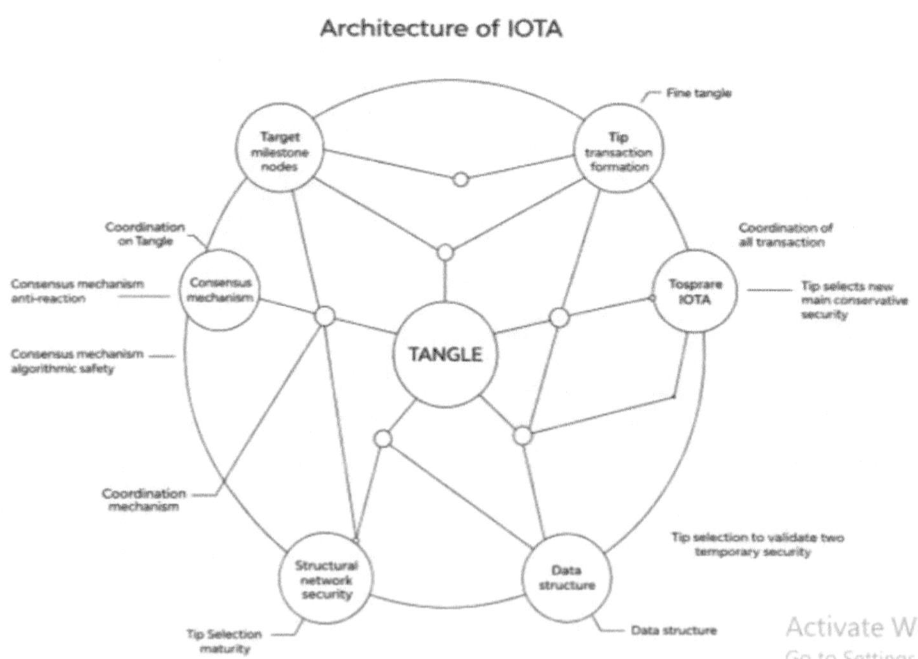

Figure 14: Architecture of IOTA.

iii. VeChain is a blockchain platform designed to enhance business processes and supply chain management. VeChain's architecture is built to provide a reliable, scalable, and efficient enterprise-ready environment suited to business customers. Below is an overview of the key elements of the VeChain architecture as shown in Figure 15. VeChain architecture was created with the needs of a business customer in mind and is designed as a safe, extensible and efficient platform suitable for blockchain-based applications.

a. **VeChainThor Blockchain**: The central component of the VeChain ecosystem, illustrated in the Venn diagram, is a data model that provides multiple blockchain capabilities.

b. **Smart Contracts**: The content of these self-executing contracts is written directly into the blockchain code to facilitate, validate, and enforce the negotiation or performance of a contract. These are based on the VeChainThor Blockchain.

c. **Consensus Mechanism**: The PoA consensus provides transactions and smart contracts with a mechanism for validation or approval by validators. Validators perform validation, accept specific transactions or smart contracts, and produce an official block to add to the blockchain, thereby creating new transactions against previous blocks. This encourages accountability, security, consistency, and regulatory compliance.

d. **Programmable Transactions**: Programmable transactions are transactions that can be programmed to autonomously perform specific tasks in response to a trigger based on predefined conditions set at the time of creation. This introduces flexibility and intelligence to software algorithms available to the blockchain.

e. **IoT Integration**: The VeChain design architecture allows for integration with IoT devices composed of sensors and nanosensors, enabling VeChain to independently store and transfer data in real time. This facilitates intelligent services and applications capable of interacting with the physical world.

f. **Data Services**: The Data Services handle the management, extraction, and storage of data accepted into the blockchain. They ensure that the data is secure, accessible, and properly managed.

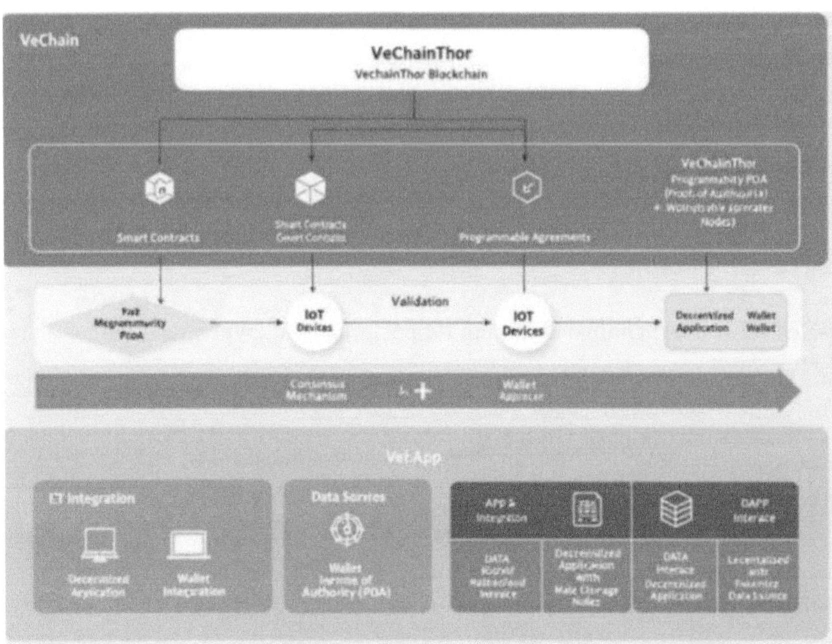

Figure 15: Architecture of VeChainThor.

g. **API Integration**: APIs (Application Programming Interfaces) allow external applications to contribute data on behalf of certain stakeholders into the VeChainThor Blockchain. Permissions for external applications are administered by approved validators, enhancing efficiency, compliance, and accountability of contributed data to uphold stakeholder commitments to the blockchain.

h. **VeChain ToolChain**: A complete BaaS (Blockchain-as-a-Service) solution that encompasses all service and product areas necessary to design, develop, deploy, and manage blockchain applications.

i. **Enterprise Applications**: Applications designed for business purposes where VeChainThor Blockchain properties can be leveraged to create better business processes with improved reliability, integrity, and transparency.

j. **User Interfaces**: The primary way users interact with the blockchain and its applications. These include mobile and web applications that provide broader access to blockchain-based services for a wider user base.

5. Conclusion

The intersection of Big Data architecture, Blockchain, and IoT offers powerful possibilities for transforming secure, scalable, and decentralized data ecosystems. This chapter has examined key architectural paradigms—from Lambda/Kappa-based frameworks for stream processing to distributed ledger constructs—to emphasize the urgency of addressing the fundamental challenges accompanying the scale, velocity, and veracity of IoT data.

The significant contributions are listed as follows:

Hybrid Architectures: The hybridization of batch/stream processing systems (i.e., Spark + Flink) with a blockchain layer (i.e., Hyperledger, Ethereum) can enable real-time analytics without sacrificing immutability or auditability.

Decentralized Governance: Smart contracts in blockchain can support autonomous coordination of devices and eliminate single points of failure in IoT networks.

Scalability Solutions: Scalability solutions such as sharding with NoSQL databases, off-chain storage using IPFS, and stateless microservices can significantly enhance performance in resource-constrained IoT edge devices.

Security Imperatives: Zero-trust models and light-weight cryptographic protocols (e.g., BLS signatures) are vital for protecting data integrity across distributed nodes.

5.1 Future Directions

Quantum-Resistant Blockchains: Preparing for post-quantum cryptography in IoT data pipelines.

AI-Driven Consensus: Use adaptive algorithms (e.g., federated learning) to minimize compute overload.

Regulatory-Aware Designs: Address challenges of balancing rates of GDPR/CCPA compliance with the transparent nature of blockchains.

Address challenges of balancing rates of GDPR/CCPA compliance with the transparent nature of blockchains. This chapter serves as a guiding framework for researchers and industry professionals, helping them design systems that are robust, efficient, and reliable as global connectivity continues to expand.

References

Al Jawarneh, I. M., Bellavista, P., Corradi, A., Foschini, L., and Montanari, R. (2021) QoS-Aware Approximate Query Processing for Smart Cities Spatial Data Streams. Sensors, 21: 4160.

Al Jawarneh, I.M., Foschini, L., and Bellavista, P. (2023). Efficient Integration of Heterogeneous Mobility-Pollution Big Data for Joint Analytics at Scale with QoS Guarantees. *Future Internet*, 15: 263.

Asch, M., Moore, T., Badia, R., Beck, M., Beckman, P., Bidot, T. et al. (2018). Big data and extreme-scale computing: Pathways to convergence-toward a shaping strategy for a future software and data ecosystem for scientific inquiry. *The International Journal of High Performance Computing Applications*, 32(4): 435–479.

Azarmi, B. (2016). *Scalable Big Data Architecture: A Practitioner›s Guide to Choosing Relevant Big Data Architecture.* Apress. https://doi.org/10.1007/978-1-4842-1326-1.

Bux, M., and Leser, U. (2013). Parallelization in scientific workflow management systems. arXiv, arXiv:1303.7195.

Chaimov, N., Malony, A., Canon, S., Iancu, C., Ibrahim, K. Z., and Srinivasan, J. (2016) Scaling spark on hpc systems. In Proceedings of the 25th ACM International Symposium on High-Performance Parallel and Distributed Computing, Kyoto, Japan, 31 May–4 June; pp. 97–110.

Chandan, A., John, M., and Potdar, V. (2023). Achieving UN SDGs in food supply chain using blockchain technology," Sustainability, vol. 15, no. 3, p. 2109, Jan. , doi: 10.3390/su15032109.

Gorton, I., and Gracio, D. K. (2012). *Data-Intensive Computing: Architectures, Algorithms, and Applications*; Cambridge University Press: Cambridge, UK.

Hlaing, Nyo Me et al. (2025). The impact of business environment towards entrepreneurs' performance: a bibliometric analysis. *International Journal of Management and Enterprise Development*, 24.3: 343–365.

Haven, B. (2024). Cloud-Native Big Data Frameworks: Trends and Challenges.

Isah, Haruna, and Farhana Zulkernine. (2018). A scalable and robust framework for data stream ingestion. *2018 IEEE International Conference on Big Data (Big Data)*. IEEE.

Kraetz, D. and Morawski, M. (2021). Architecture patterns—batch and real-time capabilities. pp. 89–104. *In:* The Digital Journey of Banking and Insurance, Volume III: Data Storage, Data Processing and Data Analysis. Cham: Springer International Publishing.

Laroui, M., Nour, B., Moungla, H., Cherif, M. A., Afifi, H., and Guizani, M. (2021). Edge and fog computing for IoT: A survey on current research activities & future directions. *Comput. Commun.*, 180: 210–231.

Shvachko, K., Kuang, H., Radia, S., and Chansler, R. (2010). The hadoop distributed file system. In Proceedings of the 2010 IEEE 26th Symposium on Mass Storage Systems and Technologies (MSST), Incline Village, NV, USA, 3–7 May; pp. 1–10.

Thantharate, P., and Thantharate, A. (2023). ZeroTrustBlock: Enhancing security, privacy, and interoperability of sensitive data through ZeroTrust permissioned blockchain, Big Data Cognit. *Comput.*, 7(4): 165, Oct. doi: 10.3390/bdcc7040165.

Tian, F. (2018). An information system for food safety monitoring in supply chains based on HACCP, blockchain and Internet of Things, Doctoral Dissertation, WU Vienna Univ. Econ. Bus., Vienna, Austria [Online]. Available: https://research.wu.ac.at/en/publications/an-information-system-for-food-safety-monitoring-in-supply-chains-3.

Wang, Y., Goldstone, R., Yu, W., and Wang, T. (2014). Characterization and optimization of memory-resident MapReduce on HPC systems. pp. 799–808. *In: Proceedings of the 2014 IEEE 28th International Parallel and Distributed Processing Symposium, Phoenix, AZ, USA, 19–23 May.*

Yildiz, O., Zhou, A. C., and Ibrahim, S. (2017). Eley: On the effectiveness of burst buffers for big data processing in HPC systems. *In Proceedings of the 2017 IEEE International Conference on Cluster Computing (CLUSTER), Honolulu, HI, USA, 5–8 September; pp. 87–91.*

Zaharia, M., Chowdhury, M., Franklin, M. J., Shenker, S., and Stoica, I. (2010). Spark: Cluster computing with working sets. *HotCloud*, 10: 95.

Zaharia, M., Xin, R. S., Wendell, P., Das, T., Armbrust, M., Dave, A. et al. (2016). Apache spark: A unified engine for big data processing. *Commun. ACM*, 59: 56–65.

Zheng, M., Zhang, S., Zhang, Y., and Hu, B. (2021). Construct food safety traceability system for people's health under the Internet of Things and big data. *IEEE Access*, 9: 70571–70583, , doi: 10.1109/ACCESS.2021.3078536.

7

AI-Driven Data Processing and Decision-Making

Machine Learning in Internet of Things

S. Anitha Elavarasi, J. Jayanthi* and *G. Kirubasri*

1. Introduction

Internet of Things: The Internet of Things (IoT) refers to the group of physical objects or devices connected via the internet that exchange data among themselves without human intervention. Devices are embedded with sensors, software, and other technologies. These devices range from home appliances, vehicles, and wearables to sophisticated industrial machines. The first IoT application started in 1982, when Carnegie Mellon connected a vending machine to the Internet to check the availability of sodas. The term "Internet of Things" was coined by Kevin Ashton in 1999, and it gained popularity with advances in technologies such as cloud computing, data analytics, and wireless communication by 2010. With the rapid growth of low-cost microprocessors, IoT has expanded to a vast range of applications such as manufacturing, agriculture, healthcare, wearable technology, sports, environmental monitoring, defense, and smart cities.

IoT technology experienced rapid growth after 2004 with advancements in microprocessor technology and communication. By 2030, over 80 billion devices were connected worldwide (Alam and Tanweer 2018, Hernández et al. 2021). Figure 1 represents the number of devices connected between 2000 and 2025, based on the statistics provided by Alam and Tanweer (2018). With billions of interconnected devices generating continuous streams of data, the potential to extract valuable insights from this data has grown significantly. Continuous development in the fields of Artificial Intelligence (AI), Machine Learning (ML), 5G networks, and edge computing offers various tools and techniques to analyze, interpret, and act upon IoT data effectively.

Four main components of IoT, depicted in Figure 2, are: (1) **Sensors**: They are used to gather data from the physical world, such as temperature, motion, pressure, light, heat, moisture, smoke

Department of Computer Science and Engineering, Sona College of Technology, Salem, Tamil Nadu.
Emails: jayanthij@sonatech.ac.in, https://orcid.org/0000-0002-2843-5693; kirubasri.cse@sonatech.ac.in, https://orcid.org/0000-0002-2086-8064
* Corresponding author: anithaelavarasi@sonatech.ac.in, https://orcid.org/0000-0003-4049-6162

DOI: 10.1201/9781003655190-9

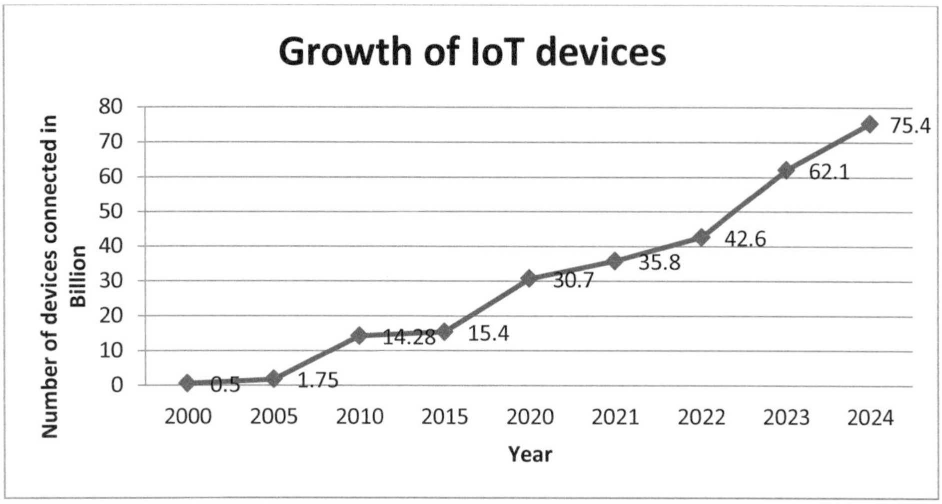

Figure 1: Growth of IoT devices. Source (Alam and Tanweer 2018).

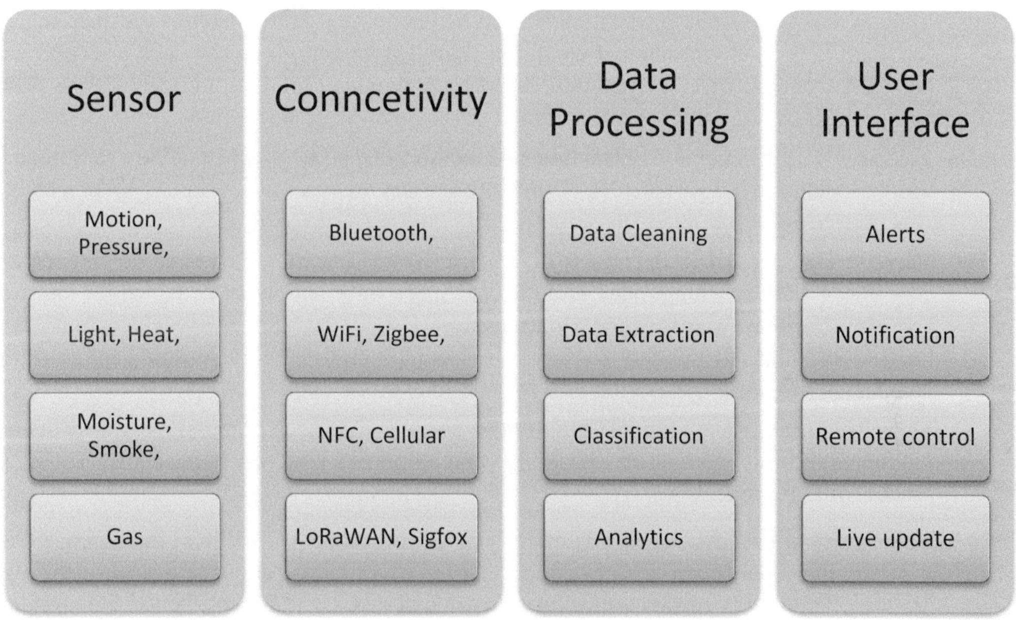

Figure 2: Components of IoT.

and so on. (2) **Connectivity**: Data collected using sensors is sent via the Internet through several communication protocols such as mobile networks, Bluetooth, Wi-Fi, ZigBee, LoRaWAN, etc. Protocols are selected based on range, power, and cost. Figure 3 represents the diverse types of protocols along with their distance coverage. Table 1 represents the range coverage of various communication protocols. (3) **Data processing**: Raw data is processed, and useful insights are derived through different techniques such as data cleaning, data extraction, aggregation, classification, and pattern identification. (4) **User interface**: Final processed information is made available to the end user in the form of dashboards, mobile applications, alerts, notifications, and text messages.

Features of IoT Data: Data is collected from various real-time devices such as sensors, which need to be analyzed immediately to gain useful insights and enable decision-making. IoT data differs from conventional data in terms of volume, velocity, variety, and veracity. Several types of data such

Table 1: Hierarchy of connectivity.

Communication Technology	Coverage Range
Bluetooth	10–100 meters
Wi-Fi	50–100 meters (indoor)
Near Field Communication (NFC)	Up to 10 cm
ZigBee	10–100 meters
Z wave	30–100 meters
Wi-Fi HaLow	Up to 1 KM
Cellular (3G, 4G,5G)	1–35 KM
Sigfox	3–50 KM
Narrowband IoT (NB IoT)	1 to 10 KM
LoRa	2–15 KM

as text, images, and videos are received from various devices, and each presents unique challenges in terms of storage, processing, and security.

- **Volume:** IoT data collected from various sensors across diverse applications continuously leads to a huge volume of data. Data such as occupancy details from smart thermostats for smart homes, environmental conditions like temperature, humidity, and pressure, and heat data from industrial automation, logistics, and patient vitals from healthcare systems produce massive amounts of information.
- **Velocity:** Data is generated in real time and requires fast processing. Data from surveillance cameras, patient vital parameters, or pressure sensors are examples where rapid processing is needed to take appropriate decisions.
- **Variety:** Data from IoT devices is diverse in nature. Location data such as geolocation coordinates (latitude, longitude), altitude, and speed information are collected for logistics and asset tracking. Cameras and drones capture visual data such as live video feeds, still images, and infrared imagery for analysis. Microphones or audio sensors capture voice commands and noise monitoring data for real-time processing.
- **Veracity:** Data generated in real time can be noisy, incomplete, or unreliable due to sensor malfunctions or network issues. Ensuring veracity plays a vital role in the decision-making process of IoT applications.

Machine Learning Process: Machine Learning (ML) is a subset of Artificial Intelligence (AI) where the system or model learns from examples and improves its performance over time through rigorous training. Training is carried out using thousands of labeled examples; hence, ML systems are often described as systems that "learn from examples" and adapt automatically. The major phases of the Machine Learning process are depicted in Figure 3.

The six major processes involved in building an ML model are: (1) Data collection: This is the most important step because the accuracy of the designed ML model depends on the quality of the data. Thousands of tagged data points gathered from various sources, including sensors, cameras, and microphones, are used to train the models. (2) Data pre-processing: Real-world data contains missing, repeated, unwanted, duplicate, and randomized data. Hence, the data needs to be cleaned, evenly distributed, and normalized. (3) Model selection: Selecting a model that is pertinent to the work at hand is crucial. Based on the input data and task, a model is chosen. Input data can be either numerical or categorical in nature. Similarly, tasks can include speech recognition, image detection, regression, classification, and prediction. Depending on the type of problem, a variety of techniques, including CNN, k-Nearest Neighbors (KNN), Support Vector Machines (SVM), and

Figure 3: Phases of Machine Learning.

decision trees, can be used. (4) Training: To identify trends and generate predictions, the data is fed into the machine learning model. Thousands of tagged data points are used to train the machine. The dataset is divided into training and testing sets. (5) Evaluation: A number of metrics, including accuracy, precision, recall, and F1 score, are used to assess the ML model. (6) Deployment: The learned model is used in real time to make the right decisions.

A vehicle number plate detection scenario is taken as an example use case to represent the various phases of ML, as depicted in Figure 4 (Anitha et al. 2019). Data collection includes image or video collection and annotation of bounding boxes around number plates. Data preprocessing includes resizing the image, flipping, rotating, and changing brightness/contrast. In the model selection phase, suitable ML algorithms such as YOLO and Faster R-CNN are chosen, and the number plate and its text are recognized. The dataset is divided into training and test sets during the training phase. During the evaluation phase, metrics like accuracy, precision, and recall are used to assess the model.

Taxonomy of Machine Learning Algorithm: Supervised learning, unsupervised learning, and reinforcement learning are the three primary classifications of machine learning algorithms.

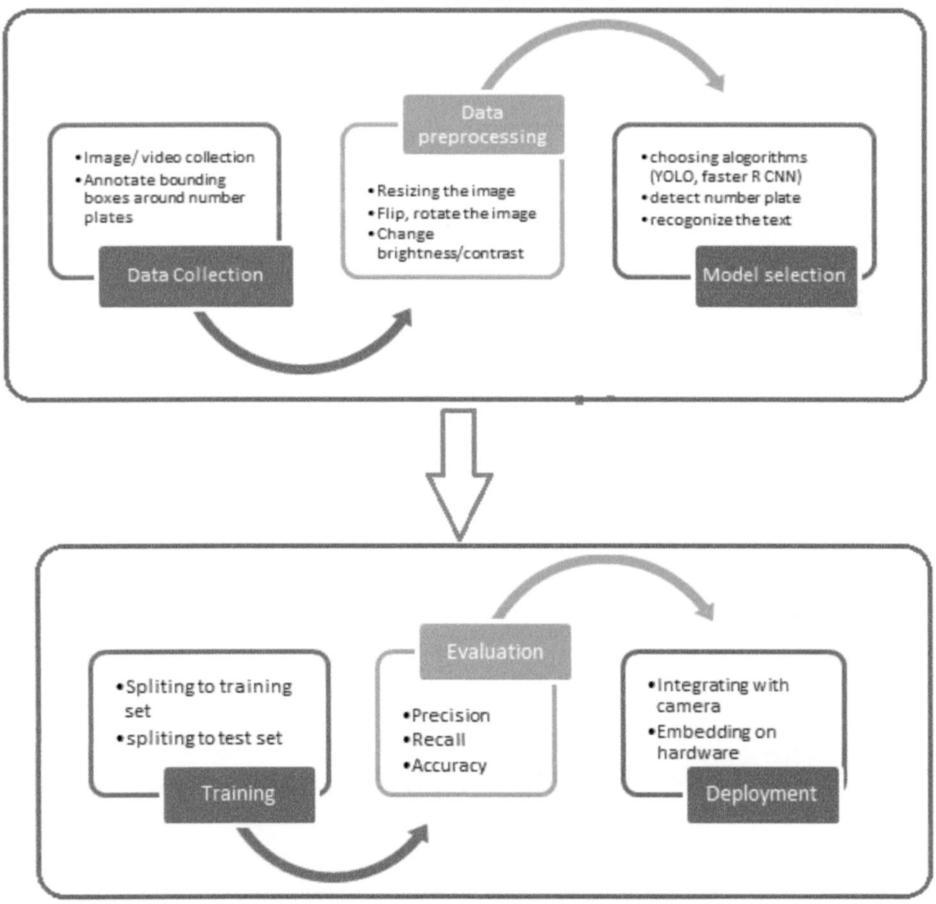

Figure 4: Vehicle number plate detection using ML.

- **Supervised Learning:** In this type of learning algorithm, labeled data is used for training a model. For new or unseen data, this training helps identify the relationship between input and output variables. A few example instances include spam detection, image classification, health prediction, and house price prediction. Algorithms include Linear Regression, Decision Trees, and the KNN classifier.
- **Unsupervised Learning:** In this type of learning algorithm, the goal is to find hidden patterns, structures, and relationships that exist among unlabeled data. A few example use cases include customer segmentation, market basket analysis, and fraud detection. Algorithms include K-Means, DBSCAN, Hierarchical Clustering, and Principal Component Analysis.
- **Reinforcement Learning:** In this type of learning, an agent learns by interacting with its environment through trial-and-error interactions and receives feedback in the form of rewards or penalties. A few example instances include robotics, gaming, and autonomous vehicles. Algorithms include Q-learning, the Actor-Critic model, and Monte Carlo Tree Search (MCTS).

2. Machine Learning Algorithms for IoT

Decision Tree Algorithms for IoT: A hierarchical tree structure with a root node at the top, internal nodes, and leaf nodes at the bottom is used to illustrate a decision tree method. It is an algorithm for supervised learning. A feature is represented by a node, a decision rule by a branch, and an outcome by a leaf node. This approach is applicable to both regression and classification techniques.

Why Decision Trees for IoT?

- High interpretability, i.e. for healthcare domains, the decision is easily visualized and easily understood.
- It requires very low computational overhead, making it suitable for resource-constrained IoT environments.

Decision tree in Scikit learn: A decision tree's objective is to construct a model that uses basic rules derived from the data features to forecast the value of a target variable (scikit-learn Developers 2025a). The DecisionTreeClassifier() is used for classification jobs (scikit-learn Developers 2025b), and DecisionTreeRegressor() is used for regression (scikit-learn Developers 2025c). These functions are imported from sklearn.tree. These classes support various parameters such as tree depth (max_depth), and sample split (min_samples_split). The quality of the split is based on the criterion function such as Gini or squared_error. The metric "Gini" or "entropy" is used for information gain in classification process, "squared_error" for regression process. Both numerical and categorical data can be handled by it. Trees can be visualized using plot_tree() or export_graphviz()methods. Figure 5 represent the implementation details using scikit-learn.

Trade off

- Decision trees algorithm suffers from overfitting issues
- Decision trees are sensitive to noise, incomplete and missing values. IoT devices are prone to such faults due to sensor or network failure.
- Most of the IoT devices are battery-powered systems and need energy-harvesting mechanisms, as decision logic computation consumes more power.
- Many IoT devices operate with kilobytes of RAM, making it difficult to store and execute large models.
- On-device tree updating is not possible. Remote training and updating needs efficient communication protocols.

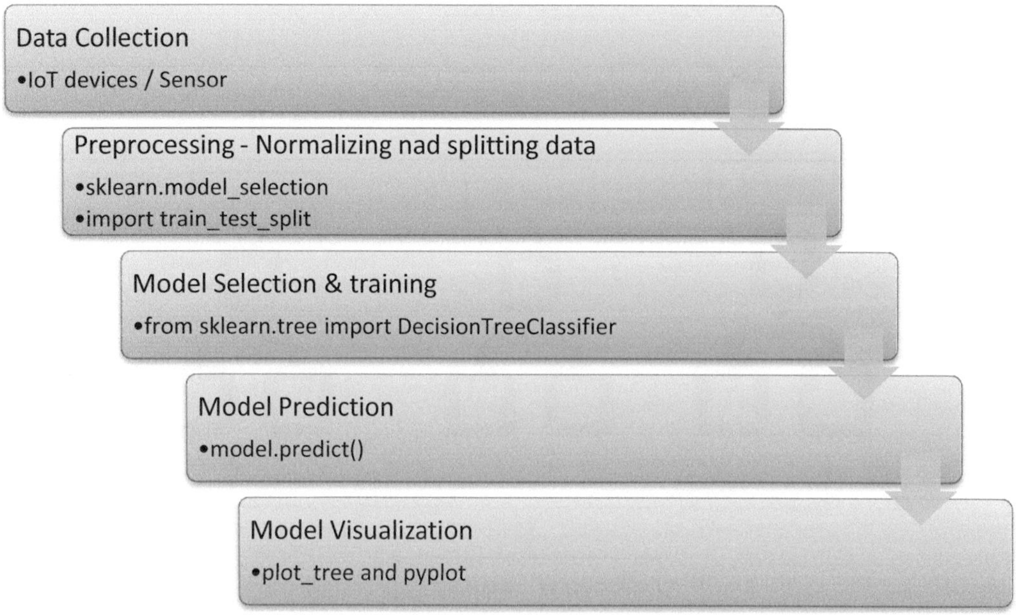

Figure 5: Decision Tree Classifier using Scikit.

Random forest for IoT: Random Forest, an ensemble learning technique depicted in Figure 6, creates several decision trees and combines them to produce more accurate predictions. It is a supervised learning approach used to solve complex problems, handle larger datasets, manage missing values, and address overfitting issues. This approach is applicable to both regression and classification techniques.

Working of Random Forest Algorithm

- Multiple decision trees are created using random feature selection, with each tree being unique.
- Every tree gives its own prediction based on what it learned from its data.
- Combine the predictions by choosing one final value based on majority voting that everyone agrees upon.
- Combining predictions from different sub-trees can lead to more accurate results.

Why Random Forest for IoT?

- By generating multiple decision trees, it reduces overfitting problem
- Able to handle noisy and unreliable data, which is common in IoT sensor networks.
- Compressed and deployed on edge devices with limited memory
- Fit the models into microcontroller units (MCUs), enabling deployment in edge-cloud architectures.

Random forest in scikit-learn: The sklearn.ensemble module includes RandomForestClassifier (scikit-learn Developers 2025d) and RandomForestRegressor classes. RandomForestClassifier creates varied set of classifiers by bringing together randomness in constructing classifier. The final prediction is by averaging the individual classifiers. It uses two arrays (X and Y), where X holds the training sample (sample and features) and Y holds the target values for class labels.

Figure 7 represents the implementation details of random forest using scikit-learn.

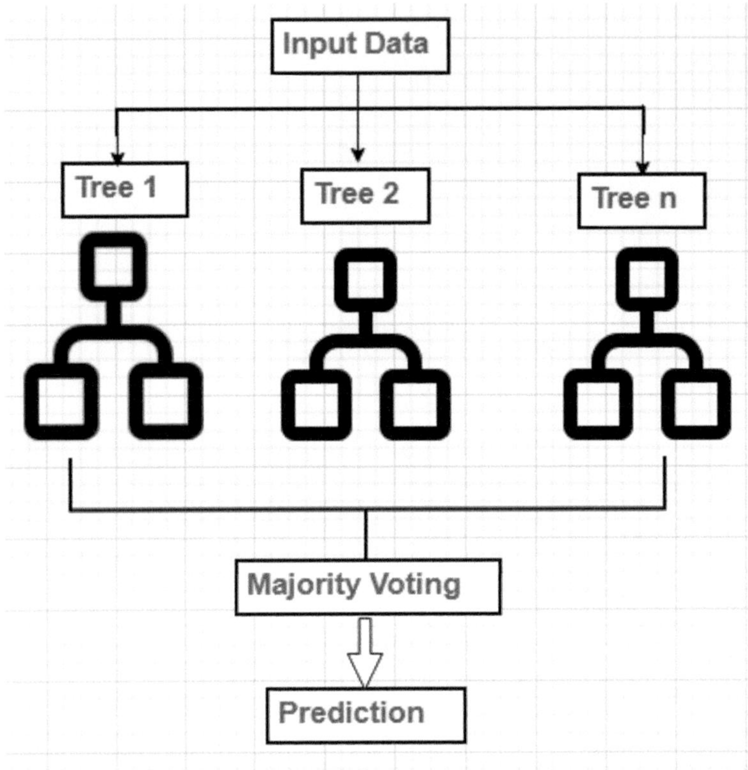

Figure 6: Random forest.

Trade off

- Random Forests handle high-dimensional data, but the performance and complexity of the algorithm highly depends on the feature selection process.
- For time-sensitive applications, tree depth must be optimized in order to reduce computational latency.
- Dynamic IoT environments need model retraining from time to time to adapt to new data patterns

Support Vector Machines for IoT: SVM is a supervised learning method used for outlier detection, classification, and regression tasks. The primary objective of this method is to identify the ideal hyperplane. The margin between data points is maximized to separate them into different classes. When there are more features than data samples, this method is resistant to overfitting problems and performs well in high-dimensional domains.

Why SVM for IoT Applications?

- To handle high dimensional data generated from numerous IoT devices and sensors and perform in an efficient manner, SVM is chosen.
- Provide high accuracy even on a small to medium-sized datasets.
- Detects unusual patters from IoT sensor, i.e., anomaly detection, fault diagnosis, etc.
- SVM is robust as it works well with non-linear decision boundaries using kernel functions like radial basis function (RBF), polynomial, and sigmoid.

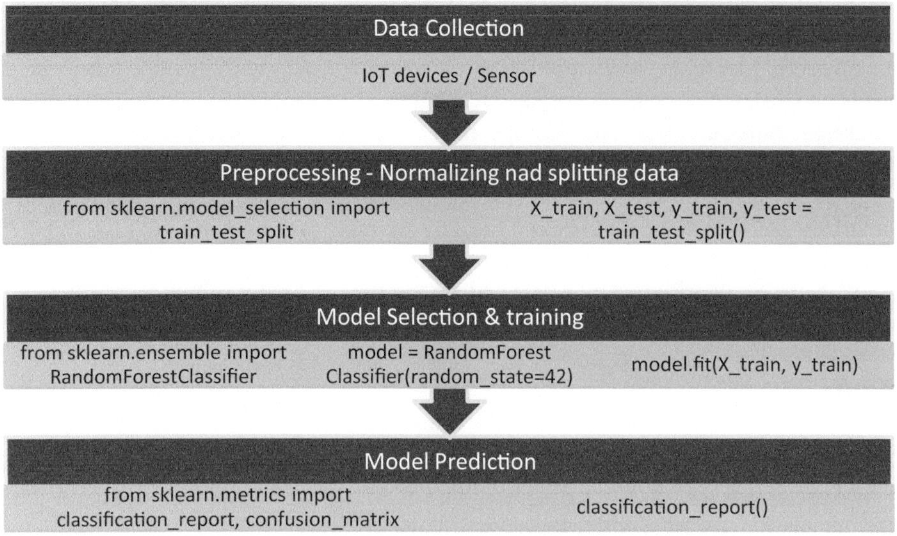

Figure 7: Random forest using Scikit.

SVM in scikit-learn: SVM in scikit-learn uses SVC, NuSVC and LinearSVC library for performing binary and multi-class classification using linear kernel function (scikit-learn Developers 2025e). To support regression SVR, NuSVR and LinearSVR library are used.

Trade off

- Radial Basis Function (RBF) is commonly used to handle non-linear data in IoT and needs more computational power and training data to achieve high accuracy.
- Less expensive linear kernel function cannot handle complex patterns in IoT data.
- Performance of SVM highly depends on hyper parameter tuning such as C (regularization) and gamma (for RBF kernel).
- SVM models (especially with linear kernels) can be optimized for deployment on microcontrollers or edge gateways using model compression.

k-Nearest Neighbors (k-NN): k-NN is a supervised learning algorithm, otherwise known as a lazy learning model because it doesn't explicitly learn from the training data during training but rather stores the data and performs calculations when making predictions. It works on the assumption that similar points can be found near one another. This can be used for both classification and regression applications. Every time a new data point is received, the distance metric—such as the Manhattan or Euclidean distance—between it and the other data points is calculated. By majority vote, the class label that appears most frequently among the new data point's *k* nearest neighbors is assigned to it. The algorithm's relevance is determined by the value of "k." A few applications include handwritten digit and satellite image classification and regression.

Why k-NN for IoT?

- It is a simple algorithm and does not require any training phase. It computes the distance and makes decision only during the time of prediction.
- It is more adaptable, i.e., it can be easily retrained or updated by simply adding new data points.
- It is a non-parametric method, i.e., it works well with irregular boundaries, and it doesn't make any assumptions about the underlying data distribution.

KNN in scikit learn: Neighbors-based functionality is used by sklearn.neighbors for both supervised and unsupervised learning techniques. Regression works for data with continuous labels in supervised learning, while classification works for data with discrete labels. scikit-learn implements two different classifiers such as : KNeighborsClassifier (scikit-learn Developers 2025f) and RadiusNeighborsClassifier. The user specifies the value of k in the former, which identifies k neighbors. Later learning is based on how many neighbors each user-specified training point has within a given radius, "r." Scipy or NumPy arrays are handled by the sklearn.neighbors class.input sparse matrices. Both distance metrics and Minkowski metrics were supported for dense matrices and sparse matrices. NearestNeighbors implements unsupervised nearest neighbors learning. Different versions of algorithm can be selected among the list of algorithms represented as auto or kd_tree and so on.

Trade off

- KNN is scalable as it can handle growing data.
- Needs more memory to hold the entire training dataset.
- Less suitable for real-time application constraints.

k-Means Algorithms: An unsupervised machine learning algorithm called K-Means groups data points that are similar to one another. The data points are grouped using a similarity or distance metric. The objective of this method is to divide "n" data points into "k" separate, non-overlapping clusters, with each point belonging to the cluster that has the closest centroid. The general steps followed are:

1. Initializing k cluster centroids (k is a user defined value).
2. Assigning each data point to the nearest centroid by applying the distance or similarity metrics.
3. Updating the centroids and repeating steps 2 and 3 until there is no change in the centroid.

Why Use K-Means for IoT Applications:

- Lightweight & Efficient: K-Means is lightweight, and requires low computational and memory requirements, making it suitable for IoT devices with limited resources.
- Data generated by IoT devices is mostly unlabeled, making K-Means more suitable as this algorithm works with unlabeled data.
- It helps identify patterns or group similar devices/events.
- IoT ecosystems are designed to accommodating substantial numbers of devices and data points as required, thereby ensuring high scalability.
- K-means is applicable to versatile applications such as smart homes, predictive maintenance, environmental monitoring, traffic control, and more.

k-Means in scikit learn: k-Means algorithm is used for cluster purpose. It is an unsupervised ML model. It is implemented in the scikit-learn library via sklearn.cluster.KMeans (scikit-learn Developers 2025 g). It tries to minimize the within-cluster sum-of-squares criterion. Scikit-learn's K-Means class allows customization of parameters like n_clusters, init, and max_iter. It's efficient and commonly used for pattern recognition, market segmentation, and image compression.

Trade off

- It is challenging to run iterative algorithms like K-means locally on IoT devices because they frequently have limited processing power, memory, and battery life.

- K-means may require multiple iterations to converge, making it unsuitable for time-sensitive tasks without optimization or approximation.
- K-means may require centralized data processing for clustering, necessitating frequent data transfers from sensors to a cloud/server. This results in high communication costs, bandwidth usage, and latency, especially for large-scale deployments.
- The quality of K-means results is sensitive to initial cluster centroid selection. Poor initialization can lead to local minima, which is problematic when retraining is expensive or infeasible in IoT.
- Traditional K-means assumes centralized data, so applying it in a distributed setting requires complex synchronization or adaptation like federated learning or distributed K-means variants.

Deep Learning in IoT: One subset of machine learning algorithms is deep learning. To maximize outcomes and improve learning capacity, it is based on a multi-layered artificial neural network (ANN). To solve real-world or real-time problems, DL uses neural networks to mimic how the human brain works. The majority of applications for Convolutional Neural Networks (CNNs) and Recurrent Neural Networks (RNNs) include object detection and image classification. Convolutional layers, pooling, dropout, and a fully connected layer make up a typical CNN model.

Why Deep Learning for IoT?

- Automatically extracts relevant features from images, audio, spatial and temporal data.
- Deep learning approach outperforms traditional ML methods on large and complex datasets and produces high accuracy
- Deep learning applies to a wide range of data types such as images, text, time series, video, and audio.
- TinyML: Specialized lightweight deep learning models that can run on microcontrollers (e.g., TensorFlow Lite for Microcontrollers).

Implementing CNN in Keras/PyTorch

1. Import all the necessary libraries.
2. Preprocessing includes normalization and labeling the data.
3. Split the data into training and testing set.
4. Creation of a convolution network model (CNN) consists of the following phases:

- **Convolutional layers:** The term convolution refers to a dot product between input (image) data arrays and filters/kernels. Stacking of multiple neural network layers is built using Keras's Sequential API. It makes use of the Conv2D () method. It can make use of different activation function such as ReLU, etc.
- **Pooling:** The process of pooling is responsible for decreasing the size of the input image after convolution. Two variants of pooling are available, i.e., Max pooling and Average pooling. Max Pooling function returns the maximum pixel value of the image considered under the Kernel function. It makes use of tf.keras.layers.MaxPool2D function. Average Pooling function returns the average pixel value of the image considered under the Kernel function. It makes use of tf.keras.layers.AveragePooling2D function.
- **Dropout layers:** It randomly reduces interconnecting neurons within the network.

- **Fully Connected layer:** It makes use of tf.keras.layers.Dense() function to create densely connected neural network. SoftMax Activation Function is used to represent the probability of an occurrence of a class or event. Dense () method makes use of SoftMax Activation function.

5. Fit the model and define the epochs and batch size.
6. Run and evaluate the model.

3. ML implementation using Scikit learn

Use case: Smart farming - Crop recommendation

Dataset: Kaggle Crop Recommendation Dataset downloaded from (Atharva Ingle., n.d.). This dataset is used to build a predictive model to recommend suitable crops using seven data fields: (1) nitrogen, (2) phosphorus, (3) potassium content in soil, (4) temperature, (5) humidity, (6) pH value, and (7) rainfall value. The dataset is tailored for Indian agricultural conditions. Figure 8 represents the loading of the Crop Recommendation Dataset and lists the first five tuples. Figure 9 represents the attribute 'Nitrogen,' and data representation of various features is shown in Figure 10.

Stages of pre-processing, model selection and evaluation using sklearn package were depicted in Figure 11, 12 and 13, respectively.

Table 2 gives a comprehensive overview of the functions employed in implementing the KNN model using scikit-learn package.

4. Literature on the use of Ml In IoT

Different machine learning methods were assessed to enhance human experiences, services, and challenges in IoT data by considering smart cities as the main use case (Mahdavinejad et al. 2018). The author chose smart city applications because nearly 60% of reviewed articles focused on this domain. Smart city use cases include smart energy (reducing power consumption and leak

```
# 1. Data Collection and loading
import pandas as pd

inputdata = pd.read_csv('/content/Crop_recommendation.csv')

# View first 5 data elements
inputdata.head(5)
```

	N	P	K	temperature	humidity	ph	rainfall	label
0	90	42	43	20.879744	82.002744	6.502985	202.935536	rice
1	85	58	41	21.770462	80.319644	7.038096	226.655537	rice
2	60	55	44	23.004459	82.320763	7.840207	263.964248	rice
3	74	35	40	26.491096	80.158363	6.980401	242.864034	rice
4	78	42	42	20.130175	81.604873	7.628473	262.717340	rice

Figure 8: Loading dataset and viewing first 5 values from dataset.

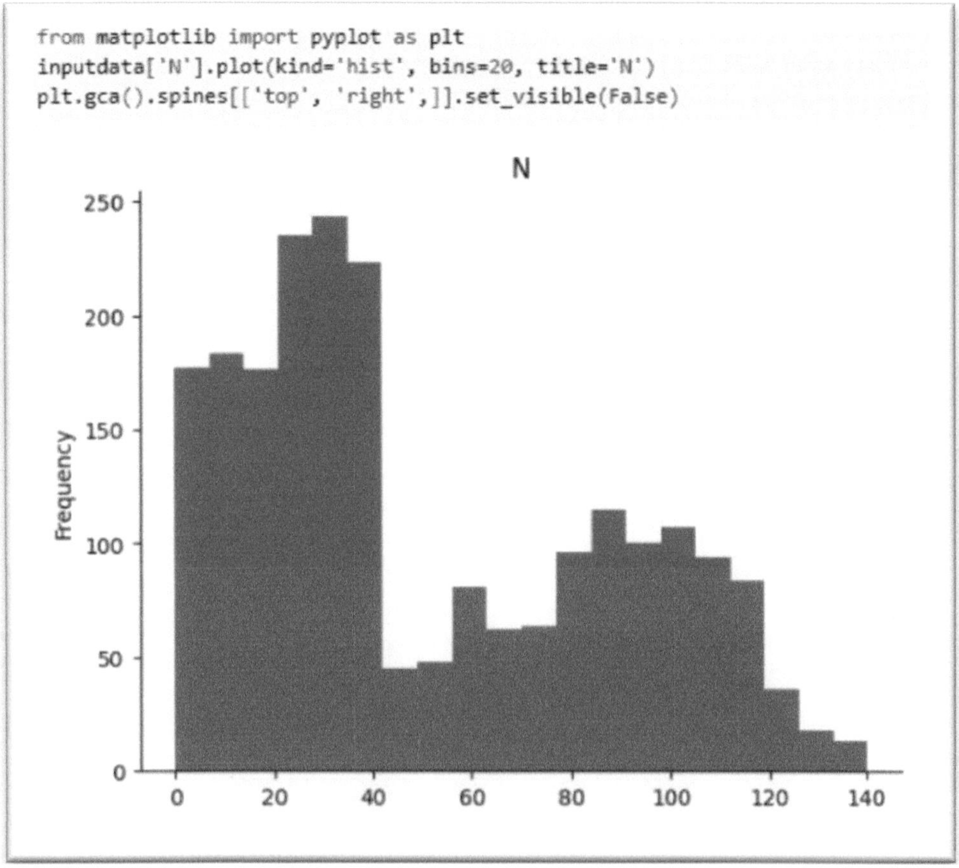

Figure 9: Data representation of feature 'Nitrogen'.

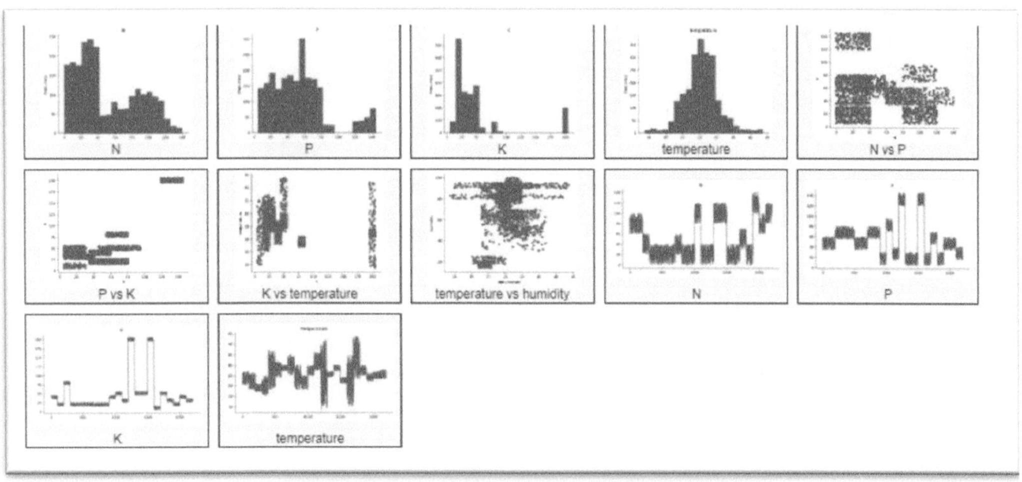

Figure 10: Data representation of various data features.

```
# 2. Data Preprocessing
from sklearn.model_selection import train_test_split
from sklearn.preprocessing import StandardScaler

# Train-test split
X_train, X_test, y_train, y_test = train_test_split(X, y, test_size=0.3, random_state=42)

# Feature scaling
scaler = StandardScaler()

X_train_scaled = scaler.fit_transform(X_train)
X_test_scaled = scaler.transform(X_test)
```

Figure 11: Data pre-processing.

```
# 3. Model Selection
from sklearn.neighbors import KNeighborsClassifier
knn_model = KNeighborsClassifier(n_neighbors=3)

# 4. Training
knn_model.fit(X_train_scaled, y_train)

# 5. Evaluation
from sklearn.metrics import accuracy_score, classification_report, confusion_matrix

y_pred = knn_model.predict(X_test_scaled)
print("Accuracy:", accuracy_score(y_test, y_pred))
print("Classification Report:\n", classification_report(y_test, y_pred))
print("Confusion Matrix:\n", confusion_matrix(y_test, y_pred))
```

Figure 12: Model Selection and Evaluation.

Table 2: Overall summary of function used in KNN model using scikit.

ML Phase	**Scikit function**
Data Collection	**load_iris()**
Preprocessing	**train_test_split(), StandardScaler()**
Model Selection	**KNeighborsClassifier(n_neighbors=3)**
Training	**knn_model.fit()**
Evaluation	**accuracy_score(), classification_report()**
Deployment	**joblib.dump(), joblib.load()**

```
Accuracy: 1.0
Classification Report:
              precision    recall  f1-score   support

           0       1.00      1.00      1.00        19
           1       1.00      1.00      1.00        13
           2       1.00      1.00      1.00        13

    accuracy                           1.00        45
   macro avg       1.00      1.00      1.00        45
weighted avg       1.00      1.00      1.00        45

Confusion Matrix:
 [[19  0  0]
 [ 0 13  0]
 [ 0  0 13]]
```

Figure 13: Model Evaluation Results.

monitoring), urban planning (city infrastructure), mobility and traffic management (autonomous vehicles and traffic control), health monitoring, and environmental management. The author used KNN and Naive Bayes for classifying traffic prediction and data abbreviation. Linear Regression and SVM were used for regression tasks. K-Means and DBSCAN were common clustering algorithms for analyzing citizen behavior and traffic patterns. Principal Component Analysis (PCA) and Canonical Correlation Analysis (CCA) were used for feature extraction. Anomaly detection methods were used to identify unusual patterns in sensor data. Future research focuses on deep learning and semi-supervised learning techniques.

To identify threats at an earlier stage, an Intrusion Detection System (IDS) searches for suspicious activity (Chaaboun et al. 2019). A study comparing supervised, unsupervised, and reinforcement learning techniques for intrusion detection in IoT systems used the IoT-23 dataset (Vitorino et al. 2021). Support Vector Machine (SVM), Extreme Gradient Boosting (XGBoost), and Light Gradient Boosting Machine (LightGBM) were used under the supervised approach. Isolation Forest (iForest) and Local Outlier Factor (LOF) represented the unsupervised approach, while the Double Deep Q-Network (DDQN) represented the reinforcement learning approach. The IoT-23 dataset is a labeled dataset consisting of 23 scenarios broadly classified as malicious and benign IoT network traffic from 2018–2019. ML algorithms played a vital role in monitoring and mitigating threats in IDS. Six machine learning techniques were used to analyze nine malware scenarios in both binary and multi-class classification contexts. Metrics such as accuracy, precision, recall, and F1-score were used to assess model performance. Experiments were conducted on an 8-core CPU, 6GB GPU, and 16GB RAM. The NumPy and Pandas libraries, Scikit-learn, TensorFlow, and Python were utilized in the implementation.

The author performed a survey of ML and DL for enhancing the security of IoT systems (Al-Garadi et al. 2020). The study defines a detailed taxonomy of ML/DL for IoT security, addressing three layers: Perception, Network, and Application. It explores security threats (both cyber and physical), threat surfaces, ML/DL algorithms for each layer, and challenges including the development of lightweight models suitable for IoT devices, federated learning to preserve data privacy, and the creation of standardized datasets for benchmarking. The Perception layer includes physical objects and connectivity, with challenges such as assigning unique IP addresses to all devices, maintaining low-power communication channels, and implementing effective routing protocols. The Network layer is vital for data transmission and routing in IoT systems. The Application layer encompasses user interfaces and services. Security solutions were presented for each layer. For the Perception layer, DL methods are used to provide physical-layer authentication and helps in achieving better spoofing detection. Reinforcement learning (RL) is effective against jamming attack, thereby enhancing the network reliability. For the Network layer, a DL model helps in finding routing protocol attacks. Smart devices are protected using ML-based intrusion detection frameworks. Unsupervised ML approaches help in finding anomalies. For the Application layer, the DL models (CNN) effectively detect Android malware automatically by eliminating manual feature engineering process. Security architecture with GANs detects abnormal behavior.

The author presents the amalgamation of DL with the IoT ecosystem (Lakshmanna et al. 2022). The various algorithms presented in the literature are Recurrent Neural Networks (RNNs), LSTM, CNN, Autoencoders, and DBN. The basic working of these algorithms, along with their IoT applications and popular datasets used, were summarized by the author. RNN is mainly used for serial and time-series data. CNN is preferred for processing spatial data. RNN and LSTM are suitable for handling sequential data, making them ideal for time-series analysis. Various applications where RNN models are applied include smart cities, energy, education, sports, and retail, as well as image recognition, along with the datasets used for analysis. LSTM models were used in traffic prediction, autonomous driving, localization, smart homes, energy, education, and sports applications. CNN models are used in healthcare, smart homes, agriculture, sports and retail, localization, and government applications. Autoencoders were used in fault assessment, the public sector, IoT infrastructure, energy, and image recognition. The DBN model is applied in transportation, intelligent transportation systems, security, energy, and the health sector. Major challenges in integrating DL into IoT are resource constraints, data privacy, and security.

An ample inspection of the convergence between deep learning (DL) methodologies and Internet of Things (IoT) systems was presented by the author. He explores how DL techniques can be coupled to process, analyze, and derive significant insights from data (Elhanashi et al. 2023). The author compares CNN, Recurrent Neural Networks (RNNs), and Deep Belief Networks (DBNs). The survey concludes that integrating DL techniques into IoT systems holds significant potential for enhancing data analysis and decision-making processes. Addressing the outlined challenges through continued research and innovation is essential for realizing intelligent, efficient, and secure IoT systems.

The integration of ML in IoT for healthcare applications (H-IoT) is presented by the author (Vaigandla et al. 2025). In other words, the Internet of Medical Things (IoMT). He presented the five layers of IoT architecture: (1) Perception: holds sensors and actuators and is responsible for collecting data from the physical environment; (2) Network: contains network infrastructure and is responsible for communication; (3) Processing: includes storage, analytics, and decision-making, responsible for handling data; (4) Application: applications or services delivered to the end user; and (5) Business: provides business models and analytics. Applications and benefits of H-IoT include enhanced patient care through personalized treatment, continuous monitoring, and tele-services. It improves patient experience through engagement and convenience. The author also discusses H-IoT technologies such as wearable devices, implantable sensors, edge and fog computing, mobile health apps and telemedicine platforms, predictive analytics, NLP, surgical robots, and automated medical dispensers. The key application areas include disease diagnosis and prediction, personalized medicine, drug discovery and development, CDSS, remote patient monitoring and telemedicine, operational efficiency, and resource management. While the benefits are substantial, the paper also addresses the challenges associated with implementing IoT and ML in healthcare. Data security and privacy are major issues since sensitive health data must be protected from breaches during transmission and storage. The study also emphasizes the necessity of device interoperability and standardized protocols to ensure seamless integration and communication within healthcare systems.

To improve real-time monitoring in automotive manufacturing, the author offers an all-inclusive framework that combines sensors, big data processing, the Internet of Things (IoT), and predictive modeling (Syafrudin et al. 2018). The system supports proactive decision-making, reduces unexpected downtimes, and improves overall production efficiency. Manufacturing processes generate vast amounts of data through IoT devices. For real-time monitoring and decision-making, this data must be processed and analyzed effectively. The amount of data is often too large for traditional systems to handle. Three primary components make up the real-time monitoring system: (1) Internet of Things (IoT)-based sensors: These are used to gather real-time data from the automotive assembly line, including temperature, humidity, acceleration, and gyroscopic movements, (2) Big data processing platform: To manage large amounts of data using MongoDB, Apache Storm, and Apache Kafka. This combination ensures low-latency processing and scalability, enabling the system to manage unstructured and streaming data effectively. (3) Prediction model: It uses DBSCAN for outlier detection and Random Forest for detecting faults in manufacturing process. The integrated system was deployed and tested in an automotive manufacturing assembly line in Korea. The evaluation focused on the system's ability to process sensor data in real-time and accurately detect faults. The results showed that the system effectively managed high-throughput data, with the hybrid prediction model outperforming traditional models in fault detection accuracy. The integration of DBSCAN improved the classifier's performance by eliminating noise and irrelevant data, leading to more reliable predictions. Future studies might explore applying the system to different manufacturing domains and integrating more sophisticated machine learning models, such as deep learning.

Implementing outlier detection using Hadoop, a big data framework, and the K-Means machine learning algorithm using LinkSmart middleware in an IoT environment (Souza and Amazonas 2015). Traditional outlier detection methods often fall short in handling the large-scale and complex IoT data present in a distributed environment. The Apache Spark big data framework is selected for its in-memory processing capabilities for real-time analysis. The system architecture proposed by the author comprises four layers: (1) Data Collection Layer: comprises various sensors through which real-time data are collected and sent to the next higher layer; (2) Data Processing Layer: uses big data frameworks such as Hadoop and Spark for processing the distributed data; (3) Outlier Detection: this layer uses machine learning algorithms to detect the outliers; and (4) Feedback: in this phase, the detected anomalies are reported back to the system for further analysis, thereby enabling an adaptive learning environment. The author discusses ensemble isolation forest approaches, K-Means, DBSCAN, and Local Outlier Factor strategies for outlier detection. The system is validated using real data from the Smart Santander Project. The ROC curve shows 0.8967 area under the curve, representing good quality of the proposed outlier detection results. The system is tested with the values obtained during February 2014 (all days). The results show 0.8456 as the minimum AUC value, 0.9465 as the maximum AUC value, and 0.8762 as the mean AUC value.

A survey is presented on various articles related to IoT security solutions since January 2016 (Banerjee et al. 2018). During this process, the author highlights a critical issue—the scarcity of publicly available IoT datasets. According to the paper, blockchain technology presents viable solutions to IoT security issues. Blockchain is an immutable, decentralized ledger. It can make it easier for stakeholders and IoT devices to share data securely. By ensuring data integrity and traceability, blockchain can help detect and prevent unauthorized data modifications. Two conceptual frameworks were proposed by the author, integrating blockchain into IoT security: (1) Blockchain-

Enabled Secure Data Sharing: Creates a tamper proof data transaction using blockchain. By recording data exchanges on the blockchain, stakeholders can verify the authenticity and integrity of the data, fostering trust in collaborative environments. (2) Blockchain-Based Intrusion Detection Systems (IDS): Uses blockchain to enhance the reliability of IDS in IoT networks. By logging security events and alerts on the blockchain, the system ensures that intrusion data remains unaltered and accessible for analysis, thereby improving the detection and response to security threats. In addition to this, the author posted nine research questions.

- How can blockchain scalability be addressed to accommodate the high transaction volume in IoT networks?
- What consensus mechanisms are most suitable for resource-constrained IoT devices?
- How can privacy be preserved when storing sensitive IoT data on public blockchain?
- What are the implications of integrating blockchain with existing IoT architectures?
- How can interoperability between different blockchain platforms and IoT devices be achieved?
- What are the cost implications of implementing blockchain solutions in IoT?
- How are regulatory and compliance issues addressed in blockchain-based IoT systems?
- What are the potential security vulnerabilities introduced by blockchain integration?
- How can user trust and acceptance of blockchain-enabled IoT solutions be fostered?

5. Role of ML in IoT

- ➢ Handling huge volume of data with diverse data type support makes ML an ideal support system for IoT application.
- ➢ Development of sophisticated algorithms makes ML suitable for IoT Applications.
- ➢ ML helps in taking decisions without human intervention.
- ➢ Deep learning, a subset of machine, plays a pivotal role in fields ranging from healthcare and finance to transportation and entertainment.
- ➢ ML models detect early signs of various component faults in connected vehicles/applications.
- ➢ ML helps IoT devices in more personalization. It adapts to individual user behaviors and preferences.
- ➢ Real-time data received from IoT devices can be analyzed to detect and predict early failure/insights.

6. Challenges and Consideration

ML based IoT ecosystem poses various challenges under four main verticals such as data, hardware, security, and ML algorithm.

- Data Issues:
    - o To handle continuous stream of real time data, there is a need for scalable storage and processing infrastructure.
    - o Data formats (structured, unstructured and semistructured data) and data standardization.
- Hardware issues:
    - o To handle noisy, incomplete, or erroneous data due to hardware failure.
    - o Sensor degradation.
    - o Limited processing, memory, and power.

- Security issues:
  - o Data privacy issues, especially in healthcare applications.
  - o Ensuring the security of confidential data remains a key concern.
  - o Can lead to various types of network attacks.
- ML issues:
  - o Need for on-device ML.
  - o Real time fast and reliable decision-making system for applications such as autonomous vehicles, industrial control, and healthcare.
  - o Label scarcity.
  - o Model interpretability.

7. IoT Applications and ML Algorithms

With billions of devices connected to the Internet, there is a need for faster processing and analysis to make decisions, which has paved the way for using ML algorithms. Table 3 represents various IoT applications with their purposes and the ML algorithms used in the literature.

Table 3: IoT applications with their purpose.

S.no	IoT Applications	Purpose	ML Algorithms suggested in literature
1.	Smart Home Automation	• To monitor and control electrical gadgets such as light, fan, AC, heater, and security cameras • Air quality checking	Decision tree
2.	Industrial Automation	• Monitor the condition of machinery • Forecast equipment failure • Air quality checking	Random Forest Deep learning Decision tree Autoencoders or LSTMs
3.	Healthcare and Wearables	• Analyzing the vitals (heart rate, ECG, oxygen, blood) • Clinical Decision Support System • Medicine dispenser • Fitness trackers, diet suggester • Voice controlled device	Deep learning SVM
4.	Smart Cities	• Traffic prediction • Energy optimization • Environment monitoring	Deep Learning
5.	Smart Agriculture	• Monitor crop growth • Predicting crop diseases • Analyzing the type of soil • Analysis of weather condition • Water management • Fertilization schedules	SVM. Random Forest
6.	Anomaly Detection	• Detecting abnormal events in temperature, pressure, or motion sensors in smart homes or industrial setups. • Intrusion Detection	SVM KNN
7.	Object detection	• Object detection using microarchitecture for real time object detection • Single short object detection for detecting human activity • Rainfall prediction	Tiny YOLO FastGRNN TinySSD TinyRainNet (CNN with BiLSTM)

Table 4: Taxonomy of ML algorithms and their respective roles in IoT application.

ML Algorithm	Smart Home Automation	Industrial Automation	Health Care Application	Smart Agricultural	Smart City Application
Linear Regression	Forecasting environmental parameters	Predicting energy consumption	Estimating the recovery time of the patient	Estimating likelihood of pest infestation	Air quality prediction, Utility load forecasting
Decision Tree	Appliance usage prediction	Fault detection	Diseases prediction	Crop disease classification	Waste categorization
Random forest	Demonstrating user behavior, Estimating energy consumption	Predictive maintenance	Diseases prediction	Crop yield prediction	Traffic analysis
SVM	Intrusion detection	Classification of defect	Tumor classification, ECG signal classification	Classifying crop health	Intrusion detection in surveillance data
KNN	Predicting user preferences	Similarity-based anomaly detection	Early diagnosis based on symptoms	Soil-based crop prediction	Traffic analysis
Deep learning algorithms (CNN)	Gesture recognition using image-based system	Identifying defective products using image data	MRI, X-ray, CT Scan image analysis	Drone image-based weed detection	Surveillance detection using video inputs, Vehicle number plate detection

Table 4 represents a brief taxonomy of various ML algorithms and their respective roles in IoT application.

8. Emerging Trends

TinyML frameworks allow ML models to be embedded on low-power devices (Syed et al. 2022). It provides ML as a Service (MLaaS). Traditional cloud-based applications require centralized servers where data from different sensors are collected and processed. In contrast, TinyML is beneficial in environments with limited connectivity and power resources. The author presents the design trade-offs of the TinyML-as-a-Service architecture and provides a case study on gesture recognition, demonstrating how it can be implemented, deployed, and maintained in IoT applications (Syed et al. 2022). Deep Neural Network (DNN) ML is used for gesture recognition. A lightweight web application designed using React, a JavaScript framework, connects to the cloud through Wi-Fi or LoRa.

A survey on the literature review of various TinyML applications published until 2023 (Abadade et al. 2023) presents a taxonomy of TinyML techniques across applications such as (1) Healthcare: blood pressure monitoring, emotion detection, smart ambulatory systems, etc., (2) Smart Farming: crop management, smart irrigation; and (3) Anomaly Detection: predictive maintenance. A few open research questions posed by the author include limited memory, hardware heterogeneity, privacy issues, and reliability.

A survey on the application of ML in different layers of communication networks such as the physical layer, access layer, network layer, Multi-access Edge Computing (MEC), Software-Defined Networking (SDN), and Network Functions Virtualization (NFV) was presented from research articles between 2008 and 2020 (Ahmad et al. 2020). The author further discussed the usage of ML for network security.

With the rapid growth of connected devices leading to potential network congestion, this can be addressed by incorporating ML at the edge of the device, where computations are performed close to the data sources. The author presented a survey on various ML algorithms, architectures, and communication standards used for Edge AI in IoT environments (Merenda et al. 2020). A total of 6,342 papers were reported using both search engines and libraries for the term "Edge Computing." Among these, only 100 papers written in English between 2014 and 2020 were reviewed. ML algorithms used in the literature include CNN, RNN, DNN, ANN, KNN, and SVM. Three architectures were focused on by the researcher: (1) On-device: ML is executed on the device itself, (2) Edge server-based architecture: Computation is performed on the edge server, (3) Joint computation: Combining both on-device and edge server architecture. Wireless standards used in the literature include Zigbee (Wireless Personal Area Network), ANT protocol (supports up to 65,533 nodes in star, tree, or mesh topology), 6LoWPAN for open IoT networks, RFID, NFC, LoRa, SigFox, 4G LTE, and 5G. The author also presented communication technologies along with data rates and distance coverage.

9. Conclusion

Cyber IoT brings a new dimension to various business models across industries. Devices are embedded with sensors, software, and other technologies. These devices range from home appliances and vehicles to wearables and sophisticated industrial machines. Data is collected from various real-time devices such as sensors, which must be analyzed immediately to gain useful insights for decision-making. Machine learning is transforming how computers interact with data and make decisions. Its applications are vast and growing, and as technology advances, it will continue to shape industries and societies. In this chapter, the reasons for choosing specific machine learning algorithms and their implementation using Scikit-learn packages were discussed. A brief taxonomy of ML algorithms and their respective roles in IoT applications was presented, along with advancements in recent trends.

References

Abadade, Y., Temouden, A., Bamoumen, H., Benamar, N., Chtouki, Y., and Hafid, A. S. (2023). A comprehensive survey on tinyml. *IEEE Access*, 11: 96892–96922.

Ahmad, I., Shahabuddin, S., Malik, H., Harjula, E., Leppänen, T., Loven, L. et al. (2020). Machine learning meets communication networks: Current trends and future challenges. *IEEE access*, 8: 223418–223460.

Alam, Tanweer. (2018). A Reliable Communication Framework and Its Use in Internet of Things (IoT). 3.

Al-Garadi, M. A., Mohamed, A., Al-Ali, A. K., Du, X., Ali, I., and Guizani, M. (2020). A survey of machine and deep learning methods for internet of things (IoT) security. *IEEE Communications Surveys & Tutorials*, 22(3): 1646-1685.

Anitha Elavarasi, S., Jayanthi, J., and Basker, N. (2019). Trajectory Object Detection using Deep Learning Algorithms. *International Journal of Recent Technology and Engineering (IJRTE)*, ISSN: 2277–3878 (Online), Volume-8 Issue-3, September 2019. Page No.: 7895–7898.

Atharva Ingle. (n.d.). Crop Recommendation Dataset. Kaggle. Retrieved from https://www.kaggle.com/datasets/atharvaingle/crop-recommendation-dataset?resource=download&select=Crop_recommendation.csv.

Banerjee, M., Lee, J., and Choo, K. K. R. (2018). A blockchain future for internet of things security: a position paper. Digital Communications and Networks, 4(3): 149v160.

Chaabouni, N., Mosbah, M., Zemmari, A., Sauvignac, C., & Faruki, P. (2019). Network intrusion detection for IoT security based on learning techniques. *IEEE Communications Surveys & Tutorials*, 21(3): 2671–2701. doi: 10.1109/COMST.2019.2896380.

Elhanashi, A., Dini, P., Saponara, S., and Zheng, Q. (2023). Integration of deep learning into the iot: A survey of techniques and challenges for real-world applications. *Electronics*, 12(24): 4925.

Hernández, Noelia, Corrales, Héctor, Izquierdo, Ruben, Ballardini, Augusto, Salinas, Carlota and García, Iván. (2021). WiFiNet: WiFi-based indoor localisation using CNNs. 10.48550/arXiv.2104.06768.

Lakshmanna, K., Kaluri, R., Gundluru, N., Alzamil, Z. S., Rajput, D. S., Khan, A. A. et al. (2022). A review on deep learning techniques for IoT data. *Electronics*, 11(10): 1604.

Merenda, M., Porcaro, C., and Iero, D. (2020). Edge machine learning for ai-enabled iot devices: A review. *Sensors*, 20(9): 2533.

scikit-learn Developers. (2025a). 1.10 Decision Trees (scikit-learn 1.7.2). In scikit-learn: Machine Learning in Python. Retrieved from https://scikit-learn.org/stable/modules/tree.html scikit-learn.

scikit-learn Developers. (2025b). sklearn.tree.DecisionTreeClassifier (API reference). In scikit-learn: Machine Learning in Python. Retrieved from https://scikit-learn.org/stable/modules/generated/sklearn.tree.DecisionTreeClassifier.html scikit-learn

scikit-learn Developers. (2025c). Decision Tree Regression Example — plot_tree_regression. In scikit-learn Examples. Retrieved from https://scikit-learn.org/stable/auto_examples/tree/plot_tree_regression.html scikit-learn

scikit-learn Developers. (2025d). sklearn.ensemble.RandomForestClassifier (API reference). In scikit-learn: Machine Learning in Python. Retrieved from https://scikit-learn.org/stable/modules/generated/sklearn.ensemble.RandomForestClassifier.html scikit-learn

scikit-learn Developers. (2025e). 1.4 Support Vector Machines (scikit-learn user guide). In scikit-learn: Machine Learning in Python. Retrieved from https://scikit-learn.org/stable/modules/svm.html scikit-learn

scikit-learn Developers. (2025f). sklearn.neighbors.KNeighborsClassifier (API reference). In scikit-learn: Machine Learning in Python. Retrieved from https://scikit-learn.org/stable/modules/generated/sklearn.neighbors.KNeighborsClassifier.html scikit-learn

scikit-learn Developers. (2025g). sklearn.cluster.KMeans (API reference). In scikit-learn: Machine Learning in Python. Retrieved from https://scikit-learn.org stable/modules/generated/sklearn.cluster.KMeans.html scikit-learn

Souza, A. M., and Amazonas, J. R. (2015). An outlier detect algorithm using big data processing and internet of things architecture. Procedia Computer Science, 52: 1010–1015.

Syafrudin, M., Alfian, G., Fitriyani, N. L., and Rhee, J. (2018). Performance analysis of IoT-based sensor, big data processing, and machine learning model for real-time monitoring system in automotive manufacturing. Sensors, 18(9): 2946.

Syed Ali Raza, Zaidi, A. M. Hayajneh, M. Hafeez and Q. Z. Ahmed. (2022). Unlocking Edge Intelligence Through Tiny Machine Learning (TinyML). *In IEEE Access*, 10: 100867–100877, doi: 10.1109/ACCESS.2022.3207200.

Vaigandla, K. K. (2025). Role of IoT and ML in Healthcare. Babylonian Journal of Artificial Intelligence, 2025, 23–36.

Vitorino, J., Andrade, R., Praça, I., Sousa, O., and Maia, E. (2021, December). A comparative analysis of machine learning techniques for IoT intrusion detection. In International Symposium on Foundations and Practice of Security (pp. 191–207). Cham: Springer International Publishing.

8

Leveraging AI to Examine Cultural Familiarity as a Moderator in the Price–Purchase Link

Insights from Vietnamese Consumers of Korean Cosmetics in the Era of Hallyu

Tran Thi Ngoc Oanh and *Bui Huy Khoi**

1. Introduction

The global cosmetics market is experiencing robust growth, with its size projected to reach USD 354.68 billion by 2025 and a compound annual growth rate (CAGR) of approximately 6.64% (Grand View Research 2023). The Asia-Pacific region dominates the market, accounting for around 40% of the total share, with China, South Korea, and Japan contributing approximately 35% of global revenue (Grand View Research 2023). In Vietnam, the cosmetics market is forecasted to reach USD 2.80 billion in 2025, with over 90% of products being imported. Among these, South Korea holds the largest share at 30% (Statista 2025). This growth is driven by rising income levels, the strong influence of Korean beauty culture (K-Beauty), and the development of e-commerce. The Vietnam-Korea Free Trade Agreement has reduced import tariffs on cosmetics by 10%, facilitating the expansion of Korean brands in the Vietnamese market. As a result, the Hallyu wave has significantly influenced Vietnamese consumer preferences. Vietnamese youth are heavily influenced by the Hallyu wave, viewing the flawless, radiant skin of K-pop idols as the ideal beauty standard, which naturally and positively drives their acceptance of Korean cosmetics. Furthermore, the role of

Industrial University of Ho Chi Minh City Ho Chi Minh City, Vietnam.
Email: Tranthingocoanh1412003@gmail.com, https://orcid.org/0009-0008-0633-3531
* Corresponding author: buihuykhoi@iuh.edu.vn, https://orcid.org/0000-0002-4976-7435

DOI: 10.1201/9781003655190-10

celebrity endorsements by popular idols such as Jennie, Lisa (BLACKPINK), and Jungkook (BTS) has significantly stimulated consumer demand. Fans tend to purchase products used or promoted by their idols as a way of expressing support. Thanks to this influence, many products sell out within hours of release, accompanied by a surge in website traffic for these brands. According to the Korea Economic Research Institute, Hallyu contributed approximately USD 23.4 billion to the consumer goods sector in Vietnam.

Theoretically, Hallyu, through the widespread popularity of K-dramas, K-pop, and other Korean cultural exports, has deeply impacted the aesthetic preferences and consumer behavior of young Vietnamese individuals (Yuen 2021), with beauty ideals such as "White like Koreans" or "Glamorous like Korean girls" (Gammon and Phan 2024). Although previous research has examined the reception of Korean popular culture in Vietnam (Gammon 2022, Gammon and Phan 2024) and behavioral studies on cosmetics (Boonmee 2015), the gap between purchase intention and actual buying behavior under the influence of the Korean Wave (Hallyu) has not yet been clearly explained (Young et al. 2009, Rausch and Kopplin 2021). Additionally, those studies emphasized that the influence of Hallyu and social media on Korean product purchase decisions in Vietnam requires further investigation, particularly regarding factors such as discounts, brand image, and advertising. Moreover, the moderating role of Hallyu familiarity in the relationship between product price and purchasing behavior has not been sufficiently explored in the Vietnamese context.

Accordingly, this study investigates the moderating role of familiarity with Hallyu (Korean Wave) in two critical relationships: between product price and purchase intention and between purchase intention and actual purchase behavior. It further examines the influence of key marketing variables—brand image, product quality, perceived price, and advertising—on consumer buying behavior and explores the mediating role of consumer attitudes in the relationship between these antecedents and purchase intention toward Korean cosmetics. The theoretical foundation of this research integrates the Theory of Planned Behavior (TPB), the Elaboration Likelihood Model (ELM), and the Push-Pull-Mooring (PPM) framework, enabling a multidimensional understanding of consumer decision-making in a cross-cultural context. The findings are expected to offer both theoretical contributions to consumer behavior literature and practical implications for firms, particularly those seeking to design effective marketing strategies in Vietnam amidst the increasing cultural influence of Hallyu and leveraging AI for examination.

2. Literature Review

2.1 Theories

The Theory of Planned Behavior (TPB) (Ajzen 1991) explains behavior through three components: attitude, subjective norms, and perceived behavioral control. TPB posits that stronger behavioral intention leads to a higher likelihood of actual behavior. This theory has been widely applied in studies on consumer behavior related to Korean cosmetics, as evidenced by the works of (Kovács and Lukovics 2022, Cayaban et al. 2023). In the present study, TPB is adopted as a foundational theoretical framework to explain the formation of purchase intention and to predict the actual buying behavior of imported cosmetics, with a particular focus on the transition from attitude to intention and, subsequently, to purchase decision.

The Elaboration Likelihood Model (ELM) explains attitude formation through two routes of information processing: the central and peripheral routes. Factors such as country of origin, brand, quality, price, and advertising influence consumers via the central route, while familiarity with Hallyu triggers the peripheral route. ELM has been widely applied in the context of cosmetics, as seen in studies by Jaini et al. (2020), and is utilized in this study to analyze consumer behavior under the influence of Hallyu. In Vietnam, the ELM has also been applied in the context of Korean cosmetics.

The Push-Pull-Mooring (PPM) model (Ravenstein 1885, Moon 1989) interprets behavior through three factors: push (dissatisfaction with the current situation), pull (attraction to new alternatives), and mooring (barriers or catalysts). In this study, the push factor refers to consumers' dissatisfaction with domestic cosmetics that fail to meet their needs. Pull factors include the attractiveness, novelty, and appeal of elements such as country of origin, brand image, product quality, price, and advertising. Attitude and familiarity with Hallyu are considered mooring factors that either facilitate or inhibit behavior change. The PPM model provides a comprehensive explanation of the behavior of purchasing Korean cosmetics, consistent with prior studies such as those by Kurniawati and Savitri (2020).

2.2 Related Concepts

The Concept of Cosmetics

Cosmetics are products applied to the human body for the purposes of cleansing, beautifying, perfuming, or altering appearance, including skincare, makeup, and hair care products (Hassali et al. 2015). They defined cosmetics as products that are applied, sprayed, or otherwise used on the skin, teeth, or mucous membranes to cleanse, perfume, beautify, or modify appearance.

The Concept of Hallyu Familiarity

The term *Hallyu*, also known as the Korean Wave, originated in the late 1990s when Korean television dramas and music began to captivate audiences in Southeast Asia and subsequently expanded to other regions in Asia and global markets (Moon and Park 2012, Sung 2014). Hallyu encompasses the global dissemination of Korean cultural products such as films, music, fashion, cosmetics, games, and cuisine (Jeong 2012, Jin 2016, Ju 2017). Familiarity with the Korean cultural wave is defined as the extent to which consumers are exposed to and experience distinct cultural values, such as television dramas, music, and cuisine, through various channels (Sun and Jun 2022). In marketing, the concept of "familiarity" refers to the amount of information or experience consumers have accumulated about a product, which in turn influences their evaluations and purchase decisions (Alba and Hutchinson 1987, Bettman and Park 1980).

2.3 Hypotheses development

Country of Origin

In the imported cosmetics industry, the country of origin is the image or stereotype that consumers form based on the culture, politics, economy, and history of the producing country, which strongly influences their evaluation of product quality, reliability, and value (Camacho et al. 2021, Nagashima 1977, Samiee 1994). Studies have shown that the country of origin not only shapes brand image but also positively influences perceptions of product quality through the halo effect (Chen 2004, Diamantopoulos et al. 2011, Gao and Knight 2007, Hamzaoui-Essoussi et al. 2011). In particular, Korean cosmetics are highly regarded by Vietnamese consumers for their quality and credibility, with the image of "beautiful skin like Korean women" serving as a typical illustration (Costa et al. 2016). Furthermore, Korean cosmetics are positioned as high-quality yet reasonably priced products (Nebenzahl and Jaffe 2014, Simonson and Drolet 2004). The country of origin also acts as an anchor point that helps consumers compare and make purchasing decisions. In advertising, Korean cosmetic brands actively leverage the country-of-origin cue by incorporating imagery, language, and cultural symbols into marketing campaigns to strengthen trust in product quality and authenticity (Eagly and Chaiken 1993, Leclerc et al. 1994, Aichner 2014). The impact of the country of origin is even greater when consumer involvement in advertising is low (Maheswaran 1994). Based on this foundation, the study proposes four hypotheses:

H1: Country of origin has a positive (+) effect on brand image.

H2: Country of origin has a positive (+) effect on product quality.

H3: Country of origin has a positive (+) effect on product price.

H4: Country of origin has a positive (+) effect on advertising.

Brand Image

Brand image is understood as the positive associations and perceptions formed in consumers' minds based on brand experiences such as logos, packaging, and communication (Keller 2001). In the context of Korean cosmetics, brand image refers to the extent to which consumers recognize, trust, and perceive the brand as suitable for their personal needs. A strong brand image significantly influences purchase intention, enhancing brand recall and emotional attachment in the realm of K-Beauty (Buil et al. 2013, Romadon 2014, Mohamed et al. 2019, Prabowo et al. 2019). Specifically, a positive brand image not only increases perceptions of product quality but also effectively drives purchasing behavior, especially as Vietnamese consumers increasingly prioritize Korean cosmetics for their reputation and style (Chung et al. 2016, Tjoe and Kim 2016). Furthermore, brand image plays a crucial role in shaping positive attitudes by building trust, favorability, and brand loyalty (Aaker 1991, Keller 2003, Laroche et al. 1996, Suh and Youjae 2006). Therefore, the study proposes the following two hypotheses:

H5: Brand image has a positive (+) influence on the intention to purchase Korean cosmetics.

H6: Brand image positively (+) affects consumer attitudes.

Product Quality

Product quality is defined as the ability to meet or exceed consumer expectations, reflected through performance, durability, appearance, and personal suitability (Chinomona et al. 2013, Girsang et al. 2020, Kotler and Keller 2016). In the Korean cosmetics industry, product quality is not only determined by natural ingredients and dermatological safety but also by its ability to evoke positive emotions, trust, and favorability from consumers (Chae et al. 2001, Hong et al. 2007, Widayat et al. 2023). Previous studies have shown that high perceived quality increases purchase intention, as consumers tend to prefer products they perceive as reliable and safe (Lee and Tai 2009, Saleem et al. 2015). Although Cayaban et al. (2023) found that in the context of Hallyu, emotional factors may overshadow quality evaluations, practical evidence suggests that trust in product quality remains a fundamental factor in sustaining long-term purchasing behavior. Moreover, product quality plays a critical role in shaping positive attitudes toward the brand, especially when consumers perceive the product as reliable and aligned with their personal needs (Anderson et al. 1994, Fornell 1992, Garg and Joshi 2018). Based on the above analysis, the following hypotheses are proposed:

H7: Product quality has a positive (+) effect on the intention to purchase Korean cosmetics.

H8: Product quality has a positive (+) effect on consumer attitudes.

Product Price

Product price is defined as the amount a customer is willing to pay in exchange for the expected value of a product (Cayaban et al. 2023, Kotler and Armstrong 2014). In the cosmetics industry, price not only reflects affordability but also symbolizes perceived product quality and prestige. Latha (2017) reported that among various factors influencing the intention to purchase cosmetics, price was the most dominant, accounting for 63.95% of consumer decision-making. Similarly, studies in Vietnam show that price is a crucial determinant of purchase intention for Korean cosmetics. A reasonable price enhances purchase intention, as consumers perceive the product as offering good value for money (Summers et al. 2006, Kumar et al. 2014). Furthermore, according to the Theory of Reasoned Action by Ajzen and Fishbein (1980), a positive evaluation of price contributes to the formation of a favorable attitude toward the product. Empirical studies in the cosmetics sector have shown that when price is perceived as reasonable relative to the benefits received, it not only fosters purchase intention but also generates a strong positive attitude (Kim 2010, Kim and Rous 2012). Based on this analysis, the following hypotheses are proposed:

H9: Product price has a positive (+) effect on the intention to purchase Korean cosmetics.

H10: Product price has a positive (+) effect on consumer attitude.

Advertising

Advertising is defined as a paid message intended to convey information or persuade customers, thereby influencing consumer perception and behavior (Belch and Belch 2018, Kotler and Armstrong 2014, Zikmund and D'Amico 1999). In the Korean cosmetics industry, advertising not only provides product-related information but also evokes emotions and builds brand trust, especially through emotional advertising, celebrity endorsements, and the use of key opinion leaders (KOLs) (Asgari and Hosseini 2015, Nurunnisha et al. 2021). Empirical studies by Belch and Belch (2013) and Moriarty et al. (2014) found that effective advertising captures attention, creates a positive impression, and enhances purchase intentions. According to the Cultivation Theory and the Theory of Reasoned Action, repeated exposure to consistent advertising can shape positive attitudes toward a brand (Ajzen and Fishbein 1975). This is particularly relevant in the cosmetics industry, where emotional factors play a central role in consumer decision-making (Kim et al. 2012, Singh and Banerjee 2018). Therefore, the study proposes the following hypotheses:

H11: Advertising has a positive (+) effect on the intention to purchase Korean cosmetics.

H12: Advertising has a positive (+) effect on consumer attitude.

The Mediating Role of Attitude

In the context of Korean cosmetics consumption, attitude is identified as a key mediating factor linking brand image, product quality, price, and advertising to purchase intention. According to the Theory of Planned Behavior (Ajzen 1991) and the Theory of Reasoned Action (Ajzen and Fishbein 1977), positive attitudes formed from favorable evaluations significantly drive consumer behavior. Several studies have shown that attitude can serve as a full or partial mediator, particularly in the cosmetics sector, where emotional influences amplified by Hallyu are substantial (Garg and Joshi 2018, Islam and Hani 2021, Widayat et al. 2023). Indirect experiences through advertising, brand imagery, or perceived quality can also strengthen attitudes, even in the absence of direct product usage (Mazodier and Merunka 2014). A positive attitude not only increases purchase intention but also encourages brand advocacy and loyalty (Ajitha and Sivakumar 2017, Shimul et al. 2022, Widayat et al. 2023). Therefore, this study proposes the following hypotheses:

H13a: Attitude mediates the relationship between brand image and the intention to purchase Korean cosmetics.

H13b: Attitude mediates the relationship between product quality and the intention to purchase Korean cosmetics.

H13c: Attitude mediates the relationship between product price and the intention to purchase Korean cosmetics.

H13d: Attitude mediates the relationship between advertising and the intention to purchase Korean cosmetics.

H14: Attitude has a positive (+) effect on the intention to purchase Korean cosmetics.

Purchase intention

Purchase intention is considered a strong predictor of actual consumer purchasing behavior (Kim et al. 2008, Rachbini 2018). According to Ajzen (1991) and Keller (2001), before making a purchasing decision, consumers form intentions based on their perceptions, attitudes, and interest in the product. They argue that purchase interest is the transitional stage that directly leads to purchase decisions. Similarly, purchase interest is positively and significantly associated with actual purchasing behavior (Jonsson et al. 2019). In the context of Vietnam, Nguyen Xuan Truong (2018) also confirmed that purchase intention plays a critical role in consumer behavior toward Korean cosmetics. Therefore, this study proposes the following hypothesis:

H15: Purchase intention has a positive (+) effect on the decision to purchase Korean cosmetic products.

The proposed research model is shown in Figure 1.

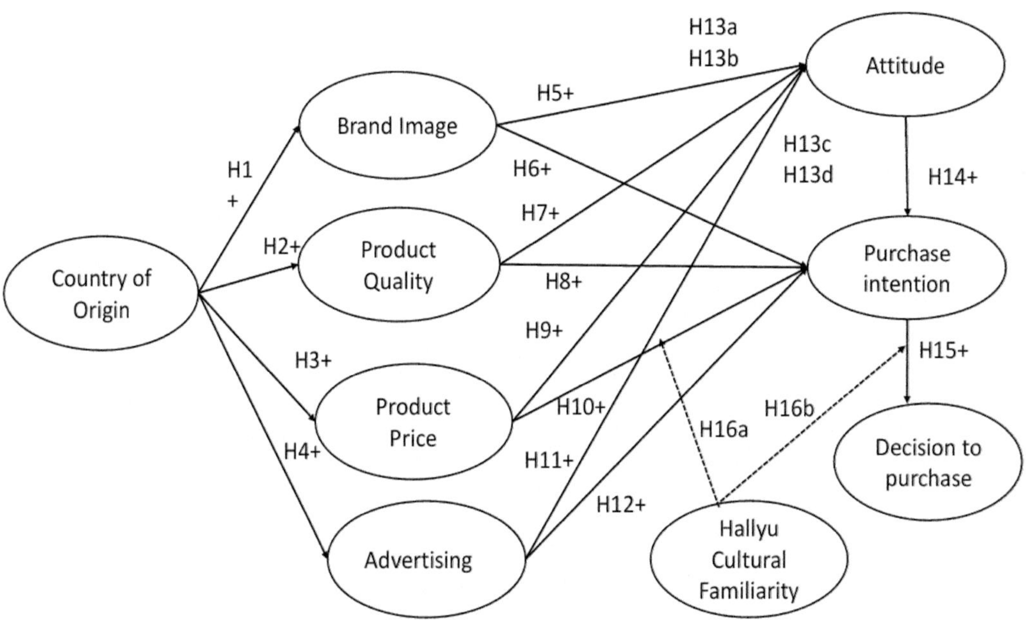

Figure 1: Research Model.

The Moderating Role of Hallyu Cultural Familiarity

Exposure to Korean cultural products such as K-dramas and K-pop enables consumers to develop positive emotions and a sense of attachment to products originating in Korea (Kim and Bae 2017). Particularly in the cosmetics sector, admiration for Korean idols has driven imitation of beauty styles and consumption habits (Lee and Nornes 2015), thereby increasing demand for Korean cosmetics. While price is commonly seen as a major factor influencing purchasing decisions (Kumar et al. 2014, Lichtenstein et al. 1993), Cayaban et al. (2023) have indicated that in contexts where consumers are strongly influenced by Hallyu, the impact of price tends to diminish. Consumers are more likely to prioritize emotional factors, and cultural attachment becomes a primary driver of consumption. Moreover, familiarity with Hallyu may also strengthen the link between purchase intention and actual purchase decision, as consumers with positive attitudes are more capable of overcoming economic or informational barriers. Based on this, the study proposes the following hypotheses:

H16a: Hallyu familiarity moderates the relationship between product price and the intention to purchase Korean cosmetics.

H16b: Hallyu familiarity moderates the relationship between purchase intention and the decision to purchase Korean cosmetics.

Based on the research objectives, literature review, and hypothesis development, the research model can be illustrated as follows:

3. Methods

3.1 Research Design

This study was conducted in two phases: qualitative and quantitative. The qualitative phase involved an extensive literature review on the Korean Wave (Hallyu) and consumer behavior toward Korean cosmetics, which informed the development of a context-specific theoretical model and the adaptation of measurement scales from prior research. These scales were refined with input from three experts—a Ph.D. scholar based in South Korea, a Vietnamese Ph.D. holder, and an industry entrepreneur—ensuring content validity. The questionnaire utilized a five-point Likert scale to maintain measurement reliability. In the quantitative phase, a pilot survey with 50 respondents was first conducted, followed by the main survey, which collected 398 valid responses through both online and in-person methods. Data analysis began with reliability assessment using SPSS 20, after which Partial Least Squares Structural Equation Modeling (PLS-SEM) was applied via SMARTPLS 4.0 to test the proposed model. Measurement reliability and convergent validity were confirmed through Cronbach's Alpha and Composite Reliability values exceeding 0.7, Average Variance Extracted (AVE) above 0.5, and Heterotrait-Monotrait Ratio (HTMT) below 0.9. Finally, the structural model was evaluated using PLS-SEM, with hypothesis testing based on a significance threshold of $p < 0.05$.

3.2 Measurement Scales

Measurement items are presented in Table 1.

Table 1: Measurement Scales.

Factor	Code	Scale Item	Source
Country of Origin (QG)	QG1	When buying imported products, the country of origin is information I consider.	Camacho et al. (2021)
	QG2	I search for information about the country of origin to choose the best imported products on the market.	
	QG3	I understand that the country of origin determines the quality of imported products.	
	QG4	When I buy expensive imported products, I always find out which country the product is made in.	
	QG5	If I have little experience with a type of imported product, I will look for information about the country of origin to help me make a better decision.	
Brand Image (HA)	HA1	When buying Korean cosmetics, I know which brands suit my needs.	Asgari and Hosseini (2015)
	HA2	I feel more secure when buying Korean cosmetics from familiar brands.	
	HA3	I know and am familiar with many Korean cosmetic brands.	
	HA4	Korean cosmetic brands have a positive image in my mind.	
	HA5	In the future, I will prioritize buying Korean cosmetic brands over brands from other countries.	
Product Quality (CL)	CL1	I choose to buy Korean products because of their superior quality.	(Cayaban et al. 2023)
	CL2	I choose to buy Korean products because of the practical benefits they offer.	
	CL3	I choose to buy Korean products because they suit my preferences and tastes.	
	CL4	I choose to buy Korean products because of their convenience and high effectiveness.	
Product Price (GSP)	GSP1	I buy Korean cosmetics because they are easy to purchase.	(Cayaban et al. 2023)
	GSP2	I buy Korean cosmetics because they offer value for the money spent.	
	GSP3	I buy Korean cosmetics because they are truly worth the price	
	GSP4	I buy Korean cosmetics because the price fits my budget.	
Advertising (QC)	QC1	The more Korean ads I watch, the more I like to buy Korean cosmetics.	Asgari and Hosseini (2015)
	QC2	I think Korean cosmetic advertisements influence my purchase decisions.	
	QC3	I like to buy Korean cosmetic brands that I know through advertisements.	
	QC4	I often remember the messages conveyed in advertisements when I see cosmetic products.	
	QC5	I often watch TV ads or pay attention to advertisements around me.	
	QC6	Korean cosmetic advertisements are effective enough to attract customers.	
Hallyu Cultural Familiarity (QT)	QT1	I am familiar with Hallyu culture.	Sun and Jun (2022)
	QT2	I have experience with Hallyu culture.	
	QT3	I have knowledge about Hallyu culture.	
Attitude (TD)	TD1	I like cosmetics from Korea.	Cayaban et al. (2023)
	TD2	I feel comfortable using Korean cosmetics.	
	TD3	I enjoy using Korean cosmetics.	
	TD4	I like Korean culture in general.	
Purchase Intention (YD)	YD1	I support the import of cosmetics from Korea.	Nguyen et al. (2023)
	YD2	Whenever possible, I will buy cosmetics from Korea.	
	YD3	I will buy Korean cosmetics if they are available.	
	YD4	I will buy Korean cosmetics if I have a need for a specific product.	
	YD5	I will share with family and friends the good qualities of Korean cosmetics.	
	YD6	I will recommend Korean cosmetics to others.	
Purchase Decision (QD)	QD1	I feel satisfied with my decision to buy Korean cosmetics.	Hanaysha (2018)
	QD2	I will actively recommend Korean cosmetics to others.	
	QD3	I regularly purchase cosmetics from Korean brands.	
	QD4	I intend to repurchase from these Korean cosmetic brands in the future.	
	QD5	Overall, I am satisfied with purchasing Korean cosmetics.	

3.3 Sampling Method

This study employed a non-probability sampling method using the convenience sampling technique to save time and costs while also facilitating easy access to the target population—Vietnamese consumers who have purchased or are currently purchasing Korean cosmetics. To ensure that the right respondents were selected, a screening question was included at the beginning of the questionnaire. Following Tabachnick and Fidell (2007), the minimum sample size for a multiple regression model is calculated using the formula: N = 8 * number of independent variables + 50. Based on this formula, the minimum required sample size was determined to be 90 observations. According to Comrey and Lee (1992) and Hair and Alamer (2022), the larger the sample size, the greater the reliability: a sample size of 50 is considered very poor, 100 is poor, 200 is fair, 300 is good, 500 is very good, and 1,000 is excellent. Therefore, to ensure representativeness and objectivity in the study, the estimated sample size used was 452 respondents, including 152 collected online and 300 through direct surveys. The data were gathered using Google Forms, targeting individuals aged 18 to 50 years old, primarily residing in major cities in Vietnam, with a focus on Ho Chi Minh City. After eliminating invalid responses, a total of 398 valid observations were retained for model testing.

4. Results

4.1 Descriptive Statistics

The survey results show that among the 398 participants, females were the majority (64.8%), mainly concentrated in the age groups of 18–25 (65.8%) and 26–30 (17.1%), reflecting a young consumer segment that is highly receptive to Korean cosmetics. In terms of occupation, students and recent graduates accounted for 51.3%, while office workers made up 28.1%. Regarding income, 45.5% earned between VND 4 to 7.5 million, which aligns with the mid-range cosmetics segment. As for shopping frequency, 63.8% reported purchasing cosmetics 1–2 times per month, indicating a relatively high consumption rate of Korean cosmetics among the survey group. Respondent demographics are summarized in Table 2.

Table 2: Descriptive statistics (n = 398).

Sample Characteristics (n = 398)		Frequency	Percentage
Gender	Male	140	35.2%
	Female	258	64.8%
Age	18–25 years old	262	65.8%
	26–30 years old	68	17.1%
	31–40 years old	39	9.8%
	41–50 years old	17	4.3%
	Over 50 years old	12	3.0%
Occupation	Student/Recent Graduate	204	51.3%
	Office Staff	112	28.1%
	Civil Servant/Worker	38	9.5%
	Business Owner	21	5.3%
	Others	23	5.8%
Income	From 4 to under 7.5 million VND	181	45.5%
	From 7.5 to under 15 million VND	105	26.4%
	From 15 to 30 million VND	90	22.6%
	Over 30 million VND	22	5.5%

4.2 Measurement Model Assessment

The Cronbach's Alpha values for the constructs ranged from 0.786 to 0.869, while the Composite Reliability (CR) values fell between 0.862 and 0.904. Additionally, the Composite Reliability (rho_a) values ranged from a minimum of 0.788 to a maximum of 0.870, indicating strong internal consistency for each scale. All these indices exceed the recommended threshold of 0.70 suggested by HaHair and Alamer (2022), confirming high internal consistency and reliability of the measurement scales. Furthermore, the Average Variance Extracted (AVE) for each construct was above 0.50, ranging from 0.526 (Product Quality – QC) to 0.732 (Perceived Value – QT). This implies that each latent variable explains at least 50% of the variance in its observed indicators (Hock and Ringle 2010). Reliability and convergent validity results are shown in Table 3.

Table 3: Results of Reliability and Convergent Validity Assessment.

Scale	Cronbach's alpha	Reliability coefficient (rho_a)	Composite Reliability (rho_c)	Average Variance Extracted (AVE)
Product Quality (CL)	0.837	0.838	0.891	0.672
Product Price (GSP)	0.786	0.788	0.862	0.609
Brand Image (HA)	0.826	0.826	0.878	0.590
Advertising (QC)	0.820	0.823	0.869	0.526
Purchase Decision (QD)	0.868	0.868	0.904	0.654
Country of Origin (QG)	0.836	0.839	0.884	0.603
Hallyu Cultural Familiarity (QT)	0.817	0.821	0.891	0.732
Attitude (TD)	0.798	0.812	0.868	0.624
Purchase Intention (YD)	0.869	0.870	0.902	0.604

The data extracted from Table 4 show that the HTMT values of all construct pairs in the model are below the threshold of 0.9, with the highest value being 0.869. Therefore, all measurement scales demonstrate discriminant validity (Henseler et al. 2015).

Table 4: Heterotrait - Monotrait (HTMT) criterion correlation.

Unit	CL	GSP	HA	QC	QD	QG	QT	TD	YD	QT x YD
Product Quality (CL)										
Product Price (GSP)	0.669									
Brand Image (HA)	0.621	0.679								
Advertising (QC)	0.625	0.592	0.578							
Purchase Decision (QD)	0.662	0.756	0.768	0.652						
Country of Origin (QG)	0.673	0.556	0.785	0.582	0.625					
Hallyu Cultural Familiarity (QT)	0.523	0.591	0.661	0.517	0.730	0.522				
Attitude (TD)	0.647	0.840	0.731	0.652	0.833	0.580	0.671			
Purchase Intention (YD)	0.690	0.849	0.750	0.644	0.869	0.640	0.735	0.846		
QT x YD	0.328	0.550	0.456	0.239	0.549	0.321	0.466	0.527	0.611	
QT x GSP	0.305	0.633	0.384	0.230	0.470	0.298	0.346	0.493	0.519	0.800

4.3 Structural Model Assessment

The bootstrapping results (with 5,000 resamples) for the relationships among constructs in the proposed research model indicate that hypotheses H1, H2, H3, H4, H5, H6, H7, H8, H9, H10, H11, H12, H13a, H13c, H13d, H14, H15, H16a, and H16b are statistically significant at the 0.05 level. Only hypothesis H13b was rejected due to a p-value greater than 0.05.

Based on the results in Table 6, the model explains purchase intention (R^2adj = 68.6%), purchase decision (60.9%), and attitude (55.7%) well. Factors such as brand image (42.2%), product quality (30.9%), and product price (19.8%) have lower explanatory power, indicating that the influence of Hallyu is stronger than product quality in shaping consumer behavior toward Korean cosmetics. The model results are shown in Figure 2.

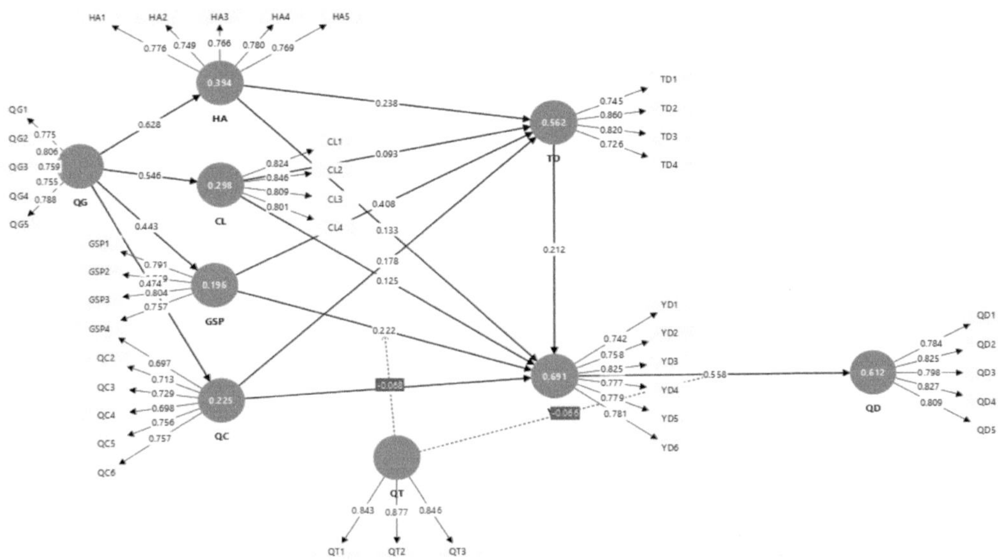

Figure 2: Hypothesis Testing Model Results.

4.4 Testing Differences in Qualitative Variables on Korean Cosmetic Purchasing Behavior

The Independent-Samples T-Test results indicate a statistically significant difference in the purchasing behavior of Korean cosmetics between male and female consumers (Levene's Test Sig. = 0.00 < 0.05; t-test Sig. = 0.00 < 0.05), highlighting gender as a significant influencing factor. For age, the ANOVA analysis confirms the assumption of homogeneity of variances (Levene's Test Sig. = 0.130 > 0.05) and reveals a statistically significant difference among age groups (ANOVA Sig. = 0.030 < 0.05), suggesting that age meaningfully impacts consumer behavior in the context of Korean cosmetics.

In contrast, ANOVA tests conducted for income and occupation show homogeneity of variances (income: Levene's Test Sig. = 0.659 > 0.05; occupation: Levene's Test Sig. = 0.163 > 0.05) but fail to detect any statistically significant differences in purchasing behavior across different income levels (ANOVA Sig. = 0.052 > 0.05) or occupational categories (ANOVA Sig. = 0.289 > 0.05). These results indicate that while gender and age significantly influence Korean cosmetic purchasing behavior, income and occupation do not serve as meaningful differentiators in this study.

Table 5: Results of Hypotheses Testing.

Hypotheses	Relationship	β	Sample mean	Standard deviation	T statistic	P-value	VIF	Conclusion
Direct effects								
H1	QG -> HA	0.651	0.652	0.040	16.259	0.000	1.000	Accepted
H2	QG -> CL	0.557	0.559	0.042	13.272	0.000	1.000	Accepted
H3	QG -> GSP	0.447	0.448	0.052	8.615	0.000	1.000	Accepted
H4	QG -> QC	0.480	0.482	0.050	9.680	0.000	1.000	Accepted
H5	HA -> TD	0.133	0.133	0.048	2.743	0.006	1.919	Accepted
H6	HA -> YD	0.237	0.239	0.053	4.462	0.000	1.657	Accepted
H7	CL -> TD	0.124	0.125	0.045	2.760	0.006	1.754	Accepted
H8	CL -> YD	0.093	0.094	0.047	1.982	0.048	1.721	Accepted
H9	GSP -> TD	0.222	0.221	0.057	3.902	0.000	2.470	Accepted
H10	GSP -> YD	0.409	0.407	0.058	7.046	0.000	1.704	Accepted
H11	QC -> TD	0.094	0.098	0.042	2.240	0.025	1.653	Accepted
H12	QC -> YD	0.178	0.179	0.047	3.755	0.000	1.546	Accepted
H14	TD -> YD	0.213	0.211	0.054	3.940	0.000	2.401	Accepted
H15	YD -> QD	0.558	0.557	0.046	12.219	0.000	1.999	Accepted
Indirect effects								
H13a	HA -> TD -> YD	0.050	0.050	0.016	3.169	0.002		Accepted
H13b	CL -> TD -> YD	0.020	0.019	0.011	1.852	0.064		Not Accepted
H13c	GSP -> TD -> YD	0.087	0.086	0.027	3.263	0.001		Accepted
H13d	QC -> TD -> YD	0.038	0.038	0.015	2.465	0.014		Accepted
Moderating effects								
16a	QT x GSP -> YD	–0.068	–0.068	0.026	2.661	0.008	1.505	Accepted
16b	QT x YD -> QD	–0.066	–0.066	0.029	2.323	0.020	1.493	Accepted

Table 6: Adjusted Coefficient of Determination (R^2adj) Results.

Unit	R^2	R^2 adjusted
Product Quality (CL)	0.311	0.309
Product Price (GSP)	0.200	0.198
Brand Image (HA)	0.423	0.422
Advertising (QC)	0.230	0.228
Purchase Decision (QD)	0.612	0.609
Country of Origin (QG)	0.562	0.557
Hallyu Cultural Familiarity (QT)	0.691	0.686

5. Discussion

Hypotheses H1, H2, H3, and H4 were supported, confirming that country of origin significantly influences brand image ($\beta = 0.651$; $T = 16.259$; $p < 0.001$), product quality ($\beta = 0.557$; $T = 13.272$; $p < 0.001$), price perception ($\beta = 0.447$; $T = 8.615$; $p < 0.001$), and advertising effectiveness ($\beta = 0.480$; $T = 9.680$; $p < 0.001$). The "Made in Korea" label creates an immediate positive impression, aligning with previous studies by Diamantopoulos et al. (2011), Gao and Knight (2007)

(Eagly and Chaiken 1993, Leclerc et al. 1994, Aichner 2014), as well as local research by Pham and Ton (2023), Nguyen (2018), Vu (2018).

Hypotheses H5, H7, H9, and H11 were also accepted, indicating that brand image ($\beta = 0.237$; $T = 4.462$; $p < 0.001$), product quality ($\beta = 0.093$; $T = 1.982$; $p = 0.048$), product price ($\beta = 0.409$; $T = 7.046$; $p < 0.001$), and advertising ($\beta = 0.178$; $T = 3.755$; $p < 0.001$) have positive effects on consumer attitudes. These findings are consistent with prior studies such as Kim (2010) (Buil et al. 2013, Kim and Rous 2012) and Romadon (2014). However, the impact of product quality on consumer attitude is relatively limited, as supported by H10. In the context of Korean cosmetics, quality is often considered a baseline expectation, shifting consumer focus toward emotional appeal and the K-Beauty trend. Compared to studies by Chae et al. (2001) (Hong et al. 2007, Widayat et al. 2023), the influence of quality appears significantly weaker.

Hypotheses H6, H8, H10, and H12 were supported, suggesting that brand image ($\beta = 0.133$; $T = 2.743$; $p = 0.006$), product quality ($\beta = 0.124$; $T = 2.760$; $p = 0.006$), price ($\beta = 0.222$; $T = 3.902$; $p < 0.001$), and advertising ($\beta = 0.094$; $T = 2.240$; $p = 0.025$) positively influence purchase intention. However, the impact of advertising primarily lies in capturing initial attention and is not strong enough to drive actual purchase behavior. This contrasts with findings by Asgari and Hosseini (2015). Notably, H6 yields different results compared to Cayaban et al. (2023), who argued that product quality is not a key driver of purchase intention due to the overwhelming influence of the Hallyu wave on consumer perceptions.

Hypotheses H13a, H13c, H13d, and H14 were accepted, indicating that attitude positively influences purchase intention ($\beta = 0.213$; $T = 3.940$; $p < 0.001$) and that purchase intention strongly predicts purchase decision ($\beta = 0.558$; $T = 12.219$; $p < 0.001$). However, H13b was not supported ($p = 0.064 > 0.05$), suggesting that attitude does not significantly mediate the relationship between product quality and purchase intention. This result aligns with Cayaban et al. (2023), who observed that consumers prioritize emotional connection and brand resonance over product quality, particularly under the influence of Hallyu.

Hypothesis H15 was confirmed, indicating that purchase intention strongly drives actual purchase behavior ($\beta = 0.558$; $T = 12.219$; $p < 0.001$), which emphasizes that once a clear intention is formed, only minor external stimuli—such as promotions or product availability—are needed to trigger the purchase.

Finally, both H16a and H16b were supported, demonstrating that Hallyu familiarity moderates the relationship between price and purchase intention ($\beta = -0.068$; $p = 0.008$) and between purchase intention and actual purchase behavior ($\beta = -0.066$; $p = 0.020$). However, the effect is negative, implying that higher familiarity with Hallyu weakens the influence of price on purchase intention and increases consumer expectations. These consumers are less susceptible to cultural "halo effects" and more critical of marketing tactics. If the product is highly priced but lacks superior performance, they are likely to refrain from purchasing and instead exhibit disappointment, skepticism, and reduced buying behavior.

6. Conclusion

This study evaluates the influence of Hallyu on consumer behavior regarding Korean cosmetics in Ho Chi Minh City, using a PLS-SEM model based on 398 valid survey responses. The model examined five independent variables—country of origin, product quality, brand image, product price, and advertising—while incorporating attitude as a mediating variable and familiarity with Hallyu as a moderating variable. The results indicate that the country of origin is the most significant determinant, exerting strong positive effects on brand image ($\beta = 0.651$), product quality ($\beta = 0.557$), product price ($\beta = 0.447$), and advertising ($\beta = 0.480$). Furthermore, product price, brand image, and product quality positively influenced consumer attitudes and purchase intentions. Attitude played a mediating role, linking perception with behavior, as reflected in the significant paths from attitude to purchase intention ($\beta = 0.213$) and from intention to purchase decision ($\beta = 0.558$). Notably, familiarity with Hallyu exhibited a negative moderating effect on the relationship between product price and purchase intention, as well as between intention and purchase decision. This suggests that consumers with a higher level of emotional attachment to Korean culture may not finalize purchases as easily, potentially due to heightened expectations or selective behavior. Demographic analysis further revealed that gender and age significantly influenced consumer behavior toward Korean cosmetics, whereas income and occupation did not present statistically significant differences.

7. Managerial Implications

First, the country of origin factor ($\beta = 0.651$) was identified as the most influential determinant in the purchase behavior of Korean cosmetics. The related variables had mean scores ranging from 3.76 to 3.91, indicating that Vietnamese consumers place significant importance on product origin. Therefore, businesses should emphasize transparent "Made in Korea" labeling, incorporate Korean cultural symbols into packaging and retail environments, and enhance authentic brand experiences through well-trained consultants, pop-up experience stores, and collaborations with KOLs to strengthen consumer trust in Korea-linked quality.

Second, product price ($\beta = 0.409$) had a substantial impact on purchase decisions, with variable means ranging from 3.81 to 4.08. This suggests that consumers highly value the perceived balance between price and product value. Brands should implement competitive pricing strategies, offer clearly segmented product lines, and promote mini-sized products and promotional campaigns. Simultaneously, they should optimize a diverse distribution network, including both online and offline channels, to increase convenience and enhance perceived value for money.

Third, brand image ($\beta = 0.237$) positively influenced consumer perceptions and attitudes, with variable means between 3.66 and 3.9. Companies are advised to invest in building a professional and consistent brand identity that aligns with their brand story. Maintaining product quality, continuously updating technology, and applying integrated and consistent communication strategies are essential for long-term brand equity.

Fourth, familiarity with Hallyu played a moderating role in consumer behavior, with variable means ranging from 3.71 to 3.97. Firms should strategically leverage Hallyu by organizing cultural experience events, incorporating K-Beauty trends into marketing content, and activating Fandom UGC (user-generated content) campaigns. Initiatives such as "K-Culture Pop-Up" events led by Korean students can help transform consumer familiarity into repeated purchases and positive word-of-mouth within communities.

Fifth, advertising ($\beta = 0.178$; direct impact on purchase intention $\beta = 0.094$) mainly influenced attitude formation, with variable means ranging from 3.5 to 3.8. Businesses should design multisensory advertising campaigns, refresh content regularly in line with trending topics, increase interactive experiences, and incorporate scientific evidence into messaging. Innovative strategies such as social media challenges can enhance persuasive power and effectively convert interest into purchasing behavior.

Sixth, product quality (impact on purchase intention $\beta = 0.124$; on attitude $\beta = 0.093$) yielded mean scores ranging from 3.8 to 4.03. This indicates that consumers appreciate effective product experiences but show varied expectations. Businesses should prioritize product personalization—through tools like co-creation labs and AI-powered applications such as StyleSync AI and Skindeai for skin analysis and tone matching—to ensure transparency in quality information (e.g. blockchain-based traceability) and optimize product efficacy and convenience to foster consumer trust and drive sustainable repeat purchases.

8. Lessons for Vietnam

To cope with the appeal of Hallyu and shifts in cosmetic purchasing behavior, Vietnamese enterprises need to focus on enhancing product origin transparency through technologies such as QR codes and blockchain. It is essential to disclose product quality assessments tailored to Vietnamese skin and climate conditions. At the same time, businesses should develop reasonable pricing strategies, ensure authentic user experiences, and obtain credible certifications. Initiating a "Vietnam Wave" as a cultural communication campaign could improve competitiveness against Korean cosmetics. Companies are advised to build brand images associated with Vietnamese culture by incorporating national symbols such as lotus flowers, Đông Sơn drums, and cultural landmarks, and use slogans like "Vietnamese Goods, High Quality." Moreover, enterprises should adopt omnichannel models combining online and offline purchasing experiences, including pop-up stores. The application of AI for personalized products, AR for enhanced product trials, gender-diverse product lines, and the expansion into specialized cosmeceutical segments are also recommended.

Vietnamese consumers, on the other hand, should become more conscious of evaluating products based on origin and quality certification instead of relying solely on advertisements or the influence of Hallyu idols. Proactively researching, trying out products, and carefully weighing real value against price can help consumers make more informed decisions. Prioritizing the use of high-quality domestic cosmetics not only protects individual interests but also strengthens the Vietnamese cosmetic industry in its competition with imported products, particularly from South Korea.

9. Limitations and Future Research Directions

This study has several limitations. First, the sample was primarily drawn from Ho Chi Minh City, which limits the generalizability of the findings to a national level. Second, the study did not incorporate psychological or emotional variables, nor did it account for the influence of emerging digital technologies that may affect consumer behavior. Third, the cross-sectional research design limits the ability to examine behavioral changes over time, particularly within the rapidly evolving context of Hallyu and digital marketing.

Future research should expand the sample to include respondents from various regions across Vietnam to enhance the study's representativeness. It is also recommended to include psychological constructs such as emotional attachment to the Korean Wave (Hallyu) to better understand its influence on consumer attitudes and behavior. Furthermore, future studies should investigate the emotional and psychological impacts of Hallyu on consumer behavior across social media platforms such as TikTok, Instagram, and Facebook. Finally, researchers should assess the effects of omnichannel shopping experiences and trust in digital platforms to gain deeper insights into consumer behavior in today's digital environment.

References

Aaker, D. (1991). Brand equity. *La gestione del valore della marca*, 347: 356.

Aichner, T. (2014). Country-of-origin marketing: A list of typical strategies with examples. *Journal of Brand Management*, 21: 81–93.

Ajitha, S., and Sivakumar, V. (2017). Understanding the effect of personal and social value on attitude and usage behavior of luxury cosmetic brands. *Journal of Retailing and Consumer Services*, 39: 103–113.

Ajzen, I. (1991). The theory of planned behavior. *Organizational behavior and human decision processes*, 50(2): 179–211.

Ajzen, I., and Fishbein, M. (1975). A Bayesian analysis of attribution processes. *Psychological Bulletin*, 82(2): 261–277. https://doi.org/10.1037/h0076477.

Ajzen, I., and Fishbein, M. (1977). Attitude-behavior relations: A theoretical analysis and review of empirical research. *Psychological bulletin*, 84(5): 888.

Alba, J. W., and Hutchinson, J. W. (1987). Dimensions of consumer expertise. *Journal of consumer research*, 411–454.

Anderson, J. C., Rungtusanatham, M., and Schroeder, R. G. (1994). A theory of quality management underlying the Deming management method. *Academy of management Review*, 19(3): 472–509.

Asgari, O., and Hosseini, M. S. (2015). Exploring the antecedents affecting attitude, satisfaction, and loyalty towards Korean cosmetic brands. *Journal of Distribution Science*, 13(6): 45–70.

Belch, G. E., and Belch, M. A. (2018). *Advertising and promotion: An integrated marketing communications perspective*. mcgraw-hill.

Belch, M. A., and Belch, G. E. (2013). The future of creativity in advertising. *Journal of Promotion Management*, 19(4): 395–399.

Bettman, J. R., and Park, C. W. (1980). Effects of prior knowledge and experience and phase of the choice process on consumer decision processes: A protocol analysis. *Journal of Consumer Research*, 7(3): 234–248.

Boonmee, J. (2015). Study of Thai women purchase decision of Korea cosmetic in Bangkok area.

Buil, I., Martínez, E., and De Chernatony, L. (2013). The influence of brand equity on consumer responses. *Journal of Consumer marketing*, 30(1): 62–74.

Camacho, L. J., Ramírez-Correa, P. E., and Salazar-Concha, C. (2021). Consumer ethnocentrism and country of origin: Effects on online consumer purchase behavior in times of a pandemic. *Sustainability*, 14(1): 348.

Cayaban, C. J. G., Prasetyo, Y. T., Persada, S. F., Mariñas, K. A., Nadlifatin, R., Borres, R. D. et al. (2023). Factors affecting filipino consumer behavior with Korean products and services: an extended theory of planned behavior approach. *Sustainability*, 15(5): 4477.

Chae, Y. M., Lee, J. H., Ho, S. H., Kim, H. J., Jun, K. H., and Won, J. U. (2001). Patient satisfaction with telemedicine in home health services for the elderly. *International Journal of Medical Informatics*, 61(2–3): 167–173.

Chen, H.-L. (2004). Testing the role of country of origin in consumer adoption of new products. *International Advances in Economic Research*, 10(3): 245–246.

Chinomona, R., Okoumba, L., and Pooe, D. (2013). The impact of product quality on perceived value, trust and students' intention to purchase electronic gadgets. *Mediterranean Journal of Social Sciences*, 4(14): 463–472.

Chung, K.-H., Yu, J.-E., Kim, W., and Shin, J.-I. (2016). The antecedent and consequences of brand image in a low-priced cosmetic brand of South Korea: The moderating effect of gender. *International Journal of u-and e-Service Science and Technology*, 9(2).

Comrey, A., and Lee, H. (1992). Interpretation and application of factor analytic results. *Comrey AL, Lee HB. A first course in factor analysis*, 2: 1992.

Costa, C., Carneiro, J., and Goldszmidt, R. (2016). A contingent approach to country-of-origin effects on foreign products evaluation: Interaction of facets of country image with product classes. *International Business Review*, 25(5): 1066–1075.

Diamantopoulos, A., Schlegelmilch, B., and Palihawadana, D. (2011). The relationship between country-of-origin image and brand image as drivers of purchase intentions: A test of alternative perspectives. *International Marketing Review*, 28(5): 508–524.

Eagly, A. H., and Chaiken, S. (1993). *The Psychology of Attitudes*. Harcourt brace Jovanovich college publishers.

Fornell, C. (1992). A national customer satisfaction barometer: The Swedish experience. *Journal of Marketing*, 56(1): 6–21.

Grand View Research. (2023). *GVR Report coverCosmetics Market Size, Share & Trends Report Cosmetics Market Size, Share & Trends Analysis Report By Product (Skin Care, Hair Care), By End-user (Men, Women), By Distribution Channel (Offline, Online), By Region (North America, Europe), And Segment Forecasts, 2024 - 2030.* Grand View Research. https://www.grandviewresearch.com/industry-analysis/cosmetics-market

Gammon, T. (2022). I'd Have Divorced My Husband If Not for Korean Dramas–Vietnamese Women's Consumption of Television Romance and Melancholia. *Studies in Gender and Sexuality*, 23(3): 207–223.

Gammon, T., and Phan, A. N. Q. (2024). Too black to be The Little Mermaid? Backlash against Disney's 2023 The Little Mermaid–continuity of racism, white skin preference and hate content in Vietnam. In *Feminist Media Studies*. Routledge. https://doi.org/10.1080/14680777.2024.2344102.

Gao, H., and Knight, J. (2007). Pioneering advantage and product-country image: evidence from an exploratory study in China. *Journal of Marketing Management*, 23(3–4): 367–385.

Garg, P., and Joshi, R. (2018). Purchase intention of "Halal" brands in India: the mediating effect of attitude. *Journal of Islamic Marketing*, 9(3): 683–694.

Girsang, N. M., Rini, E. S., and Gultom, P. (2020). The effect of brand image and product quality on re-purchase intention with customer satisfaction as intervening variables in consumers of skincare oriflame users–a study on students of north sumatra university, faculty of economics and business. *European Journal of Management and Marketing Studies*.

Hair, J., and Alamer, A. (2022). Partial Least Squares Structural Equation Modeling (PLS-SEM) in second language and education research: Guidelines using an applied example. *Research Methods in Applied Linguistics*, 1(3): 100027.

Hamzaoui-Essoussi, L., Merunka, D., and Bartikowski, B. (2011). Brand origin and country of manufacture influences on brand equity and the moderating role of brand typicality. *Journal of Business Research*, 64(9): 973–978.

Hanaysha, J. R. (2018). An examination of the factors affecting consumer's purchase decision in the Malaysian retail market. *PSU Research Review*, 2(1): 7–23.

Hassali, M. A., Al-Tamimi, S., Dawood, O. T., Verma, A. K., and Saleem, F. (2015). Malaysian cosmetic market: Current and future prospects. *Pharmaceutical Regulatory Affairs*, 4(4): 1–3.

Heilbroner, R. L., Ajzen, I., Fishbein, M., and Thurow, L. C. (1980). *Understanding attitudes and predicting social behavior*. Prentice Hall.

Henseler, J., Ringle, C. M., and Sarstedt, M. (2015). A new criterion for assessing discriminant validity in variance-based structural equation modeling. *Journal of the Academy of Marketing Science*, 43(1): 115–135. https://doi.org/10.1007/s11747-014-0403-8.

Hien, N. N., Phuong, N. N., Tran, T. V., and Thang, L. (2020). The effect of country-of-origin image on purchase intention: The mediating role of brand image and brand evaluation. *Management Science Letters*, 10(6): 1205–1212.

Hock, M., and Ringle, C. M. (2010). Local strategic networks in the software industry: An empirical analysis of the value continuum. *International Journal of Knowledge Management Studies*, 4(2): 132–151.

Hong, J., Kim, J. M., Jung, W. K., Kim, S. H., Bae, W., Koo, H. C. et al. (2007). Prevalence and antibiotic resistance of Campylobacter spp. isolated from chicken meat, pork, and beef in Korea, from 2001 to 2006. *Journal of Food Protection*, 70(4): 860–866.

Hwang, H. (2014). The influence of the ecological contexts of teacher education on South Korean teacher educators' professional development. *Teaching and Teacher Education*, 43: 1–14.

Islam, M. M., and Hani, F. F. (2021). Hopes and reality: consumers' purchase intention towards whitening cream. *Future Business Journal*, 7: 1–12.

Jeong, J.-H. (2012). Ethnoscapes, mediascapes, and ideoscapes: Socio-cultural relations between South Korea and China. *Journal of international and area studies*, 77–95.

Jin, D. (2016). *New Korean Wave: Transnational cultural power in the age of social media*. University of Illinois press.

Jonsson, E., Källström, P., and Wallander, E. (2019). Why do we purchase from e-retailers?: An explanatory study of the e-loyalty antecedents impact on consumers purchase intentions on the e-commerce market. In.

Jaini, A., Quoquab, F., Mohammad, J., and Hussin, N. (2020). I buy green products, do you…? The moderating effect of eWOM on green purchase behavior in Malaysian cosmetics industry. *International Journal of Pharmaceutical and Healthcare Marketing*, 14(1): 89–112. https://doi.org/10.1108/IJPHM-02-2019-0017.

Ju, H. (2017). National television moves to the region and beyond: South Korean TV drama production with a new cultural act. *The Journal of International Communication*, 23(1): 94–114.

Keller, K. L. (1993). Conceptualizing, measuring, and managing customer-based brand equity. *Journal of Marketing*, 57(1): 1–22.

Keller, K. L. (2001). Building customer-based brand equity: A blueprint for creating strong brands.

Keller, K. L. (2003). Understanding brands, branding and brand equity. *Interactive Marketing*, *5*: 7–20.

Kim, D. J., Ferrin, D. L., and Rao, H. R. (2008). A trust-based consumer decision-making model in electronic commerce: The role of trust, perceived risk, and their antecedents. *Decision Support Systems*, 44(2): 544–564.

Kim, K.-D., and Bae, S.-J. (2017). Hallyu and the traditional cultural genes of Korea. *Kritika Kultura*, 1(29): 16.

Kim, M. (2010). Influence of leaming Stress, depreSSion and dySfunctional atitude On School performance among middle School boyS. *Unpublished master's thesis, Yonsei University, Seoul, Korea.*

Kim, Y. S., and Rous, J. J. (2012). House price convergence: Evidence from US state and metropolitan area panels. *Journal of Housing Economics*, 21(2): 169–186.

Kim, Y. T., Choi, B. Y., Lee, K. O., Kim, H., Chun, J. H., Kim, S. Y. et al. (2012). Overview of Korean community health survey. *Journal of the Korean Medical Association*, 55(1): 74–83.

Kotler, P., and Armstrong, G. (2014). Principles of Marketing Fifteenth Edition-Global Edition.

Kotler, P., and Keller, K. L. (2016). *A framework for marketing management*. Prentice Hall.

Kovács, P., and Lukovics, M. (2022). Factors influencing public acceptance of self-driving vehicles in a post-socialist environment: Statistical modelling in Hungary. *Regional Statistics*, 12(2): 149–176. https://doi.org/10.15196/RS120206

Kumar, A. H., John, S. F., and Senith, S. (2014). A study on factors influencing consumer buying behavior in cosmetic products. *International Journal of Scientific and Research Publications*, 4(9): 1–6.

Kurniawati, D. A., and Savitri, H. (2020). Awareness level analysis of Indonesian consumers toward halal products. *Journal of Islamic Marketing*, 11(2): 531–546. https://doi.org/10.1108/JIMA-10-2017-0104.

Laroche, M., Kim, C., and Zhou, L. (1996). Brand familiarity and confidence as determinants of purchase intention: An empirical test in a multiple brand context. *Journal of Business Research*, 37(2): 115–120.

Latha, D. (2017). The role of perceived occupational stress on organizational effectiveness. *IJAR*, 3(1): 783–789.

Leclerc, F., Schmitt, B. H., and Dubé, L. (1994). Foreign branding and its effects on product perceptions and attitudes. *Journal of Marketing Research*, 31(2): 263–270.

Lee, J. W., and Tai, S. W. (2009). Determinants of product quality perceptions and their application to marketing standardisation: The case of the automobile in Kazakhstan. *International Journal of Emerging Markets*, 4(2): 119–136.

Lee, S., and Nornes, A. M. (2015). *Hallyu 2.0: The Korean Wave in the age of social media*. University of Michigan Press.

Lichtenstein, D. R., Ridgway, N. M., & Netemeyer, R. G. (1993). Price perceptions and consumer shopping behavior: a field study. *Journal of Marketing Research*, 30(2): 234–245.

Maheswaran, D. (1994). Country of origin as a stereotype: Effects of consumer expertise and attribute strength on product evaluations. *Journal of Consumer research*, 21(2): 354–365.

Mazodier, M., and Merunka, D. (2014). Beyond brand attitude: Individual drivers of purchase for symbolic cobranded products. *Journal of Business Research*, 67(7): 1552–1558.
Mohamed, R. N., Mohamad, B., Borhan, H., Osman, I., and Kamaralzaman, S. (2019). The determinant factors of supply chain management on purchase intention of an international branded apparels status quo. *International Journal of Supply Chain Management*, 8(3): 677–684.
Moon, H.-J., and Park, S.-H. (2012). A study the relation between popular factors and likability of Hallyu and the national image. *Journal of Public Relations Research*, 16(4): 246–279.
Moriarty, S., Mitchell, N. D., Wells, W. D., Crawford, R., Brennan, L., and Spence-Stone, R. (2014). *Advertising: Principles and practice*. Pearson Australia.
Nagashima, A. (1977). A comparative "made in" product image survey among Japanese businessmen. *Journal of Marketing*, 41(3): 95–100.
Nebenzahl, I. D., and Jaffe, E. D. (2014). Estimating demand functions from the country-of-origin effect. pp. 159–178. *In: Product-Country Images*. Routledge.
Nguyen, N. H., Dao, T. K., Duong, T. T., Nguyen, T. T., Nguyen, V. K., and Dao, T. L. (2023). Role of consumer ethnocentrism on purchase intention toward foreign products: Evidence from data of Vietnamese consumers with Chinese products. *Heliyon*, 9(2).
Nguyen, X. T. (2018). The impact of Hallyu 4.0 and social media on Korean products purchase decision of generation C in Vietnam. *The Journal of Asian Finance, Economics and Business*, 5(3): 81–93.
Nguyen, X. T. (2019). Factors impacting on Korean consumer goods purchase decision of Vietnam's generation Z. *Journal of Distribution Science*, 17(10): 61–71.
Nurunnisha, G. A., Setiawan, D. W., Wijayanti, F., and Sinaga, O. (2021). The effect of product, distribution, and digital marketing toward female's purchase intention (case study: Bts album on Weverse shop). *Review of International Geographical Education Online*, 11(7): 3429–3440.
Pham, N. V. T., and Ton, H. N. N. (2023). Marketing Strategy for Korean Consumer Goods among Young Consumers in Ho Chi Minh City–Identifying Determinants of Purchase Decision. *Review of Integrative Business and Economics Research*, 12(3): 230–250.
Prabowo, P. K., Brahmasari, I. A., and Suryani, T. (2019). The effect of country of origin on brand image, perceived quality, and purchase intention of hydraulic excavator for limestone quarry in east java Indonesia. *Archives of Business Research*, 7(12): 83–98.
Rachbini, W. (2018). The impact of consumer trust, perceived risk, perceived benefit on purchase intention and purchase decision. *International Journal of Advanced Research*, 6(1): 1036–1044.
Ravenstein, E. G. (1885). *The laws of migration*.
Romadon, Y. (2014). *Pengaruh Green Marketing Terhadap Brand Image Dan Struktur Keputusan Pembelian (Survei pada Followers Account Twitter@ PertamaxIND Pengguna Bahan Bakar Ramah lingkungan Pertamax Series)* Brawijaya University].
Statista. (2025). *Beauty & Personal Care - Vietnam*. Statista Market Insights.
Saleem, A., Ghafar, A., Ibrahim, M., Yousuf, M., and Ahmed, N. (2015). Product perceived quality and purchase intention with consumer satisfaction. *Global Journal of Management and Business Research: E Marketing*, 15(1): 21–28.
Samiee, S. (1994). Customer evaluation of products in a global market. *Journal of International Business Studies*, 25(3): 579–604.
Shimul, A. S., Cheah, I., and Khan, B. B. (2022). Investigating female shoppers' attitude and purchase intention toward green cosmetics in South Africa. *Journal of Global Marketing*, 35(1): 37–56.
Simonson, I., and Drolet, A. (2004). Anchoring effects on consumers' willingness-to-pay and willingness-to-accept. *Journal of Consumer research*, 31(3): 681–690.
Singh, R. P., and Banerjee, N. (2018). Exploring the influence of celebrity credibility on brand attitude, advertisement attitude and purchase intention. *Global Business Review*, 19(6): 1622–1639.
Suh, J.-C., and Youjae, Y. (2006). When brand attitudes affect the customer satisfaction-loyalty relation: The moderating role of product involvement. *Journal of Consumer Psychology*, 16(2): 145–155.
Summers, T. A., Belleau, B. D., and Xu, Y. (2006). Predicting purchase intention of a controversial luxury apparel product. *Journal of Fashion Marketing and Management: An International Journal*, 10(4): 405–419.
Sun, L., and Jun, J.-W. (2022). Effects of Hallyu on Chinese consumers: a focus on remote acculturation. *Sustainability*, 14(5): 3018.
Sung, S.-Y. (2014). K-pop reception and participatory fan culture in Austria. *Cross-Currents: East Asian History and Culture Review*, 3(1): 56–71.
Tabachnick, B. G., and Fidell, L. S. (2007). *Experimental designs using ANOVA* (Vol. 724). Thomson/Brooks/Cole Belmont, CA.
Tjoe, F. Z., and Kim, K.-T. (2016). The effect of Korean Wave on consumer's purchase intention of Korean cosmetic products in Indonesia. *Journal of Distribution Science*, 14(9): 65–72.
Widayat, W., Sari, H. M. K., and Masudin, I. (2023). Determinants of consumer purchasing behavior toward Korean cosmetic products: Evidence from Indonesia. *Innovative Marketing*, 19(2): 129.

Young, S. E., Friedman, N. P., Miyake, A., Willcutt, E. G., Corley, R. P., Haberstick, B. C. et al. (2009). Behavioral Disinhibition: Liability for Externalizing Spectrum Disorders and Its Genetic and Environmental Relation to Response Inhibition Across Adolescence. *Journal of Abnormal Psychology*, 118(1): 117–130. https://doi.org/10.1037/a0014657.

Yuen, Y. L. E. (2021). Fashion and Beauty in the Time of Asia. *Dress*, 47(2): 217–219. https://doi.org/10.1080/03612112.2021.1929401.

Zikmund, W. G., and D'Amico, M. (1999). Marketing. Cincinnati. *Ohio: Southwestern College Publishing*. Vu Thi Phuong. *The Factors Affecting Purchase Intention of Foreign Products: Focusing on Korean Cosmetic Products in Vietnam* 울산대학교 일반대학원 경영학과].

9

Industry 5.0 and Smart Economy

A Managerial Perspective

Archana Ambekar,[1] *Leelavathi S.,*[2] *Kiran Kumar M.*[3]
and *Amit Kumar Tyagi*[4,*]

1. Introduction

The world is experiencing a major transformation driven by technological advances. Technology has influenced industrial change, and as economies now operate through technology, we must redefine how we think about economic activities. Industrialization began with the First Industrial Revolution, where mechanization provided an alternative means of human labor, but we have since seen multiple waves of industrialization that have changed how goods are produced, consumed, and delivered. Each wave, from steam to making buildings, cars, planes—everything automated—has raised technological capability. As industrialization progresses deeper into automation, we must rethink and redefine a human-machine paradigm to explore sustainable, creatively driven innovation that addresses needs for personalization—Industry 5.0 (Tóth et al. 2023). Moving from fully automated manufacturing to a human-influenced model of prioritizing innovations by creativity and intellectual value, Industry 5.0 identifies that while machinery and automation provide efficiency, productivity can still flourish through resonance with human experience, fulfilling tangible needs while improving quality of life. This paradigm shift is significant because it introduces a new reality: machines are partners and not just tools in creating unique products. New decision-making can include the development of nuanced experiences driven by AI and human input.

Alongside the development of Industry 5.0, the concept of the Smart Economy is also gathering steam. A smart economy is an economic model that uses digital technologies like AI, the Internet of Things (IoT), and blockchain, as well as big data, to facilitate productivity, innovation, and

[1] Jain College, Bengaluru City University, Vasavi Campus Bangalore.
Email: archana.ambekar@jaincollege.ac.in, 0009-0004-5767-9381

[2] JSS Centre for Management Studies, JSS Science and Technology University, Mysore.
Email: leelavathi360@gmail.com, 0009-0005-2330-2524.

[3] CMS Business School, Jain (Deemed-to-be University), Bangalore, India.
Email: kiranm287@gmail.com, 0000-0003-2084-9418.

[4] National Forensic Sciences University, Gandhinagar, Gujarat, India.
* Corresponding author: amitkrtyagi025@gmail.com, 0000-0003-2657-8700

DOI: 10.1201/9781003655190-11

sustainability. In a smart economy, decisions are made based on up-to-date information, AI-assisted reasoning, and automated processes that represent a collaborative approach toward optimizing economic and social outcomes. The smart economy's promise lies in embedded intelligence added to the design of production, distribution, and consumption of goods to establish a more sustainable, efficient, and inclusive economic system (Özdemir and Hekim 2018). The connection between Industry 5.0 and the Smart Economy comes with new management challenges and opportunities. With business complexity on the rise, there are vital opportunities available to managers who can grasp the transforming power of technologies to influence growth, innovation, and sustainability. This chapter will explore the connection between Industry 5.0 and the Smart Economy from a managerial perspective and examine the role of technology involved in business strategies, organizational structures, and decision-making processes. The chapter will also investigate the key elements that managers ought to consider in navigating organizations through a new era of technological transformation.

It is entirely reasonable to characterize the transitional periods of Industry 1.0 to Industry 4.0 as the technological cycles of industrial transformation. The First Industrial Revolution began with the introduction of mechanical production powered by steam engines in the late 18th century, which ushered in the factory system and mass production, fundamentally altering the manufacturing industry and providing economies of scale. The next industrial revolution, which involved the introduction of electricity, saw the development of assembly lines and automation. Industry 3.0 involved the use of information technology (IT), which set the foundation for digitalization, as computers and robotics began to take over most repeatable tasks on production lines (Nahavandi 2019). Industry 4.0 added cyber-physical systems (CPS), IoT, AI, and big data to previous technological advancements to respond to unprecedented demand variations, tight turnaround times, and complexity, which allowed industry to adopt smart factories where machines—and even intelligent machines—can connect and communicate with one another and adjust processes autonomously and in real time. All the above advancements introduced efficiencies, shortened time windows, and optimized resources, but legitimate concerns arose about displacing workers, dehumanizing manufacturing systems, and lacking any personal touch.

With the introduction of Industry 5.0, we have an opportunity to take a human-centered approach to the evolution of Industry 4.0. Industry 5.0, as part of the Industry 4.0 continuum, involves the combination of collaborative robots (cobots), AI, and machine learning alongside human endeavor and expertise. Unlike traditional robots, cobots are not intended to replace human labor but to work with humans and enhance their capabilities. With cobots that possess these qualities, customization empowers innovative, creative, and personalized solutions with humans at the core of the process, which potentially acts as an answer to current pressing issues of modern times such as sustainability, social inequalities, and resource shortages (Dalenogare et al. 2018). Industry 5.0's explosion of knowledge and information coincides with the growing desire for customized products and services. Consumers increasingly expect greater customization and responsiveness to their individual needs, primarily driven by digital platforms providing vast amounts of data. As consumer desires evolve, businesses must also evolve to remain sustainable and efficient producers.

The emergence of Industry 5.0 presents an opportunity, and it is necessary to have the economic framework—provided by the smart economy—to take full advantage of the promise that Industry 5.0 holds. A smart economy features a high degree of digital technologies in all facets of production, distribution, and consumption. Digital technologies enable resource optimization, better decision-making for economic actors, and ultimately allow businesses to deliver increasingly personalized and compelling services. The smart economy utilizes data analytics, AI, blockchain, and related technologies so that manufacturers and consumers can make increasingly smart consumption and production choices. This could include smarter supply chain management, enhanced personalized consumer service and experience, and real-time financial and value transactions. The Smart Economy, at its foundation, seeks to improve the productivity of both the private and public sectors while maintaining focus on sustainable development. Smart economies help businesses allocate resources more efficiently, make better decisions, and drive economic policy through innovative pathways that can lead to greater economic resilience and engagement while reducing the environmental impact associated with economic activity. Thus, Industry 5.0, with its focus on human-centric innovation, aligns with the overall vision of the Smart Economy. Managers in the "new normal" must learn to navigate an ever-changing and complex economic space where organizations must balance the efficiencies of digital technologies with the stimulation of human creativity, innovation, and purpose (Tyagi 2023a, Tyagi 2023b). Within this continuum, finding equilibrium between human capital and digital transformation will be critical to effective leadership.

As organizations move into Industry 5.0, managers will face a new range of unique challenges related to technological integration, organizational change management, and workforce transformation. In adapting to this new context, managers will need to reconsider conventional business models, leadership standards, and decision-making practices. One of the most complex challenges will be managing human and machine interactions. AI and robots can provide greater productivity and precision, but they still require human supervision, creativity, and ethics. This type of technology calls for a new style of leadership that embodies collaboration, agility, and continuous learning. Additional issues to consider include the ethics of AI and automation and how to mitigate their impact on the workforce. Another major management challenge is data governance. To support smart systems, organizations must collect vast amounts of data. Given the sheer scale of our data environments and the consequences of mismanagement, ease of data movement in and out of systems, data security, and regulatory compliance remain critical as our society becomes more digitally dependent. Managers must ensure that data flows in a timely and accurate manner through systems while implementing sufficient security to avoid potential cyber threats. In addition, managers must address technological capabilities to close the digital divide and ensure workforce members can successfully operate in technology-based environments. This may include upskilling, reskilling, and transforming organizational learning and culture. Table 1 provides a comparative summary of the industrial revolutions, showing how human roles and technological focus evolved toward Industry 5.0.

This research chapter seeks to examine the managerial implications of Industry 5.0 in the smart economy context. The following research questions will be addressed:

- What are the prominent managerial aspects associated with the adoption of Industry 5.0 technologies and how can businesses tackle them?
- How does incorporating AI, IoT and other digital technologies affect managerial decision-making and the formal structure of organizations?
- What are the workforce implications in set-up and management under industry 5.0 practices?
- Where is the role of managers in accessing the smart economy to bring forth innovation and sustainable growth?

Table 1: Evolution from Industry 1.0 to 5.0.

Industrial Revolution	Key Technology	Human Role	Focus	Output Impact
Industry 1.0	Steam & Mechanization	Manual Labor	Mechanized Production	Efficiency
Industry 2.0	Electricity & Assembly Line	Specialized Labor	Mass Production	Scalability
Industry 3.0	Computers & Automation	Supervisory Role	Automation	Speed
Industry 4.0	IoT, AI, Big Data	Minimal Input	Optimization & Autonomy	Precision
Industry 5.0	Cobots, AI, Digital Twins	Collaborative Co-creation	Human-Centered Innovation	Personalization & Sustainability

The remainder of this chapter has been organized as follows: Section 2 reviews the literature on the evolution of Industry 1.0 to 5.0, with a particular emphasis on human-centric innovation. Section 3 reviews the leading technologies that are enabling Industry 5.0 and their implications for the smart economy. Section 4 considers some of the key managerial implications of Industry 5.0, including changes to leadership, upskilling the workforce, and ethical issues. Section 5 considers some of the key challenges and risks faced by organizations in relation to Industry 5.0, including data privacy, cybersecurity, and the digital divide. Section 6 reviews a number of case studies of leading organizations to highlight issues of application and key learnings. Section 7 develops a strategic and performance framework that can guide managers in effectively adopting Industry 5.0 principles. Finally, Section 8 concludes this research by outlining various future directions.

2. Literature Review

The notion of Industry 5.0 has emerged from technological advancements and the increasing use of technology in business processes, especially manufacturing and production. While the earlier industrial revolutions involved mechanization, automation, and digitalization, Industry 5.0 is about a human-centric approach that uses advanced technologies and combines them with human ingenuity, skills, and decision-making. Industry 1.0 was characterized by the use of mechanical production powered by steam, which changed manufacturing by enabling large-scale and efficient production of goods. Industry 2.0 took this a step further by harnessing electrical power to mass-produce goods through assembly line production. The Third Industrial Revolution brought computers and IT to automate repetitive tasks and create a connected, data-driven world.

The Fourth Industrial Revolution, which commenced in the early twenty-first century, introduced CPS, IoT, AI, and big data technologies to make operations more autonomous and reliant on data-driven decision-making. Industry 4.0 focused primarily on efficiency, optimization, and autonomy of systems, yet it also resulted in concerns about human labor devaluation and marginalization, as machines began to assume cognitive and physical tasks. Industry 5.0 represents the reintroduction of the human aspect of industry into the technological realm. The focus is on cobots, AI, quantum computing, digital twins, and advanced technologies that support and enhance humans' cognitive and physical capabilities. The goal of Industry 5.0 is to support and amplify collaboration between humans and machines so that results can be achieved that neither could accomplish independently. Humans are no longer viewed as impediments to be automated out of production but as innovators and co-creators who use machines' computational capabilities to create, solve problems, and make strategic decisions.

AI plays a central role in the decision-making process of Industry 5.0. AI capabilities allow organizations to engage in intelligent automation of operations and autonomous decision-making. AI can optimize processes, predict future trends, and offer customized solutions to solve problems. Consequently, managers must become more data-literate, make decisions based on data, and formulate strategies that integrate AI insights. Quantum computing may radically change the industrial landscape by solving complex problems at speeds that traditional systems cannot match. In the context of Industry 5.0, quantum computing has the potential to optimize supply chain operations, enhance manufacturing processes, and simulate predictive outcomes. Managers must remain informed about developments in quantum computing and invest in training and infrastructure to capitalize on these advancements (Van Erp et al. 2024).

Digital twins are digital representations of physical assets, processes, or systems that allow businesses to emulate physical world dynamics before taking action in the physical system. It collects and uses data in real-time from sensors embedded in the assets to create predictive and the unwanted intelligence. Digital twins allow managers to maximize performance, minimize downside, and increase efficiency. Because businesses can use digital twins to inform their strategic and decision-making process, organizations can prevent costly mistakes, reduce threats and risks, and optimize efficiencies. The Smart Economy is an economic model utilizing digital technologies such as AI, IoT, blockchain, and big data, incorporating them into every aspect of production, consumption, supply chain management, and financial transactions. It builds on digital technologies to access real-time data, utilize autonomous control systems, and apply intelligent decision-making to better allocate resources, increase productivity, and reduce costs. Key aspects include: (1) Data informed decision making - Organizations can leverage real time data and real-time AI to make better informed and more timely decisions based on the analysis of patterns across the demand and supply chain. (2) Automation and personalization - Automation and AI play a keyrole in personalizing the experience, leverage the productivity of each actor in supply chain to reduce costs, and reduce the manual labor to utilize these human resources more effectively. (3) Sustainability - A key pillar of the Smart Economy. Human labor often generates pollution and waste; thus, the Smart Economy seeks to reduce waste, optimize human resource utilization, and deploy advanced technologies such as AI and digital twins to improve energy efficiency, enhance supply chain operations, and mitigate carbon impact. (4) A decentralized approach - Using blockchain or similar protocols enables transparency and security, allowing peer-to-peer transactions across services, goods, and finance.

Industry 5.0 combines human-centered technology with data-based economic models involving human-robot collaboration. It aims to meet the objectives of smart economic development through collaborative problem-solving, personalization, and efficiency. At a practical level, Industry 5.0 enables companies to scale responsibly while maintaining personal relationships with customers. It also provides real-time data and intelligent resource management structures that help managers balance supply and demand effectively (Raisa et al. 2025). Incorporating blockchain as part of the value supply chain process provides opportunities for decentralized economic activity, making it easier to automate numerous business-related processes such as payments to suppliers and resource management. Sustainability is also a central focus of Industry 5.0, enhancing resource efficiency, reducing carbon footprints, and strengthening the economy to be more sustainable. This often involves balancing corporate efficiency with ethical decision-making and social responsibility (Angelidou et al. 2022). Ultimately, Industry 5.0 is a new framework for engaging in the modern economy. Research on Industry 4.0, digital transformation, and automation has been extensive, but there is significantly less research on Industry 5.0, human-machine collaboration, and their incorporation into the Smart Economy. Moreover, much of the current literature is dispersed across various fields, and few interdisciplinary studies combine management, economics, and technology to provide a comprehensive understanding of Industry 5.0. This study addresses these gaps by presenting a unified approach to the managerial implications of Industry 5.0—specifically, how technologies like cobots, AI, and blockchain are transforming business strategy, decision-making, and leadership. The study also provides a perspective on how Industry 5.0 can be incorporated into the Smart Economy while leveraging human creativity and machine capabilities for sustainable development..

3. Industry 5.0 Technologies and Smart Economic Transformation

Industry 5.0 represents a transformation toward a more efficient and sustainable business model. This transformation depends on advances in AI and ML, where machines can perform repetitive tasks, follow instructions to make informed decisions, learn, and autonomously improve performance. Some applications of AI include predictive maintenance, production schedule optimization based on customer demand, and analyzing customer purchase preferences to provide personalization. Cobots can work alongside humans in the workplace, helping them perform tasks with precision, speed, and repetitive movement. The precision of creating cobots generates great efficiencies in many industries, particularly manufacturing, since they can optimally partner with humans to create highly productive workflows. Managerial decision-making must evolve so that managerial practice can interpret AI insights effectively while balancing them with human instinct and reasoning (Esposito et al. 2018). IoT connects physical objects to the internet and allows them to collect and exchange data. In an

Industry 5.0 context, IoT enables smart factories and smart supply chains that use real-time data to enhance operational efficiencies, reduce consumption, and optimize resources. Managers should consider how to develop and adapt traditional business operations to incorporate IoT systems into their work, allowing for performance monitoring systems and asset tracking with near-immediate feedback for data use. Blockchain is a secure and verifiable method of recording transactions or exchanges of data that allows for the decentralization of business models. It provides unique opportunities to reduce costs in areas such as supply chains, where traceability and transparency are needed to operate effectively (Dong et al. 2024). While blockchain technology remains in its infancy, business managers must consider its capabilities for automating business processes and verifying that transactions are authentic and secure. Big data analytics uses sophisticated algorithms to analyze vast amounts of structured and unstructured data, identifying patterns and trends that were previously not possible. Managers must create a data-driven culture in their organizations and use big data to make better-informed decisions. By understanding customer behavior and market trends, managers can improve customer satisfaction, manage supply chains more effectively, and increase overall operational efficiency.

Digital twins are digital representations of physical objects or systems that, when used in Industry 5.0, simulate real-life processes effectively. They allow managers to identify potential bottlenecks, predict failures, and optimize workflows that can be replicated in physical environments. Quantum computing is still in its infancy, but it will change the world when the technology reaches a point where it can solve problems that classical computers never could. The processing of information occurs at speeds many orders of magnitude faster than classical computing, making it capable of significantly enhancing artificial intelligence. Figure 1 illustrates the technology stack of the smart economy, showing interconnected layers of AI, IoT, blockchain, big data, digital twins, and user interfaces.

The inclusion of technology into industrial operations and economic architectures has transformed global economies into smart economies. These economies are driven by real-time data and based on economic models built on automation and knowledge systems. They enable smart manufacturing and production systems, sustainable resource management, decentralized finance, and data-driven decision-making. Some of the Industry 5.0 technologies creating smart factories include real-time

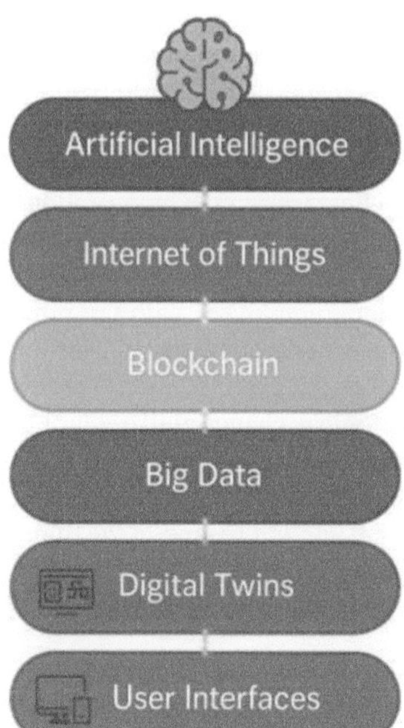

Figure 1: Smart Economy Technology Stack.

data, AI, and IoT. Beyond smart factories, these technologies automate and optimize manufacturing processes based on real-time data. Accordingly, smart factories seek to transition toward adaptable manufacturing rather than linear production. This transformation allows companies to reconfigure production depending on demand, reducing waste and energy consumption (Morrar et al. 2017). Companies can utilize predictive analytics to forecast equipment failures, thereby reducing resource loss and downtime. AI and blockchain are used for decentralized finance (DeFi), which allows for peer-to-peer transactions without intermediary financial institutions such as banks. Smart contracts enable peer-to-peer transactions automatically once predefined conditions have been met, making them more efficient than traditional agreements. Smart contracts significantly reduce the need for human intervention in transactions, lowering transaction friction and consequently reducing transaction costs. In summary, it is evident that these technologies have revolutionized the modern world and provide opportunities for improved efficiency and a more sustainable future. The use of these technologies will fundamentally change the role of managers and managerial decisions, as managers must evolve beyond traditional command-and-control leadership to adopt more agile, collaborative, and data-driven leadership styles (Boughzala and De Vreede 2015). AI and big data will require managers to understand and make decisions based on algorithmic results. For example, the use of collaborative robots and digital twins will necessitate new organizational structures and workforce strategies. Beyond direct coaching, managers need to build digital competencies within their teams so employees can effectively leverage new technologies. Managers are also confronted with ethical issues surrounding AI, data privacy, and the displacement of human workers. As managers adapt to increasingly automated systems, they must pay close attention to accountability, transparency, and ethics throughout the deployment processes.

4. Managerial Implications of Industry 5.0

As Industry 5.0 continues to integrate human creativity with advanced technologies such as AI, collaborative robots, and quantum computing, leadership styles are undergoing a fundamental shift. The traditional method of command and control is not necessarily being replaced, but the environment is changing. Hierarchical, top-down decision-making structures are no longer suitable as the rate of technological change and the future of work evolve rapidly. To lead organizations effectively, managers in Industry 5.0 must become transformational leaders who empower people to combine human creativity with advanced technologies to drive innovation and new thinking. One of the main components of this leadership transformation is the development of digital skills. More than ever, managers are expected not only to understand emerging technologies but also to comprehend how they influence organizations, strategies, and decision-making (Demir et al. 2019). For example, the use of AI and machine learning in organizations presents challenges for managers, who must understand the technology well enough to make informed decisions, guide their teams, and evaluate the success of new systems. Equally important, managers must develop the ability to lead human-robot teams, providing employees with the right resources to remain satisfied and productive while overcoming the gaps between worker capabilities and robotic efficiency. In today's environment, managers cannot merely be experts in management; they must also adopt a learning mindset. The fast pace of technological change will require managers to continuously learn to remain relevant. This includes understanding advancements in AI, blockchain, quantum computing, and IoT, as well as how these technologies can fit into different business models. Organizations that fail to develop these skills will risk being left behind in the digital future.

4.1 Decision-Making in the Smart Economy: AI-Supported vs. Human Intuition

Perhaps the most significant change that will arise from Industry 5.0 is the influence of AI on decision-making. Traditionally, in business settings, decision-making was based largely on intuition, experience, and human reasoning. With AI, however, decision-making is now supported through real-time data analysis, predictive algorithms, and machine learning models that can sift through massive datasets and produce insights faster and more accurately than humans (Mittal et al. 2017). Managers now need to find the balance between using AI-enabled insights and relying on human intuition. AI helps optimize processes, predict future performance, and create better customer experiences. For example, AI algorithms can identify consumer behavior patterns, estimate product demand, and recommend services or products to customize experiences, thereby improving the customer journey. AI is a powerful tool for analyzing large datasets and identifying patterns and trends, but it cannot use uniquely human capabilities such as empathy, ethics, or judgment in complex and ambiguous situations. For managers, the biggest challenge will be determining when to trust AI-generated insights and when to rely on personal or team instincts. AI enables a data-driven process for decision-making, while human leaders should guide decisions involving ethics, creativity, or long-term vision. Managers must initiate processes that integrate data-driven insights with human judgment through hybrid decision-making approaches, creating value by combining AI capabilities with human intuition (Tyagi 2023a). Figure 2 illustrates a visual spectrum demonstrating decision-making scenarios where AI is preferred, where human intuition is crucial, and the hybrid middle ground.

The latest technologies, such as AI, robotics, and quantum computing, result in significant organizational change. Industry 5.0 requires businesses to move beyond traditional workflows and adopt new ways of working that emphasize a collaborative workforce between humans and machines, data-driven decision-making, and organizational agility. The rapid pace of these advances will undoubtedly create some degree of resistance among employees, especially if they do not understand the changes or fear job loss (Bokhorst et al. 2022). Effective organizational change management in the age of Industry 5.0 requires managers to anticipate such resistance and remove barriers to implementing new technologies by fostering a culture of continuous learning and adaptability. Employees will need reassurance that while technology may seem to threaten job security, these developments will instead expand their capabilities and create opportunities in more creative and strategic roles. Agility is now more imperative than ever. Change is inevitable in the age of Industry 5.0, and managers have a duty to remain agile amid new technologies, markets, and customer demands, making their organizations lean, flexible, and responsive. The best way to support this transformation is to involve employees directly in the process. Managers should encourage inclusivity, allowing employees to interact with new technologies, understand their implications, and participate in redesigning future workflows (Nambisan et al. 2019). This creates engagement from the bottom and reduces resistance, employees feel a sense of ownership and, therefore, will be more willing to embrace that technology and be innovative. Additionally, the level of agility required in Industry 5.0 might require a change in organizational form. In a fast-paced data-driven environment, traditional hierarchical organizational structures may not be as

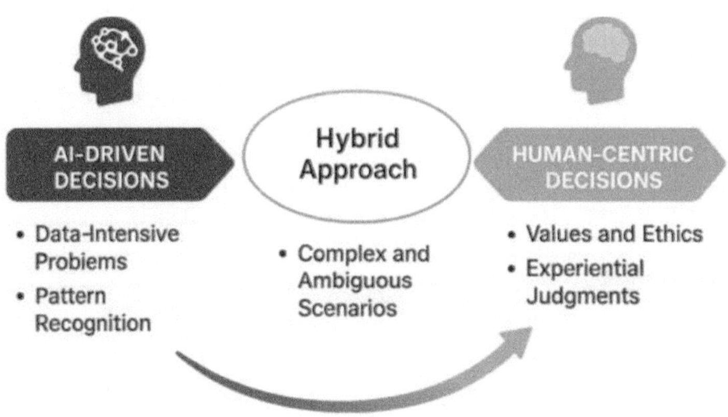

Figure 2: Decision-Making Balance: AI vs. Human Intuition.

effective. Organizations may need to consider adopting "flatter" organizational structures to support communication and decision-making across levels. The ability to pivot quickly, experiment with new ways of working, and modify the approach given findings from data will be important when finding competitive advantage. In Industry 5.0, innovation extends beyond product development or R&D—it is embedded in every aspect of the business, including production processes, customer service, and operations (Wei and Shen 2025). While technology remains a key driver of innovation, the differentiating factor in the Industry 5.0 environment is how managers leverage both human creativity and machine intelligence to drive new business models and gain competitive advantage.

Managers will also need to rethink traditional methods of strategy development in the Industry 5.0 world. Conventional strategic planning, which relies on static forecasts and long-term projections, will be less effective in this dynamic environment. Strategy must now be flexible and adaptive to technological, market, and consumer behavior shifts (Nosova et al. 2021). Organizations will need to continuously monitor technological trends, market movements, and consumer preferences, adjusting their strategies based on real-time insights. Data analytics plays a crucial role in the management decision-making process, particularly in innovation planning and system monitoring. By using data analytics, managers can make informed, real-time decisions that influence innovation and strengthen competitive positioning. The manager's focus should be on cultivating a culture of innovation—encouraging employees and technology to work collaboratively to solve complex business challenges creatively. To achieve this, managers should promote behaviors centered around experimentation, cross-functional collaboration, and entrepreneurial mindsets that foster an environment where innovation is continuous.

As organizations adopt technologies such as Artificial Intelligence, the Internet of Things, and blockchain, ethical issues and data governance become increasingly important. In decision-making, AI and machine learning algorithms raise concerns about bias, transparency, and accountability. Managers are responsible for ensuring that these technologies are used ethically—not only to protect the brand's reputation but also to meet consumer expectations that technology use will comply with or exceed ethical standards related to human rights. For example, auditing and monitoring algorithms are prudent management practices, ensuring that organizations do not develop systems that unintentionally discriminate against specific groups (Dignum 2022). Data governance is another major concern. As organizations collect and analyze vast quantities of data, including personal information, they must implement robust policies on data privacy, security, accessibility, and regulatory compliance, such as GDPR. Companies should clearly outline data practices regarding collection, storage, sharing, and disposal while respecting stakeholders' rights over their data. Managers in the age of Industry 5.0 must balance resource allocation between technological innovation and ethical decision-making to ensure that advances in artificial intelligence and automation align with societal norms, legal standards, and expectations. Furthermore, managers bear responsibility for sustainability-oriented decisions, leveraging technology to minimize environmental impact and contribute to long-term societal well-being. Table 2 provides key managerial challenges in the Industry 5.0 context along with proposed mitigation strategies for successful integration.

Table 2: Key Managerial Challenges and Mitigation Strategies in Industry 5.0.

Managerial Challenge	**Description**	**Recommended Strategy**
Data Governance	Ensuring secure and ethical data use	Strong compliance, AI audits, policy reforms
Human-Machine Collaboration	Resistance to new workflows	Reskilling, inclusive culture, change agents
Ethical AI Deployment	Bias, transparency, decision traceability	Algorithm auditing and stakeholder oversight
Organizational Inertia	Hesitancy toward disruptive models	Leadership development and culture change
Legal & Regulatory Ambiguity	Uncertainty around responsibility and compliance	Proactive internal governance structures

5. Challenges and Risks

One of the most pressing challenges in the era of Industry 5.0 and smart economic systems is ensuring data privacy and cybersecurity. The use of AI, IoT, and interconnected systems increases the number of devices and the amount of data generated, transferred, and stored over platforms that were not originally designed to process such information. While data is critical for information-based automated decisions and intelligent automation, it has also become an attractive target for hackers. In smart factories, cobots and digital twins operate on cloud-based systems that contain sensitive information about processes, intellectual property, and personnel (Alves et al. 2023). A cyber breach of either of these systems could lead to major operational breakdowns, reputational loss, and financial damage. AI platforms with automated decision-making processes also pose serious privacy concerns when they process personal information such as consumer behaviors, biometrics, and financial data. Weak or improperly implemented data governance protocols can prove fatal to user privacy. Managers must commit to robust cybersecurity frameworks to protect their digital assets. This requires financial investment in encryption, intrusion detection systems, multi-factor authentication, and continuous network vulnerability monitoring. There is also a growing ethical need to manage data in a transparent and clear manner that allows users to make informed choices about their data. Moreover, companies must adhere to the respective legal frameworks for data protection and management, such as the European GDPR data management directives and the U.S. California CCPA (Bauer et al. 2015). The potential of Industry 5.0 is promising, but its development depends on the readiness of industries, organizations, and countries. Advanced technologies require complex infrastructures, technical capacity, and significant investment. Not every organization—particularly SMEs—will have the ability to adapt quickly or easily. This situation will widen the digital divide both between organizations within the same economy and between countries at different stages of development. Economies with limited access to high-speed internet, digital skills, and capital investment plans will likely struggle to industrialize effectively in the Industry 5.0 era. Similarly, workers in traditional industries, even in developed countries, may lack access to the upskilling required to remain relevant. Various stakeholders in government and business must collaborate to develop inclusive digital infrastructures that promote technological democratization (Tyagi and Abraham 2021). This includes providing financial incentives for digital transformation, investing in education and vocational training in technology, and building open-source tools to lower barriers to entry. If left unaddressed, the digital divide will not only slow down the diffusion of Industry 5.0-related innovations but also reinforce existing global and local economic inequalities. The effective transition to Industry 5.0 requires more than the application of technology—it demands a cultural shift within organizations. Many organizations experience inertia, where people resist changes to existing structures, processes, systems, and mindsets. Inertia may stem from fear of job loss, lack of technical expertise, or diminished managerial control. Management resistance is particularly relevant to organizational inertia. Leaders unfamiliar with flat organizational structures—characterized by collaborative, real-time decision-making and human-machine integration—may resist such changes. Decision-makers may also hesitate to rely on AI for strategic planning or to invest in costly technologies without a clear return on investment. Overcoming inertia requires a forward-looking, inclusive organizational vision. Management must be educated on the strategic benefits of Industry 5.0 and develop the leadership skills necessary to effectively guide digital transformation (Huang et al. 2023). Change management is essential to mitigate fear and foster a culture of innovation and experimentation. Stakeholder engagement, education and training, pilot projects, and other change initiatives can help familiarize personnel with new strategies and systems. Industry 5.0 technologies are evolving faster than regulatory frameworks can keep pace. Legal uncertainty surrounds many AI-related issues, including AI accountability, liability in human-cobot interactions, data ownership, algorithmic transparency, and cross-border data flows. The lack of uniform and comprehensive legal standards creates uncertainty in the risk-averse environments of managers, investors, and developers. For example, if an AI algorithm recommends a course of action that leads to financial or operational losses, who is liable—the software developer, the data provider, or the organization implementing the AI? Similar ambiguity exists in defining legal responsibility for autonomous robotics operating in public or workplace environments, where rapidly evolving technologies outpace existing regulations. This regulatory void slows adoption and increases risks for early adopters. To mitigate these challenges, regulatory authorities must proactively engage with industry leaders, researchers, and ethicists to design technology-neutral, adaptable, and scalable regulatory frameworks. Meanwhile, organizations should establish internal governance structures and ethical frameworks to minimize risk exposure while awaiting comprehensive legislation. Alongside the benefits of Industry 5.0, there are also drawbacks related to economic inequality and workforce displacement. While collaboration between humans and machines is a central goal of Industry 5.0, higher levels of automation could reduce jobs or widen existing income inequalities, particularly for low-skill roles (Ali et al. 2022). The challenge is especially severe in labor-intensive sectors where repetitive tasks are easily automated or performed by AI and cobots. The economic benefits of digital transformation are unlikely to be evenly distributed. High-skilled workers with digital expertise will become more prosperous, while low-skilled workers will likely face greater marginalization. Regions without access to training and reskilling opportunities will experience deeper inequalities. If these trends are not addressed, economic transformation may result in increased social instability, reduced economic participation, and growing tensions between workforce segments. Therefore, companies and governments must invest in reskilling and upskilling initiatives to ensure that workers displaced by Industry 5.0 can reenter the economy successfully. Institutions should also adapt educational programs to emphasize digital literacy, ethical reasoning, interdisciplinary knowledge, and hybrid work competencies. Universities will need to recognize that future workers will assume new roles—combining human insight with technical understanding—to thrive in the evolving landscape of Industry 5.0.

6. Case Studies

The transition to Industry 5.0 is not something that might happen—many progressive companies are already implementing collaborative robots, AI, digital twins, and other cutting-edge technologies. These real-world applications provide valuable insights into the operational, tactical, and strategic changes needed to succeed in a smart economy. This section reviews case studies from leading companies such as Siemens, Bosch, and Tesla and includes analyses across various sectors such as healthcare, manufacturing, and finance. In addition, insights from interviews and secondary data illustrate how current managers are addressing Industry 5.0 demands, what they have learned, and the key success factors for future strategic initiatives (Lezzi et al. 2018).

Siemens is often cited as the leader in the adoption of Industry 5.0 technologies. In their manufacturing plants, they have adopted digital twins, artificial intelligence, and collaborative robotics to bolster productivity, streamline costs, and ultimately allow for a higher degree of product customization. For example, Siemens' Amberg Electronics Plant in Germany today has more than 75% of its value chain processes automated and connected to digital systems. However, what Siemens has that other companies lack in implementing Industry 5.0 is the emphasis on the human centredness of the workforce. Employees are trained to work with machines and not in conflict with machines. It is worth noting that Siemens' managers took the lead in the introduction of Industry 5.0 technologies by focusing on the workforce and developing change management plans that ensure employees are engaged with Industry 5.0 technologies while minimizing resistance. They were able to achieve grand levels of productivity enhancements without taking away from human creativity and decision-making.

Bosch is another global manufacturing leader that has developed an Industry 5.0 strategy within its smart factories in Europe and Asia. Bosch's model focuses not only on smart factories but also on how humans and machines can best adapt together. By integrating AI-derived analytics from production lines and involving humans for oversight, Bosch has designed adaptive production systems. In its automotive component factories, digital twins are used to simulate manufacturing processes, enabling ongoing optimization in real time and allowing managers to perform predictive maintenance. Managers at Bosch have fostered a culture of continuous learning and digital experimentation. The company has also integrated ethical considerations into its digital strategy; for instance, in AI-driven quality control and process optimization systems, Bosch ensures that algorithms are auditable and transparent (Bankins and Formosa 2023). Trust with stakeholders is built on a balance between innovation and ethics. Bosch effectively leverages the benefits of Industry 4.0 while creating legitimate business and social value.

Tesla is another case, just as unique but informative at the same time. Known for its hyper-aggressive strategy of automation, Tesla also had clear intentions of creating a fully automated production line at its Gigafactories. Unfortunately, Tesla found itself faced with challenges because of a heavy reliance on robotic automation. Even Elon Musk himself publicly stated that they made a mistake with the amount of automation they had in place, and reinforced how human workers were going to be needed at the production level. In light of this, Tesla completely changed their position regarding automation and decided to go with a mixed Industry 5.0 style model that is taking advantage of both machine precision and human labor flexibility. Tesla had begun using industrial cobots to perform physically demanding tasks while the human workers were responsible for design changes, quality inspections, and strategic responsibilities. This position underscores a key managerial insight: we must achieve a level of automation where human judgment complements—and elevates—the tasks performed by machines. Nothing drives teams to embrace true agility and realign strategies for technological integration more effectively than hands-on experience with innovators like Tesla.

Across industries, the adoption of Industry 5.0 appears to have unique characteristics and challenges. In manufacturing, for example, companies such as Siemens and Bosch are focused on process optimization, workforce augmentation, and customization of products. In comparison, the healthcare industry has applied Industry 5.0 to enhance diagnostics, patient monitoring, and individualized treatment. For example, in some advanced hospitals, AI algorithms assist radiologists by reading imaging data and making suggestions about possible diagnoses, while cobots help in taking care of laboratory functions. Meanwhile, digital twins are being developed of human organs to model treatment so that doctors can choose which therapies will be most effective and least invasive. Here, managers must navigate the challenge of clinical effectiveness alongside patient trust and data privacy. They must also contend with tight regulations and ethical considerations, especially given the implications of decisions driven by AI on human lives.

In the financial services space, we see Industry 5.0 technologies being utilized to strengthen fraud detection, boost risk management, and foster personalized banking experiences. AI algorithms can rapidly sift through large datasets to identify suspicious transactions, thus enabling real-time fraud alerts. Additionally, chatbots and virtual assistants can engage in dialogue with customers geared toward customized financial recommendations. Moreover, the simplified integration of Industry 5.0 technologies still demands a new type of leadership competence. Identifying leaders who can comprehend the algorithmic processes and their anticipated impact on the various dimensions of customer experience, compliance, and competitive positioning will be an ongoing challenge. Leaders also need to manage organizational change, including how to train finance professionals to work hand in hand with intelligent systems. The financial services industry demonstrates that there is a tangible application of Industry 5.0, and although it might differ vastly from manufacturing or healthcare, the manager's responsibilities are still fundamentally framed in balancing innovation with human intervention and ethical governance.

The interview-based research, alongside the secondary data analysis, provides more empirical insights into how managers are addressing various facets of the transition. We conducted interviews with middle- and senior-level managers in Industry 5.0 pilot organizations and encountered an ongoing emphasis on urgency, agility, digital literacy, and cross-functional collaboration. Interestingly, managers revealed that successful implementation of Industry 5.0 technologies has less to do with the technology itself and more to do with the existing culture in the organization. Companies that encourage experimentation and innovation find employees more willing to embrace new tools and methodologies.

We learn multiple lessons from these cases and empirical understandings. First, technology adoption will require framing strong change management strategies. Technology alone cannot create transformation; it needs to work within an organizational context and culture that enable its implementation. Second, leadership matters. Leaders must not only sponsor digital initiatives but also model values such as adaptability, transparency, and collaboration. Third, workforce development is important. If the nature of work is evolving, then continuous learning and digital skills development are essential strategic activities. Fourth, ethical and legal considerations should be built into managerial decision-making from the start; this is even more important when working with AI and individual data.

7. Strategic Roadmap for Managers in Smart Economy

As organizations move from traditional digitalization efforts to a more sophisticated, human-centric perspective of Industry 5.0, the nature of management as a profession is becoming much more complex and strategic. Management has moved beyond just enforcing established systems and practices to facilitating intelligent transformation, enabling interdisciplinary collaboration, and stewarding ethical, sustainable practices (Tyagi and Sreenath 2021). A roadmap for organizations to follow through this change is critical—one that has clear, actionable recommendations, a structured plan for adoption, realistic and actionable steps for sustainable change, and clear requirements for evaluating transition progress.

The first step in this roadmap will be to align the strategic plan with the principles of Industry 5.0. Business leaders must understand that this phase of industrial evolution is not simply driven by automation or cost savings; it is characterized by curated human-machine collaboration, personalization, resilience, and sustainability. Managers need to actively drive an organizational culture that embraces change, supports digital literacy, and celebrates opportunities for experimentation. Organizational leadership should foster, support, and promote interdisciplinary teams that include technologists, human resource practitioners, operational employees, and strategists who co-design the processes, policies, and supportive systems to achieve smart organizational goals. Meanwhile, public policymakers play a key role as well. By way of incentives, infrastructure development, clarity of regulation, and education reform, public policy can help create an enabling environment to support enterprises transitioning to Industry 5.0.

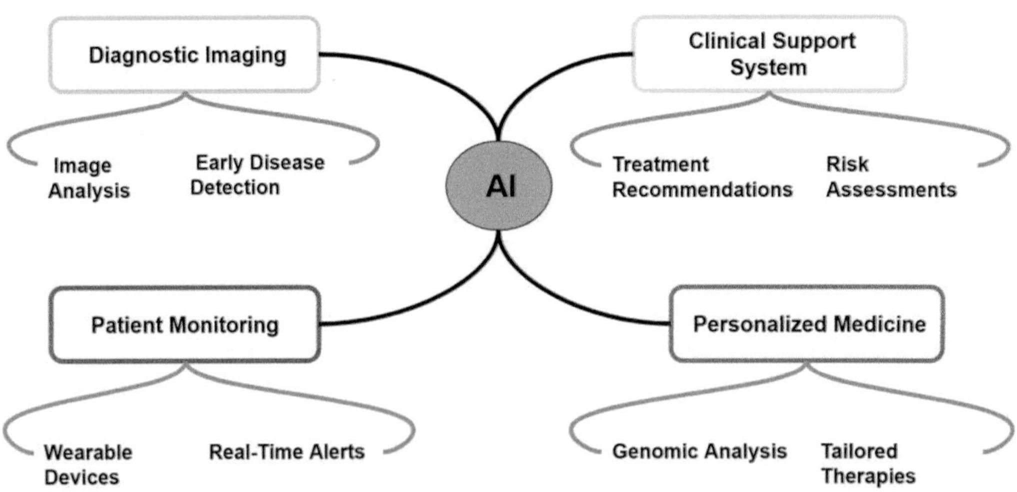

Figure 3: Case Study Insights in Siemens, Bosch, Tesla.

To implement these strategic objectives, managers could focus on a specific Industry 5.0 Managerial Adoption Framework (I5MAF) that is organized within a four-faceted model: Vision Alignment, Technological Integration, Human Enablement, and Governance & Ethics. Under the first facet, Vision Alignment, an organization must articulate what Industry 5.0 means for them over the long term. Organizations need to define the value propositions and a roadmap that connects these propositions with their business objectives. This facet involves stakeholder mapping to ensure employees, customers, partners, and communities are all included in the transformation journey. The second facet, Technological Integration, establishes the role of infrastructure, interoperability, and data ecosystems. Managers will be responsible for selecting, implementing, and integrating technologies aligned to collaborative robots, AI analytics, digital twins, and IoT technologies, to name a few. However, it is crucial that managers not only assess technology for efficiency but also for how technology choices can bolster human input and decision-making. Collaboration with IT and engineering teams will be highly relevant in this facet to ensure that new systems are aligned with business needs and can scale in the future.

Human Enablement, which is the third pillar, is the fundamental foundation for Industry 5.0. Managers must invest in reskilling and upskilling for their workforce to co-create value with intelligent systems. Workforce development courses must be individualized, modular, and continuous, helping workers advance alongside innovation. Equally, managerial positions must foster high-trust, high-psychological safety, and inclusive workplaces where employees can engage in trials and give feedback without fear of reprisal or putting their job at risk. Governance & Ethics, the fourth pillar, ensures that advanced technologies are adopted in ways that are transparent, ethical, accountable, and legally compliant. Managers should have strong data governance processes related to privacy, security, and AI responsibility (Mutlag et al. 2018). They should also provide ethical AI audits, transparency in algorithms, transparent data-sharing policies requiring consent, and embed environmental and social sustainability into operational processes and supply chain decisions, accounting for broader ESG considerations.

Transitioning to a resilient and human-centric model also consists of sequential stages for managing the transition. The first stage will be diagnostic assessment, where managers perform a baseline evaluation of their organization's current digital maturity, current capabilities across human resource systems, and their technological assets, and then must understand their readiness gaps before beginning to develop a realistic transformation roadmap. This is followed by the pilot implementation stage, where the manager will use small-scale implementations in low-risk environments to test technologies, processes, and training models they want to explore in their organization. Piloting technology is very useful for generating quick wins, enhancing internal advocacy, and iterative improvement before rolling ideas out to scale.

The third stage is scaling and institutionalizing, where managers must move early adopters across the organization and task them to learn how to optimize the existing technology, develop new processes, and redesign accordingly. In this stage, managers may choose to invest in the necessary digital infrastructure to support the changes they are making to their workforce system—upgrading existing technology, designing, and piloting new tools to support processes. Finally, in the feedback and iterate stage, managers need to continually assess and improve their efforts (Wamba et al. 2016). Managers should use their diagnostic evaluation and agile reviews to form iterative improvements, design employee feedback opportunities, and develop data maturity in their performance reviews to enable real-time adjustments to strategies while recalibrating technology and workforce models within fiscal and human resource constraints.

To effectively gauge and measure progress and maturity, managers need to develop relevant tailored key performance indicators (KPIs) against the goals of Industry 5.0. The KPIs should cover four principal areas. First, belonging to technology capability, we have system uptime, AI accuracy, process automation levels, and data interoperability measures. Second, in relation to people and machine collaboration, examples include employee-tech adoption rates, task co-performance, and user experience satisfaction measures. Third, under Sustainability and Resilience, organizations

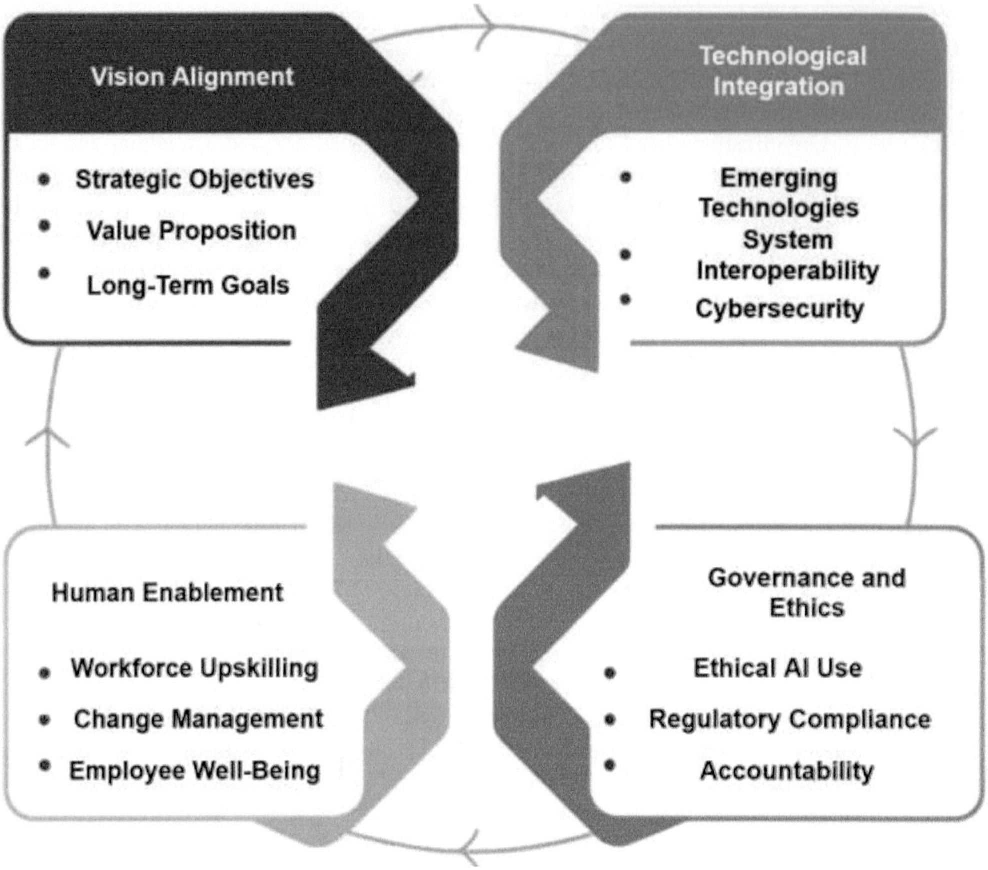

Figure 4: The Industry 5.0 Managerial Adoption Framework (I5MAF).

can measure resource reductions, resource circularity index, or operational recovery times from disruption. Finally, we have ethical and good governance measures, including regulatory compliance for data, bias audits related to AI, and stakeholder trust measures. This will provide a more thorough view of how to adopt responsibly. Figure 4 visualizes a four-pillar framework showing vision alignment, technological integration, human enablement, and governance & ethics for effective Industry 5.0 managerial adoption.

Although no two organizations' transformative journeys are ever the same, a core principle of the 5.0 principles is universal: technology must serve human values, operational actions must be based on resilience rather than rigidity, and organizational strategies must focus on adaptability and long-term value creation. The smart economy will reward organizations that innovate quickly in an inclusive, ethical, and sustainable way. Table 3 presents an overview of core Industry 5.0 technologies and corresponding managerial responsibilities for value creation.

Figure 5 illustrates various KPI categories including technological capability, people-machine collaboration, sustainability metrics, and governance effectiveness.

8. Conclusion and Future Directions

Industry 5.0 represents a new language and a new context for how the human experience of work, labor, value, and creativity plays out. No longer are we fixing things solely for automation and efficiency; Industry 5.0 is about the "collaborative human," personalization of production and consumption, and sustainable and responsible innovation. Now, when combined with the opportunities of the

Table 3: Industry 5.0 Enabling Technologies and Managerial Roles.

Technology	Role in Smart Economy	Managerial Responsibility
Artificial Intelligence (AI)	Predictive analytics, decision automation	Ensure interpretability and integrate insights
Collaborative Robots (Cobots)	Augment human labor	Facilitate collaborative workflows and training
IoT	Real-time monitoring and feedback	Infrastructure integration and security management
Blockchain	Trust, traceability, smart contracts	Implement compliance and decentralization protocols
Digital Twins	Simulations and proactive optimization	Scenario planning and virtual validation
Quantum Computing	High-speed decision computation	Strategy alignment and skill forecasting

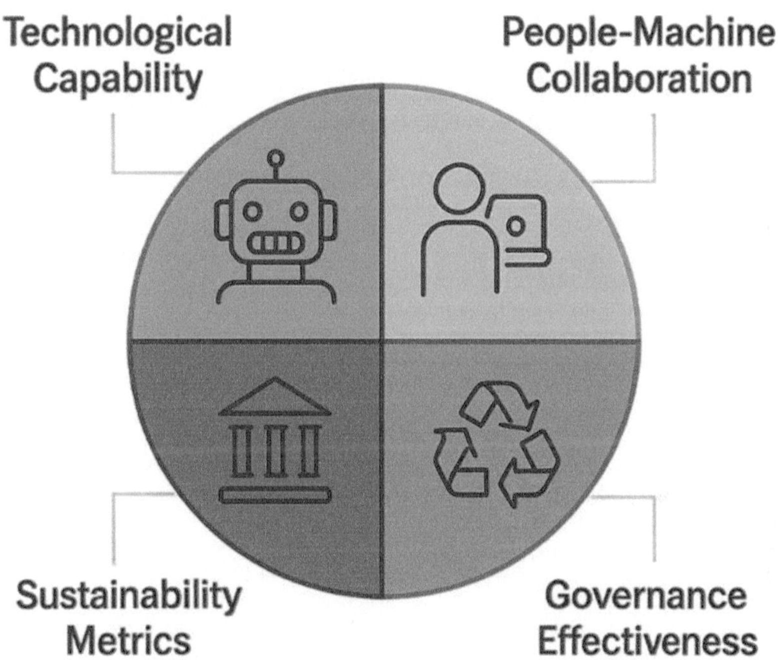

Figure 5: Managerial KPI Dashboard for Industry 5.0.

smart economy in connected and smart technologies, organizations are going to require managers with agile leadership skills for employees, customers, communities, stakeholders, and partners, with an inclusive and ethically oriented approach. The success of Industry 5.0's transformation will not lie strictly in technology adoption but in optimizing cultural transformation, learning transformation, and governance transformation in organizations—cultures that are supportive rather than siloed. Moving forward, researchers and managers need to focus on how adaptive organizations can build tools to facilitate ongoing, real-time decision-making, provide data ethics, and enable functional collaboration. Policymakers and education consortia will need to prioritize digital literacy, reskilling, and digital infrastructure development to address prevailing issues of the digital divide. Policymakers will also need to focus on dynamic regulatory frameworks to consider issues of accountability, data privacy, and workforce displacement due to technological change. Managers, as Industry 5.0 continues to take shape, will need to be advocates for innovation while being socially responsible and using technology to improve human value and performance outcomes (meaning less displacement). Interdisciplinary, collaborative, and manager-led research can be leveraged to continue understanding new use cases, improving strategic patterns, and enhancing sustainable best-practice promotion around meaningful human economic transformations.

References

Ali, N., Isa, Z. M., Bakar, S. A., Jali, F. a. @. A., and Shaharruddin, S. (2022). Industrial Revolution (IR) 4.0: Opportunities and Challenges in online business. *International Academic Symposium of Social Science 2022*, 85. https://doi.org/10.3390/proceedings2022082085.

Alves, J., Lima, T. M., and Gaspar, P. D. (2023). Is Industry 5.0 a Human-Centred Approach? A systematic review. *Processes*, 11(1): 193. https://doi.org/10.3390/pr11010193.

Angelidou, M., Politis, C., Panori, A., Bakratsas, T., and Fellnhofer, K. (2022). Emerging smart city, transport and energy trends in urban settings: Results of a pan-European foresight exercise with 120 experts. *Technological Forecasting and Social Change*, 183: 121915. https://doi.org/10.1016/j.techfore.2022.121915.

Bankins, S., and Formosa, P. (2023). The ethical implications of artificial intelligence (AI) for meaningful work. *Journal of Business Ethics*, 185(4): 725–740. https://doi.org/10.1007/s10551-023-05339-7.

Bauer, W., Hämmerle, M., Schlund, S., and Vocke, C. (2015). Transforming to a hyper-connected society and economy – towards an “Industry 4.0.” *Procedia Manufacturing*, 3: 417–424. https://doi.org/10.1016/j.promfg.2015.07.200.

Bokhorst, J. A., Knol, W., Slomp, J., and Bortolotti, T. (2022). Assessing to what extent smart manufacturing builds on lean principles. *International Journal of Production Economics*, 253: 108599. https://doi.org/10.1016/j.ijpe.2022.108599.

Boughzala, I., and De Vreede, G. (2015). Evaluating Team Collaboration Quality: The development and field application of a collaboration maturity model. *Journal of Management Information Systems*, 32(3): 129–157. https://doi.org/10.1080/07421222.2015.1095042.

Dalenogare, L. S., Benitez, G. B., Ayala, N. F., and Frank, A. G. (2018). The expected contribution of Industry 4.0 technologies for industrial performance. *International Journal of Production Economics*, 204: 383–394. https://doi.org/10.1016/j.ijpe.2018.08.019.

Demir, K. A., Döven, G., and Sezen, B. (2019). Industry 5.0 and Human-Robot co-working. *Procedia Computer Science*, 158: 688–695. https://doi.org/10.1016/j.procs.2019.09.104.

Dignum, V. (2022). Responsible Artificial intelligence: recommendations and lessons learned. pp. 195–214. *In: Social and cultural studies of robots and AI*. https://doi.org/10.1007/978-3-031-08215-3_9.

Dong, M., Bonnefon, J., and Rahwan, I. (2024). Toward human-centered AI management: Methodological challenges and future directions. *Technovation*, 131: 102953. https://doi.org/10.1016/j.technovation.2024.102953.

Esposito, C., De Santis, A., Tortora, G., Chang, H., and Choo, K. R. (2018). Blockchain: a panacea for healthcare Cloud-Based data security and privacy? *IEEE Cloud Computing*, 5(1): 31–37. https://doi.org/10.1109/mcc.2018.011791712.

Huang, K., Wang, K., Lee, P. K., and Yeung, A. C. (2023). The impact of industry 4.0 on supply chain capability and supply chain resilience: A dynamic resource-based view. *International Journal of Production Economics*, 262: 108913. https://doi.org/10.1016/j.ijpe.2023.108913.

Lezzi, M., Lazoi, M., and Corallo, A. (2018). Cybersecurity for Industry 4.0 in the current literature: A reference framework. *Computers in Industry*, 103: 97–110. https://doi.org/10.1016/j.compind.2018.09.004.

Mittal, S., Khan, M. A., Romero, D., and Wuest, T. (2017). Smart manufacturing: Characteristics, technologies and enabling factors. *Proceedings of the Institution of Mechanical Engineers Part B Journal of Engineering Manufacture*, 233(5): 1342–1361. https://doi.org/10.1177/0954405417736547.

Morrar, R., Arman, H., and Mousa, S. (2017). The Fourth Industrial Revolution (Industry 4.0): A Social Innovation perspective. *Technology Innovation Management Review*, 7(11): 12–20. https://doi.org/10.22215/timreview/1117.

Mutlag, A. A., Ghani, M. K. A., Arunkumar, N., Mohammed, M. A., and Mohd, O. (2018). Enabling technologies for fog computing in healthcare IoT systems. *Future Generation Computer Systems*, 90: 62–78. https://doi.org/10.1016/j.future.2018.07.049.

Nahavandi, S. (2019). Industry 5.0—A Human-Centric solution. *Sustainability*, 11(16): 4371. https://doi.org/10.3390/su11164371.

Nambisan, S., Wright, M., and Feldman, M. (2019). The digital transformation of innovation and entrepreneurship: Progress, challenges and key themes. *Research Policy*, 48(8): 103773. https://doi.org/10.1016/j.respol.2019.03.018.

Nosova, S., Norkina, A., Makar, S., and Fadeicheva, G. (2021). Digital transformation as a new paradigm of economic policy. *Procedia Computer Science*, 190: 657–665. https://doi.org/10.1016/j.procs.2021.06.077.

Özdemir, V., and Hekim, N. (2018). Birth of Industry 5.0: Making Sense of Big Data with Artificial Intelligence, “The Internet of Things” and Next-Generation Technology Policy. *OMICS a Journal of Integrative Biology*, 22(1): 65–76. https://doi.org/10.1089/omi.2017.0194.

Raisa, J. F., Rahman, M. S., Mahmud, I., Kaiser, M. S., and Han, D. S. (2025). Transition towards Healthcare 5.0: Impact of COVID-19 in the healthcare industry. *ICT Express*. https://doi.org/10.1016/j.icte.2025.04.002.

Tóth, A., Nagy, L., Kennedy, R., Bohuš, B., Abonyi, J., and Ruppert, T. (2023). The human-centric Industry 5.0 collaboration architecture. *MethodsX*, 11: 102260. https://doi.org/10.1016/j.mex.2023.102260.

Tyagi, A. K. (2023a). Decentralized everything: Practical use of blockchain technology in future applications. pp. 19–38. *In: Elsevier eBooks*. https://doi.org/10.1016/b978-0-323-96146-2.00010-3.

Tyagi, A. K. (2023b). Dew Computing. pp. 332–345. *In: Advances in systems analysis, software engineering, and high performance computing book series*. https://doi.org/10.4018/978-1-6684-8531-6.ch017.

Tyagi, A. K., and Abraham, A. (2021). Recent trends in blockchain for information systems security and privacy. In *CRC Press eBooks*. https://doi.org/10.1201/9781003139737.

Tyagi, A. K., and Sreenath, N. (2021). Cyber Physical Systems: Analyses, challenges and possible solutions. *Internet of Things and Cyber-Physical Systems*, 1: 22–33. https://doi.org/10.1016/j.iotcps.2021.12.002.

Van Erp, T., Carvalho, N. G. P., Gerolamo, M. C., Gonçalves, R., Rytter, N. G. M., and Gladysz, B. (2024). Industry 5.0: A new strategy framework for sustainability management and beyond. *Journal of Cleaner Production*, 461: 142271. https://doi.org/10.1016/j.jclepro.2024.142271.

Wamba, S. F., Gunasekaran, A., Akter, S., Ren, S. J., Dubey, R., and Childe, S. J. (2016). Big data analytics and firm performance: Effects of dynamic capabilities. *Journal of Business Research*, 70: 356–365. https://doi.org/10.1016/j.jbusres.2016.08.009.

Wei, J., and Shen, Y. (2025). Impact and mechanism of digital transformation on performance in manufacturing firms. *Innovation and Green Development*, 4(1): 100205. https://doi.org/10.1016/j.igd.2025.100205.

Section 3
Applications and Use Cases

10

Forecasting Humidity Saturation for Climate-Resilient and Sustainable Agriculture in Africa's Arid Farmlands Using Machine Learning

*Tamsangqa Benson Lubasi** and *Nikodemus Angula*

1. Introduction

Sub-Saharan Africa (SSA) faces existential threats from climate change, with arid regions like Botswana, Namibia, and Niger experiencing 1.5°C above global averages and growing seasons shortened by up to 25 percent (Tefera 2025). Over 75 percent of producers are smallholder farmers vulnerable to technological gaps and erratic rainfall (Ndung'u 2023). In western Namibia, drought intervals have shrunk from 10 to 3 years, while desertification in Niger and Botswana accelerates groundwater depletion and undermines irrigation viability (Tefera 2025). Dew, contributing up to 30 percent of dry season moisture in places like Botswana, remains excluded mainly from agricultural planning despite its importance (Wei 2024). Dew point variability influences fungal outbreaks, irrigation efficiency, and soil retention, yet remains under-forecasted. Traditional coping strategies, including dew harvesting cowpeas and nocturnal tillage, are outpaced by climate extremes. The 2023 heat dome desiccated soils beyond traditional adaptation capacity, and indigenous indicators like Acacia bloom or termite activity are eroding (Tefera 2025). Technological gaps further compound risks; smallholders often lack even basic moisture sensors (Ndung'u 2023).

Machine learning (ML) offers precise dew point forecasting by decoding complex atmospheric variables (Wei 2024). LSTM, XGBoost, and Random Forest models trained on ERA5 data each

Department of Commerce and Management, Namibia University of Science and Technology, Windhoek, Namibia.
Email: chcangula@gmail.com, https://orcid.org/0000-0002-3963-0267
* Corresponding author: thamulubasi@gmail.com, https://orcid.org/0009-0000-3589-8269

DOI: 10.1201/9781003655190-13

match region-specific humidity dynamics. For example, LSTM suits Namibia's marine layers, XGBoost excels in Botswana's sparse data zones, and RF handles Niger's dusty conditions. These models offer transformative utility. SMS-based advisories such as "80 percent dew expected, skip irrigation" conserve groundwater and optimise disease prevention. In Zinder, Niger, abandoned lands regain viability through dew-informed irrigation; in Botswana's Kgalagadi, farmers revive melon crops without pumped water. ML-guided forecasts enable desert regeneration by fusing traditional indicators with algorithmic accuracy, reducing ambiguity and building civilisational resilience in arid climates. The future of SSA farming lies in unlocking the water that is invisibly suspended in the air.

Recent discussions highlight how artificial intelligence and machine learning are becoming crucial in tackling environmental challenges and improving agricultural resilience in climate-affected areas such as sub-Saharan Africa. The rapid expansion of big data through Internet of Things devices, satellite images, and local sensor systems now allows for timely analysis of microclimatic conditions like dew point, which is especially important in regions with unpredictable rainfall. These tools support irrigation strategies and disease prediction methods that are both adaptable and grounded in the local context (Tyagi and Chahal 2020). As the field advances, new combinations with technologies such as blockchain and edge computing may help make localised forecasts more secure and accessible, particularly in remote farming communities. These innovations are shaping a new era in agriculture, where smart and affordable solutions can work alongside traditional knowledge to give farmers the tools they need to adapt and succeed despite growing environmental pressures (Pramod et al. 2021).

1.1 Research Questions

1. Can machine learning models (Random Forest, XGBoost, LSTM) accurately forecast humidity saturation across diverse arid agroecologies in Botswana, Namibia, and Niger using meteorological data?
2. Which model architecture delivers optimal performance (RMSE, MAE, R^2) for dew point prediction in each country's unique microclimate?
3. How can forecasted dew point values be operationalized into SMS-based irrigation advisories to enhance water use efficiency and climate resilience?

1.2 Hypotheses

4. LSTM models will outperform other ML approaches in forecasting accuracy due to their capacity to capture temporal dependencies in humidity saturation patterns.
5. Model performance will vary regionally, with superior results in Niger due to more stable desert climates compared to Namibia's coastal variability.
6. Integration of dew point forecasts into irrigation scheduling will reduce agricultural water use by $\geq 25\%$ while maintaining crop yields.

1.3 Objectives

1. Data Synthesis & Preprocessing: Curate and clean hourly meteorological data (ERA5 reanalysis) for Botswana, Namibia, and Niger (2018–2025), featuring dew point, temperature, precipitation, pressure, and wind components.
2. Model Development & Optimization: Train and hyperparameter tune three ML architectures (RF, XGBoost, LSTM) using time-series cross-validation and sequence windowing.
3. Performance Validation: Evaluate models using RMSE, MAE, and R^2 metrics, with country-specific benchmarking.

2. Literature review

2.1 Climate Vulnerabilities and Agricultural Fragility in Arid African Regions

Botswana, Namibia, and Niger face growing agricultural fragility due to climate change, including rising temperatures, erratic rainfall, droughts, and desertification (Spear et al. 2015). In Southern Africa, Botswana and Namibia face high evapotranspiration and reduced seasonal predictability, resulting in frequent crop failures. Although adaptation methods, such as water harvesting, exist, their adoption remains low due to inadequate institutional support (Spear et al. 2015).

Niger, in the Sahel, exemplifies compounded vulnerability where marginal soils, limited irrigation, and sociopolitical fragility drive resource conflict. Climate-induced stress contributes to tensions, but underlying causes include governance and exclusion (Akinyetun et al. 2024).

Widespread land degradation further threatens resilience, especially in African drylands (El-Beltagy and Madkour 2012). Efforts must integrate traditional knowledge, modern technology, and decentralised services (Speranza 2010).

Addressing root causes of vulnerabilities, such as poverty and inadequate infrastructure, is essential. Without this, adaptation remains superficial in regions where agricultural failure risks social and ecological collapse.

2.2 Dew Point, Humidity Saturation, and Microclimatic Dynamics in Agriculture

The dew point, the temperature at which air becomes saturated and condensation begins, plays a vital role in agriculture, particularly in arid regions. When relative humidity hits 100%, further cooling results in dew formation, a passive water source during calm, clear nights (Villalobos et al. 2017). Dew can alleviate crop water stress and replenish topsoil moisture.

Studies in Hungary show that vegetative barriers enhance dew accumulation by altering wind dynamics (Somfalvi-Tóth et al. 2019).

2.3 Machine Learning Applications in Climate and Agricultural Hydrology

Machine learning (ML) is changing and improving agricultural hydrology by predicting future readings of different weather aspects and other natural predictable figures like dew point, precipitation, and groundwater levels, which are critical for irrigation and decision-making for sustainable and improved farming. ML is exceptional at detecting non-linear patterns in big datasets from sensors and satellites (Lange and Sippel 2020), demonstrating superiority over traditional models due to its flexibility and ability to scale (Lange and Sippel 2020). Combining ML with process-based models improves parameterisation, bias correction, and surrogate modelling, improving accuracy and producing better results (Xu and Liang 2021). In the field of agriculture, machine learning (ML) has become quite helpful for managing water use and tracking pollution in real time. Some models, like deep learning and quadratic reinforcement networks, have already been used to sort irrigation data with good accuracy (Lin et al. 2024). Other models, such as multilayer perceptrons, have managed to reflect changes between weather and water use in underground water systems (Sahoo et al. 2017). In most cases, ML has performed better than the old methods, especially when it comes to predicting floods and rainfall (Mosaffa et al. 2022).

Taken together, these tools are changing how farmers and planners handle water. ML doesn't just help with making better forecasts; it also helps scale up solutions, accelerate innovation, and manage resources more wisely. This has made it an important part of farming and water systems, especially in places affected by changing weather patterns like the three target countries in this research (Namibia, Botswana, and Niger).

2.4 Regional Model Transferability and Local Adaptation of Forecasting Systems

Predictive models usually do not perform well when used in new regions. This usually happens due to differences in models' original tuning, the quality of data, and the following factors: weather conditions and land features. In particular, areas with fewer or no weather stations or unusual climate patterns tend to show this problem more. A model trained in a different environment may miss important signals when moved to another different environment. For example, a machine learning predictive model that works well in tropical areas might fail to capture the dry winds and temperature changes found in semi-arid zones like some parts of Southern Africa, including the three main case countries in this study (Spear et al. 2015, El-Beltagy and Madkour 2012). In such cases, relying on the model without adjusting or fine-tuning it can lead to wrong predictions and misleading forecasts. It's worth noting that before applying any model to a new location, it's important to look closely and study local weather behaviour, land features, currently available research knowledge in that area, and how farmers in the area are already responding to climate stress (Speranza 2010, Tefera 2025). Generationally passed-on knowledge of the local people is also important to consider.

In African agroecologies, these issues are particularly acute, necessitating the adaptation of hyperlocal models. Model transferability, the ability to retain predictive performance in new settings, varies widely, with perfect prognosis methods being more transferable than model output statistics (Baño-Medina et al. 2024), which are GCM dependent. Wang et al. (2023) showed that a regionally adapted hydrological forecast system outperformed GloFAS in China, stressing local tuning. Yang et al. (2019) revealed that regionalization methods cause high errors, especially in low precipitation zones common in Africa, while noting that quantile mapping helps reduce initialization errors. Additionally, Priyatikanto et al. (2023) found that domain adaptation techniques such as DANN and KLIEP improve crop model reliability across zones. Sangiorgio and Guariso (2024) emphasised evaluating LSTM model transferability in data-scarce areas using station clustering, a relevant strategy for Africa's under-instrumented regions.

To quantitatively assess model transferability and adaptation, this research will employ three core evaluation metrics:

- Root Mean Square Error (RMSE): To quantify the average magnitude of forecasting errors, with sensitivity to larger errors.
- Mean Absolute Error (MAE): To measure the average absolute deviation between predicted and observed values, providing a straightforward interpretation of forecast accuracy.
- Coefficient of Determination (R^2): To indicate the proportion of the variance in the observed data that is predictable from the model, thereby reflecting model explanatory power.

By basing on these metrics, with a comprehensive analysis of domain adaptability, the study aims to analyse and improve the performance of predictive systems for arid African agroecological regions.

2.5 Digital Advisory Systems and Sustainable Irrigation for SDG Aligned Agriculture

Digital advisory tools like SMS systems and smart irrigation setups have become very important, especially when trying to link farming activities to the Sustainable Development Goals (SDGs), mainly Goals 6 (clean water), 13 (climate action), and 15 (life on land). These tools help farmers make better decisions, particularly around when and how much to irrigate. In many cases, smart irrigation uses technologies such as the Internet of Things (IoT), making it possible to monitor water use closely and reduce waste, which is key in dry places where every drop counts (Priyambada et al. 2023).

In addition, geospatial systems have helped by giving farmers maps that show which parts of the land need attention, and with this knowledge, farmers can seek advice or research on what kind of attention exactly should be given to that part of the land. This has reduced overuse of chemicals, improved yields, and protected biodiversity at the same time (Pandey and Pandey 2023). When this is combined with modern practices like precision farming and ideas from the bioeconomy, food production improves without harming the land.

It's also worth noting that in some Islamic countries, new approaches like digital farming and Islamic digital finance are helping to grow agriculture in a fair and inclusive way (Tlemsani et al. 2023). These systems are being aligned with the SDGs to support wider goals like reducing poverty and promoting justice.

Overall, mixing digital advice with local irrigation methods makes it easier to build farming systems that are not only efficient but also fair and strong enough to cope with climate changes.

3. Methodology

3.1 Study Area and Data Sources

The study targeted dryland farms in Botswana, Namibia, and Niger, using ERA5 reanalysis data (2018–2025) to extract key meteorological variables relevant to humidity, which were then converted to interpretable units and averaged over national farmland zones.

The drylands of southern Africa, including Botswana, Namibia, and Niger, are among the most climate-stressed regions on Earth. These semi-arid zones face very seasonal and highly variable rainfall, frequent droughts and flash floods, and rising temperatures (Spear et al. 2015, El-Beltagy and Madkour 2012). In short, it's what our rural neighbours often know from bitter experience: rain-fed farming here is a gamble. Scientists note that people in dry areas are likely to be most seriously hit by shifts in moisture and temperature regimes due to climate change. In such marginal settings, traditional farming is risky and yields can collapse when rains fail or storms intensify (Speranza 2010, Somfalvi-Tóth et al. 2019). We therefore set out to analyse meteorological data for these countries, using it to understand humidity trends and help design better adaptation tools. Our methodology is built on rich ERA5 reanalysis climate data (2018–2025) and aims to translate that raw information into the meaningful variables that farmers and agronomists care about (see Table 1). We extracted key weather variables from ERA5 and converted them into familiar units (e.g., °C and mm), then averaged them over the main agricultural zones of each country. The idea is to bridge global climate data and local farm realities. After all, precision tools like this support the Sustainable Development Goals on food and water security.

Table 1: Variables used in the ML models.

Variable	Description	Unit	Source
u10	10 m eastward wind component	m/s	ERA5
v10	10 m northward wind component	m/s	ERA5
d2m	2 m dew point temperature	°C (converted from K)	ERA5
msl	Mean sea level pressure	Pa	ERA5
sp	Surface air pressure	Pa	ERA5
tp	Total precipitation	mm (converted from m)	ERA5
t2m	2 m air temperature	°C (converted from K)	ERA5
longitude	Longitude of the data point	Degrees	ERA5
latitude	Latitude of the data point	Degrees	ERA5
valid_time	Timestamp of the observation	YYYY-MM-DD HH:MM:SS	ERA5

Our data sources were global reanalyses (ERA5) that assimilate observations into a consistent climate record. ERA5 provides hourly and daily estimates of weather fields on a high-resolution grid. We used data from 2018 through 2025 to cover recent conditions. From ERA5, we selected variables known to govern humidity and dew formation, as well as overall climate stress. These include the 10-metre wind components (*u*10, *v*10) in m/s, which capture the horizontal wind vectors; the 2-metre dew point temperature (*d2m*) in °C, our proxy for atmospheric moisture; mean sea level pressure (*msl*) and surface air pressure (*sp*) in pascals; total precipitation (*tp*) in millimetres; and 2 meter air temperature (*t2m*) in °C. We also recorded the geographic latitude/longitude and timestamp for each grid point. In practical terms, *u*10 and *v*10 tell us how air masses and moisture are moving across the landscape, while *d2m* directly indicates how saturated the air is. The pressure fields (*msl*, *sp*) influence humidity via condensation and evaporation processes. Rainfall (*tp*) is obvious: it's the primary source of water for crops. Air temperature (*t2m*) drives evaporation and plant stress; for example, higher temperatures increase vapour pressure demand and often call for higher humidity in the root zone. Each of these variables thus plays a role in the crop water balance and microclimate that farmers know so well. Somfalvi-Tóth et al. (2019) found that capturing this microclimate requires exactly these kinds of measurements; they instrumented both an agroforestry plot and an open field and recorded 10 m wind speed and direction, 2 m temperature and relative humidity, rainfall amount/intensity, dew formation, and soil moisture (volume %). We followed a similar logic, albeit on a broader national scale.

In technical terms, the ERA5 data come with native units that we converted to more interpretable ones. Temperatures (*d2m*, *t2m*) are given in Kelvin, so we subtracted 273.15 to report them in Celsius (°C). Precipitation (*tp*) is output in metres per time interval, so we multiplied by 1000 to get millimetres (mm). The pressure values (*msl*, *sp*) are already in pascals (Pa), which we left unchanged. Wind components (*u*10, *v*10) are in m/s and needed no conversion other than possibly changing sign conventions. In essence, we prepared a dataset where each row is a specific location in a country's farm areas at a given time, with columns for eastward/northward wind, temperature, dew point (humidity), precipitation, and pressure. Including longitude, latitude, and timestamp lets us average or filter by region and season. We then grouped these gridded data by each country's agricultural zones—for example, national farm belts or biome regions—and computed mean values. In other words, we took the raw ERA5 grid (which might have thousands of points per country) and distilled it into country-level time series for each variable. This step is crucial because individual grid cells can be noisy, but national agricultural averages are more robust and meaningful for policy and planning. As one review of precision agriculture notes, merging satellite or model data at relevant scales helps "map and monitor soil and crop conditions" and boosts yields while conserving resources.

Because our focus is humidity (dew point), we paid special attention to those variables. Dew point (*d2m*) itself is an intrinsic measure of moisture: it is the temperature at which air becomes saturated with water vapor, and it rises when humidity is high. This means dew point and relative humidity are closely linked; our use of *d2m* provides a direct method to control on the actual moisture content of the air. As Shamshiri et al. (2018) emphasize from greenhouse studies, dew point (and related vapour pressure deficit) is key to understanding plant water stress. They write that *"humidity alone does not provide any information about plant transpiration"* and that the vapour pressure deficit, which depends on both temperature and humidity, is the driving force for water movement in plants. In practical terms, if the day is hot (high *t2m*) but dew point is also high, plants can maintain leaf moisture. But if hot *t2m* combines with low *d2m*, the drying force is huge. Our methodology thus models how *t2m*, *d2m*, and pressure interact to dictate humidity levels. One tip from agronomy guides is that plants under high temperature often require higher humidity to avoid stress. This is precisely why we include all these factors: dew point can't be interpreted in isolation if we want to advise farmers about irrigation or forecasts.

Wind components (*u*10, *v*10) also matter for humidity, even if it's not obvious at first glance. For instance, strong winds can evaporate moisture from both soil and foliage, lowering humidity and raising vapour pressure deficit. In our drylands, afternoon winds may bring in dry air from the desert. Conversely, evening breezes from wetter areas or from higher latitudes may cool the land and allow humidity (and dew) to form. Pandey and Pandey (2023) note that geospatial data (like ERA5) are invaluable for mapping such patterns over time. They highlight that modern agriculture increasingly relies on remotely sensed climate inputs to optimise yields while conserving water. In fact, climate-smart agriculture models often combine wind, humidity, and rain data to fine-tune irrigation—for example, triggering watering only when both temperature and vapour deficit exceed certain thresholds. By including *u*10 and *v*10, we capture the physical transport of air masses; with *msl* and *sp*, we implicitly capture barometric influences on moisture as well. Somfalvi-Tóth et al. (2019) likewise tracked wind at 10m together with humidity and rain, underscoring that these factors go hand-in-hand in dryland microclimates.

The coordinates (latitude, longitude) and timestamps in our dataset let us tie these variables to place and season. For example, October in Niger's farmlands may look very different from May in Botswana or January in Namibia. We kept the full timestamp (YYYY-MM-DD HH:MM:SS) so that later analyses, such as machine learning models, can learn diurnal cycles and seasonal shifts. We averaged each variable across all farm grid cells in a country to get a national profile. In effect, each country yields a time series like *"average dew point in Botswana on 2020–05–15 at 12:00 UTC"* and so on. This approach smooths out local noise (like one grid cell seeing a thunderstorm) but retains broad patterns. It also means our results directly inform national planning. Studies of smallholder adaptation often emphasise this kind of scale-up: while farmers vary plot-to-plot, policy needs broad indicators like "*Botswana's mean growing-season humidity*" to make resilient decisions (Ndung'u 2023, Pandey and Pandey 2023).

To process and summarise these variables, we combined ERA5 with simple Python routines. We wrote scripts to apply the unit conversions (Kelvin→C, m→mm) and to compute country-level means weighted by farmland area. In practice, we masked ERA5 grids to national agricultural zones (for example, excluding high mountains or cities) before averaging. Then, for each date/time, we recorded the aggregated *u*10, *v*10, *t2m*, *d2m*, *tp*, *msl*, *sp*. The workflow was: download ERA5 for each country, transform units, spatially aggregate, and stack the variables. (This is akin to building a "digital twin" of each country's farming climate.) In our team, we like the idiom: *"garbage in, garbage out"*, so careful data preprocessing was key. We double-checked that the converted dew point and temperature values matched known weather station readings where possible, and we ensured rainfall sums from ERA5 were realistic (e.g., tens of mm per storm, not hundreds).

Ultimately, these processed variables feed into our deeper analyses (e.g., regression or machine learning to predict dew point). But even as raw descriptors, they tell an important story. For instance, if Botswana's *tp* is persistently low and *t2m* persistently high, we know the atmosphere is often thirsty (high vapour pressure deficit). If *d2m* moves up and down more than expected, that may signal unusual humidity swings that can stress crops. By quoting values like R^2 or *RMSE* from model results, we measure how well our inputs explain actual moisture patterns. It's worth remembering why we care: each variable in our table is a piece of the climate puzzle that farmers live and farm by.

To a farmer or extension officer, these numbers translate into actionable insights. For example, a high mean *t2m* combined with low *d2m* might suggest irrigation is urgently needed that week. High *tp* readings signal a good rain event. Sharp drops in *msl* could forecast storms. By mapping these ERA5-derived variables onto dryland realities, we aim to offer "smart" guidance for agriculture. Indeed, the literature shows that integrating such data into precision farming—for example, through automated irrigation systems or advisory services—can significantly advance food security goals.

In summary, our study area is the dry farm lands of Botswana, Namibia, and Niger, and our data comes from ERA5 reanalysis (2018–2025). We extracted variables—wind (*u10*,*v10*), temperature (*t2m*), dew point (*d2m*), precipitation (*tp*), pressure (*msl*,*sp*)—that jointly describe humidity conditions. These were converted to °C or mm, aligned by time, and averaged over agricultural zones. As Speranza (2010) notes, rural dryland communities live with *"high temperatures and high rainfall variability"* that make rain-fed farming risky. By carefully handling these variables, we aim to turn climate data into actionable knowledge, blending modern technology with on-the-ground wisdom. After all, as Tefera (2025) shows, combining traditional water management with machine learning can build *"resilient agricultural systems"* in sub-Saharan Africa. Our methodology is a step in that direction, using geospatial climate data to sustain farming livelihoods under harsh conditions.

3.2 Data Preprocessing

ERA5 data were preprocessed into model-ready features by converting units, deriving wind metrics, and engineering cyclical time features. Interpolation and carry-forward techniques addressed missing values, preserving temporal consistency and enhancing model performance on climate patterns.

3.3 Machine Learning Models and Training

We developed three models for predicting humidity via dew point: Random Forest (RF), XGBoost, and Long Short-Term Memory (LSTM) neural networks. Scikit-learn handled RF, XGBoost's native library implemented gradient boosting, and TensorFlow/Keras supported LSTM.

RF: Trained using RandomizedSearchCV for hyperparameter tuning, RF captured non-linear patterns via an ensemble of decision trees.

XGBoost: Sequential trees minimized residuals, with early stopping on validation loss to prevent overfitting. Key parameters like learning rate and depth were optimized.

LSTM: Keras-based LSTM modelled temporal dependencies using sliding windows of hourly climate data. Features included t2m, tp, msl, sp, wind metrics, and cyclical time features. The model was trained with the Adam optimiser, tuned via dropout, batch size, and sequence length, and validated using early stopping.

All models were trained per country, with 2018–2023 for training and 2024–2025 for testing, ensuring causality and avoiding data leakage.

3.4 Hyperparameter Tuning and Validation

We used time-aware methods for tuning and validation. For Random Forest and XGBoost, we applied scikit-learn's TimeSeriesSplit to preserve temporal order. This walk-forward validation trained models on earlier periods and validated on later ones, avoiding future data leakage. Optimal hyperparameters were selected based on average performance across folds.

For LSTM, the last segment of the training data served as a validation set for early stopping. Grid or randomised search explored variations in LSTM depth, units, dropout, and learning rate, with validation loss guiding optimal configurations.

Evaluation relied on Root Mean Squared Error (RMSE), Mean Absolute Error (MAE), and R^2 (coefficient of determination). RMSE penalises large deviations, MAE captures average error magnitude, and R^2 measures explained variance. These metrics were computed per model and country, including the final hold-out year (2024–2025), enabling robust performance comparison across models and regions.

3.5 Deployment Approach

Our hybrid deployment system delivers forecasts to smallholder farmers via: (1) SMS alerts (e.g., "High humidity expected tomorrow—delay irrigation") from centrally hosted Random Forest/XGBoost models, and (2) on-device LSTM (pruned/quantised to TensorFlow Lite) for offline use. The SMS system, sent in local languages through automated gateways, provides actionable advice like "Dew at 80% tonight—skip watering". This balances accuracy with accessibility in low-connectivity areas while optimising water use.

3.6 Model Interpretability and Regional Tuning

To improve interpretability and local relevance, models were calibrated to reflect regional climate dynamics. In Namibia's coastal zones, marine boundary layer effects were emphasised due to sea spray and wind-driven humidity. In Niger, Harmattan winds (October–February) cause humidity drops; models incorporated seasonal indicators and emphasised longwave radiation signals via msl and tp.

Feature importance was analysed per country, and post-training diagnostics included partial dependence plots and Shapley values. These tools revealed variable influence on humidity, for instance, how rising temperature or falling pressure affected dew point.

Insights guided regional retraining: if the model under-predicted during Botswana's rainy season, those months received added weight. This iterative tuning improved accuracy and contextual performance.

Overall, interpretability methods and domain knowledge ensured that each country—coastal Namibia, savannah Botswana, and Sahelian Niger—benefited from a locally adapted, context-sensitive forecasting model.

3.7 Tools and Libraries

The analysis pipeline was built in Python using scikit-learn, XGBoost, and TensorFlow/Keras. Data handling used pandas, xarray, and GeoPandas; visualisation used Matplotlib, Seaborn, and Cartopy. Parallel processing and reproducibility practices ensured efficient, consistent model training and results.

4. Results

In this study, we evaluated the performance of Random Forest (RF), XGBoost, and Long Short-Term Memory (LSTM) models for forecasting dew point (a proxy for humidity) in desert farmlands of Botswana, Namibia, and Niger. Table 2 summarises the root mean squared error (RMSE), mean absolute error (MAE), and coefficient of determination (R^2) for each model and region. Lower RMSE and MAE values indicate more accurate forecasts, and higher R^2 indicates better model fit. The Table 2 below shows that performance varies by algorithm and country. We discuss each model's results in turn.

Table 2: Model performances.

Model	Country	RMSE	MAE	R^2
RandomForest	Botswana	3.8171	2.7375	0.7886
XGBoost	Botswana	3.6172	2.6082	0.8102
LSTM	Botswana	3.0799	2.3233	0.8623
RandomForest	Namibia	4.8281	3.7231	0.7344
XGBoost	Namibia	4.7145	3.6658	0.7467
LSTM	Namibia	3.9306	3.0629	0.8240
RandomForest	Niger	3.6579	2.8605	0.8320
XGBoost	Niger	3.5162	2.7620	0.8448
LSTM	Niger	3.1331	2.4289	0.8768

4.1 Random Forest Results

Random Forest captured dew point trends moderately but struggled in Namibia's arid climate, showing reduced accuracy due to data sparsity and regional climatic variability.

4.2 XGBoost Results

XGBoost performed slightly better than Random Forest in all the regions, capturing non-linear relationships, even though accuracy gains were modest and still limited by the local climate complexities of the regions.

4.3 LSTM Results

LSTM obtained the best results overall, leveraging temporal patterns to predict dew point more accurately despite higher training complexities in the subject countries.

4.4 Forecast Visualization and Model Performance

Result-based generated visual plots display model forecasts in comparison with actual dew points, clearly highlighting temporal accuracy, error fluctuations, and performance differences across Botswana, Namibia, and Niger over six days.

4.4.1 Botswana Forecasts

The dew point forecast analysis for Botswana in the Figure 1 below shows important insight into how different machine learning models perform in a semi-arid inland climate region. The study compared the performance of the predictive models as they were tested against actual observed dew point values over a fixed period to evaluate their effectiveness in capturing atmospheric moisture changes. These results match what has been observed in similar environments where LSTM models adapt better to weather changes, therefore producing better output. While both Random Forest and XGBoost use tree-based systems that are good for finding general patterns, they were not sensitive enough to detect and flag fast changes in dew. Their outputs were often smoothed out, failing to catch short-term peaks and drops in humidity. This makes them less useful for real-time decisions in agriculture.

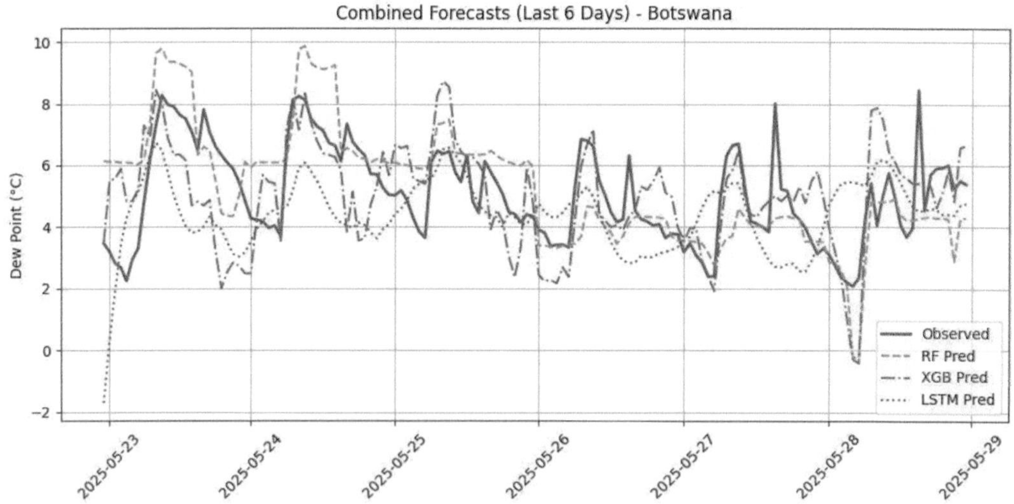

Figure 1: Botswana combined model forecasts.

In fact, Random Forest was the least responsive. It missed several peaks, flattening its predictions and making the dew point curve less realistic. For farmers depending on accurate alerts, this kind of output may lead to wrong irrigation choices. XGBoost did a bit better; it followed the trend but was late in picking up quick shifts. For instance, when dew increased rapidly one evening, XGBoost reacted slowly, which would have delayed any action taken based on its advice.

These issues are not just academic. In places like Botswana, a few hours can make or break a crop. Wrong timing in watering leads to nutrient loss in soil or even crop stress. In contrast, the LSTM model's memory-based design helped it learn patterns across time and respond better. It understood how earlier weather patterns affected later changes, which is a big plus for climate-smart farming.

One more thing that makes LSTM helpful is its ability to project into the future. After training, the model can use what it has learned to predict dew several days ahead. This can support planning, like when to plant, spray, or harvest. A farmer with access to three-day dew forecasts might save water or avoid using chemicals when nature already provides the moisture.

Also, dew point in Botswana is affected by more than just temperature. Factors like air pressure, wind, and soil dryness also play a part. LSTM can take all these into account and find patterns that other models miss. Because of this, it becomes a strong tool for regions with mixed or shifting climate conditions.

The Botswana results clearly show the strength of LSTM in forecasting dew point. While XGBoost and Random Forest have their uses, they fall short when it comes to tracking short-term weather changes. For dry regions where every drop matters, a model like LSTM gives farmers a chance to adapt better, save water, and achieve higher crop yields.

4.4.2 Namibia Forecasts

Namibia's weather is more complex than the other countries. One day it is hot and dry, the next it is cool and damp, especially near the western coast. We tested how well three models, Random Forest, XGBoost, and LSTM, could follow dew point changes in this shifting climate.

Between May 24 and May 26, we saw sharp jumps and drops in dew point readings. These were not minor swings; they happened fast, within hours. That is typical in Namibia. The country sits between the cold Atlantic and the warm inland plateau. This mix creates unusual weather where the air can be dry in the afternoon and suddenly moist at night. Therefore, we needed models that could handle fast changes.

Out of the three, LSTM clearly did better. It followed the quick ups and downs very closely. This makes sense because LSTM remembers past weather values and adjusts over time. Dew does not just show up; it builds from factors such as earlier temperature, wind, and air pressure. That is what LSTM understands best.

Both XGBoost and Random Forest missed the sharp dew peaks. Their predictions looked too smooth (see Table 2). For example, on May 25, when the dew rose, there was only a slow increase. The same thing occurred when the dew dropped; there was a lag.

This may seem like a minor matter, but it is actually a significant issue. In real farming, missing a dew event can waste water. Imagine a farmer waters the field, thinking it is dry, yet the dew has already handled it. That is a waste of limited resources.

In coastal Namibia, where water is highly valuable, many smallholder farmers collect dew using nets or sheets. However, dew forms for only a few hours early in the morning. If they get the timing wrong by even an hour, they lose that chance. That is where LSTM helps. It provides better short-term predictions so that farmers can act quickly.

LSTM also opens opportunities for early alerts. If combined with a mobile system, farmers could receive messages such as: "Expect high dew tomorrow morning, irrigation not needed." Simple information like this helps protect crops and manage water more efficiently.

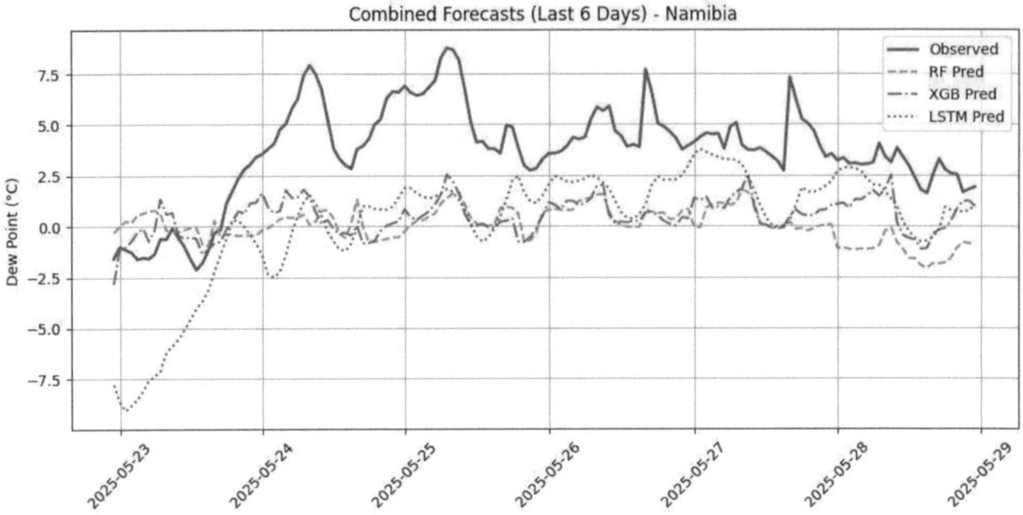

Figure 2: Namibia combined model forecasts.

This does not mean Random Forest and XGBoost are without value. In more stable areas with slow climate shifts, they might actually perform better. They are also faster to train and easier to interpret. However, in a place like Namibia, where weather changes rapidly and with intensity, they struggle to adapt.

Therefore, this is not simply about choosing the best model. It is about selecting the right one for the specific environment. In Namibia's case, LSTM is not only more accurate; it is also more practical. It aligns with how weather behaves in the region and supports real decisions on the ground. For small farmers who rely on nature's rhythm, that kind of accuracy is not a luxury. It is essential.

4.4.3 Niger Forecasts

The forecast plot for Niger compares how three machine learning models, RF, XGB, and LSTM, performed when predicting dew point values over six days. The plots in Figure 3 visually display the trends of each model's predictions, which can help us to compare and analyse the model behaviours. The plots in the figure also show how well these models captured Niger's unique climate. Unlike coastal or semi-arid regions, Niger is dominated by the Sahara Desert, which means it experiences extreme dryness with only occasional changes and fluctuations in humidity. This makes forecasting dew point a little harder but very important and necessary for farming, especially in areas where irrigation is limited and farmers depend on natural moisture.

During the observed period from May 23 to May 28, Niger's dew point stayed relatively steady compared to places like Namibia, where humidity swings are more dramatic. Still, small dips and rises occurred, probably due to night-time cooling, changing winds, or brief moist air moving in from nearby areas. The LSTM model did the best job of following these subtle changes and matched the observed data closely. Because LSTM can remember previous sequences, it picked up on gradual shifts that the other models missed. For example, on May 25, there was a small drop in dew point likely caused by dry winds. While RF and XGBoost smoothed over this dip, LSTM captured it well, showing its sensitivity to changes over time.

Random Forest, which works well in many situations, had trouble with Niger's small humidity changes. Its predictions were generally flatter and missed the short-term dips. This happens because tree-based models like RF average many decisions, which can oversimplify fluctuations. In a dry place like Niger, even tiny dew point changes affect how well crops stay hydrated. If a model misses a natural dew event, farmers might water crops unnecessarily and waste scarce water.

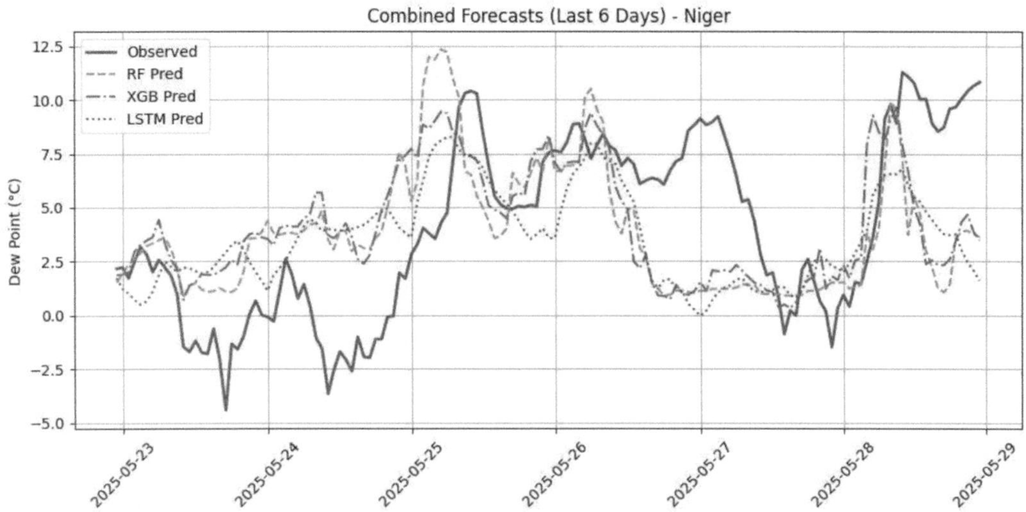

Figure 3: Niger combined model forecasts.

XGBoost did better than RF but still didn't match LSTM. It caught broader trends, like the steady rise in dew point from May 26 onwards, but missed the finer details. Its predictions also lagged a bit during day-to-day changes. This finding shows that while XGBoost performs well with complex patterns, it does not learn real-time sequences as effectively. In Niger, where temperature differences between day and night can be quite wide, this delay can affect key farming decisions. A missed dew event might mean failing to benefit from natural moisture at the right time.

Such differences have real-life consequences. In Niger, farming depends largely on crops like millet and sorghum. These crops need precise water management. Accurate dew point predictions help farmers know the best time to irrigate.

5. Conclusion

The comparison of Random Forest, XGBoost, and LSTM across Botswana, Namibia, and Niger has produced one clear fact: LSTM handles dew point changes more accurately. In Botswana, it recorded an RMSE of 3.08 and R^2 of 0.86, capturing both gradual and sudden humidity shifts that other models missed. In Namibia, where coastal and inland air mix rapidly, LSTM's RMSE of 3.93 outperformed Random Forest's 4.83 by tracing sharp ups and downs closely. Niger's mostly stable but crucial minor dew point variations were best tracked by LSTM, with an RMSE of 3.13 and R^2 of 0.88, enabling precise irrigation timing in very dry zones. These results show that models which learn from past sequences and adapt over time deliver superior forecasts in semi-arid and arid climates. While Random Forest and XGBoost may serve when fast training and interpretability are priorities, they smooth out key fluctuations and risk poorer water-management decisions. The choice of forecasting tool must therefore match local weather dynamics. In regions where every drop counts, LSTM's ability to learn temporal patterns offers farmers the best chance to optimise irrigation, conserve scarce water, and build resilience against drought and desertification.

References

Akinyetun, T. S., Fatai-Abatan, A., and Ogunbodede, N. (2024). Heated Environment, Armed People: Between "Climate Change Conflict" and "Fragility Conflict" in the Sahel. *Journal of Asian and African Studies*, 00219096241285108.

Baño-Medina, J., Iturbide, M., Fernández, J., and Gutiérrez, J. M. (2024). Transferability and explainability of deep learning emulators for regional climate model projections: Perspectives for future applications. *Artificial Intelligence for the Earth Systems*, 3(4): e230099.

El-Beltagy, A., and Madkour, M. (2012). Impact of climate change on arid lands agriculture. *Agriculture & Food Security*, 1: 1–12.

Lange, H., and Sippel, S. (2020). Machine learning applications in hydrology. In *Forest-water Interactions* (pp. 233–257).

Lin, Y. C., Alorfi, A. S., Hasanin, T., Arumugam, M., Alroobaea, R., Alsafyani, M. et al. (2024). Water agricultural management based on hydrology using machine learning techniques for feature extraction and classification. *Acta Geophysica*, 72(3): 1945–1955.

Mosaffa, H., Sadeghi, M., Mallakpour, I., Naghdyzadegan Jahromi, M., and Pourghasemi, H. R. (2022). Application of machine learning algorithms in hydrology. *Computers in Earth and Environmental Sciences*, 585–591.

Ndung'u, S. N. (2023). A data-driven model for sustainable deployment and adoption of climate smart agriculture practices among smallholder farmers in Kakamega County. *MMUST Repository. f.*

Pandey, P. C., and Pandey, M. (2023). Highlighting the role of agriculture and geospatial technology in food security and the Sustainable Development Goals. *Sustainable Development*, 31(5): 3175–3195.

Pramod, A., Naicker, H. S., and Tyagi, A. K. (2021). Machine learning and deep learning: Open issues and future research directions for the next 10 years. pp. 463–490. *In: Computational Analysis and Deep Learning for Medical Care: Principles, Methods, and Applications*. John Wiley & Sons, Inc.

Priyambada, I. B., Hardyanti, N., Budihardjo, M. A., Puspita, A. S., and Cahyati, A. P. (2023). A review of smart agricultural transition to achieving Sustainable Development Goals (SDGs): Smart irrigation system. *IOP Conference Series: Earth and Environmental Science*, 1268(1): 012010.

Priyatikanto, R., Lu, Y., Dash, J., and Sheffield, J. (2023). Improving generalisability and transferability of machine-learning-based maize yield prediction model through domain adaptation. *Agricultural and Forest Meteorology*, 341: 109652.

Sahoo, S., Russo, T. A., Elliott, J., and Foster, I. (2017). Machine learning algorithms for modeling groundwater level changes in agricultural regions of the US. *Water Resources Research*, 53(5): 3878–3895.

Sangiorgio, M., and Guariso, G. (2024). Transfer learning in environmental data-driven models: A study of ozone forecast in the Alpine region. *Environmental Modelling & Software*, 177: 106048. 21

Shamshiri, R. R., Jones, J. W., Thorp, K. R., Ahmad, D., Che Man, H., and Taheri, S. (2018). Review of Optimum Temperature, Humidity, and Vapour Pressure Deficit for Microclimate Evaluation and Control in Greenhouse Cultivation of Tomatoes. *International Agrophysics*, 32(2): 287–302.

Somfalvi-Tóth, K., Burucs, Z., and Nyisztor, B. (2019). Online soil moisture and microclimate investigations on agroforestry and agricultural field plots. *Regional and Business Studies*, 11(1): 29–37.

Spear, D., Haimbili, E., Angula, M., Baudoin, M. A., Hegga, S., Zaroug, M. et al. (2015). *Vulnerability and adaptation to climate change in the semi-arid regions of Southern Africa.*

Speranza, C. I. (2010). Resilient adaptation to climate change in African agriculture. *Studies. econstor.eu.*

Tefera, M. L. (2025). Sustainable solutions to land degradation and rainfall variability in Sub-Saharan Africa: Integrating traditional water management, agricultural intensification, and machine learning. *Università degli studi di Sassari.*

Tlemsani, I., Zaman, A., Hashim, M. A. M., and Matthews, R. (2023). Digitalization and sustainable development goals in emerging Islamic economies. *Journal of Islamic Accounting and Business Research.*

Tyagi, A. K., and Chahal, P. (2020). Artificial intelligence and machine learning algorithms. pp. 188–219. *In:* Challenges and Applications for Implementing Machine Learning in Computer Vision (). *IGI Global Scientific Publishing.*

Villalobos, F. J., Mateos, L., Testi, L., and Fereres, E. (2017). Air temperature and humidity. pp. 55–67. *In: Principles of Agronomy for Sustainable Agriculture*. Springer.

Wang, H., Zhong, P., Zsoter, E., Prudhomme, C., Pappenberger, F., and Xu, B. (2023). Regional adaptability of global and regional hydrological forecast system. *Water*, 15(2): 347. 18

Wei, S. (2024). Towards sustainable irrigation management: A machine learning approach to monitoring and optimization. *Arizona State University.*

Xu, T., and Liang, F. (2021). Machine learning for hydrologic sciences: An introductory overview. *Wiley Interdisciplinary Reviews: Water*, 8(5): e1533.

Yang, X., Magnusson, J., and Xu, C.-Y. (2019). Transferability of regionalization methods under changing climate. *Journal of Hydrology*, 568: 67–81.

11

Emerging Technologies for Smart Banking and Financial Systems for Modern Era

Prakruthi N. Udupa,[1] *Anusha K. M.*,[2] *Suma B. R.*[3] and *Mansi Kukreja*[4,*]

1. Introduction

Banking and finance have always been foundational to economic development and social progress. From early barter and coin-based economies to digital banking and payments with a mobile device, the history of financial services is a story of the human quest for greater efficiency and security in the manner we manage money, while also increasing convenience and accessibility. The past two decades have seen a rapid change in the way financial services are provided, accessed, and regulated; the driver of this evolution has been increasingly rapid technological change. Banking has predominantly happened in bank branches, on paper and by hand, managing accounts with a bank representative and face-to-face with customers. Reliance upon banks to perform these banking functions created opportunities for instrument error, operational inefficiencies, and inherent limitations of geography and proximity. The introduction of internet banking and ATMs began the first wave of digital transformation for the financial services sector, providing opportunities for remote transactions, account access in real time, faster processing, and remote bank account access through public networks of communication technology, namely telecommunication networks. The start of the 21st century ushered in a digital banking revolution, where information and communication technologies (ICT) were ubiquitous in service delivery—smartphones, mobile applications, and internet-based services had emerged as the interface between consumers and banks. Digital banking not only offered convenience and around-the-clock availability but also helped pave the way for new technologies such as artificial intelligence (AI), blockchain, cloud computing, big data analytics, and biometrics (Kou and Lu 2025). In terms of a technological paradigm shift, this has created a new set of expectations for customers. Nowadays, customers expect not only digital access but intelligent, friendly, personalized financial experiences. The modern customer wants to onboard seamlessly,

[1] Surana College Autonomous, Kengeri, India, Email: prakruthin.udupa@gmail.com, 0000-0001-8891-9465
[2] Community Institute of Management Studies, Bengaluru City University, Bengaluru, India. Email: anusha.km1@gmail.com, 0009-0007-9638-0069
[3] S-Vyasa Deemed to be University, Bengaluru, India, Email: Sumaramamurthy@gmail.com
[4] CMS Business School, Jain (Deemed-to-be University), Bangalore, India.
* Corresponding author: mansikukreja80@gmail.com

DOI: 10.1201/9781003655190-14

have their transactions processed in real time, receive proactive and timely alerts on fraud, have voice-enabled assistants to carry out their requests, and receive tailored product recommendations. All these expectations are indicative of smart banking—the next stage in banking that incorporates systems that are driven by data, intelligent, secure, and responsive to customer changes.

Digital transformation has proven to be of immense value and has created enormous challenges for traditional financial institutions. Legacy systems, regulatory response, cybersecurity threats, and limited adaptability are only a few of the pressing challenges that keep traditional banks from innovating at the pace of technology (Zalan and Toufaily 2017). Broad-based innovation will be needed to overcome the challenges of emerging, entrenched FinTech startups and neobanks built solely on digital foundational infrastructures. Emerging technologies such as AI and blockchain are not simply evolutionary technologies—they are potentially disruptive technologies. They are reimagining and omnichanneling back-office tasks, decisions, customer engagement, compliance processes, and the way in which financial transactions can be delivered. For example, AI is enabling smarter customer and credit scoring and predictive analytics, while blockchain is facilitating decentralized finance (DeFi) models to bypass authorities. In light of this context, a few pertinent questions come to mind: What challenges are traditional financial systems facing in adapting to the new technological landscape? Which technologies are having the greatest impact on the banking and financial ecosystem? How can financial institutions efficiently employ these technologies to remain competitive, secure, and customer-centric in the modern day? This research will seek to answer these questions by identifying the technological, operational, and social elements of smart banking. This inquiry will also address how these various technological elements are being implemented across the world, what regulatory frameworks are evolving in tandem, and what the future of financial services will look like (Neves et al. 2023).

The motivation of this research is to look at the science of innovation in today's world and the role of technology as part of the wider process of innovation in the transformation of banking and financial services. The project to achieve this will provide a common description of the process of innovation for the financial services industry through technological advancement, which is affecting both frontend and backend innovations and services. The study will explore the contributions of many of the key technologies of interest such as AI, machine learning, blockchain, IoT, cloud, and biometrics toward customer service automation, fraud detection, digital identity management, decentralized finance, and real-time payments; open banking ecosystems and platforms, API architecture, and programmable and trustless contracts. In addition to evaluating technological advancements, the research will analyze the opportunities and threats that this change will bring. This encompasses operational efficiencies, enhancements in financial inclusion, improvements in user experience, and reductions in cost on one side of the coin, and data privacy, cyber threats, ethics, and productive constraints from regulatory agencies on the other. The research will have a global context and will include illustrative case studies from developed/advanced and developing economies to demonstrate the diverging adoption paths for technology in contrasting socioeconomic contexts. The research will provide a balanced overview by including innovation, policy implications, and sociotechnical contexts. This research explores qualitative, exploratory methodology through a review of academic literature, industry studies, and case studies. The main approach is to summarize secondary data, compiling evidence of growth and trends related to smart banking technologies. The formulated literature consists of peer-reviewed journal articles, whitepapers from banks and technology providers, government position papers, and regulatory guidance. Case studies will allow for an understanding of how banking technologies are used in practice in different regions of the world, including examples from India, the US, Brazil, Kenya, and members of the EU, allowing for theoretical ideas to be understood with practical applications and also allowing for more reliable comparisons of different innovation models. This research will examine statistical data and market research from reputable sources like McKinsey, Accenture, BIS, IMF, and the World Bank where available to substantiate claims about trends in adoption, customer behavior, and outcomes in the market. A thematic analysis approach will be used to identify and classify the key themes, challenges, and opportunities that emerge from these technologies. While the research does not involve gathering firsthand empirical data—deemed not appropriate given its conceptual and exploratory nature—it offers frameworks and models to inform strategic initiatives and guides for further empirical research in this area.

The remainder of the chapter has been organized as follows: In Section 2, we introduce the conceptual evolution of digital and smart banking and assess the transition from traditional banking, where branch networks drove service delivery, to mobile-first, data-centric, intelligent banking delivery methods. In Section 3, we illustrate the emerging technologies driving smart banking, discussing the relevant innovations of Artificial Intelligence and Machine Learning (AI/ML), Blockchain and Distributed Ledger Technology (DLT), IoT, Cloud Computing, Edge Computing, and Quantum Computing. Each subsection details the influence of the technology on service automation, personalization, fraud detection, transparency of transactions, scalability, and security in relation to cryptographic principles. In Section 4, we present digital financial services within the constructs of digital payments and mobile wallets, DeFi, inclusive open banking APIs, and regulatory technology (RegTech), and we position how these digital financial services challenge and change access and integration in financial technology and compliance in banking. Finally, Section 5 provides case studies and global scenarios comparing how each of the smart banking technologies in developed economies compares to emerging markets and the role of public infrastructure, innovation ecosystems, and regulatory values. Section 6 expresses the benefits, possibilities, and obstacles to smart banking in terms of better decision-making, financial inclusion, accountability, and compliance, but also risks in terms of data privacy, regulatory shortcomings, digital exclusion, lack of interoperability, and inertia within the financial system. Section 7 conveys the outlook on smart banking, predicting developments such as hyper-personalization, autonomous banking using AI, digital currency use, and ecosystems of collaboration with traditional banks and FinTechs, influenced by changing consumer behavior and policy direction. Section 8 uses a summary of insights to close the chapter while reiterating the idea that the future of banking will depend on how well it can balance innovation with inclusion, security, and accountable governance while managing rapid technological change.

2. The Evolution of Banking in Digital Era

Banking has always been a conservative industry. The banking industry has relied on physical infrastructure, paper-based processes, and personal interaction. Traditional banking systems were stable, well-regulated, and had many resources but were often slow in processing transactions, limited in access and operational abilities, and burdened by significant bureaucratic processes. Customers were required to go to a bank branch to deposit and withdraw money, open accounts, and access services like loans and insurance. Even when ATM machines and phone banking were introduced, how banks operated remained centrally located and dependent on manual processes. Transitioning out of this traditional workflow and process to smart banking involves a journey that is both incremental and transformative (Arner et al. 2015). The term smart banking refers to the application of data in banking operations, the use of digital technologies and automation, and the adoption of customer-centric digital design in banking services. Smart banking focuses on the delivery of seamless, secure, and intelligent financial services using various digital delivery channels. The rise of online access to information, the proliferation of smartphones, and the availability of mobile apps changed how banks engage with customers. Banks have moved away from being limited to physical branches to becoming digital platforms that provide financial services as customers need them. Figure 1 illustrates the historical progression from traditional banking methods to smart, AI-driven financial services.

One of the core elements of smart banking is real-time availability. Customers can move money, apply for loans, get investment guidance, and even complete KYC without visiting a physical bank (French Chamber of Commerce and Industry in Hong Kong 2022). While it seems like a seamless

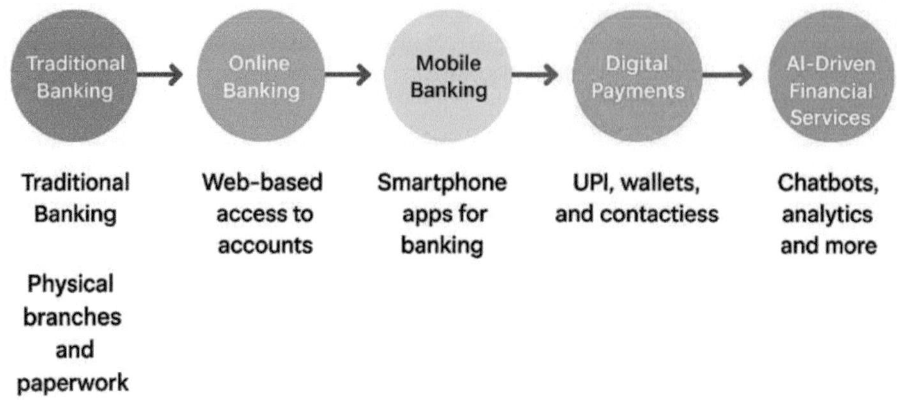

Figure 1: Timeline of Banking Technological Evolution.

user experience, there is an intricate ecosystem of cloud computing, artificial intelligence, APIs, and secure databases behind it. The operational processes have changed from manual data input and standalone systems to automated workflows and integrated platforms with real-time analytics. The improvements in technology have enabled banks to enhance efficiencies, decrease operational costs, and react quickly to market conditions. Smart banking is not limited to digital interfaces, as it incorporates adaptive and predictive service delivery. Banks are also utilizing artificial intelligence to anticipate customer needs and provide advanced support. For example, chatbots can address basic questions, while predictive analytics can find opportunities for loans or investments based on customer behavior patterns. This indicates a major shift in how banks operate—from being transaction-focused to experience-focused (Tyagi 2023a, Tyagi 2023b). Table 1 reinforces distinctions such as physical infrastructure vs. digital platforms, manual processing vs. automation, and delayed vs. real-time access.

In the digital age, financial technology or FinTech companies have emerged as an entirely new category of financial service provider. Many of these companies are startups that use modern technology to provide financial products and services that are faster, simpler, user-friendly, and cheaper than those of established banks (Deloitte Insights 2025). FinTech companies create value across a number of financial services, from digital payments to peer-to-peer lending, robo-advisory to insurtech, wealth management to crowdfunding, and RegTech. Perhaps the most noticeable aspect of FinTech's contribution to the advancement of banking is the disruption of the value chain. FinTech companies have carefully focused on specific problems in financial services and unbundled bank services, reformulating them into a customer-friendly format. FinTech companies are much

Table 1: Core emerging technologies enabling smart banking, their financial applications, and potential benefits.

Artificial Intelligence (AI) & Machine Learning (ML)	Chatbots, robo-advisors, fraud detection, credit scoring	Personalized services, enhanced decision-making, 24/7 customer support
Blockchain & DLT	Smart contracts, cross-border payments, secure data storage	Transparency, immutability, cost reduction, elimination of intermediaries
Internet of Things (IoT)	Wearable banking, ATM monitoring, contextual offers	Real-time engagement, predictive service delivery, improved infrastructure efficiency
Cloud & Edge Computing	Core banking as a service, CRM, real-time analytics, open banking support	Scalability, faster deployment, cost efficiency, reduced latency
Quantum Computing	Post-quantum cryptography, risk modeling, derivative pricing	Massive computational power, improved financial modeling, futureproof encryption

more agile in the way they develop and communicate with customers, allowing them to iterate and rework their ideas quickly compared to traditional banks with longer innovation cycles.

The notion of neobanks signals the next stage of evolution in the FinTech story. Neobanks—virtual-only banks, also referred to as digital-only banks—operate entirely online with no physical branches. They offer banking through mobile apps and online web services only. Neobanks have gained widespread interest for their simplified interfaces, lower fees, real-time support, and quicker transaction times. While traditional banks are often hindered by legacy infrastructure, neobanks are built cloud-native, which means they can scale easily, be redesigned quickly, and are already designed to integrate with other digital services in a digital banking ecosystem. Most neobanks even have features around automated budgeting, spending analytics, and goal-based saving—optimally suited for product-adopting and growth-oriented users who are younger and more digitally native. Neobanks such as Revolut, N26, and Jupiter Bank in India exemplify the digital banking experience as neobanks. The emergence of neobanks has also compelled traditional banks to accelerate costly digital transformations or pursue partnerships and acquisitions for competitive differentiation in a rapidly evolving digital banking environment (Javaid et al. 2022). Governments and regulators, on their end, have done their part by creating regulatory sandboxes and digital banking licenses that allow enough boundary flexibility and stability for innovative financial actors within the same space.

FinTech and neobanks also reflect a deeper, more organic transition from institution-centered banking to customer-centered banking in consumer finance, redefining a client-relevant value proposition around user experience, timing, and personalization. Consumer behavior has been a major driving force behind banking's digital transformation. Historically, banking customers placed value on trust, security, and stability—three qualities that traditionally belonged to legacy players and physical branches. The digital generation, shaped by the on-demand economy and continual connectivity, demands more than trust; they demand access, personalization, and immediacy. The modern customer has no time for long queues or paper-based processes. Instead, they expect frictionless onboarding with instant transactions and 24/7 access to services. Digital services have raised the bar across all industries, and banking is no different. Companies like Amazon, Uber, and Netflix have conditioned users to expect sophisticated user interfaces, curated and intelligent recommendations, and a seamless signup and user experience. As a result, financial institutions must use UX and design elements with similar visual technologies and design paradigms. Additionally, digital literacy and mobile penetration have proliferated globally across developed and developing markets. In India, for example, the coalescence of Aadhaar-based eKYC, the Unified Payments Interface (UPI), and mobile wallets created a digital payments ecosystem that reaches even the rural and unbanked (Tyagi 2023a). In the case of Kenya, M-Pesa harnessed mobile-based financial services to create an economy around credits, savings, and insurance where formal banks had yet to exist.

Today's consumers are also interested in personalization. They anticipate banking experiences that are tailored to their financial habits, life aspirations, and risk tolerances. Many banks are now in a position to meet these expectations through advanced analytics and AI that personalize insights, financial planning tools, and contextualized product recommendations. For example, a bank app will notify a customer when spending is out of character, suggest saving plans for a goal, or offer investment opportunities based on their objectives. Security has also become an important concern as financial data is increasingly stored and transmitted online (Gomber et al. 2018). Consumers want the digital experience and also expect to be protected against growing cyber threat activity. Banks must provide trust in their digital platforms by offering a range of security options such as biometric authentication, real time fraud detection, and blockchain security devices. In conclusion, the key transformation in how consumers engage with banking services—from being passive users of banking infrastructure to active collaborators in digital financial ecosystems—has shifted the success parameters in banking. Not only do banks need to focus on technology investments, but they also need to drive foundational changes in culture, organizational agility, and their agenda for customer engagement.

3. Key Emerging Technologies for Smart Banking

Advances in technology have driven change in banking at a pace never experienced before. The cost-effective application of various emerging technologies has transformed the way banks provide services, connect with customers, deal with risk, and offer value-added services to their customers. New technologies improve institutional efficiency, scalability, and innovation while responding to changing customer expectations, evolving regulatory requirements, and changing market conditions. There are currently six emerging technological areas driving change in the banking industry: artificial intelligence, blockchain technology, the Internet of Things, cloud technology, quantum computing, and cybersecurity plus biometric credentialing.

3.1 Artificial Intelligence (AI) and Machine Learning (ML)

AI and ML represent one of the foundations of the transformation of smart banking. The systems are designed to imitate human cognitive functions like learning, analyzing data, and making decisions, allowing financial services firms to automate complex activities, improve decision-making accuracy with research and market data, and provide highly personalized services at scale. In customer service, chatbots have characteristics of AI because they respond to customers' requests in real time. Bank customers can interact with chatbots and communicate with an electronic assistant through voice. Chatbots use Natural Language Processing (NLP), which allows them to resolve frequently asked inquiries, help users complete common banking functions, and escalate any complex issues during a customer interaction with a human agent. Implementing chatbots presents opportunities to create more automation within the call center, reduces costs associated with inbound volume, and most importantly, improves customer satisfaction because chatbots drive availability and 24/7 serviced chatting operational contact. Bank of America's Erica, one of the first of its kind, as well as HDFC's Eva, have demonstrated that conversational AI can engage customers meaningfully and effectively. Robo-advisors are another large area of AI applications (Kar 2015). Robo-advisors and marketplaces are service providers that enable customers to create and manage investments at reduced fees since their investment management service is algorithm-driven instead of relying on resource-intensive human approaches. By utilizing customer preferences focused on risk appetite, investment objectives, time horizon, and market conditions, robo-advisors are able to develop investable portfolios or, in some cases, create diversified portfolios that automatically rebalance on behalf of customers. There are already several emerging firms like Betterment, Wealthfront, and Charles Schwab Intelligent Portfolios that are capturing the interest of customers who are tech-savvy, have disposable income, and want data-driven solutions to their investment objectives. Figure 2 highlights the key technologies driving smart banking, including AI/ML, Blockchain, IoT, Cloud/Edge, and Quantum Computing.

Fraud detection is one of the areas where machine learning has had a significant impact. Traditional methods of detection are rule-based, and these traditional approaches typically do not work as fraud continues to become more elaborate and complex. Machine learning algorithms, as a dynamic model, have the potential for learning about past transactions/behaviors with respect to detecting instances of anomaly in real-time transaction flows. As the machine learning systems are trained, they adapt to their environment, becoming better at reducing false positives as well as improving fraud detection more generally. Machine learning has also changed the landscape associated with credit scoring and risk assessment. Rather than just using historical financial data and static models, machine learning algorithms look at nontraditional data for assessing financial risk or credit quality. In particular, machine learning algorithms are often used to assess individuals with limited or no banking history and for providing access to credit for individuals who are underbanked,

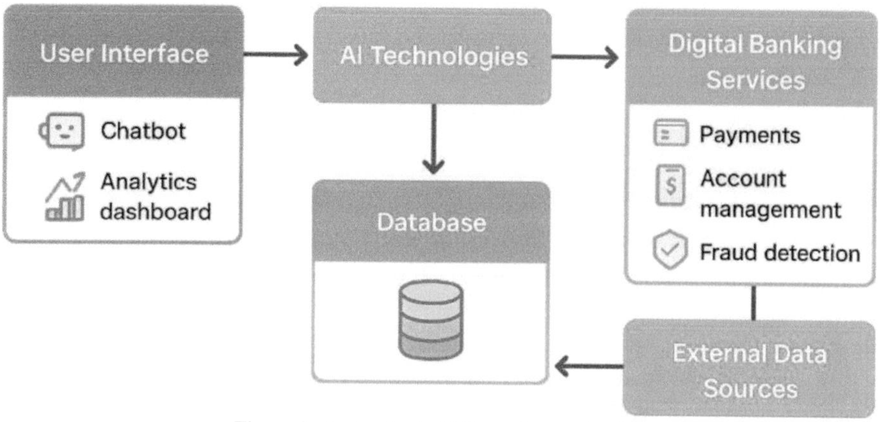

Figure 2: Technological Pillars of Smart Banking

thus enhancing financial inclusion (Omarini 2024). While there are many benefits to utilizing AI within the banking domain, using machine learning algorithms also has its disadvantages. One of the primary limitations of using machine learning algorithms, however, is that many of the models are "black boxes," which makes it very difficult for the model to explain its decisions. This lack of explainability can make it difficult to trust a model and, just as importantly, to adhere to regulated environments. In addition, ethical considerations concerning data privacy, algorithmic biases, and customer profiling need to be addressed prior to engaging users or customers for banking services, which can only be mitigated at present with model governance and ethically driven AI.

3.2 Blockchain and Distributed Ledger Technology

Blockchain and Distributed Ledger Technology (DLT) represent a fundamentally different way to record, verify, and settle financial transactions. A blockchain is a digital ledger that stores information in secure blocks that are chained together in chronological order. Because of blockchain's decentralized architecture and tamper-evident design, it does not rely on central intermediaries to store or verify data, and it creates a trust mechanism based on cryptographic consensus methods. One of the most impactful applications of blockchain in banking is the automated execution of contracts, known as smart contracts, which have underlying rules built directly into their code. Smart contracts provide a mechanism for payment to occur automatically when certain conditions are met. This eliminates the need for humans to intervene in the payment process and significantly reduces time delays and fraud risk. Blockchain creates a more secure and transparent method of storing data within a distributed ledger since it guarantees protected transactions and allows for the appropriate records to be viewed by users while inhibiting data changes. The significance of this capability begins with cross-border payments, which are usually disadvantageous since traditional payment mechanisms require substantial fees, wait times, and reconciliation processes when payments "cross over" between more than two banks. Examples of blockchain payments in innovation include scaled solution platforms such as JPMorgan's Onyx solution and RippleNet, which facilitate payment services with real-time settlement prices, drastically reduced overall costs compared to wire or ACH payments, and even improved record transparency for all involved. DeFi is an example of how blockchain emulates financial systems in a different manner, which is emerging strongly into society. DeFi gives access to financial services such as lending, borrowing, trading of assets, and yield farming without the use of intermediaries, as it utilizes decentralized protocols built on blockchains that are typically public blockchains, such as Ethereum (Tyagi and Abraham 2021). DeFi also changes the existing financial system the most toward democratized finance, but it does present regulatory and systemic risks that will need to be addressed through adaptive law, especially since DeFi operates in time-critical regimes at very high speeds. Although the transformation of traditional banking to include blockchain is underway, there still remain obstacles to blockchain integration such as scalability, energy consumption, regulatory clarity, and the feasibility of adapting existing infrastructure to include blockchain solutions. In the meantime, without regulatory or systemic resolution of these obstacles, continued innovations in blockchain for improved security, efficiency, and customer empowerment will undoubtedly keep appearing.

3.3 Internet of Things (IoT) and Wearable Banking

The IoT is the connection of physical objects—the things surrounding us—each of which has a unique identifier and is able to collect, transmit, and process data via the Internet. In the context of banking, the use of IoT allows easy interfaces from consumers to services by layering banking capabilities into daily objects. Wearable banking gives users the ability to perform transaction activity through their smartwatches or wristbands. This extends the convenience of traditional banking to mobile and contextual use cases. Banking and FinTech companies are already partnering with companies involved in wearable motion technologies to create solutions for contactless transactions with only a simple gesture of the user's hand. IoT also fits nicely with banking in the real-time data collection and behavioral measurement areas. Banking services can observe the pattern of customer spending, geolocation, and even monitor customers' driving habits to alter insurance premiums or provide customer context for promotions. As an example, when a user enters a car showroom, a banking app may allow for instant preapproved auto loan options, depending on the IoT data available. Likewise, ATMs and branches enabled by IoT can monitor environmental conditions, consolidate maintenance schedules, and even recognize unauthorized access or vandalism (Deloitte 2025). By creating connected ecosystems, banks can both increase operational efficiencies and offer aspects of more intelligent management of infrastructure as well. The proliferation of IoT into banking has also raised concerns around data privacy, data interoperability, and cybersecurity. Each device that connects to a network represents a potential point of vulnerability or attack; therefore, banks must follow strong encryption, secure firmware updates, and identity authentication protocols.

3.4 Cloud Computing and Edge Technologies

Cloud computing has been a significant enabler of digital banking, offering scalability, cost-effectiveness, and agility. Upon moving core banking processes to cloud services, banks leverage a broad range of services, including data storage, computing capacity, and applications, where the financial institutions do not have to rely on their own physical infrastructure. Software-as-a-Service (SaaS) is one model used, where a bank deploys a software application streaming over the cloud with minimal implementation. The time to market for new products is significantly reduced using SaaS. Popular, readily available customer relationship management (CRM) platforms, anti-money laundering platforms, and even core banking platforms exist in the cloud today. They have allowed banks to retain operational flexibility, reduce operational costs, and allow innovation in processes to occur much faster. Cloud computing is a powerful enabler of open banking initiatives, which allow third-party developers to build on traditional banking for clients. In doing so, banks, FinTechs, and regulators are collaborating and can co-create innovative solutions for end-user benefit. Open banking frameworks, such as the revised Payment Services Directive (PSD2) in Europe, are advancing innovation in financial services, requiring financial institutions to share their data in accordance with customers' consent. Edge computing, a complementary technology to cloud computing, simply put, is the opposite—that is, moving the capacity closer to the data source, whether it be an IoT device or a remote mobile endpoint that collects data. This minimizes latency and improves the performance of time-sensitive applications. Examples include biometric authentication or real-time fraud detection. In smart banking applications, edge computing improves responsiveness and allows banks to make intelligent decisions at the point of interaction. With the benefits of cloud solutions, the banking sector must work through regulatory compliance, data residency, and vendor lock-in. Banks are going to have to partner heavily with cloud solutions providers to mitigate compliance risks around security standard requirements, utilization and backup, and appropriate audit processes with sensitive financial data (Porter and Heppelmann 2015).

3.5 Quantum Computing

While quantum computing is currently in the early stages of development, it has the potential to change banking systems even further by providing the ability to solve computationally intensive problems an order of magnitude faster than classical computers. Quantum computers use quantum bits (qubits) or subatomic particles that can be in a superposition of states, allowing them to perform high-dimensional calculations at once. Quantum computing will, in part, have huge ramifications in finance when we consider cryptography. Encryption protocols are widely based on the computational difficulty of factoring large numbers, as with RSA or ECC, which is a task that quantum computers, using Shor's algorithm, could accomplish very easily indeed. It goes without saying that this renders the current encryption protocols more or less obsolete, meaning we need crypto protocols quickly and urgently that are quantum-resistant. The ramifications of quantum computing extend beyond security. Quantum computing can lead to improved portfolio optimization, risk modeling, fraud detection, and derivative pricing once it is able to sample from vast datasets and develop new insights that were not attainable using classical algorithms. Wall Street has several institutions that have expressed varying degrees of interest in quantum computing, such as Goldman Sachs, J.P. Morgan Chase, and HSBC. However, like all things quantum, quantum computing is more of a long game. Technological barriers must be addressed before quantum systems will be commercialized. In the meantime, financial institutions must start preparing by experimenting with quantum simulators while at the same time developing post-quantum cryptographic approaches. Table 2 summarizes each technology (AI/ML, Blockchain, IoT, Cloud, Quantum) with its specific use case and impact.

Table 2: A comparison of operational, technological, and service-level aspects between traditional and smart banking systems.

Aspect	Traditional Banking	Smart Banking
Customer Interaction	In-person at branches	24/7 digital platforms (apps, chatbots, AI assistants)
Transaction Speed	Delayed; requires manual processing	Instantaneous; real-time transactions
Service Delivery	Paper-based, branch-centric	Digital-first, omnichannel delivery
Technology Use	Minimal; legacy systems	AI, ML, Blockchain, IoT, Cloud, APIs, Edge Computing
Product Personalization	One-size-fits-all banking products	Personalized offerings via analytics and AI
Access and Inclusion	Limited to banked regions and individuals	Mobile-first, inclusive, reaches rural and underbanked populations
Operational Efficiency	High cost, resource-intensive	Automated, scalable, cost-effective
Compliance and Security	Manual audits and reports	RegTech automation, biometric authentication, real-time fraud detection

4. Smart Financial Systems and Innovations

The digital transformation occurring within the financial industry is not simply limited to backend operations or client-facing aspects; it permeates the very core of future financial systems. As technology continues to advance, we will see financial systems become smart financial systems that will be nimble, interoperable, and designed for the needs of the digital-first world. Smart financial systems will expose a wide array of innovations like real-time digital payments, decentralized financial products, and data-driven open banking, and regulatory technology (RegTech) will automate compliance. Further, these innovations will reinvent financial services while providing financial inclusion, improved user experience, and more effective regulatory engagement.

4.1 Digital Payments and Mobile Wallets

Perhaps the most noticeable and consequential transformation in contemporary finance is the emergence of digital payment systems and mobile wallets. Digital payment systems have supplanted traditional cash-based interactions with money and, in some cases, redefined the very nature of interaction. Digital payment systems, often involving a mobile device but also contactless cards and, occasionally, QR codes, allow for interaction with money in a very seamless way, often eliminating any physical presence at all. One of the most significant examples of a digital payment system is the UPI deployed in India. UPI is developed by the National Payments Corporation of India (NPCI), which enables real-time interbank transactions through mobile platforms by linking bank accounts with unique identifiers. UPI uses interoperable frameworks with immediate effective speeds and is zero cost for every practical purpose, making it a vital part of the digital economy in India and advancing financial inclusion, online and offline, even in urban and rural areas. In June 2024, there were 11.6 billion UPI transactions recorded, demonstrating its widespread reach and adoption across all demographics. Near-field communication (NFC) and QR code-enabled payment systems also have a significant impact on the cashless ecosystem. NFC allows users to instantly make a payment by simply tapping their smartphone device or a contactless card at a payment terminal, frequently incorporated into systems like Apple Pay or Google Pay. QR codes are more common in developing nations due to the lower costs of implementation. Merchants can create QR codes that customers scan to wire funds directly, simplifying digital commerce for small businesses. The social and economic implications of digital payments are significant. They increase transparency for governments, deter tax evasion and costs associated with cash management by introducing digital cash, and they offer speed, convenience, and security for consumers. Digital payments also offer ways for governments to formalize an economy and better target transfers or subsidies, like welfare payments, to those in need. Nevertheless, governments must confront risks like cybersecurity and privacy issues, digital literacy gaps, and infrastructure challenges so that everyone can have equal access to the advantages of a cashless economy.

4.2 Decentralized Finance (DeFi)

Decentralized finance, commonly referred to as "DeFi," is a complete shift in approach to traditional financial intermediation. Built on blockchain platforms, DeFi removes intermediaries—e.g., banks and brokerage firms—entirely from the financial transaction process; ultimately, participants can have peer-to-peer financial interactions through "smart contracts." Smart contracts are programmable computer code that can automatically negotiate, settle, or enforce an agreement using algorithms to execute tasks without human interaction, producing automated, verifiable, and consensus-based outcomes in a transparent, permissionless, and highly customizable manner. DeFi platforms can facilitate a broad mix of financial services, including lending, borrowing, trading of assets, derivatives, insurance, and yield farming. For example, in a peer-to-peer lending protocol like Aave or Compound, users lend their crypto assets out to someone else, receive interest, or borrow crypto assets using collateral. To execute these transactions, smart contracts act as the intermediary—they facilitate near-instant transaction settlements without requiring banks to intervene. Staking in DeFi consists of locking assets in a blockchain protocol to earn rewards or to gain the right to governance and consensus (Google Cloud n.d.). Decentralized Autonomous Organizations (DAOs) are governance models that are free from central authority, where community-based decision-making is exercised through voting on a blockchain. DAOs are capable of governing DeFi projects and transparently controlling resources, all while taking part in a governance model that focuses on decentralization and communal ownership. While there are new, innovative aspects to DeFi, it is not devoid of risk and regulatory obstacles. Without central oversight, DeFi's novel structure can lead to vulnerabilities and bugs in smart contracts, flash loan attacks, and liquidity challenges.

The anonymity of market participants also presents challenges regarding anti-money laundering (AML) and Know Your Customer (KYC) protocols. Governments and regulators internationally are challenged in determining how to classify and supervise individuals and entities that employ DeFi protocols, which rarely operate in respect to individuals or jurisdictions. However, it is clear from the advent of DeFi that there is a demand for a more transparent, open, and programmable financial service. As it continues to grow, it emphasizes the demand for autonomy in finance and access to financial services for traditionally marginalized groups.

4.3 Open Banking and API Ecosystems

Open banking is a game-changing model that redefines how financial data is shared, controlled, and used. Open banking relies on the principle that individuals own their financial data and should be able to share that data securely with third-party organizations in order to access improved financial services. The sharing of this financial data is accomplished through APIs, in which external developers can connect their own services to banks' services through a standardized, secure method. Open banking is supported through regulation—examples of the regulations that foster open banking are the European Union's Revised Payment Services Directive (PSD2) and media reports of consumers moving to new banks or financial institutions for improved banking experiences. Open banking has brought in a new ecosystem of FinTechs that include, among others, account aggregators, personalized budgeting using AI and ML, alternative credit scoring, and investment advisory services. Open banking promotes innovation around the customer instead of waiting for banks to create a better product before consumers can benefit. By encouraging open banking, banks are asked to relinquish their monopoly over their customers' data. Consumers will have a unified perspective of their finances across many banks, receive tailored financial recommendations, and will not be restricted by switching brands (Maze 2025). Startups and FinTechs can deliver niche solutions that are specific to a target market, while banks can leverage those tools to build on their offerings and interact even more with their clients. Beyond the benefits to consumers, open banking increases competition, transparency, and financial inclusion. It encourages incumbent banks to continue innovating, to adopt more consistent technology, and to challenge the nature of the relationship with their customers. Countries like the UK, Australia, and India are all seeing momentum around open banking, with governments and regulators trying to stimulate interoperability and security. Implementing open banking isn't simple and requires established frameworks around a few important issues, including customer security and how to educate consumers toward understanding open banking and consent. Providing standardized APIs, encrypting data, and using effective frameworks within consent management are critical to protecting sensitive financial data and offering privacy protections to users. Figure 3 shows the data flow and interaction between banks, third-party service providers, and customers under open banking frameworks.

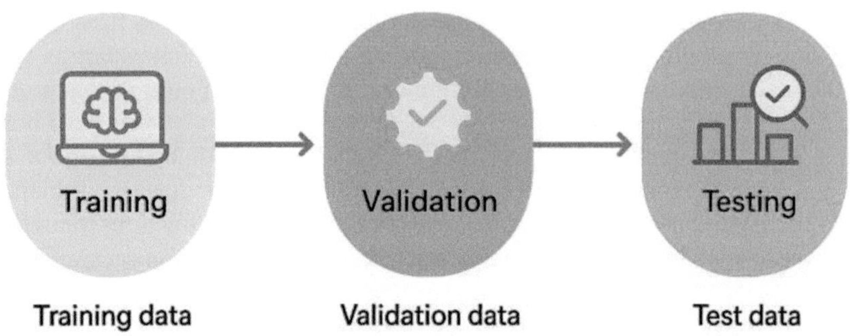

Figure 3: API-based Open Banking Architecture.

4.4 RegTech and Compliance Automation

Regulatory Technology, or RegTech, is a developing field that uses technology—specifically advanced technology—to simplify compliance processes, reduce regulatory risk, and increase transparency in the financial industry. As regulatory burden continues to increase and take on greater significance for financial institutions, especially concerning areas such as AML, KYC, data protection, and fraud prevention, RegTech enhances operations and provides complex diagnostic solutions related to compliance. A highly impactful application of RegTech is real-time KYC and AML monitoring (Suryono et al. 2020). The traditional manner of compliance is often a less structured, labor-intensive process, which creates significant exposure to human error. In contrast, RegTech offerings leverage AI and machine learning to review and analyze both structured and unstructured data that occur in large volumes and at high speeds, in addition to using real-time applicant identification clearance and review, and transaction monitoring with risk scoring. RegTech tools have the potential to identify suspicious patterns of behavior, flag irregularities, and ensure compliance with prescribed regulatory thresholds, while reducing the significant time and cost associated with client onboarding and monitoring throughout the course of their business with the recipient entity (Shanti et al. 2024).

Another important change is the use of machine-readable rules and digital rulebooks. The ability to encode compliance regulations, guidelines, and rules into machine-executable formats allows the rules to be fully automated, including interpretation and enforcement. This leads to less ambiguity, less variability, faster changes in perspectives, and better compliance. Financial firms can even simulate compliance, understand the impact, and put the right controls in place. Furthermore, RegTech allows for more accurate reporting and auditability. Reporting tools can automatically generate and aggregate compliance data from various systems and provide reports to regulators in acceptable formats. This reduces administrative burden and improves responsiveness to regulatory inspection or audits. The benefits of RegTech go beyond saving costs and improving efficiency. RegTech enhances an institution's ability to manage risk, respond to regulatory change, and maintain a sound compliance environment. It also helps eliminate some of the compliance burden for smaller financial firms and reduces the burden for their startup competitors, thereby promoting inclusiveness in financial systems. However, implementing a RegTech solution will require thoughtful integration with IT systems and regulator-specific requirements, ensuring that staff are trained to engage effectively with compliant systems. Regulators will also need to evolve regulations and frameworks designed to accommodate RegTech innovation, which may include sandbox environments to test or certify new solutions.

5. Case Studies and Global Implementations

New technologies are transforming banking and financial systems in a nonlinear fashion—how it manifests differs by geographical region, economic structure, and regulatory structure. Case studies in developed and developing economies shed light on how financial innovation is being used in different locations, applied, scaled, and institutionalized. The practitioner stories built within the case studies highlight best practices, reveal persistent challenges carried over from bureaucracies, and provide lessons to be applied more broadly on a global scale. Ultimately, the case studies shed light on the multiple trajectories of smart banking adoption from both developed financial markets and emerging economies, demonstrating the importance of (1) policy structures translating into real outcomes, (2) the availability of enabling infrastructure, and (3) innovation ecosystems.

5.1 Smart Banking in Developed Economies

In developed markets such as the United States, European Union, and Japan, smart banking technologies are offered in the context of deep digital engagement, formal regulation, and strong demand from consumer segments to receive personalized and on-demand financial services. These markets tend to be the development grounds for advanced applications of artificial intelligence, applied blockchain technologies, and digital asset development. For example, JPMorgan Chase is taking the lead in the United States with its use of blockchain technology for institutional banking (Tyagi and Sreenath 2021). Using its Onyx platform, JPMorgan has developed JPM Coin, a digital token to facilitate instant cross-border settlements with its institutional client base. With the use of DLT, JPMorgan claims they have established transaction visibility, reduced transaction times, and eliminated the need for third-party clearinghouses. The institution also boasts its investments in AI platforms to improve the efficiency of fraud detection and automate client billing advisory services. In the European Union, open banking is reaching mainstream usage with the help of the revised PSD2 regulation. PSD2 regulates banks to open their APIs to licensed third-party providers, thereby creating competition and encouraging innovation. Consequently, platforms like Tink in Sweden and Yolt in the Netherlands are now capable of providing comprehensive representations of associated financial data from multiple institutions, as well as offering insights for customers. While the EU's strong data protection laws—like its General Data Protection Regulation (GDPR)—help to form a security-focused environment, they also allow for innovation and a relative balance between data security and data innovation. Japan offers an interesting paradigm, as it uniquely combines progressive technology with a conservative culture. Cash continues to dominate as an accepted medium of exchange; however, the Japanese government is aggressively pursuing cashless and hybrid cash systems, spurred by economic digitization and the Tokyo 2020 Olympics. For example, financial giants such as MUFG Bank (Mitsubishi UFJ Financial Group) have introduced digital-only banks (e.g., MUFG Coin), are poised to earn returns on investment from blockchain-based financial platforms for interbank transaction settlements, actively pursue cashless transaction deals, and have made strides toward AI-powered risk analytics for customer risk modeling, as well as robotic process automation (RPA) for back-office processing. Hybrid use of technological paradigms and cash continues (Lim and Soesilo 2024).

5.2 Smart Banking in Emerging Markets

Emerging countries have exhibited tremendous creativity in leveraging smart banking technology to work around infrastructure deficiencies and deliver financial services to sections of society that have traditionally been excluded from the formal financial system. In many cases, these countries have "leapfrogged" the traditional banking model's legacy, moving directly to mobile, API-based, or decentralized models that allow inclusive and scalable financial services. India is a leader in digital financial inclusion. The Aadhaar biometric identity system and the UPI have created a digital public infrastructure enabling seamless financial transactions across the spectrum of social and economic status (Rysin et al. 2023). UPI provides instantaneous payments from one bank to another on mobile platforms, and it is now one of the largest real-time payment systems in the world. Supported by India Stack, which includes digital identity, eKYC, e-signature, and consent architecture, India enables an ecosystem for paperless and presence-less banking.

In addition to established government entities, private players have also played an important role in establishing India's fintech space. Paytm, PhonePe and Google Pay changed the game for retail payments, while neobanks such as Jupiter and Fi are changing the banking experience for urban millennials. Moreover, Indian regulators have introduced the Account Aggregator system, which provides access across financial institutions in a secure manner through consent, which creates a competitive financial ecosystem and a collaborative one at the same time. Kenya is a completely different example with MPesa launched by Safaricom. Initially, it facilitated money transfers via SMS. Today, it is one of the best one-stop services for financial utility, offering savings, loans, insurance, merchant payments, and more (Javaheri et al. 2023). It has changed tens of millions of lives whose only means of access to finance was informal credit. M-Pesa's success is based on one critical aspect: simplicity. It is a mobile-first design, offering m-wallet services backed by thousands of agents who facilitate cash-in and cash-out transactions even in remote areas. Table 3 provides

Table 3: Countries with new initiative of payments.

Country	Technology/Initiative	Key Features	Outcome/Impact
India	Aadhaar, UPI, India Stack	Biometric ID, real-time mobile payments, eKYC, paperless banking	Massive financial inclusion, 11+ billion UPI transactions/month (2024), digital ecosystem reach
Kenya	M-Pesa	Mobile-based P2P money transfer, agent network, offline SMS integration	Financial empowerment of unbanked rural populations, informal economy digitization
Brazil	Pix (by Central Bank of Brazil)	Real-time payments using mobile, QR, phone/email IDs	140+ million users in < 3 years; reduced transaction cost, enhanced entrepreneurship
USA	JPMorgan Onyx, Erica AI	Blockchain settlement, AI chatbot for customer service, predictive analytics	Streamlined B2B transactions, automated client servicing
EU	PSD2, Open Banking, GDPR	API sharing regulations, consumer data control, FinTech integration	Competitive FinTech ecosystem, cross-bank personalization
Japan	MUFG Coin, RPA, AI risk analytics	Blockchain-based digital yen, robotic automation, AI modeling	Hybrid cash–digital model, automation-led banking modernization

a quick cross-country snapshot of initiatives like UPI (India), M-Pesa (Kenya), and Pix (Brazil), among others.

Brazil has also distinguished itself as an innovator in the digital payments landscape with the introduction of Pix, an instant payment system launched by the Central Bank of Brazil. Pix enables instantaneous transfers and payments through mobile apps using identifiers such as telephone numbers, emails, and QR codes. The instant payment platform has rapidly scaled across Brazil, reaching more than 140 million users in less than three years. Pix is an example of how a digital payment system can accelerate financial access, stimulate micro-entrepreneurship, and reduce the costs of payment transactions for the population. The lessons learned from India, Kenya, and Brazil illustrate how public-private partnerships, regulatory vision, and user-oriented design can enable financial innovation in resource-constrained environments, as well as how it is important to have inclusive infrastructure for both digitally literate and marginalized users (Jagtiani and Lemieux 2019). Figure 4 presents a comparative chart depicting smart banking adoption in the US, EU, Japan, India, Kenya, and Brazil.

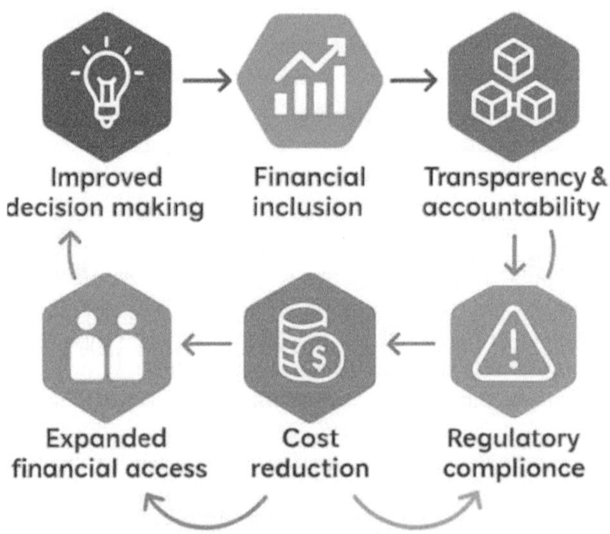

Figure 4: Global Adoption of Smart Banking Technologies.

6. Benefits, Opportunities and Challenges

Smart banking enhances decision-making capabilities because of analytics and the ability to access data in near real time (Central Bank of Ireland n.d.). ML models today process more transactional and behavioral data than humans can recognize, allowing banks to understand the behavior of customers through actionable insights with trend analysis and predict purchases when customized per user behavior. For banks, data-driven insights provide improved credit scoring, better risk assessment, and more customized products. For the customer, it improves the value of the user experience and the quality of financial advice provided, along with more timely nudges that reinforce or suggest better financial behavior. The combination of dashboards and smart reporting also enables management teams to react faster to market fluctuations, regulatory changes, or operational disruptions through real-time data insights.

Technology-enabled finance solutions also have the potential to expand financial inclusion. Emerging technologies act as bridges from the informal financial system or underserved/unbanked communities to more formal financial systems. In places where banks have limited or no traditional banking infrastructure or access to capital, financial technology offerings provide significant growth opportunities that enable access to basic services such as savings accounts, credit, and insurance through mobile banking, digital wallets, communication platforms, and biometric verification systems. Initiatives like these have enabled millions of people to engage in the digital economy, helping to reduce income disparities and support social mobility (Tyagi 2022). Additionally, open banking ecosystems enable data sharing and new forms of credit assessments, allowing individuals and small businesses that were previously excluded to gain access to formal credit markets. Transitioning to smart banking is also beneficial because it increases transparency around financial transactions and leads to accountability. Blockchain technology has the ability to create immutable records, timestamp each transaction, and eliminate the possibility for parties to commit fraud or manipulate, misrepresent, or misreport financial transactions. Furthermore, RegTech's automated compliance checks and real-time detection of suspicious activity allow financial institutions to remain compliant with regulatory obligations. Digital audit trails and machine-readable regulations can equip organizations with tools to speed up reporting, lessen administrative burden, and foster a degree of assurance and trust among regulators, financial service providers, and consumers alike. Figure 5 presents a mapping of challenges such as cybersecurity, regulation, digital inequality, and interoperability.

While these advantages are noteworthy, adopting smart systems for banking does not come without its fair share of challenges and obstacles. One of the primary challenges relates to data privacy and cybersecurity (Rauchs et al. 2018). As financial institutions begin to collect and process higher volumes of sensitive personal and transactional data, they present a greater opportunity for cybercriminals to attack. Cyber threats including phishing, ransomware, compromised data, and insider threats can erode customer confidence and lead to significant financial and reputational costs. Individual privacy, while secure, remains questionable, and collectively, AI and biometrics impose unique governance obligations and online security and incident response protocols to mitigate risk. A second barrier pertains to regulation and legislation. Financial innovation can advance faster than regulators can respond, creating ambiguity and a gray area in which new technologies move forward without any regulatory guidance. Questions remain about accountability, jurisdiction, and consumer protection with respect to decentralized finance platforms, algorithms based around AI, and cross-border digital assets. Regulation that is inconsistent and outdated creates inertia that can either hamper innovation or fail to protect against harmful practices. A main concern is ensuring that innovation is being developed within regulatory boundaries and ethical norms. Concerns persist that the promise and power of technology are being carried forward while ensuring that systemic risks and consumer rights are upheld.

Equitable adoption is also limited by issues of digital literacy and access. Smart banking solutions are designed to provide greater convenience, but that convenience is not shared equally

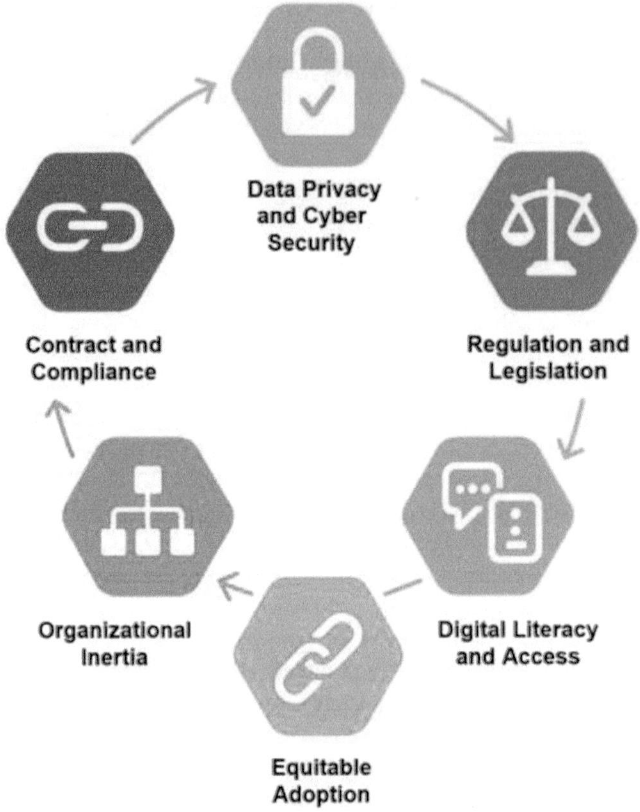

Figure 5: Key Challenges to Smart Banking Implementation.

across populations. People or groups with limited access to smart devices, the internet, or education remain excluded from digital financial services, thus exacerbating inequality. Additionally, user experience could limit engagement with a digital platform that is too complicated to use, and language can be a barrier to participation for people uncomfortable with technology. Inclusive design standards, multilingual platforms, and community-based digital literacy programs that increase an individual's capacity to engage digitally with banking services are necessary. Interoperability and standardization also represent further technical challenges. The financial services ecosystem is rapidly evolving and growing, with an increasing number of FinTechs, neobanks, payment gateways, and third-party developers. The growing complexity of integrating various systems is a challenge, particularly when using incompatible APIs or databases that are not integrated. Different cybersecurity standards can potentially lead to failures, security gaps, inefficiencies, and ultimately a poor customer experience. Financial institutions will have to commit to standards-based architecture as well as establish governance models across different platforms to provide users with a seamless and resilient experience.

Organizational inertia is another barrier in the transition to smart banking. Banks have historically been subject to inertia; legacy IT systems, hierarchical decision-making, and corporate cultures characterized as "risk averse" can slow down the pace of innovation. The move to digitally driven operating models requires a change not only in technology but also in culture. Moving to digital models requires competence and skills that are often missing in legacy banks—it isn't just about deploying new technology, it's about changing behaviors and decision-making altogether (Egger et al. 2020). If banks are to compete against FinTechs, neobanks, payment processors, and third-party developers, they must provide leadership buy-in, cross-functional collaboration, and invest in upskilling employees. Change management methodologies, continuous training, and agile working practices and tools can help financial institutions create an adaptive and sustainable working environment within the new digital paradigm.

7. Future Trends and Research Directions

The metamorphosis of banking and financial systems is still far from complete. As technology continues to develop rapidly, the financial services industry stands poised at the edge of yet another tectonic shift. Emerging paradigms like embedded finance, digital currency, autonomous banking systems, and financial sustainability are forecasted to further characterize the next 10 years of smart financial systems. In parallel, evolving regulatory environments, changing consumer expectations, and international socioeconomic trends will disrupt how financial institutions design products, manage risk, and facilitate economic development. This chapter outlines the most significant future trends and highlights where further research is required to advance safe, inclusive, and intelligent financial innovation. One of the most disruptive future trends is the introduction of embedded finance into nonfinancial platforms. Embedded finance is the integration of financial services into the user experiences of nonbank platforms. With greater expansion of digital ecosystems, especially on e-commerce platforms, telehealth services, and ridesharing services, users not only expect but want financial services at the point of need. This change moves away from the standard approach of an institution-centric model to a platform-centric model that offers context within the user experience whenever possible. Enterprise technology companies and retailers today are rapidly embedding credit options, digital payments, and insurance offerings into their platforms, requiring consumers to have less and less interaction with banks. For banks and financial services, this offers the chance to broaden their market, but it also presents the challenge of remaining relevant and resilient in a customer journey that is defined and curated by third-party ecosystems.

Another significant evolution, however, is the worldwide experimentation with Central Bank Digital Currency (CBDC). CBDCs are digital versions of a state's fiat currency that are issued and regulated by a central bank. CBDCs are not cryptocurrencies, which are decentralized sources of wealth and often highly speculative. The purpose of a CBDC is to provide a stable digital payment form of money with government backing, replacing or augmenting hard currency. China, Sweden, and India are among several countries developing CBDC models or at the pilot phase. The introduction of CBDCs will have implications for monetary policy, cross-border settlement, and financial inclusion. It raises a number of questions, including privacy, levels of interoperability, measures taken to ensure cyber resilience, and the ability of CBDCs to coexist with commercial banking environments. A significant amount of research must be carried out to analyze the potential systemic impacts of deploying CBDCs on financial stability, data protection, privacy, and legal implications. Note that AI will also rapidly evolve from simple automation and predictive activities to more advanced forms of financial decision-making. The rise of self-learning algorithms that can make complex investment, lending, or risk-aversion decisions adds new levels of efficiency but also additional ethical and regulatory challenges. The ability for AI to manage portfolios, underwrite loans, and meet consumer demands—all typically without human involvement—could reduce costs dramatically and potentially improve access. However, balancing transparency, fairness, and accountability in AI applications in finance remains an important research challenge. Issues regarding algorithmic fairness, model interpretability, and consumer consent will continue to shape the agenda in this area. Future research should focus on advancing interpretable and responsible AI models, developing frameworks for auditing AI, and gaining a better understanding of the socioeconomic implications of algorithmic financial services.

Quantum computing, while still largely in the proof-of-concept phase, is expected to be a disruptive force impacting financial modeling, cryptography, and asset optimization. Quantum algorithms could significantly improve the accuracy and speed of derivative pricing simulations, portfolio risk analysis, and scenario testing. A more immediate concern is the threat of quantum computing to existing encryption methods. The cryptographic algorithms that secure online transactions and blockchain will be compromised in a post-quantum world. Consequently, there is increasing demand for research to develop quantum-resistant cryptographic methods and strategies for migrating financial institutions to secure alternatives. Sustainability and ESG perspectives are increasingly becoming embedded within financial institutions' decision-making. Upcoming smart banking systems will likely integrate ESG analytics into credit scoring, investment advisory, and portfolio management functions. AI and big data can be used to examine the sustainability metrics of companies, projects, or individual behaviors. However, there are no standard ESG data and reliable scoring systems that are consistent and free from greenwashing or unreliable evaluations. There needs to be a research agenda to develop common and transparent ESG data taxonomies and to explore the use of blockchain for verifiable sustainability disclosures. In addition, financial innovation should support sustainable development goals by incentivizing green investments and allowing inclusive access to climate-resilient finance.

Cybersecurity will continue to be an area of increasing interest and focus as financial systems become more intertwined and reliant on digital infrastructure. With the rise of API-based ecosystems, decentralized finance behaviors, and cloud-based banking technologies, the attack surface of financial cyber threats is expanding. Future trends in financial sector cybersecurity may include the deployment of artificial intelligence to detect threats in real time, zero-trust security architecture, and authentication through biometric or behavioral means. But along with these trends come new threats, including adversarial AI technology, deepfake frauds, vulnerabilities within supply chains through provider partnerships with financial institutions, and new types of attacks that existing security systems may not yet be equipped to handle. Research in this area must be proactive amidst a rapidly evolving risk landscape and must seek to develop adaptive security frameworks, ethical hacking methods, and coordinated tactical approaches to global cyber defense for the sector. The evolution of RegTech and SupTech will also realign how compliance in the digital sector is monitored and overseen. Regulators are already moving into the space of using AI and data-driven analytics to monitor market behavior and detect anomalies in real time. The move toward machine-readable regulation, automated compliance engines, and coordinated risk reporting and exposure management will require not just technical innovation but also regulatory change that allows regulators to utilize nontraditional means of oversight. Future research will assist in developing intelligent regulatory systems that are sufficiently agile and robust to enable innovation while simultaneously safeguarding stability and consumer protection.

Interoperability and data standardization—ensuring financial data remains secure across all platforms—are the next areas of development. As open banking, data-sharing ecosystems, and digital identity evolve on a global scale, appropriately developed data protocols and formats will be critical. If efforts toward the future of financial services and financial technology are not built with standardization in mind, we will continue to add layers of risk and inefficiency that detract from firms and endanger people's security. Nations and institutions will need to work together and coordinate to develop interoperable infrastructure that allows for secure, transparent, and seamless data exchange. Currently, research may help propel the development of global financial data standards, cross-border open banking, and consent-based user verification, strengthening user autonomy. In addition, we must not forget the human side of digitized finance. As we move toward digitization and automation in financial services, we will also experience challenges tied to decreased human-to-human interactions, cognitive overload, and the danger of digital exclusion. Our innovation agenda should ensure that inclusive design, digital literacy, and user empowerment are front and center. Research should better inform how to build technology that is easy to understand, accessible, and humanized—particularly for older adults, low-income communities, and people with disabilities. We face the challenge not only to build increasingly intelligent systems but also to build systems that promote equity, cultural sensitivity, and alignment with human values.

8. Conclusion

The modern-day evolution of banking and financial systems is being extensively influenced by technology-driven innovations. Technologies of interest—all impacting the structure, delivery, or experience of financial services—include artificial intelligence, blockchain, cloud computing, IoT, and biometrics. The operational efficiency, the ability to engage consumers in real time, enable financial inclusion, and automate regulatory compliance, among other capabilities, are available to financial institutions like never before. Despite all of this, transitioning to smart banking is not without challenges. Concerns around data privacy, cybersecurity vulnerabilities, digital exclusion, and the adaptability of regulation persist. In addition to these issues, there is the challenge of unbalanced technological capability between regions. Through a global review of trends and case studies of smart banking developments in developed and developing economies, this chapter illustrated the different pathways to adopting smart banking. It calls for inclusive design and interoperability, governance of ethical AI, and regulatory agility to promote an equitable and secure financial ecosystem. The digital transformation of banking requires engagement with the entire fintech ecosystem, including—but not limited to—banks, fintechs, technology companies, regulators, and consumers. Engaging financial technology innovation responsibly is the best way to realize smart banking in a sustainable manner for inclusive and resilient financial services in the digital environment.

References

Arner, D. W., Barberis, J. N., and Buckley, R. P. (2015). The evolution of FinTech: A new post-crisis paradigm? *SSRN Electronic Journal.* https://doi.org/10.2139/ssrn.2676553.

Central Bank of Ireland. (n.d.). What is 'FinTech' and how is it changing financial products? https://www.centralbank.ie/consumer-hub/explainers/what-is-fintech-and-how-is-it-changing-financial-products.

Deloitte Insights. (2025, June 17). 2023 banking and capital markets outlook. https://www.deloitte.com/us/en/insights/industry/financial-services/financial-services-industry-outlooks/banking-industry-outlook-2023.html.

Deloitte. (2025, June 16). Financial services. https://www.deloitte.com/us/en/Industries/financial-services/about.html.

Egger, D. J., Claudio Gambell, Jakub Marecek, Scott McFaddin, Martin Mevissen, Rudy Raymond et al. (2020). Quantum computing for finance: State-of-the-art and future prospects. *IEEE Transactions on Quantum Engineering,* 1: 1–24. https://doi.org/10.1109/tqe.2020.3030314.

French Chamber of Commerce and Industry in Hong Kong. (2022, August 31). Accenture unveils 'Technology Vision 2022–Meet me in the Metaverse'. *FCCIHK.* https://www.fccihk.com/publications-news/news/n/news/accenture-unveils-technology-vision-2022-meet-me-in-the-metaverse.html.

Gomber, P., Kauffman, R. J., Parker, C., and Weber, B. W. (2018). On the FinTech revolution: Interpreting the forces of innovation, disruption, and transformation in financial services. *Journal of Management Information Systems,* 35(1): 220–265. https://doi.org/10.1080/07421222.2018.1440766.

Google Cloud. (n.d.). AI in banking: Applications, benefits and examples. https://cloud.google.com/discover/ai-in-banking.

Jagtiani, J., and Lemieux, C. (2019). The roles of alternative data and machine learning in FinTech lending: Evidence from the LendingClub consumer platform. *Financial Management,* 48(4): 1009–1029. https://doi.org/10.1111/fima.12295.

Javaheri, D., Fahmideh, M., Chizari, H., Lalbakhsh, P., and Hur, J. (2023). Cybersecurity threats in FinTech: A systematic review. *Expert Systems with Applications,* 241: 122697. https://doi.org/10.1016/j.eswa.2023.122697.

Javaid, M., Haleem, A., Singh, R. P., Suman, R., and Khan, S. (2022). A review of blockchain technology applications for financial services. *BenchCouncil Transactions on Benchmarks, Standards and Evaluations,* 2(3): 100073. https://doi.org/10.1016/j.tbench.2022.100073.

Kar, A. (2015). Retail banking in the contemporary financial environment. *Journal of Social Science Research,* 9(1): 1792–1795. https://doi.org/10.24297/jssr.v9i1.3776.

Kou, G., and Lu, Y. (2025). FinTech: A literature review of emerging financial technologies and applications. *Financial Innovation,* 11(1). https://doi.org/10.1186/s40854-024-00668-6.

Lim, W., and Soesilo, P. K. M. (2024). Factors affecting the intention to use digital banking services: Case study on Generation X in Greater Jakarta. In *Studies in Big Data* (pp. 271–281). https://doi.org/10.1007/978-3-031-71213-5_25.

Maze. (2025, April 30). Banking customer experience trends to watch in 2025. https://maze.co/collection/customer-experience/banking-trends/.

Neves, C., Oliveira, T., Santini, F., and Gutman, L. (2023). Adoption and use of digital financial services: A meta-analysis of barriers and facilitators. *International Journal of Information Management Data Insights,* 3(2): 100201. https://doi.org/10.1016/j.jjimei.2023.100201.

Omarini, A. (2024). The digital transformation in banking and the role of FinTechs in the new financial intermediation scenario. *SSRN Electronic Journal.* https://doi.org/10.2139/ssrn.4862876.

Porter, M. E., and Heppelmann, J. E. (2015, October 1). How smart, connected products are transforming companies. *Harvard Business Review.* https://hbr.org/2015/10/how-smart-connected-products-are-transforming-companies.

Rauchs, M., Andrew Glidden, Brian Gordon, Gina C. Pieters, Martino Recanatini, François Rostand et al. (2018). Distributed ledger technology systems: A conceptual framework. *SSRN Electronic Journal.* https://doi.org/10.2139/ssrn.3230013.

Rysin, V., Prokopenko, O., Muravskyi, O., Pechenko, R., Holiachuk, N., and Zinchenko, A. (2023). Personalization of banking products (services) using digitalization technologies. *WSEAS Transactions on Business and Economics,* 20: 2528–2539. https://doi.org/10.37394/23207.2023.20.216.

Shanti, R., Siregar, H., Zulbainarni, N., and Tony, N. (2024). Revolutionizing banking: Neobanks' digital transformation for enhanced efficiency. *Journal of Risk and Financial Management,* 17(5): 188. https://doi.org/10.3390/jrfm17050188.

Suryono, R. R., Budi, I., and Purwandari, B. (2020). Challenges and trends of financial technology (FinTech): A systematic literature review. *Information,* 11(12): 590. https://doi.org/10.3390/info11120590.

Tyagi, A. K., and Abraham, A. (2021). Recent trends in blockchain for information systems security and privacy. In CRC Press eBooks. https://doi.org/10.1201/9781003139737.

Tyagi, A. K., and Sreenath, N. (2021). Cyber Physical Systems: Analyses, challenges and possible solutions. Internet of Things and Cyber-Physical Systems, 1: 22–33. https://doi.org/10.1016/j.iotcps.2021.12.002.

Tyagi, A. K. (2023a). Decentralized everything: Practical use of blockchain technology in future applications. pp. 19–38. *In:* Elsevier eBooks. https://doi.org/10.1016/b978-0-323-96146-2.00010-3.

Tyagi, A. K. (2023b). Dew Computing. pp. 332–345. *In:* Advances in systems analysis, software engineering, and high performance computing book series. https://doi.org/10.4018/978-1-6684-8531-6.ch017.

Tyagi, S. (2022). Analyzing machine learning models for credit scoring with explainable AI and optimizing investment decisions. *arXiv.* https://doi.org/10.48550/arxiv.2209.09362.

Zalan, T., and Toufaily, E. (2017). The promise of FinTech in emerging markets: Not as disruptive. *Contemporary Economics,* 11(4): 415–430. https://doi.org/10.5709/ce.1897-9254.253.

12

E-Waste Management in Communication Technology

AI Integrated Bioleaching

M. Kanchana,[1,*] *Bhargava Satya Himank Chinni*[1] and *Prathap Mani*[2]

1. Introduction

Electronic waste, or e-waste, refers to discarded electronic products such as computers, smartphones, televisions, and circuit boards. Due to rapid technological advancements and constant consumer demand for new gadgets, the lifespan of electronic devices has shrunk significantly. As a result, the world generates huge amounts of e-waste every year. Over 53 million metric tons was generated in 2021, and this number is projected to continue increasing, according to the Global E-Waste Monitor. Throwing out e-waste the wrong way may be harmful for the environment and human health. Lead, mercury, and cadmium are some of the dangerous materials that can leach into the soil and water, which can harm ecosystems and people's health. Landfilling and incineration are two common ways to get rid of trash, but they pollute the environment. The rapid growth of communication technology has made the e-waste situation worse, which shows how much we need smarter systems that can sort and recover materials more efficiently. Sorting by hand is a big part of traditional recycling, but it's not very efficient, takes a lot of time, and can be bad for the environment. Scientists have been exploring ways to couple artificial intelligence with environmental technologies like bioleaching (involving bacteria leaching metals from electrical waste) to address these challenges. AI approaches have progressed tremendously in recent years in terms of computer vision and data analysis, particularly with respect to deep learning and machine learning. Computers can now distinguish between different kinds of e-waste by employing techniques such as gray-scale conversion and feature description methods like Local Binary Patterns. However, there are a series of challenges that need to be addressed to build a robust AI-assisted bioleaching system. One principal challenge is the absence of standard datasets of e-waste to train the models. This requires collecting images of e-waste from multiple locations and manually preprocessing the images for

[1] Department of Computing Technologies, SRM Institute of Science and Technology, Kattankulathur–603203, Chengalpattu District, Tamil Nadu, India, Email: bc1429@srmist.edu.in

[2] Department of Computer Science & Information Technology, American University of Kurdistan. Email: prathap.mani@auk.edu.krd

* Corresponding author: kanchanm@srmist.edu.in, https://orcid.org/0000-0002-6961-4735

DOI: 10.1201/9781003655190-15

Table 1: Global E-Waste Generation Trends.

Year	Total Global E-Waste	Communication Devices Share
2019	53.6 million tons	~ 15%
2021	57.4 million tons	~ 17%
2023	61.3 million tons	~ 20% (estimated)
2025	~ 74 million tons	~ 22% (projected)

model use. Additionally, obtaining high-quality sensor measurements for bioleaching, including pH and metal concentrations, is difficult, making it necessary to develop simulated datasets. Training deep learning models with insufficient data, such as CNN-2D, also proved difficult, leading to overfitting that required extensive regularization and hardware tuning. Despite these challenges, this research was able to produce a scalable system that builds upon image-based classifications and sensor-based predictions to assist in the sorting of e-waste, facilitating the metal recovery process. This hybrid approach could revolutionize the way e-waste is processed, enabling more efficient recycling and sustainability.

2. Statistical Report on E-Waste Management in Communication Technology

According to table 1, in the year 2019, global e-waste reached 53.6 million tons, with communication devices making up ~ 15% of the total. In 2021, e-waste increased to 57.4 million tons, and the share from communication devices rose to ~ 17%. In 2023, an estimated 61.3 million tons of e-waste was generated globally, with ~ 20% coming from communication technologies. Global e-waste is expected to hit ~ 74 million tons, with communication devices contributing ~ 22% in the year 2025.

According to Table 2, smartphones and tablets (38%) are the largest contributors due to rapid replacement cycles and mass adoption worldwide. Laptops and computers (22%) are the next major sources of e-waste, especially from work and educational sectors. From Table 3, e-waste components carry both economic value (recoverable metals) and environmental hazards, making safe recycling and proper material segregation essential. According to Table 4, developed regions like the EU and USA have better recycling infrastructure, while emerging economies like India and China face challenges in managing the sheer volume and ensuring environmentally sound disposal. From the above statistical data, e-waste management is not just a technical issue, it's a social, environmental,

Table 2: Sources of E-Waste in Communication Technology.

Device Type	Contribution to E-Waste (%)
Smartphones & Tablets	38%
Laptops & Computers	22%
Routers, Modems, Cables	15%
Telecom Equipment	10%
Accessories (earphones, chargers)	15%

Table 3: Composition of E-Waste in Communication Devices.

Component	Example Metals	Environmental Impact
Printed Circuit Boards	Gold, Copper, Lead	High recovery value, toxic if leaked
Casings & Frames	Plastic, Aluminium	Non-biodegradable
Batteries	Lithium, Cobalt	Fire hazard, toxic leachates
Displays	Mercury, Indium	Health hazard during disposal

Table 4: Country-Level Insights.

Country	Key Stat	Notes
India	~ 5 million tons (2023)	Low recycling capacity, growing telecom e-waste
USA	~ 6.9 million tons	High per capita generation, moderate recycling
China	~ 10.1 million tons	Leading e-waste generator, improving regulation
EU	> 40% of e-waste recycled	Stronger enforcement of EPR and collection

and economic necessity. Proper handling ensures a cleaner planet, healthier communities, and sustainable development. To address these issues, this work proposes AI-integrated bioleaching techniques for sustainable development and environment.

3. Related Work

E-waste handling has improved, and several research studies have examined the use of artificial intelligence (AI) to enhance processes such as image classification and bioleaching for material recovery. The following discusses some important research advancing the development of smart e-waste management systems.

The proposed method in one study can be applied in many different areas of waste management, including municipal waste systems, recycling centers, and e-waste recovery. The system uses IoT-enabled sensors and deep learning (DL) models to help people make smart, real-time decisions on how to sort waste and recover materials. According to preliminary results, the WMR-DL framework reduces operating costs and enhances resource management while increasing recovery efficiency by up to 30%. This method contributes to a cleaner and greener world by encouraging sustainability and greatly lowering the environmental impact of waste disposal systems. The first thorough and methodical analysis of the wide range of CNN applications in intelligent waste identification and recycling is presented in this work. To categorize images of different types of construction and demolition waste, a deep convolutional neural network (CNN) was created in (Peter Davis 2021). To simulate situations on construction sites where on-site sorting is impractical, waste was placed in a standard trash can. The deep CNN recognized C&DW classes with 94% accuracy. Construction waste stream operations may benefit from such deep CNN models.

Based on a data-efficient image transformer (DeiT), (Islam 2023) introduced EWasteNet, a novel two-stream method for accurate e-waste image classification. The suggested method shows great promise in tackling the worldwide issue of managing e-waste. By correctly identifying e-waste images, it reduces the health and safety risks associated with improper disposal and promotes effective waste management and recycling. To correctly categorize electrical devices, (FNU Abhimanyu 2022) presented RGBX, a multi-modal image classification technique that combines important properties from external RGB images with those produced from X-ray images. Eighty percent of the e-waste produced in wealthy nations is exported, according to (Hicks 2005). Despite international initiatives for greater control and global transboundary movement of hazardous waste, illicit trading and informal recycling have made it more difficult to recycle e-waste globally (Liu 2023).

In developing and transitional countries, e-waste typically makes up 1–2 percent of all solid waste, and this percentage is predicted to increase significantly soon (Sadeghi et al. 2022). The ideas and techniques of biomining and bioleaching were examined and addressed by the authors (Jaiswal and Srivastava 2024), who also gave a global overview of e-waste. Utilizing biomining to extract valuable metals from electronic waste is the focus of the review. To achieve sustainable goals in the management of waste from the ICT sector, the authors (Alaskar et al. 2022) investigated the existing state of electronic waste worldwide, strategies created to manage it, and approaches to be used. Additionally, data pertaining to manufacturers' strategies for resolving the problem of electronic waste management were included.

In (Singh et al. 2025), the load cell sensor, microcontroller, and IoT devices guarantee real-time data, and a middleware platform provides encryption mechanisms that facilitate seamless communication between the blockchain network, IoT gateway, and CVMS while addressing concerns about data security, scalability, and privacy. As a result, the system offers an accountable, safe, and secure platform for e-waste management procedures. This study (Santhuja and Suresh 2025) suggests an e-waste management system based on Convolutional Neural Networks (CNNs) that is coupled with the Internet of Things and Blockchain. IoT sensors and RFID tags provide real-time e-waste monitoring, while blockchain ensures transparent and secure waste transit tracking. A CNN-based image classification model achieved 96.2% accuracy in separating components into recyclables, reusables, and hazardous goods after being trained on a set of 50,000 e-waste photos. Compared to traditional methods, the system demonstrated a 32% increase in sorting efficiency and a 28% reduction in processing time through automated decision-making.

The study (Adedeji and Wang 2019) also employed ResNet-50 CNN and SVM to analyze a group of pictures of waste. During the pre-training phase, the ResNet-50 CNN model showed that it could sort images into 1,000 separate groups. A team of researchers devised a new method for identifying and sorting e-waste using the "You Only Look Once" Version 5 (YOLOv5) DL algorithm in 2021. Concurrently, the researchers created a simple online platform named Renew e-waste, helping e-waste producers and collectors communicate more effectively, thereby streamlining the collection process. As part of the categorization process, the YOLOv5 model takes in a photo of the e-waste from the producer, identifies the various types of e-waste visible in the photo, and creates separate categories. The research designed by (Kim 2023) developed an automatic AI-based system for detecting e-waste. The study used Convolutional Neural Networks (CNN-2D) features and Local Binary Patterns (LBP) to identify features in an average of 10,000 photographs, including cell phones, circuit boards, and computer peripherals. Even though the process was successful in categorization, it could not be used for predicting metal retrieval rates and therefore was not a full-cycle e-waste management process. The research done in (Mehta and Roy 2023) examined AI's ability to potentially improve bioleaching by modeling how pH, temperature, microbial activity, and metal concentration can be used to predict metal recovery from e-waste. Machine learning regression approaches like Support Vector Regression (SVR) and Random Forest were applied. While the study was effective at optimizing bioleaching parameters, it was not applicable for image classification, which meant it could not efficiently sort e-waste in real time. This study (Zhang 2022) developed a smart sorting system that uses AI-powered visual recognition and is set up on a conveyor belt. The system used CNN-based classification to sort items like resistors and PCBs effectively. However, it could not be scaled up or used flexibly because it required specialized hardware and could not be integrated with recovery methods like bioleaching.

This study (Kumar and Singh 2023) employed deep learning to automatically identify and recover components of e-waste. Their CNN classifier achieved over 94% accuracy, although it was only tested on datasets from controlled labs. It emphasized the importance of having more diverse, real-world datasets so that the results could be generalized and applied more effectively. This study used supervised learning models like Decision Trees and Gradient Boosting to predict how well bioleaching would work. The goal was to improve the recovery of metals from microbes using AI. The study successfully predicted metal recovery performance, but it did not use image-based categorization, which meant it could not identify all bioleach recovery options. This study suggested a hybrid AI platform (Nair and Banerjee 2024) that combined image-based classification (using CNNs and LBP) with sensor data analysis (using XGBoost and Decision Trees) to predict how well bioleaching would work. Even though the platform had superior metal recovery rates and operating efficiency, it wasn't good for industrial use on a big scale because it could only be tested on a limited scale and not in real time.

4. Proposed Method

- **System Architecture**

The system presents three main modules: 1. Image Classification, 2. Bioleaching Prediction, 3. Decision Engine (integrating both modules). The system is designed to classify images of e-waste for categorization and use sensor data to predict bioleaching success rates.

- **Image Classification Module**

The preprocessing stage includes resizing and converting e-waste photos to grayscale for dataset standardization. Feature extraction utilized techniques like Local Binary Patterns (LBP) and calculated statistical features such as mean, median, and variance to capture the image's critical textural characteristics. Convolutional Neural Networks (CNN-2D) and Residual Neural Networks (ResNet) were employed to classify the e-waste, with CNN-2D performing most of the classification work while ResNet improved accuracy by using deeper network structures to capture complex patterns. In the bioleaching portion, the Bioleaching Prediction Module processes sensor data in either .csv or .xlsx formats, providing essential features such as pH, temperature, metal content, and microbial activity. The sensor inputs for the XGBoost model are used to predict bioleaching efficiency by analyzing these parameters and generating an assessment of potential metal recoveries under optimal conditions. The Decision Engine combines outputs from both the image classification and bioleaching prediction modules to provide recommendations for waste classification or resource recovery, completing the automated e-waste management process. To ensure scalability and adaptability, the following advanced algorithms were utilized in the system:

- **ResNet (Residual Neural Network):** Improved image classification through the use of residual connections that allow deeper networks to be trained, enhancing the recognition of true patterns in e-waste components.
- **XGBoostf (Extreme Gradient Boosting):** Used for predicting the bioleaching efficiency of various scenarios due to its superior performance with structured data. Using the gradient boosting tree algorithm, it is able to model the complex relationship between sensor features of images and recovery success.
- **K-Means Clustering:** For visually similar e-waste images, K-Means was used to cluster them into groups, providing an unsupervised approach prior to CNN classification. This was beneficial in organizing the dataset into classified groups to enable efficient training.
- **Convolutional Neural Networks (CNN-2D)**

CNN-2D is a deep learning algorithm designed for image data. It performs classification by automatically extracting features from images through multiple layers of convolutional filtering, where features may include edges, textures, shapes, etc. The features extracted from the convolutional layer are then passed through fully connected layers, where classifications are made. CNNs are mainly useful for identifying patterns in images to recognize objects, classes, and categories. They are also used for classifying audio, time-series, and signal information. In this methodology, CNN-2D, as shown in Figure 1, is used for image classification of e-waste components. The model classifies e-waste components, such as mobile phone parts and circuit board parts, as reusable or non-reusable, based on the features detected in the images. The automated detection of e-waste components via this algorithm enhances recycling processes by allowing much faster sorting of e-waste components.

- **Residual Neural Network (ResNet)**

ResNet is a type of deep neural network that addresses the vanishing gradient problem that affects deep networks by utilizing residual connections. Residual connections bypass certain layers in a model and instead send the connections of the data from the earlier layers to the deeper layers. This allows the model to learn complex patterns. This is why ResNet can train deep networks without losing information. The methodology integrates ResNet to improve image classification. ResNet

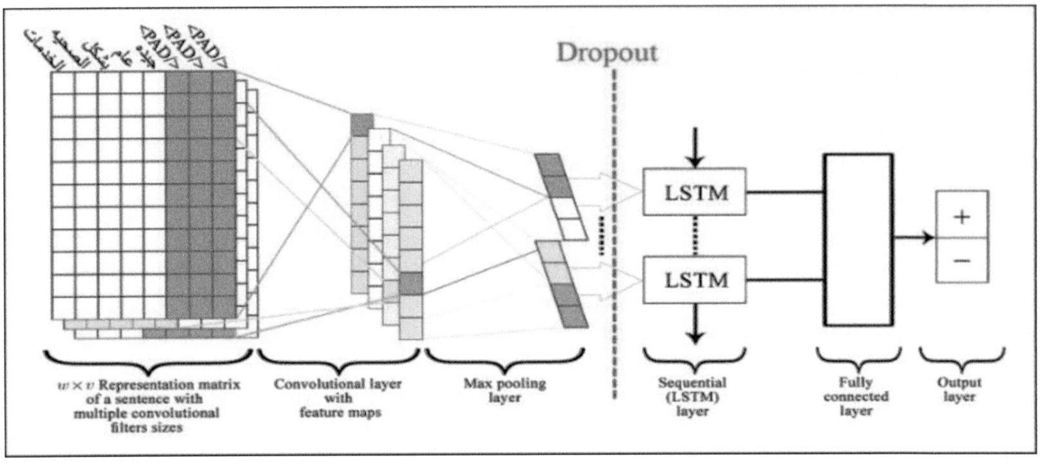

Figure 1: Architecture of CNN.

is utilized as a replacement for CNN-2D and employs deeper layers in the network to achieve improved classification performance on more complex e-waste features, as shown in Figure 2. This is especially important when the classification task involves components that are very similar to one another.

- **Extreme Gradient Boosting (XGBoost)**

XGBoost is a robust machine learning algorithm based on the gradient boosting approach, which sequentially builds decision trees to correct the errors of previous trees. It implements a boosting methodology comprising defined weak models that focus on instances that were difficult to predict successfully. XGBoost goes a step further by incorporating regularization to limit overfitting. In this example, XGBoost is used to predict bioleaching efficiency. By analyzing sensor data such as pH, temperature, and microbial activity, XGBoost learns how these variables impact bioleaching performance. It helps the system determine where conditions are optimal for metal recovery from e-waste, thereby increasing the sustainability and efficiency of recycling systems.

- **K-Means Clustering**

K-Means clustering is an unsupervised learning method that helps create clusters of related data points. The process assigns data points to a certain number of clusters (K) based on the similarity of features. K-Means clustering iteratively updates the clusters by changing the centroids to minimize variance within each cluster. In this methodology, K-Means clustering serves as a preprocessing stage to cluster visually similar e-waste images. This separates the e-waste dataset into several more manageable and organized groups before running CNN-2D for classification. By using K-Means clustering to organize the e-waste images, it also helps to reduce computational complexity by allowing the model to focus on representative subsets of data in each clustered set, which may improve overall performance.

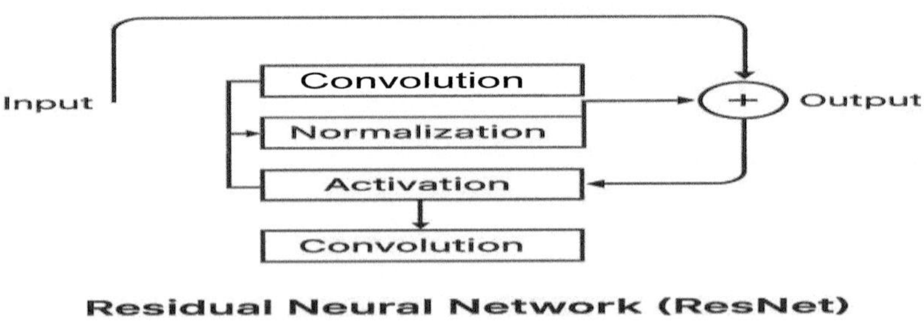

Figure 2: Architecture of ResNet.

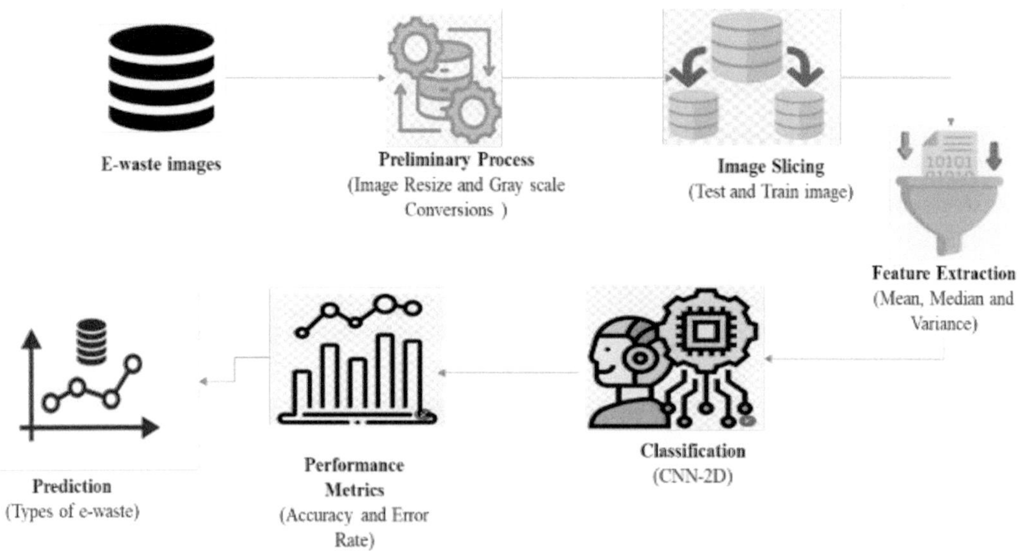

Figure 3: Ensemble Model Architecture.

- **Ensemble Model Architecture**

The Ensemble Model Architecture in Figure 3 combines both image classification and bioleaching prediction tasks in an integrated model so that classification predictions and recovery predictions by bioleaching can be completed in real time and seamlessly. The ensemble model architecture illustrates how CNN-2D, ResNet, XGBoost, and K-Means Clustering have been integrated into the model. It depicts the cohesive interaction of these methods to enhance both the accuracy of image classification and the optimization of predictions for bioleaching recovery in e-waste management.

- **Implementation**

The stage of the system is broken down into various steps in order to achieve proper e-waste management. The implementation of the system proceeds through an Input and Preprocessing step, where the e-waste images are accepted in .jpg or .png formats and resized, then converted to grayscale to maintain a consistent dataset. The next step is Feature Extraction, which evaluates the statistical features of the image (mean, median, and variance) as a means to characterize the image. Additionally, Local Binary Patterns (LBP) are implemented to capture texture features and enhance the system's ability to distinguish between different parts of e-waste. The next step is the Model Generation phase, during which the model is trained with the associated feature sets of the images. CNN-2D or ResNet supports the classification of e-waste material. The system then proceeds to the Prediction phase, where it classifies new images not previously processed through the system. Additionally, real-time sensor data are processed to predict the potential success of the bioleaching process and estimate the recovery of metal from e-waste (especially lithium) simultaneously.

- **Input and Preliminary Process**

Part 1: Image sampling and processing. Before classification, images of e-waste are collected, and preprocessing is performed on these images. The dataset contains images of e-waste in .png or .jpg format from the Kaggle dataset or equivalent; the images are preprocessed by resizing for consistency in size across all input data. Figures 4 and 5 show the original and resized images, respectively. The resize() method enables images to be resized. Images are then converted from RGB to grayscale, as shown in Figure 6. Converting to grayscale minimizes redundancy in the dataset using the formula: $imgGray = 0.2989 \times R + 0.5870 \times G + 0.1140 \times B$. The degradation lessens computational load, while retaining significant features for classification.

Figure 4: Original Image.

Figure 5: Resized Image.

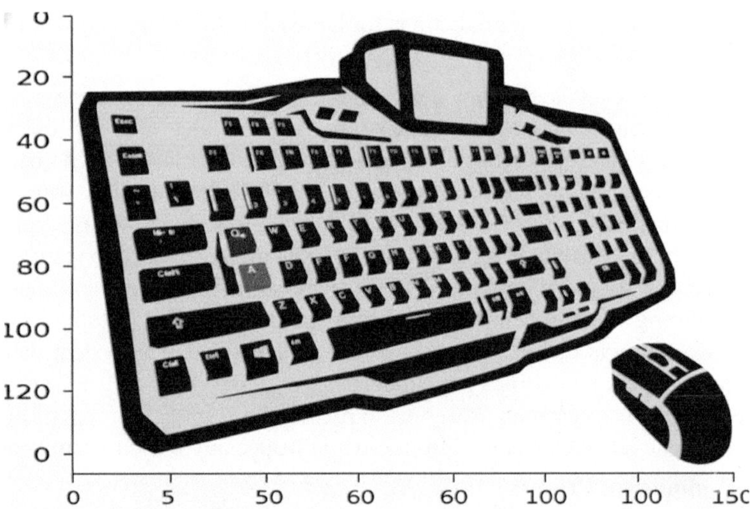

Figure 6: Gray Scale Image.

- **Feature Extraction**

This is a primary stage in image preclassification, in which the images consist of multiple statistical features of pixel values to characterize them. The mean filter quantifies the average intensity of all pixel intensities of the image or within a Region of Interest (ROI), providing a first approximation of the lightness of the image. The median filter removes noise while preserving the definitive characteristics of an image, such as edges or finer details, while also being effective at maintaining the integrity of images affected by noise. The variance, in terms of pixel intensities, quantifies the contrast—i.e. the variability of pixel intensities—which differentiates between intensity values.

Variance is also important for texture analysis, where the model tries to identify areas in the pixels that vary from surrounding pixels; hence, allowing the model to determine multiple textures during image classification.

In image processing, mean, median, and variance are also widely used to describe the statistical features of pixel values in a picture. We are employing the same methods in our work. **Mean:** In image processing, the mean can be found by averaging the pixel values in a Region of Interest (ROI) or the whole image. It shows the average brightness level of the pixels and is widely used for filtering, adjusting contrast, and normalizing images. Mean filtering, for instance, replaces the value of each pixel with the average of the surrounding pixel values. This helps smooth the image and reduce noise. **Median:** The median is a great way to eliminate noise in images. In median filtering, the value of each pixel is replaced with the median value of its neighbors. This process works well to preserve edges and fine details while reducing salt-and-pepper noise and other types of impulse noise.

Variance: Variance measures the variability or contrast in pixel values within an image. High variance indicates a wide range of pixel intensities, while low variance suggests a more uniform distribution of intensities. Variance is often used in texture analysis, where it can help distinguish between regions of uniform texture and regions with varying textures.

- **Image Splitting**

The image splitting diagram in Figure 7 represents the division of the dataset into training (70%)—used to develop the model—and testing (30%) subsets—used to evaluate the model's performance. This step ensures that the model is trained on a large portion of the data while reserving a portion for performance evaluation to avoid overfitting.

- **Result Generation**

The final system output and performance is evaluated through various metrics, as shown in Table 5.

The system's functionality is verified through several test cases designed to ensure its performance across different scenarios. Test Case 1 focuses on image classification of a valid e-waste item, where the objective is to verify the system's ability to correctly classify an e-waste image. Test case examples here help illustrate how the system is intended to function. The first test case, Test Case 1, tests for a successful e-waste identification classification. The input includes a .jpg circuit board image, and the expected output is "E-Waste." Test Case 2 verifies that input images, which do not belong to e-waste, are not incorrectly classified as e-waste. The input includes a .png image of an ordinary cardboard box, with an expected output of "Not E-Waste." Test Case 3 verifies the bioleaching efficiency prediction. In this example, the user will enter pH = 2.5, Temperature = 35 degrees Celsius, Copper concentration = 250 ppm, and Microbial activity = High. The intention is to evaluate the prediction for bioleaching success, so the expected output is "High recovery efficiency

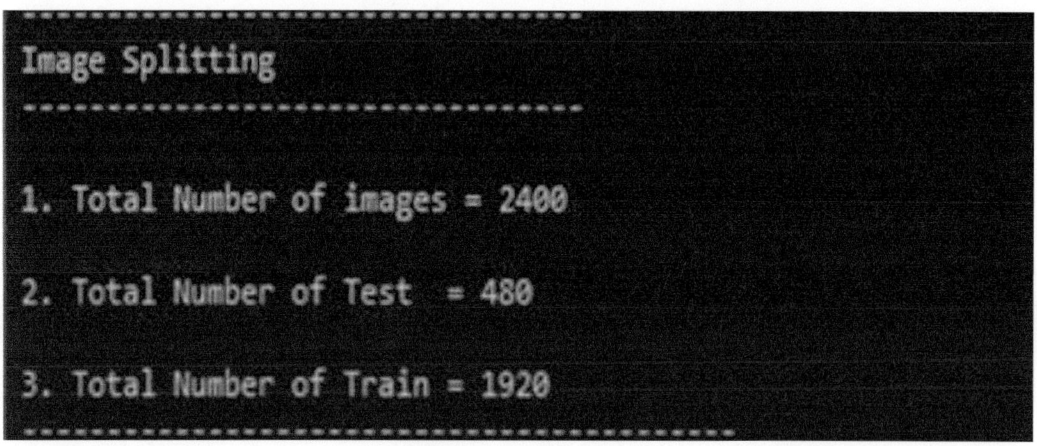

Figure 7: Image Splitting.

Table 5: Evaluation Matrices Description.

S. No	Metric Name	Description	Formula
1	Accuracy	Ratio of correct predictions to the number of predictions made.	(TP + TN)/(TP + TN + FP + FN)
2	Error Rate	The rate of incorrect predictions made by the model.	(FP + FN)/(TP + TN + FP + FN)
3	Precision	The proportion of accurately forecasted positive observations to the total forecast positives	TP/(TP + FP)
4	Recall	The proportion of positively predicted observations that were correct to all true positives.	TP/(TP + FN)
5	F1-Score	The weighted average of precision and recall, offering a balance between the two.	2 * (Precision * Recall)/(Precision + Recall)

expected." Test Case 4 evaluates input not with sensor data but with an image input to the system that produces a corrupted image file. The input, a corrupted .jpg image that holds no value for the experiment, has a preset expected output acknowledging the corruption: "Error: Invalid image format." Call this Test Case 2: input_2_corrupted_data. Lastly, Test Case 5 is to evaluate the invalid scenario variation for sensor data format (text instead of numeric value). The case addresses the pH input with text, for example, the pH is "acidic." The expected output is "Error: Invalid data type in sensor input".

5. Dataset Details

The datasets utilized for this project consist of two types—e-waste pictures and sensor data—which are both necessary for the classification and prediction work for the AI-integrated bioleaching model. The e-waste picture dataset contains a variety of e-waste images, including Printed Circuit Boards (PCBs), mobile phone components, and miscellaneous electronic parts. Images in this dataset were collected from online repositories, including Kaggle, and supplemented with curated images to ensure representation and diversity across categories of e-waste. The initial images collected had to be preprocessed to extract features and classify images suitable for CNN-2D tasks. E-waste pictures' prominent features include image resolution, texture, and component shape, which are important for developing the CNN-2D model. This dataset was developed primarily for training and testing the e-waste image classification model to distinguish usable from non-usable e-waste components.

Figure 8 illustrates the relative significance of various attributes commonly analyzed in electronic waste (e-waste). Metal concentration (mg/kg) dominates the assessment, accounting for 84.6%, highlighting its critical role in determining the value and environmental impact of e-waste. Total metal weight (9.3%) also contributes substantially, reflecting the recoverable metal mass. Other factors such as temperature (4.4%), pH (1.1%), and microbial activity (0.5%) play smaller but relevant roles, particularly in the context of metal recovery processes like thermal or biological treatments. Overall, the chart underscores the primary focus on metal content in e-waste analysis.

Figure 9 displays the distribution of various metals present in electronic waste based on their weight. Gold constitutes the largest share (38.4%), followed by zinc (23.2%) and nickel (15.6%), indicating the high presence of both precious and base metals. Silver (12.8%), lead (5.2%), and copper (4.8%) also contribute to the metal content. This distribution underscores the economic potential and environmental importance of metal recovery from e-waste.

Table 6 shows the e-waste item analysis report with five key parameters related to an e-waste sample, likely to assess environmental or processing impacts. The pH value 5.4 indicates the sample

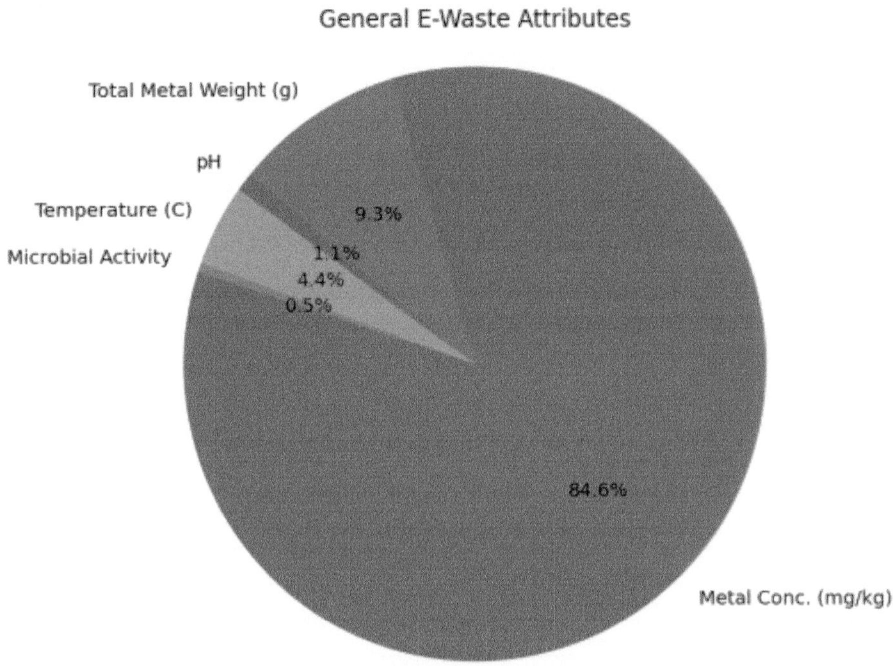

Figure 8: General E-Waste Attributes.

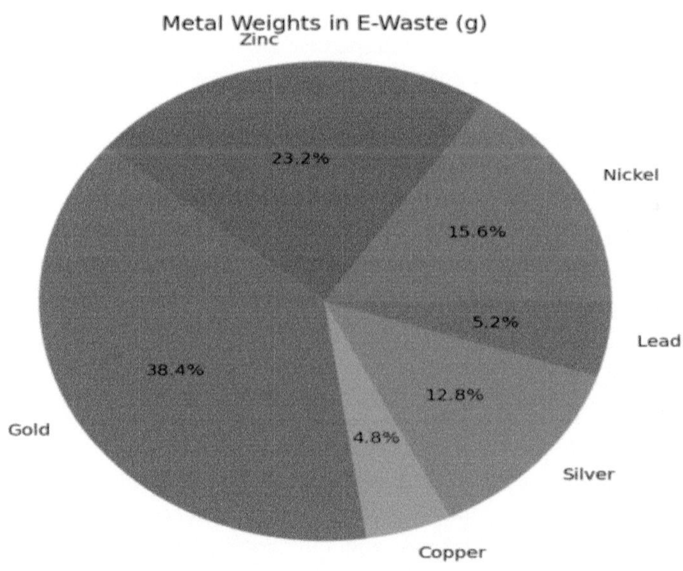

Figure 9: Metal Weights in E-Waste (g).

is slightly acidic, which can influence metal solubility and microbial behavior. Temperature (22.1°C) represents normal room temperature, which helps maintain typical microbial activity levels. A moderate level (2.67) of biological activity, possibly related to biodegradation or bioleaching processes, was observed. Metal concentration (420.45 mg/L or ppm) shows a high concentration of metals, which may be valuable for recovery or harmful if untreated. Total metal weight (46.21 g) indicates the total extractable or measurable metal content in the analyzed e-waste sample.

Table 6: E-Waste Item Analysis Report.

E-Waste Item Analysis Report		
1)	pH Value	= 5.4
2)	Temperature (C)	= 22.1
3)	Microbial Activity	= 2.67
4)	Metal Concentration	= 420.45
5)	Total Metal Weight (g)	= 46.21

Table 7 specifically focuses on the **weights of individual metals** recovered or measured in grams and lists the mass of six different metals found in the e-waste sample. The **total metal weight** of these six metals adds up to approximately **46.21 g**, matching the value reported in the table.

Table 7: Weight of the Different Metals.

Metal Weights (g)		
6)	Gold weight	= 17.75
7)	Copper Weight	= 2.23
8)	Silver Weight	= 5.91
9)	Lead Weight	= 2.39
10)	Nickel Weight	= 7.19
11)	Zinc Weight	= 10.74

Table 8 presents the metal composition percentages of an e-waste sample, showing how much each metal contributes to the total sample mass in percentage terms. The percentages indicate the proportion of each metal in the overall sample, likely based on total sample weight.

Table 8: Metal Composition Percentages.

Metal Percentage		
12)	Gold Percentage (%)	= 4.13
13)	Copper Percentage (%)	= 0.52
14)	Silver Percentage (%)	= 1.38
15)	Lead Percentage (%)	= 0.56
16)	Nickel Percentage (%)	= 1.67
17)	Zinc Percentage (%)	= 2.5

The sensor dataset contains structured data specific to the bioleaching process alongside the image dataset. This structured dataset contains relevant data essential to the bioleaching process, such as pH, temperature, metal concentration, and microbial activity, which are necessary to predict bioleaching success. Through the analysis of these selected data parameters, the system can predict the efficiency of metal recovery from e-waste. The sensor data was involved in the training and testing of the bioleaching efficiency prediction model (XGBoost), which is vital for optimizing the recovery process of various metals. The data sourced from both image and sensor datasets was separated into a 70%-30% split for evaluating both models. This ensured the models were sufficiently trained on the data while still holding enough data to evaluate performance and quantify the accuracy and generalizability of the model. Table 9 summarizes the specifics of the e-waste image dataset and the sensor dataset developed for this study. The table outlines the source, format, features included, and use of each dataset, highlighting their role in image classification and bioleaching prediction.

Table 9: Data sets used in the project.

Dataset Type	Source	Format	Features Included	Used For	Number of Images Used
E-Waste Images	Kaggle, Online Sets	.jpg, .png	Image resolution, texture, component shape	CNN-2D classification	10,000+ images (from Kaggle and curated sources)
Sensor Data	Simulated + Public CSVs	.csv, .xlsx	pH, temperature, metal concentration, microbial activity	Bioleaching efficiency prediction	N/A

6. Results and Discussion

The AI-powered e-waste management technical system offered strong recognition performance on both image classification and bioleaching prediction. The recommended dual-model approach (CNN-2D-based image classification + bioleaching prediction by analysis of sensor data) performed very well in the management of e-waste.

A) Image Classification Results

The CNN-2D model, utilizing Local Binary Pattern (LBP) as a texture-based feature extraction technique, demonstrated an overall classification accuracy of 94.5% after 20 epochs of training. Other e-waste categories, including printed circuit boards (PCBs), mobile components, and other electronic hardware, exhibited relatively consistent performance as well. The use of LBP significantly improved texture recognition, especially in distinguishing between reusable and non-reusable e-waste components. We have used precision, recall, and F1-score metrics to evaluate the model's performance. The results indicated strong generalization capabilities, making the model suitable for real-time applications. Figure 10 compares the accuracy, precision, recall, and F1-score of the proposed CNN-2D model against ResNet and other baseline models. It visually emphasizes the superior performance of the CNN-2D model in terms of classification metrics. The proposed model produces a precision of 94.5%, recall of 92%, and F1-score of 93.2%.

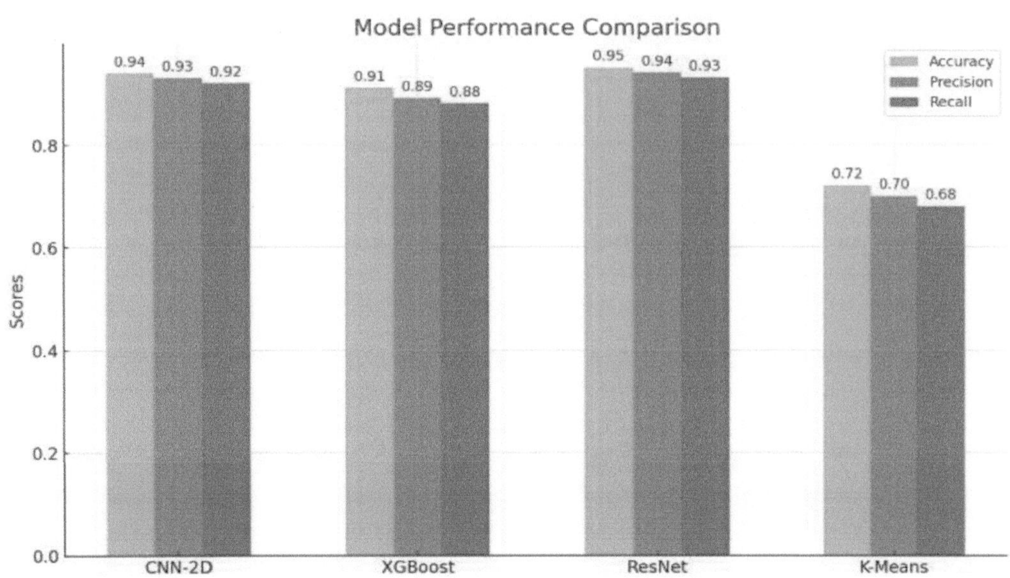

Figure 10: Various Model Performance Comparison

Figure 11 plots the accuracy of the CNN-2D model against the number of epochs during training. It shows the model's gradual improvement in accuracy as training progresses, demonstrating the convergence of the model's performance. The correlation between training time and accuracy during training is depicted in Figure 12. It illustrates the trade-off between computation time and performance by showing how the model's accuracy rises with each epoch while training time progressively increases.

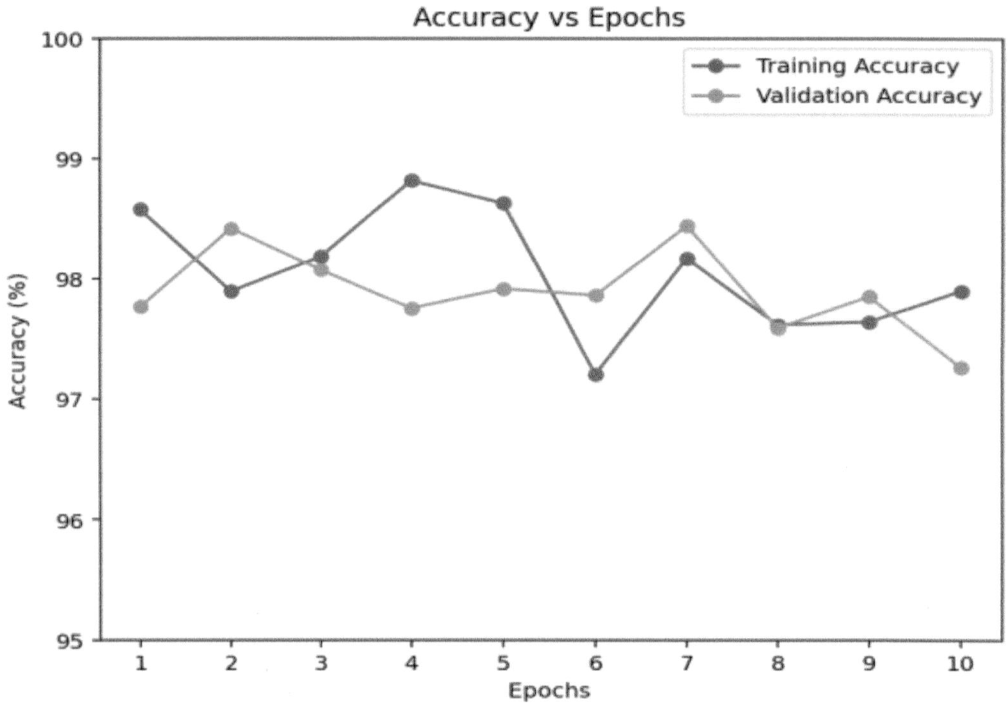

Figure 11: Accuracy vs. Epoch.

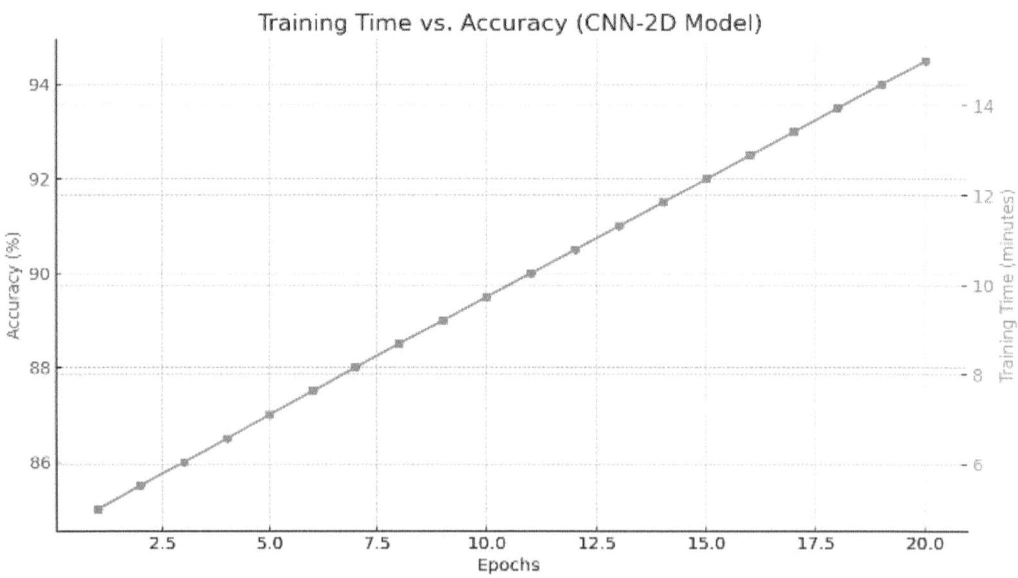

Figure 12: Timeline Vs Accuracy.

B) Sensor Data and Bioleaching Prediction Results

The XGBoost model used for predicting bioleaching efficiency showed excellent performance by accurately predicting metal recovery rates based on historical sensor data, which included parameters like pH, temperature, metal concentration, and microbial activity. The model demonstrated strong alignment with expected outcomes, particularly under optimal environmental conditions. The predictions made by the sensor-based model aligned with the outcomes of real-world bioleaching experiments, supporting the system's potential to improve resource recovery in e-waste recycling. The output achieved graph in Figure 13 provides a visual representation of the system's performance in test cases. It compares the expected versus actual outputs for various test cases, such as image classification and sensor data prediction, showing how the system correctly classified and predicted outcomes. Table 10 shows the various test case descriptions and the results achieved for better understanding purposes.

The confusion matrix also corroborates these findings by demonstrating the distribution of true positives, false positives, true negatives, and false negatives, as described in Figure 14. The model showed 94.5% e-waste categorization accuracy and 92% accuracy for non-e-waste classification. This confusion matrix graph visualizes the true positive, false positive, true negative, and false negative predictions for the CNN-2D model. It helps assess the model's performance in distinguishing between e-waste and non-e-waste, highlighting areas of misclassification.

The CNN-2D model, enriched with LBP for texture feature extraction, achieved an impressive classification accuracy of 94.5% after 20 training epochs. The model demonstrated excellent consistency across different classes of e-waste, successfully distinguishing between reusable and non-reusable components, and Table 11 shows the performance analysis of the proposed system. The confusion matrix analysis revealed that this model achieved a precision of 94.5% and 92% recall, reflecting its reliability and ability to accurately classify e-waste images in real time. Experimental results validate the efficacy of integrating deep learning models (CNN-2D) with environmental

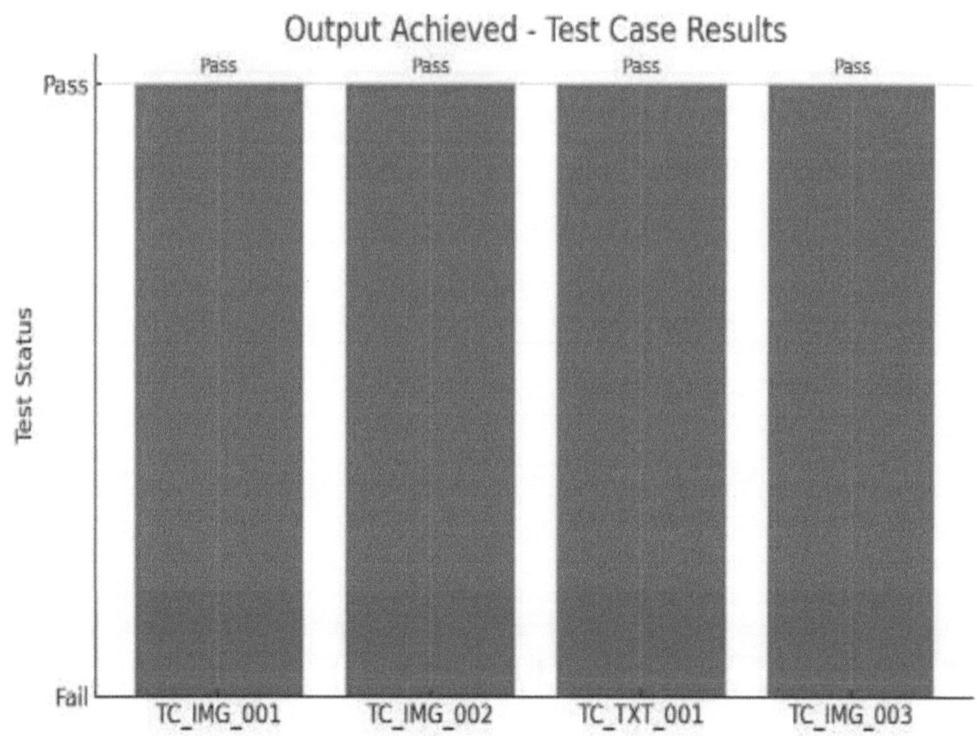

Figure 13: Output Achieved.

Table 10: Test Case Description and Results Achieved.

Test Case ID	Input Type	Expected Output	Actual Output	Status
TC_IMG_001	PCB Image	E-Waste	E-Waste	Pass
TC_IMG_002	Box Image	Not E-Waste	Not E-Waste	Pass
TC_TXT_001	Sensor Data	High Efficiency	High Efficiency	Pass
TC_IMG_003	Corrupted Image	Error	Error	Pass

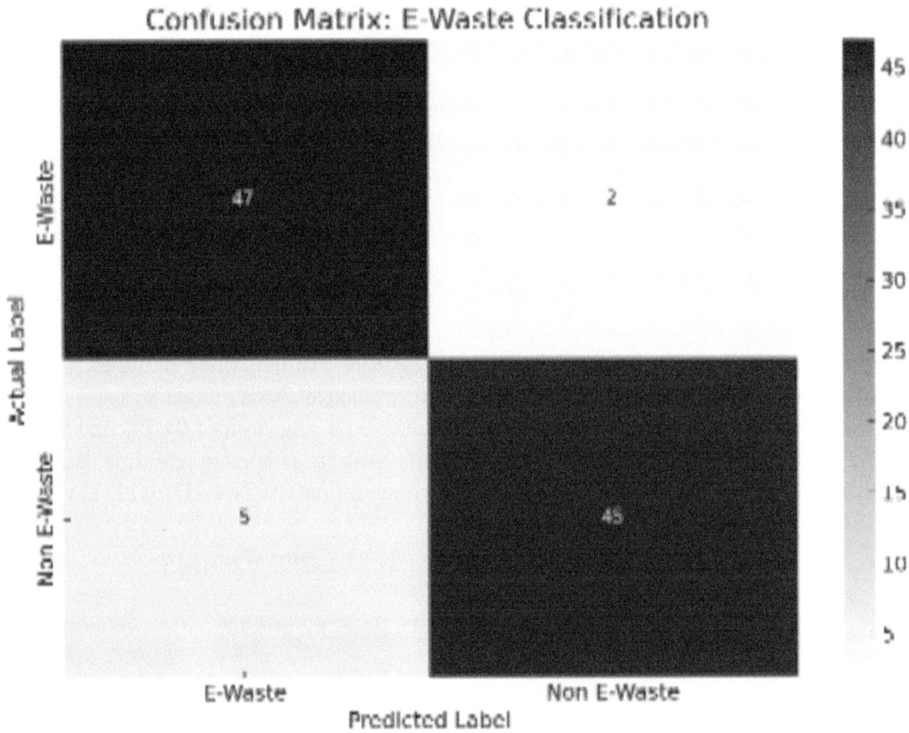

Figure 14: Confusion Matrix.

Table 11: Performance Analysis of the Proposed System.

Class Type	TP	FP	TN	FN	Precision	Recall	F1-Score	Accuracy (%)	Support
E-Waste	47	3	40	2	94.5%	95.5%	95%	94.5%	92
Non E-Waste	45	5	38	3	90.0%	92.0%	91.0%	92.0%	91

data analysis (with XGBoost being used to predict bioleaching) in the management of e-waste. This system has shown its capability to optimize and improve the process of e-waste processing with the addition of automatic classification and predictive bioleaching. The proposed framework is highly scalable and can be used in smart city and green recycling initiatives, which will ultimately help in promoting more environmentally friendly practices in the electronic waste industry. Future developments may focus on larger real-world datasets and real-time IoT sensor integration. The model's performance and generalization capacity would be further improved by increasing the amount of training with more diverse, real-world data. Integration of real-time sensor data from IoT devices could greatly improve the accuracy and responsiveness of the system, enabling it to more effectively respond to dynamic environmental conditions.

7. Conclusion and Future Enhancements

This study presented a comprehensive AI-driven prototype intended for the management of e-waste in the communication technology sector, blending image processing with deep learning and data-driven bioleaching optimization. It uses CNN-2D as the image classification system for identifying components from waste or those to be reused. Although Local Binary Patterns (LBP) are used for texture-extracted features through which the various components of e-waste can be determined, text mining functions and analysis of sensor input enable predictions of optimal bioleaching conditions, such as pH levels, temperature, and microbial activity. This integration not only optimizes the effectiveness of metal recovery but also restricts the use of chemical-intensive technologies, resulting in more environmentally friendly practices.

By combining machine-based image classification with comprehensive microbial extraction methods, the proposed model provides a green and sustainable solution to the growing global problem of electronic waste. The AI model represents the principles surrounding a circular economy and is used in both resource recovery and classification processes. It reduces waste through the recovery of precious metals, decreases landfill use, and helps avoid environmental threats posed by electronic waste. The flexibility of the method means it could be adapted for a variety of electronic components to promote sustainable urban planning and drive industry stakeholders toward environmentally responsible practices. The proposed model could be incorporated into the waste management processes of smart cities, aligning with the vision of sustainable urban development. Beyond demonstrating the environmental sustainability of e-waste management, this model serves as a foundational reference point for future interdisciplinary research, connecting the fields of computer science, environmental engineering, and microbiology.

For the future, there are several promising paths to enhance the scalability, robustness, and efficiency of the AI-powered bioleaching model. One path would be to use real-time streaming data from IoT sensors embedded in waste processing units. These can provide continuous real-time data on environmental parameters and chemical reactions, allowing the system to adapt and optimize the bioleaching process in real time. Enhancing deep learning models by incorporating ResNet or Vision Transformers (ViTs) would increase image classification accuracy—especially regarding the complex variance represented by e-waste. Incorporating more training data and an even richer, globally diverse dataset could further improve the model's ability to generalize across regions and types of electronic waste. Exploring reinforcement learning may allow the model to adjust to varying microbial behavior, leading to optimal bioleaching effectiveness over time. Introducing blockchain technology may add a new level of traceability and transparency for metal recovery, ensuring accountability and trust in the system itself.

Another area of significant potential growth is the application of the model to smart city waste treatment plants. Real-world applications could permit continual model enhancement and performance improvement. Moreover, modifying the system to accommodate different waste types such as plastics and organics would broaden its scope and application. Collaboration with recyclers could provide valuable real-world data that would help refine the system based on actual performance in the field. For sustainability, and to build upon current success, exploring synthetic biology to create superior microbes for bioleaching could significantly enhance recovery rates of economically valuable metals. Overall, the integration of AI, sensor networks, and biotechnology represents the future of sustainable e-waste management, and the model from this research provides a strong initial step toward addressing the growing global e-waste dilemma and serves as a template for continuous future improvements and sustainability-oriented solutions worldwide.

References

Adedeji, O., and Wang, Z. (2019). Intelligent waste classification system using deep learning convolutional neural network. Procedia Manufacturing, 35: 607–612. https://doi.org/10.1016/j.promfg.2019.05.086.

Arun, M. (2025). Investigation of a deep learning-based waste recovery framework for sustainability and a clean environment using IoT, *Sustainable Food Technology*, 3: 599–611.

Dassi, R., Kamal, S., and Babu, B. S. (2021). E-Waste Detection and Collection assistance system using YOLOv5. CSITSS 2021 - 2021 5th International Conference on Computational Systems and Information Technology for Sustainable Solutions, Proceedings. https://doi.org/10.1109/CSITSS54238.2021.9682806.

Fan, X., Zhang, H., Peng, W., Lü, F., and He, P. (2022). Applications of convolutional neural networks for intelligent waste identification and recycling: A review, *Resources, Conservation and Recycling*, 180: 106214.

FNU Abhimanyu, Zodage, T., Thillaivasan, U., Lai, X., Chakwate, R., Santillan, J. et al. (2022). RGB-X Classification for Electronics Sorting, *arXiv preprint* arXiv:2209.03509, 2022.

Hicks, C., Dietmar, R., and Eugster, M. (2005). The recycling and disposal of electrical and electronic waste in China-legislative and market responses. Environ. Impact Assess Rev., 25: 459–471.

Islam, N., Jony, M. M. H., Hasan, E., Sutradhar, S., Rahman, A., and Islam, M. M. (2023). EWasteNet: A Two-Stream Data Efficient Image Transformer Approach for E-Waste Classification, in *Proceedings of the 2023 IEEE 8th International Conference on Software Engineering and Computer Systems (ICSECS)*, Penang, Malaysia, 2023, pp. 435–440.

Kamal Miyalal Alaskar et al (2022). Current Status of Electronic Waste in Information Communication Technology (ICT) and its Environmental Management – A Review, Current World Environment, 17(3): 836–845.

Kim, J. (2023). AI-Based Image Classification System for E-Waste 2023.

Kumar, T., and Singh, M. (2023). Deep Learning for Electronic Waste Recovery, 2023.

Liu, K., Tan, Q., Yu, J., and Wang, M. (2023). A global perspective on e-waste recycling,Circul. Econ. (2023), Article 100028.

Meghna Jaiswal and Sudhakar Srivastava (2022). A review on sustainable approach of bioleaching of precious metals from electronic wastes, Journal of Hazardous Materials Advances, Volume 14: 2024.

Mehta, P., and Roy, A. (2023). Sustainable Bioleaching using AI Optimization,2023.

Nair, A., and Banerjee, R. (2024). Integrated AI Platform for Circular E-Waste Economy, 2024.

Patel, S. (2024). Metal Recovery via AI-Based Microbial Processes, 2024.

Peter Davis, Fayeem Aziz, Mohammad Tanvi Newaz, Willy Sher, and Laura Simon (2021). The classification of construction waste material using a deep convolutional neural network, Automation in Construction, Volume 122, 2021.

Sadeghi, M., Fadaei, A., and Ataee, M. (2020), Assessment of hospitals medical waste management in Chaharmahal and Bakhtiari Province in Iran, Arch. Agricul. Environmen. Sci, 5: 157–163.

Singh, N., Chimaniya, S., Pradhan, H., Singh, D., and Singh, K. (2025). Enabling Sustainable E-Waste Management Through Blockchain Technology, *2025 3rd International Conference on Disruptive Technologies (ICDT)*, Greater Noida, India, 2025, pp. 1–5, doi: 10.1109/ICDT63985.2025.10986526.

Santhuja, P., and Suresh, A. (2025). CNN-Based Circular Economy in E-Waste Management with Blockchain and IoT Integration, *2025 International Conference on Visual Analytics and Data Visualization (ICVADV)*, Tirunelveli, India, 2025, pp. 237–242, doi: 10.1109/ICVADV63329.2025.10961209.

Zhang, L. (2022). Intelligent Urban E-Waste Sorting System, 2022.

13

Machine Learning-Enabled Digital Twins for Diagnostic and Therapeutic Purposes

Nandini Manickam,[1] *Vijayakumar Ponnusamy,*[1,*] *Emilija Kisic*[2] and *Nemanja Zdravković*[2]

1. Introduction

Digital twin (DT) technology has become the next step in advanced data analytics. This technology is similar to creating a digital replica of a physical object. The twins created are linked to the data sources of the environment and receive real-time updates from the original physical object. When they are interconnected within one system to replicate the original, it helps in achieving better simulation results, decision-making, and improved performance. This process is called the metaverse. It uses real-world data to create simulations based on current and past data. Better outcomes are obtained when software platforms, artificial intelligence (AI), and the Internet of Things (IoT) are integrated. IoT sensors allow the transfer of data from physical objects to the digital world. With advancements in machine learning (ML) and data analytics, monitoring, diagnosis, predictive analysis, and testing services are made easier.

Healthcare industries require these advanced technologies that depend on real-world data for healthcare administration, healthcare analytics, and biomedical signal analysis. Digital twinning is best suited to provide personalized assistance by monitoring, diagnosing, and providing treatments to individual patients. This helps in the development of medical devices, and healthcare providers make use of these real-time data to update hospital details, allocate work for staff, and update resources available for patients. This improves patient care and the overall efficiency of the healthcare sector. Figure 1 depicts the structure of digital twin technology.

Depending on the application, digital twins can be divided into four categories. They are component twins, which represent individual parts of the system. These twins are used in modeling integral parts under certain conditions. Asset or product twins are the digital versions of physical

[1] Department of Electronics and Communication Engineering, SRM Institute of Science and Technology, Kattankulathur, Chengalpattu, Chennai.
Email: nm6075@srmist.edu.in, https://orcid.org/0009-0004-6853-4997

[2] Faculty of Information Technology, Metropolitan University, Serbia.
Emails: emilija.kisic@metropolitan.ac.rs, https://orcid.org/0000-0003-3059-2353;
nemanja.zdravkovic@metropolitan.ac.rs, https://orcid.org/0000-0002-2631-6308

* Corresponding author: vijayakp@srmist.edu.in, https://orcid.org/0000-0002-3929-8495

DOI: 10.1201/9781003655190-16

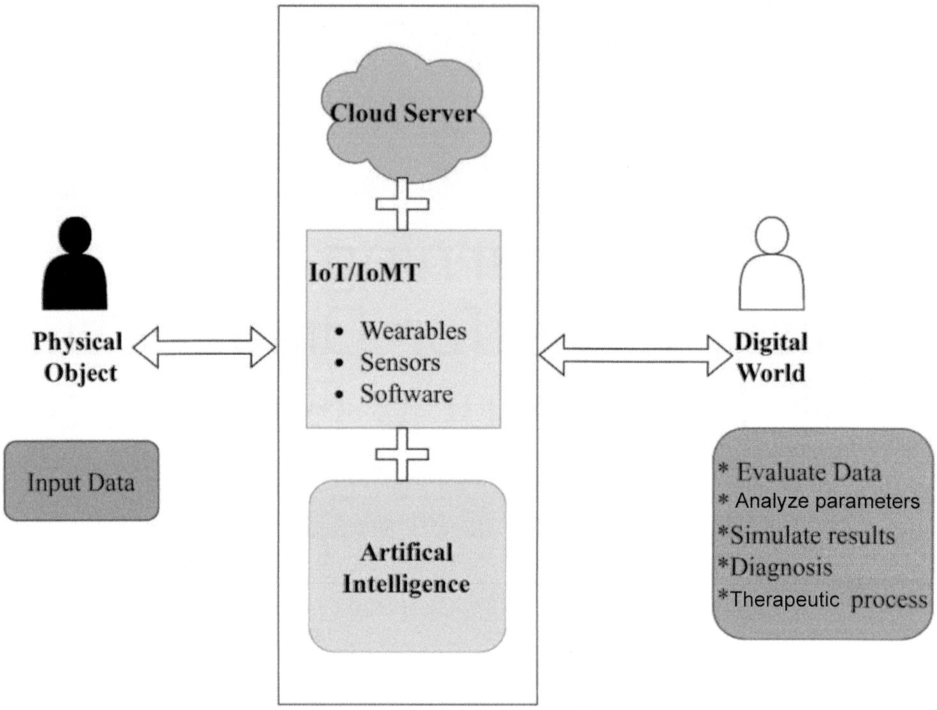

Figure 1: Structure of Digital Twin Technology.

objects composed of various component twins that integrate to form a product. This is mainly used to understand how individual parts work in a single real-world scenario. System twins or unit twins represent systems of products working in an integrated manner. This helps in understanding how products interact with one another to improve performance. Lastly, process twins represent the digital version of systems that work together. Figure 2 shows the types of digital twins.

Some of the advantages of digital twins are:

- Cost-effective and requires less time
- Improves efficiency with a proper update of information that replicates real-time changes
- Improves customer service
- Continuous monitoring, identifying, and rectifying issues.

Figure 3 portrays the flow diagram of the literature review for digital twin technologies in various applications.

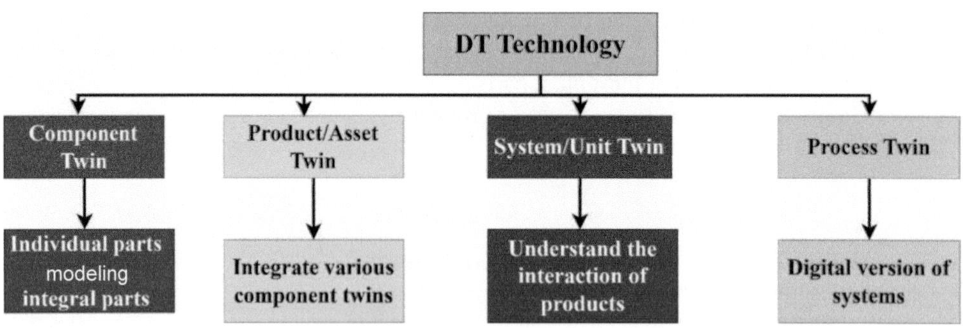

Figure 2: Types of Digital Twins.

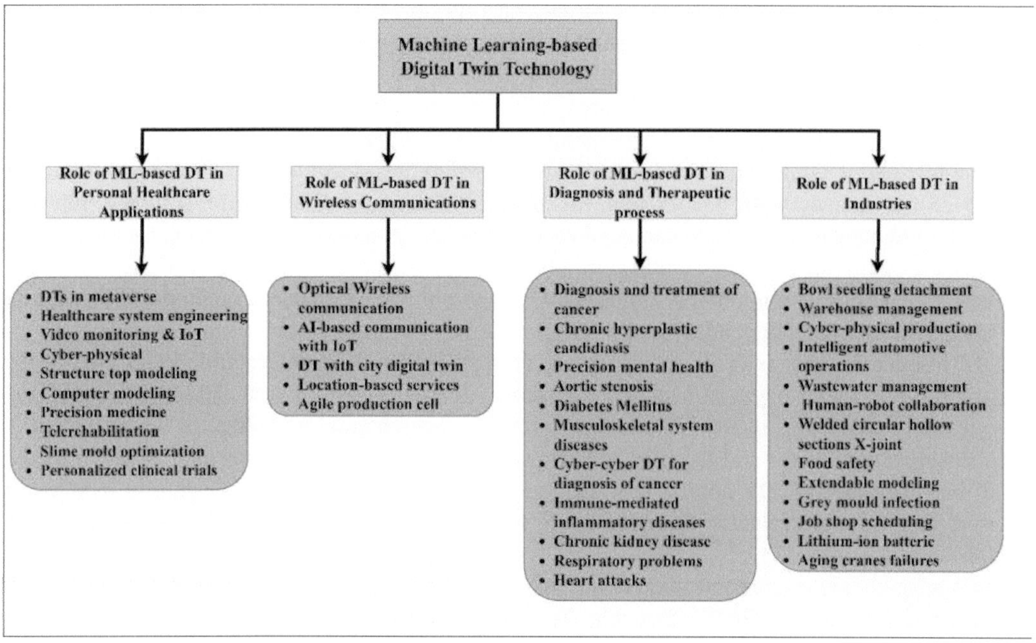

Figure 3: Flow diagram representation of literature review.

2. Related Work

Machine Learning (ML) helps the system learn from the data received and arrive at decisions without explicit programming. ML helps in the analysis of large data gathered from sensors and other sources, thereby improving the performance of digital twin technology. ML algorithms are used to assess the data obtained from the digital twin to determine whether its behavior has deviated in any way. ML models are trained to recognize patterns from past data to detect anomalies. ML algorithms can determine the model that requires repair or maintenance by comparing present and past data and predicting based on correlations. The role of integrating ML and digital twin technology in various applications is discussed below.

2.1 Role of Machine Learning-based Digital Twin in Healthcare Applications

Digital twins have become popular and widely used in various applications like industries and healthcare, especially in the metaverse, which has gained significant attention (Turab et al. 2023). As a result, a virtual world is created in which each person can interact with their environment and digital copies. Developing personalized and precision medicine to examine the working of DTs in the metaverse has been explored. The incorporation of DTs in the metaverse enhances security and privacy while offering patients individualized treatment.

DT technology has transformed the healthcare system into an enhanced personal care system where real-time data are integrated using advanced analytic techniques and virtually simulated to make accurate predictive and preventive analyses through ML algorithms (Vallée et al. 2023). By analyzing data received from various sources, such as medical history, patient characteristics, physiological data, and DTs, personalized healthcare can be provided. A safe and realistic environment can also be created for healthcare professionals.

DT technology has gained attention in industrial sectors for improving operations, productivity, reliability, quality, flexibility, and cost-effectiveness. The healthcare sector is an important domain that receives the complete benefits of DT technology (Mohamed et al. 2023), (Nader et al. 2023).

Healthcare system engineering is based on engineering analysis and design principles that help provide better healthcare services. Healthcare system engineering tools assist in continuous improvement, risk management analysis, supply chain processes, optimized resource utilization, and safety measures that improve healthcare quality and cost-effectiveness resource utilization, and safety measures that improves healthcare quality and cost-effectiveness.

The Internet of Things (IoT) has recently developed in the telecommunication sector (Stergiou et al. 2023). To connect with one another, this network makes use of hardware, software, sensors, and real-world objects. Sensor and camera data are collected, processed, and tracked. Using a cloud-based monitoring system, mobile cloud computing (MCC) processes and transmits video and image data using an algorithmic approach. Using the DT scenario, the healthcare system demonstrates its increased efficiency and dependability.

DT of microorganisms deals with analysis, recognition, and characterization in the metaverse by applying it to cyber-physical healthcare systems (CPHS) (Jamshidi et al. 2023). Taking DT of bacteria under study, the role of DT can transform the research and knowledge of bacteria since it helps in separating useful and harmful bacteria so that there can be efficient diagnosis. To create DT of bacterial species, DT approaches such as MobilenetV2, EfficientNetV2-S, and ResNet-50 are used. Training and evaluation are done using a hybrid dataset. EfficientNetV2-S performs better with an accuracy of 99.58%.

DT in healthcare is widely studied under two categories: treatment of disease and enhancement of healthcare (Sheng et al. 2023). Structure Top Modeling (STM) is a data mining method that is used as an analytical tool. DT technologies are widely used in many applications that facilitate daily life activities (Doungtap et al. 2023). DT employs mobile devices for image data collection, cloud processing, and 3D deep learning applications. Using deep learning techniques, an end-to-end pipeline is developed, and an accurate 3D model is automatically generated from the photographs gathered. The model uses ShapeNetCars and deep 3D datasets for providing end-to-end processes using VR.

Computer modeling for patients using clinical and gene data concentrates on eradicating two major issues: one is when a new drug is given, it should not affect the existing treatment, diet, or routine over time. The second is that it's a challenge to treat patients whose bodies resist the current therapies (Chebanov et al. 2023). The modeling tools are Generative Adversarial Networks (GAN) and deep learning. Drugs are used as inputs in a patient model, and survival probability estimates are produced as outputs. This makes simulating clinical trials and predicting their outcomes easier. A combination of Convolutional Neural Networks and deep learning can solve problems that interpret human intelligence (Currie et al. 2023). Integration of AI and DT for interpreting medical data in the form of signs and symbols without context in the digital space is called simulacrum (Rubeis et al. 2023). Here, DT is a collection of data points taken from the context and uses operational logic in an algorithm. Hyperreality data needs context.

AI algorithms are used in edge networks to automate information by meeting end users' needs in a technique known as mobile artificial intelligence-generated content (AIGC) (Chen et al. 2023). Mobile AIGC is an emerging technology for applications called Human Digital Twin (HDT). This enhances personalized healthcare by providing data, modeling DT, running tests, and medical services. Telerehabilitation development has become one of the emerging technologies during the pandemic situation, where patients need remote-aided healthcare assistance (Falkowski et al. 2023). The performance is evaluated using DTs to get updates and the current status of the patient in the physical world to the digital world. This is based on a questionnaire and bibliographic approach. Controlling the rehabilitation exoskeleton through DT is analyzed, and techniques for improvement are explored. Figure 4 illustrates the role of ML-based digital twins in healthcare applications.

Deep learning-based classification for IoT wearable devices has gained attention (Annamalai et al. 2023). Data encryption and classification in an IoT setting are accomplished by creating a slime mold optimization using ElGamal encryption and a deep learning-based classification model.

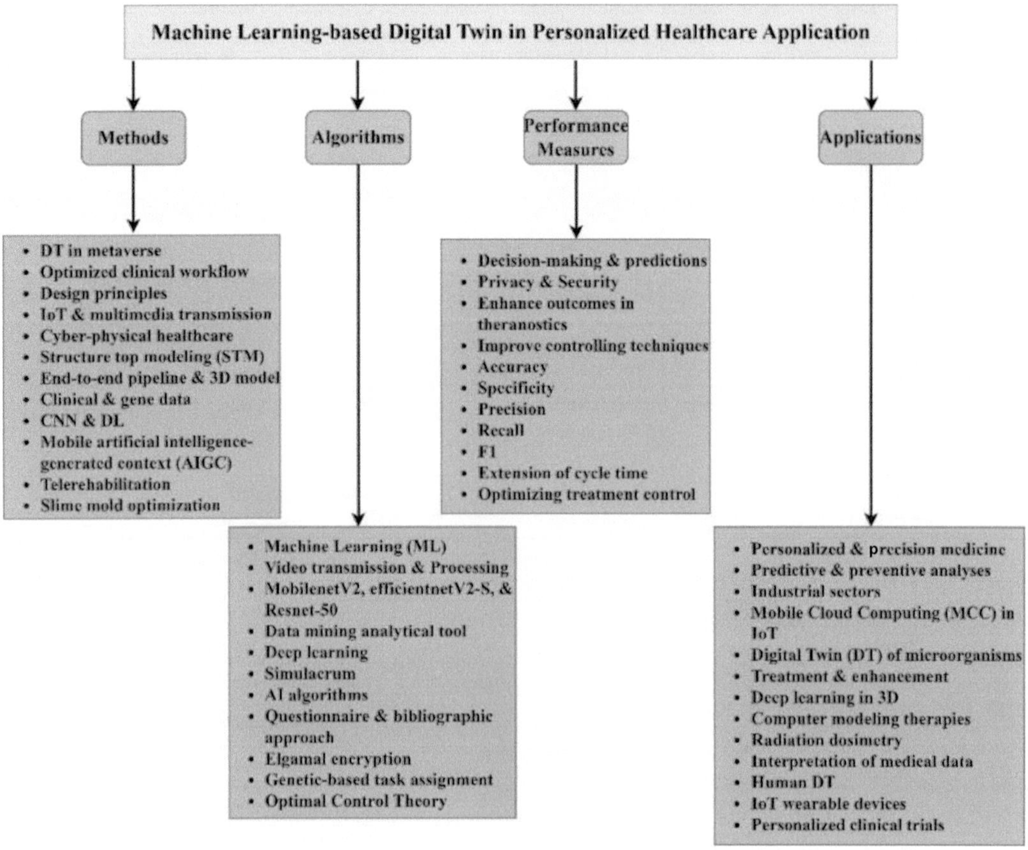

Figure 4: Role of Machine learning-based Digital twin in Personal Healthcare Applications.

The proposed method showed the following performance measures: Specificity – 98.5%, Precision – 98.75%, Recall – 98.3%, Accuracy – 98.5%, and F1 Score – 98.25%.

Interaction between humans and robots has contributed to the development of DT technology to achieve accurate monitoring and reflection of the physical world, including environmental changes (Ma et al. 2024). This study deals with the design of an environment-adaptive assignment method for DTs, where the factor-event-act mechanism is used to analyze dynamic events and observe changes in both the physical and digital worlds. The development of four mathematical techniques for personalized clinical trials enhances computational performance (Yankeelov et al. 2024). Initially, mathematical models are adjusted according to patient-specific data to produce accurate predictions. Finally, data assimilation is done where continuous updates received can reconstruct the outcomes in response to the treatment process. AI and DT technology are widely used in various applications (Kreuzer et al. 2024). Papers related to these two technologies are analyzed. Only a small proportion of papers have elaborated on DT in real time, gathering data from the physical system and implementing it in a physical-to-virtual environment. Table 1 represents the review of literature on digital twins in healthcare applications.

2.2 Role of Machine learning-based digital twins in wireless communications

Optical wireless communication (OWC) has become an important factor in the development of 6G technology, and DT technology meets the requirements of wide applications that provide good optimization in overall system performance (Eldeeb et al. 2023). These technologies are integrated as a virtual replica to reflect the characteristics of the physical object. Factors like recent

Table 1: Role of Machine Learning-based Digital Twins in Healthcare Applications.

References	Method	Algorithm used	Results
Turab et al. 2023	DTs in the metaverse	Modeling, simulation, and interfaces	Improves decision-making and predictions
Vallée et al. 2023	Predictive and preventive analysis	ML algorithms	Patient safety and efficiency
Mohamed et al. 2023	Healthcare system engineering	ML algorithms	Improve the process and outcome of healthcare
Nader et al. 2023	Healthcare system engineering	ML algorithms	Improve the process and outcome of healthcare
Stergiou et al. 2023	Video monitoring and IoT in cloud	Video transmission and processing algorithm	High latency and efficiency than conventional algorithm
Jamshidi et al. 2023	Cyber-physical healthcare	MobileNetV2, EfficientNetV2-S, and ResNet-50	V2-S performs better with an accuracy of 99.58%.
Sheng et al. 2023	Structure top modeling (STM)	Data mining method analytical tool	94 papers grouped to enhance healthcare
Doungtap et al. 2023	Deep learning in 3D	CAP–UDF model using ShapeNetCars	L2 chamfer distance of 0.089 and 0.129
Chebanov et al. 2023	Computer modelling for patients	Deep learning and GAN	Simulating clinical trials and predicting their outcomes becomes easier
Currie et al. 2023	Precision medicine-radiation dosimetry	ML and DL algorithms	Enhanced outcomes in theranostics
Rubeis et al. 2023	DT for cardiovascular disease	Simulacrum and simulation	More personalized medicine
Chen et al. 2023	Mobile AI-generated content (AIGC)	AI algorithms	Effectiveness of mobile AIGC studies through testbeds
Falkowski et al. 2023	Telerehabilitation development	Questionnaire and bibliographic approach	Controlling techniques through DT are explored
Falkowski et al. 2023	slime mold optimization-IoT security	Elgamal encryption and deep learning	Specificity-98.5% Precision-98.75% Recall-98.3% Accuracy-98.5% F1 score-98.25%
Ma et al. 2024	Environment-adaptive assignment method for DTs	Genetic algorithm-based task assignment	Extension of cycle time from 7500 to 12,800s. DT-HRC is required to reduce the cycle time
Yankeelov et al. 2024	Personalized clinical trials	Mathematical techniques and optimal control theory	Predicting and optimizing treatment control outcomes
Kreuzer et al. 2024	AI and DT technology applications	Analysis based on the articles search	Out of 149 articles, only few studies have developed DT in real-time by applying them in a physical-to-virtual environment.

advancements, challenges, and structural analysis based on DT-assisted OWC systems are studied and their performances are evaluated.

Wireless body area networks (WBANs) enable self-diagnosing capabilities in healthcare through AI-based communication with IoT (Ponnusamy et al. 2022). Augmented and virtual reality has also enhanced the diagnosis. DT technology helps to monitor, automate, and optimize their processes in the digital world. Digital twin networks (DTN) and city digital twins (CDT) are essential for network operators and planning (Turcanu et al. 2024). This is forecasted as a future in 5G and 6G networks. Developing location-based services (LBS) has a few challenges, such as providing a standardized interface when interacting with the physical world and proper intelligent services for supporting system performance (Lombardo et al. 2024). To resolve this issue, a multilayer DT-based architecture is needed that provides ML-based intelligent LBS to perform continual learning (CL). This architecture evaluates latency and complexity through analysis of execution graphs using DL. Figure 5 depicts the role of ML-based DT in wireless communication.

Replacing wired communication with wireless communication within an agile production cell can be done by maintaining the number of access points at a minimum (Szabó et al. 2024). The architecture is optimized, and access points are reduced through the application of reinforcement learning (RL). With the help of user equipment (UE) connected to the cameras and network resource management (NRM), this design helps standardize the 5G network. To increase training speed, DTs aid in quick ray tracing for radio propagation modeling. Table 2 illustrates how digital twins based on machine learning are used in wireless communications.

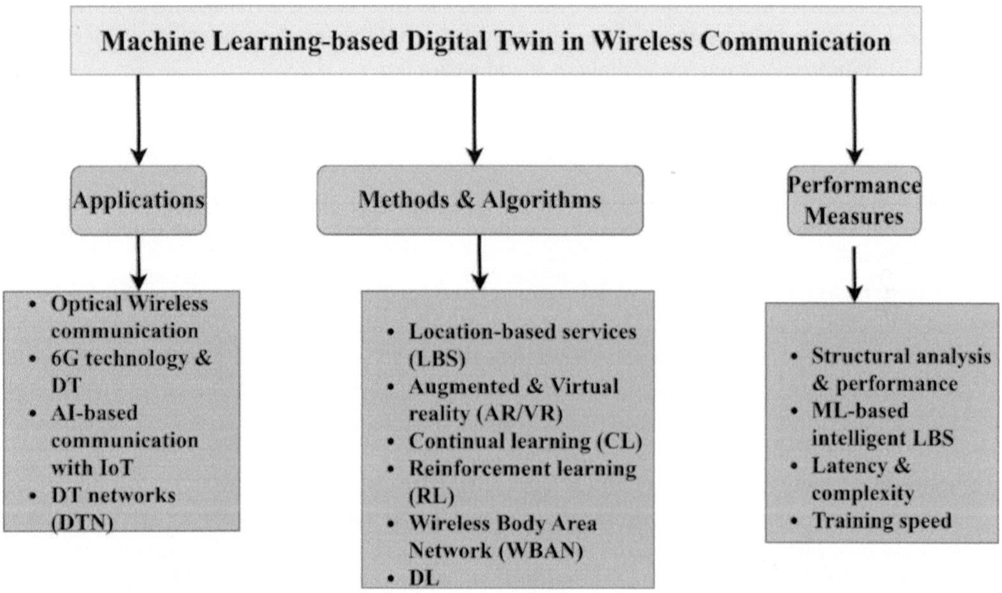

Figure 5: Role of Machine Learning-based Digital Twin in Wireless Communication.

Table 2: Role of Machine Learning-based Digital Twins in Wireless Communications.

References	Method	Algorithm used	Results
Eldeeb et al. 2023	Optical wireless communication (OWC)	Big data technology, AI/ML/ DL technologies, modelling and visualization, and intelligent reflecting surface (IRS)	Efficiency, scalability and resource allocation design methods are analyzed.
Ponnusamy et al. 2022	AI-based communication with IoT	Wireless body area networks (WBANs), augmented and virtual reality	Effective surgery and diagnosis with real-time assistance
Turcanu et al. 2024	Digital twin networks and city digital twin	Analysis for 5G and 6G networks	Improves decision-making, real-time monitoring and latency
Lombardo et al, 2024	Location-based services (LBS)	ML-based intelligent and continual learning	Adaptive AI with continual learning helps in recognizing patient's trajectories. Showed the same autoML module while training.
Szabó et al, 2024	Agile production cell with wireless communication	Reinforcement learning Network resource management (NRM) User equipment (UE)	ARIAC 2020 environment showed a speed-up rate of 3Xtimes more than setup without curriculum learning

2.3 Role of digital twins in diagnosis and therapeutic process

ML and DT technologies are used in the diagnosis and treatment of cancer (Moztarzadeh et al. 2023). The primary focus of this is breast cancer, which is the second most important early diagnosis concern. It is a complicated task to digitize this illness in the metaverse. In order to address this, real-time data is developed using four machine learning approaches, and DT is employed for both therapeutic and diagnostic purposes. Linear regression is used in the development of digital twins of cancer. A rare kind of oral infection known as chronic hyperplastic candidiasis (CHC) causes white or white-red patches on the oral mucosa (Liska et al. 2023). Leukoplakia is a related infection that is frequently confused with CHC due to its combination of red and white patches. The tongue, retro-angular mucosa, and buccal mucosa are among the regions where these lesions are most frequently observed. The gradient boosting regression (GBR) technique is utilized to identify the risk factors and forecast the site of CHC. Experiments show that the presence of CHC is mostly found in the retro-angular mucosa. Performance assessment has proved the robustness of the classification method, which improves accuracy in diagnosis and treatment.

There is a necessity for a proper system to handle precision mental health for accurate identification and diagnosis of mental health problems and to provide an appropriate therapeutic process (Spitzer et al. 2023). DTs tend to be the most innovative technology needed for the analysis of mental health. This helps in accurate prediction and treatment processes, which are continuously monitored. The process is analyzed between the therapist and the patient to evaluate predictions and treatment outcomes through regular monitoring.

Aortic stenosis (AS) is the most widely spread valvular disease and is increasing among elderly people (Rouhollahi et al. 2023). This has become a public health challenge since they lack medical treatment to prevent valve calcification. While less invasive transcatheter aortic valve replacement (TAVR) has emerged as the most popular technique, surgical aortic valve replacement (SAVR) is still the standard procedure. Patient-specific models are created and simulated for analyzing predictive outcomes. Two-dimensional U-Net produces better outcomes than other architectures. This helps in developing an AI tool for reconstructing DT geometry.

Diabetes is one common and chronic disease that needs attention to improve the patient's life (Chu et al. 2023). There are only limited healthcare resources compared to the population affected by diabetes mellitus (DM). DT is the solution that is applied in various healthcare applications as an emerging tool in disease management. More concentration on diagnosis, pathogenesis, and therapeutic processes is required due to the increasing rate of disabilities in musculoskeletal diseases (Sun et al. 2023). The idea of a mirrored reproduction of the real thing has led to the development of DT and AI in the medical field through medical image analysis. DT helps in personalized healthcare and the analysis of musculoskeletal diseases. Biomedical properties of the musculoskeletal system achieve real-time personal care.

Information about cancer patients is gathered by the Cancer Registry of Norway (CRN) via messages sent by different Norwegian hospitals and healthcare facilities (Lu et al. 2023). These messages are initially validated by an automated registry system. EvoCLINICAL uses an algorithm to select the appropriate cancer message from the candidate dataset and compare the registry system with it. Performances are evaluated through precision, recall, and F1. DTs help in treating individual patients through drugs that suit their body characteristics (Benson et al. 2023). The issue is that clinical practice and the disease's complexity are not aligned. Immune-mediated inflammatory illnesses are the primary focus of research on the clinical application of DT for disease prediction and prevention. Deep learning algorithms help in classifying different stages of musculoskeletal pains through 3D images (Ekambaram et al. 2023). Deep learning algorithms are compared with CNN and Spatial Transform Networks (STN) with Grad-CAM images. Deep learning, which correctly recognizes rehabilitation activities with training accuracy of 98.12% and validation accuracy of 95%, increases the efficiency of musculoskeletal rehabilitation therapy.

IoT and AI have contributed to Diabetes Mellitus (DM) management (Satuluri et al. 2022). Invasive and non-invasive methods are explored for managing DM. Effective invasive techniques for glucose monitoring include machine learning algorithms and near-infrared spectroscopy (NIR). DT technology is used to distinguish between cancer patients in the physical world and the digital world to arrive at an accurate prediction of diseased and non-diseased patients (Subasi et al. 2024). This manages the risk assessment, diagnosis, and therapeutic process. It also deals with drugs and assessment of the treatment. Diagnosis and treatment of thoracic disorders are done by integrating artificial intelligence and DT (Avanzato et al. 2024). IoT sensors and AI algorithms represent the respiratory health of the patient. YOLO version 8 neural architecture is adapted to predict and classify lung DT from chest X-ray images. Prediction has achieved an accuracy of 98% through continuous monitoring and diagnosis.

DT technology serves as a solution for analyzing and diagnosing mental health disorders. A dialogue system is introduced for exploring the mental status of an individual (Abilkaiyrkyzy et al. 2024). Chatbot architecture is used to interact with healthcare providers through pre-trained models for the detection of severity levels and achieved an accuracy of 84.75%. Integration of deep learning and DT technology through images and features diagnoses kidney diseases at early stages (Sasikaladevi et al. 2024). Data from radiological images that are recognized using deep learning techniques serve as the foundation for the automatic detection of chronic kidney disease (CKD). Abdominal CT scan pictures and urograms are used to construct reliable DTs for CKD diagnosis. Better prognoses for kidney diseases are done based on the validation from the DT system. This technology helps in monitoring respiratory problems, giving personalized treatment, and decision-making based on how an individual responds to the treatment given (Khan et al. 2024). Statistical time series data and correlation techniques are used in conjunction with deep learning (DL) and machine learning (ML) to accomplish this process. Principal Component Analysis (PCA) is used to measure the breaths per minute (BPM) and is found to achieve an accuracy of 92.3%.

Chronic illnesses are complex, and processing heart data is difficult since there are varied symptoms (Chai et al. 2024). Accurate heart disease diagnosis and cardiac data processing are made possible by the combination of edge, fog, and cloud computing. For feature extraction, it makes use of an optimized cascaded convolutional neural network (CCNN), and particle swarm optimization and galactic swarm optimization techniques are used to fine-tune the CCNN hyperparameters. Early heart attack prediction is achieved through the creation of DT (Upadrista et al. 2024). Through a virtual depiction of the patient and a model developed using real-time data, this aids doctors in performing tests and anticipates the onset of disease early to save the patient's life. Along with this architecture, blockchain and machine learning algorithms are used. Blockchain is responsible for making the data secure and ML algorithms are responsible for prediction. This has achieved an accuracy of 97.1%. Figure 6 portrays the role of DT in diagnosis and therapeutic processes.

Table 3 depicts the role of machine learning-based digital twins in diagnosis and therapeutic processes.

2.4 Role of Machine learning-based Digital twins in Industries

The bowl seedling detachment caused by the shaking of transplanting machinery has been analyzed by the DT plant factory transplanting system using a physical entity monitoring approach (Kaikang et al. 2023). In order to determine the ideal operating conditions, this method examines the vibrations of the machines under various operating circumstances. A comparison of the working conditions of transplantation equipment, axiomatic design for optimal deployment, demand analysis for the physical entity, and vibration power spectrum density of transplanting equipment under various conditions are all examined. This analysis helps in combining DT and exploring working conditions to achieve real-time visualization of the transplanting machine.

The warehouse management system requires the support of technological intervention since it has an unstructured layout, damage control, and order management (Maheshwari et al. 2023). A

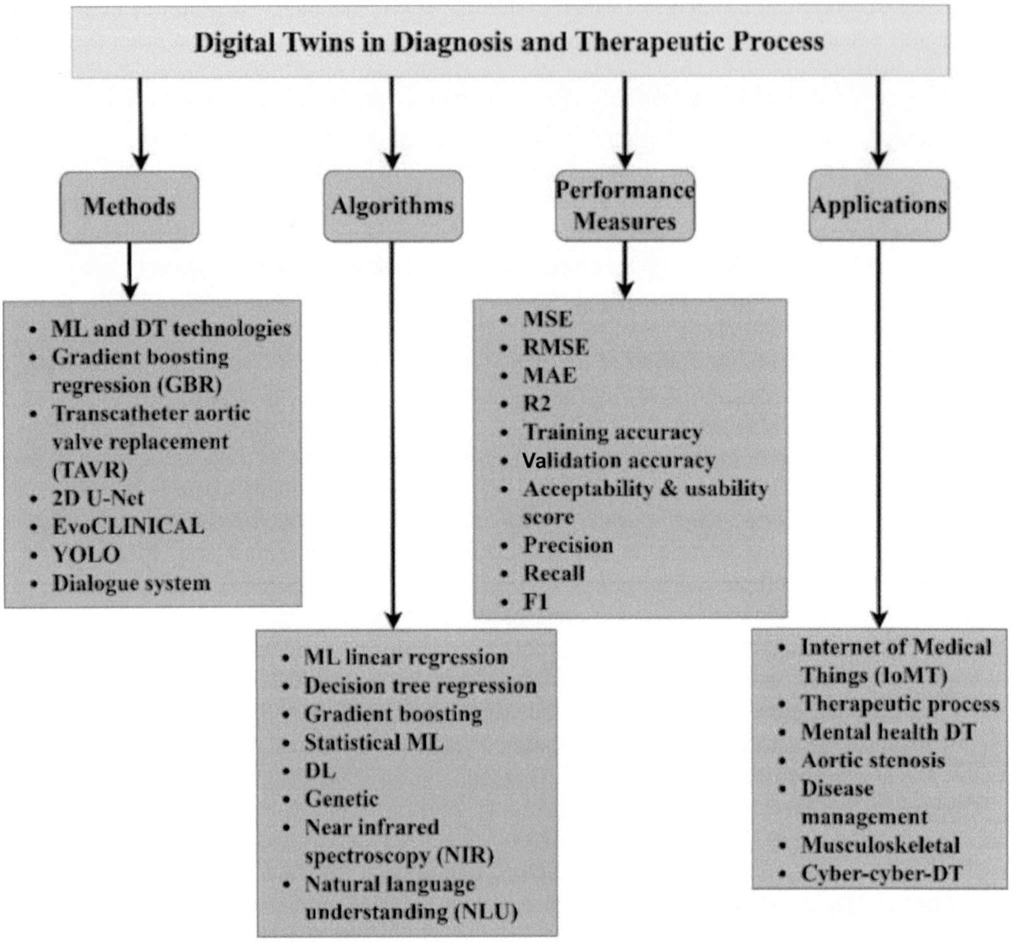

Figure 6: Role of DT in Diagnosis and Therapeutic Process.

DT-based warehouse system is proposed with knowledge graphs that are consolidated, and transfer of goods takes place from an inbound supplier to an outbound customer. This system also explores the interoperability and integrability of implementing DT. This study serves as a toolbox for practitioners by incorporating DT in the warehouse, leading to efficient supply chain management.

The most crucial elements and difficulties encountered by small and medium-sized businesses (SMEs) in Europe are identified in order to implement smart manufacturing (Nagy et al. 2023). By using secondary data analysis, large firms are able to exclude SMEs from the market because of the significant disadvantages that SMEs have in the cyber-physical production system. Deep learning and virtual simulation techniques are used in Industry 4.0 smart process planning. Business productivity is increased by the integration of autonomous robotics and machine learning technologies. Due to the emergence of digital replicas, this has altered how businesses operate (Attaran et al. 2023, ÜNAL et al. 2023). DT has improved the performance of the system and consumes less time. Data acquisition, data modeling, and data application are the three characteristics of DT using four technologies: IoT, cloud computing, artificial intelligence (AI), and extended reality (EX). Real-time data is collected and stored by these systems. This data provides useful information for producing a digital replica of the real item.

Table 3: Role of Machine Learning-based Digital Twins in Diagnosis and Therapeutic Processes.

References	Method	Algorithm used	Results
Moztarzadeh et al. 2023	DT and AI technology for the diagnosis and treatment of cancer disease	ML linear regression Decision tree regression Gradient boosting algorithm Random forest regression	Performance measured using mean square error (MSE), root mean squared error (RMSE), mean absolute error (MAE), and coefficient of determinants (R2) used to identify diseased and non-diseased patients
Liska et al. 2023	Analysis of Chronic hyperplastic candidiasis (CHC)	Gradient boosting regression technique	Shows better performance when compared with other algorithms
Spitzer et al. 2023	Precision mental health	Statistical machine learning algorithms	Improved mental health diagnosis (MHD) through psychopathology and psychotherapy.
Rouhollahi et al. 2023	Transcatheter aortic valve replacement (TAVR) has become the dominant method for Aortic stenosis (AS)	Deep learning models - U-Net used for image segmentation	Two-dimensional U-Net produces better results than others and makes DT geometry and calcification distribution easier
Chu Y et al. 2023	DT for diabetes mellitus (DM)	Medical Digital Twin (MeDigiT) for data acquisition, modeling and simulating.	MeDigiT is cost-effective, reduces animal testing, and prevents disease.
Sun et al. 2023	DT and AI for diagnosis of musculoskeletal diseases	Biomechanical performance using hybrid modeling designs	DT-based simulations show more accurate, and personalized treatments.
Lu et al. 2023	Cyber-cyber DT for diagnosis of cancer	EvoCLINICAL uses a genetic algorithm	Performance of EvoCLINICAL showed an accuracy, precision, and F1 greater than 91%
Benson et al. 2023	DT in drug treatment for immune-mediated inflammatory diseases	Predictive, preventive, personalized treatment	Improves diagnostic and therapeutic processes
Ekambaram et al. 2023	Convergence of 3D image into plane-based projection to predict musculoskeletal system diseases	Deep learning algorithms compared with CNN Spatial transform networks (STN) with grad CAM images.	Recognition of rehabilitation activities with training accuracy of 98.12% and validation accuracy of 95%.
Satuluri et al. 2022	Integration of AI and IoT in the prediction of diabetes mellitus	Near infrared spectroscopy (NIR) and ML algorithms	Study effective non-invasive method for glucose monitoring.
Subasi et al. 2024	DT for predicting cancer patients	ML algorithms	Prediction of diseases, treatment modeling, predictive analysis explored.
Avanzato et al. 2024	Lung DT for diagnosis and management of thoracic activities	YOLOv8 neural network and AI algorithms	Average accuracy of 96.6%-system performance Average accuracy of 98%
Abilkaiyrkyzy et al. 2024	Dialogue system for prediction of mental illness	Chatbot design with emotional mental intelligence (EMI) and Natural language understanding (NLU)	Accuracy in detecting mental illness –69% and acceptability and usability score of 84.75%
Sasikaladevi et al. 2024	Detection of chronic kidney disease (CKD) using DT	Deep learning and analysis through CT images	Out of 12,446 scans, 3,709 were with cyst, 5,077 normal, 1,377 with stone, and 2,283 with tumor Performance achieves validation accuracy of 99.71
Khan et al. 2024	DT for monitoring respiratory problems	ML and deep learning using channel state information (CSI) sensor Bagged tree ensemble algorithm	Breaths per minute with estimation accuracy of 92.3% Bagged ensemble showed accuracy of 89.2 and 83.7%
Chai et al. 2024	Edge-fog computing with DT for predicting chronic illness	Optimized cascaded convolutional neural network (CCNN), particle swarm optimization (PSO), and galactic swarm optimization (GSO)	Performance measures showed an accuracy of 87.4%, precision of 91.3%, recall of 85.1%, and F1 score of 90.3%
Upadrista et al. 2024	DT to predict heart attacks	Headless architecture with integration of other devices for other health conditions	Heart attack can be predicted at an accuracy of 97.1%

Recycling water is greatly aided by the wastewater treatment process. The negative consequences damage the environment. To resolve this, DT fault detection architecture using a convolutional autoencoder is used for monitoring wastewater treatment (Liu et al. 2023). By getting continuous updates from wastewater data, simulated data and failures in virtual space are detected. The sliding window method and Bayesian fusion methods are used to enhance the effects of the treatment process. The adoption of deep learning in manufacturing technologies collaborates with human robots to take care of safety and reliability (Wang et al. 2023, Kobayashi et al. 2024). DT makes use of these human operators, and robots can manage the operations to classify the manufacturing process. A convolutional neural network is trained through the data received from DT and tested between digital and physical objects.

Structural health monitoring (SHM) is one of the important technologies in industries (Macías et al. 2024, Liang et al. 2024). The application of DTs and ML algorithms helps to check the difference between the physical environment and the replica created. This is mainly used to detect damages and their locations accurately in the industrial environment. Artificial intelligence is used for decision-making. For an accurate prediction of prognostics, explainable AI (XAI) and interpretable machine learning (IML) are applied (Vodyaho et al. 2024). This integration supports DTs in accurate prediction, model interpretability, validation, and reliability.

A combination of data fabric and DT characteristics is used to resolve data management and contextualization issues (Trappey et al. 2024). This makes use of the previous proposal that has high-level data processing activities and helps in the traceability of data. Prognostics and health management (PHM) aims to build mathematical models for predicting abnormalities in ammonia critical equipment and sending warnings (Baicun et al. 2024, Aziz et al. 2024). Using sequence segmentation algorithms, Gaussian distribution, and long and short-term memory networks, real-time monitoring and early warnings are predicted. Cyber-physical systems are large-scale distributed systems that can integrate with small sub-systems using IoT (Jiao et al. 2024, Taherkhani et al. 2024). To manage continuous agile architectures, DTs are used where models are developed from the previous models of earlier stages. This set of models uses dynamic DTs to manage heterogeneous cyber-physical systems with varied structures and characteristics. Digital transformation (DT) is an emerging technology where it is divided into several domains to improve system performance (Mohammed et al. 2024). DT is employed in cloud computing, CPS, AI, deep learning, edge computing, and data analytics. Transdisciplinary engineering research focuses on complex systems using digital transformation.

Dealing with a welded circular hollow section (CHS) X-joint which is subjected to axial loading, a strain-interfaced DT is established (Cheok et al. 2024). ML algorithms are used to identify, locate, and measure the extent of the crack. These algorithms receive the strain readings to estimate the problem in joints, and remote sensing methodologies are used to avoid physical evaluation, thereby improving the diagnosis. Providing safe and secure food products is an important need for every individual (Nair et al. 2024). A potential approach that focuses on monocular depth estimation in food safety utilizes 3D reconstruction and DT analysis in mixed reality. Web service protocols and global network path networks (GLPN) are used to do this. Construction industries lack data related to construction, real-time data acquisition, and visualization (Opoku et al. 2024). Design of construction-based DTs, which is a combination of building data modeling and IoT, aims to implement DT in the university library as a living lab to make interactions with live environments like data collected from sensors related to thermal, lighting, and air quality and update the system accordingly.

Generating networks that represent the real-world network system is established through the development of extendable modeling architecture for DT-oriented complex network systems (DT-CNs) (Wen et al. 2024). Modeling is based on the characteristics of nodes, and based on the node's preference, rules are framed for creating a connection between nodes. This helps in the analysis of disaster resilience. Microverse is a task-oriented, edge-scale solution for smart cities (Qian et al. 2024). Data processing is done through on-site/near-site data and fusion of data that helps in decision-making. Authentication is done through smart public safety surveillance (SPSS). Botrytis cinerea and grape bunch morphology play a vital role in grey mould infection (Zanchin et al. 2024). In order to assess this problem, DT shape analysis is used to gauge the extent of grey mould based on grape bunch morphology. The photogrammetry technique was used to rebuild the DT. Multiple linear regression models and PCA were used to identify the descriptors that were most closely associated with the symptoms of grey mould. The parameters that were compared between the two models were the R-squared value and the root mean square error. A flexible job shop scheduling network that analyzes real-time scheduling is created by integrating cloud-edge and DT (Gao et al. 2024). There are two main goals of this architecture. First, zero-one programming is used to improve the flexible job shop time-space network's makespan and energy usage. Second, resources for scheduling tasks are allocated using a cloud-edge layout. A genetic algorithm is utilized to obtain an updated real-time timetable. Figure 7 depicts the role of ML-based DT in industries.

In plant factories, the cyber-physical-biological system (CPBS) is employed for crop production and energy management (Hu et al. 2024). By capturing biological dynamics such as temperature, humidity, illumination, and CO_2 levels, this system optimizes control factors. Utilizing DT, physics-informed deep learning (PIDL) approaches are used to evaluate crop conditions and the microclimate

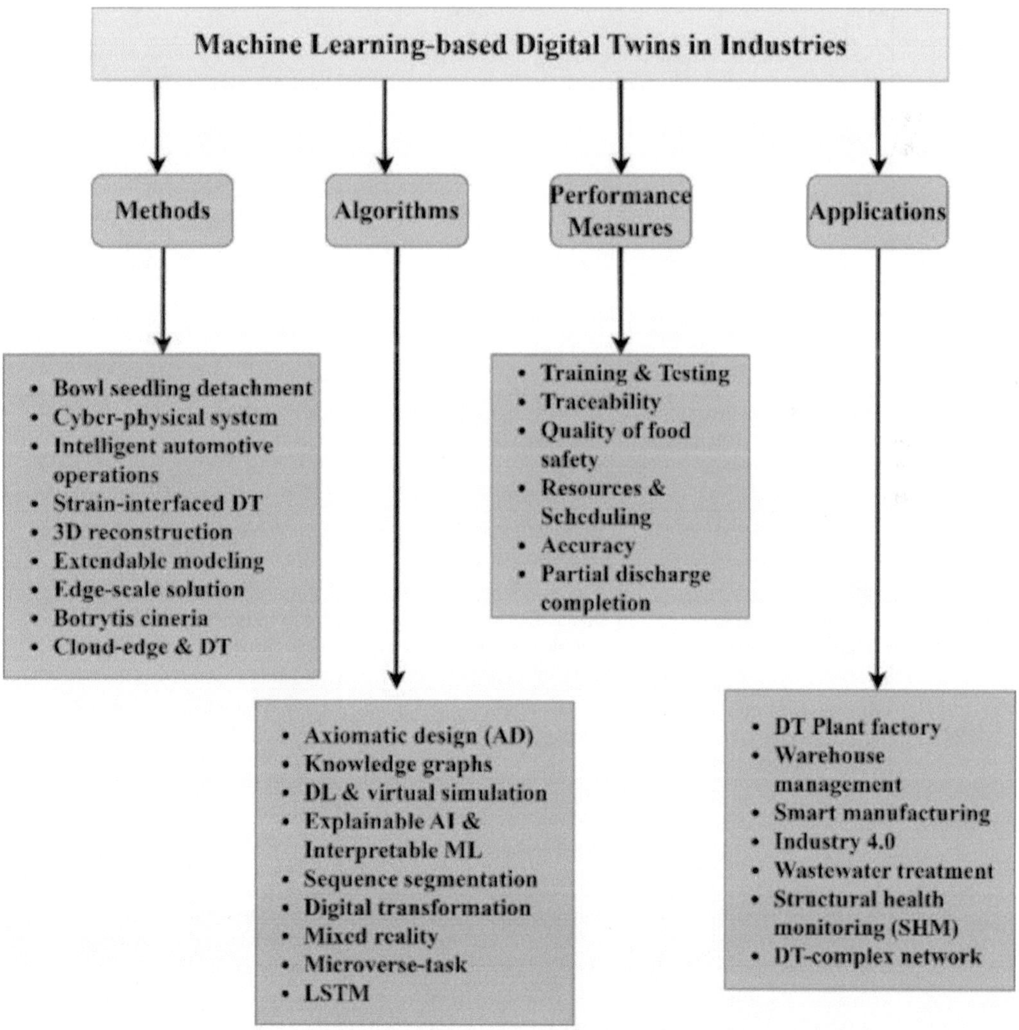

Figure 7: Role of ML-based DT in Industries.

Table 4: Role of Machine Learning-based Digital Twins in Industries.

References	Method	Algorithm used	Results
Kaikang et al. 2023	DT for monitoring bowl seedling detachment	Axiomatic design (AD) theory based on optimal deployment plan	Under operating temperature, 3000 transplants were done per hour. Highest vibration amplitude-12.03 Hz
Maheshwari et al. 2023	DT-based warehouse management system	Knowledge graphs Theoretical and systematic analysis	Serves as a toolbox for efficient supply chain management
Nagy et al. 2023	Cyber-physical production in European small and medium-sized enterprises (SMEs)	Deep learning and virtual simulation algorithm	Enhances business efficiency through ML and autonomous robotic technologies
Attaran et al. 2023 - ÜNAL et al. 2023	Intelligent automotive operations using DT	IoT, cloud computing, artificial intelligence (AI), and extended reality (EX)	Data acquisition, data modeling, and performance are measured.
Liu et al. 2023	Fault detection architecture for wastewater management	Convolutional autoencoder and Bayesian fusion methods	Sliding window and fusion methods improve the treatment process. Statistical and action points help in decision
Wang et al. 2023 – Kobayashi et al. 2024	DT utilizes human operators and robots to manage operations and classify the manufacturing process	Convolutional neural network and DT	Training and testing techniques for human-robotics collaboration
Macías et al. 2024 - Liang et al. 2024	Structural health monitoring (SHM) in industries	DT and ML algorithms	Detect damages and locations in industrial environments
Vodyaho et al. 2024	Prediction of prognostics	Explainable AI (XAI) and interpretable machine learning (IML)	Helps in prediction, validation, and reliability
Trappey et al. 2024	Data fabric and DT	Data fusion tasks	Data process activities and traceability are improved.
Baicun et al. 2024 - Aziz et al. 2024	Prognostic and health management (PHM) through mathematical models	Sequence segmentation algorithm, Gaussian distribution, long and short-term network	Early predictions through warnings
Jiao et al. 2024 - Taherkhani et al. 2024	Cyber-physical system integrating small sub-systems using IoT	Models use dynamic DTs for heterogeneous systems	Continuous agile structures to study varied structures and attributes
Mohammed et al. 2024	DT employed in cloud computing	Digital transformation (DT) Transdisciplinary engineering for complex systems	Helps to improve system performance
Cheok et al. 2024	Strain-interfaced DT to identify welded circular hollow sections X-joint	ML algorithms to measure the extent of crack	Estimate problems in joints through remote methods and improve diagnosis
Nair et al. 2024	Food safety using 3D reconstruction and DT	Mixed reality Web service protocols and global network path networks (GLPN)	Improves the quality of food safety
Opoku et al. 2024	Construction-based DT	Building data modeling and IoT	Interactions with the live environment and gets updated accordingly
Wen et al. 2024	Extendable modeling architecture for DT-oriented complex networks	Modeling based on the characteristics of nodes	Helps in the analysis of disaster resilience.

Table 4 contd. ...

...*Table 4 contd.*

References	Method	Algorithm used	Results
Qian et al. 2024	Edge-scale solution for smart cities	Microverse -task oriented Smart public safety surveillance (SPSS)	Helps in decision-making and improves quality
Zanchin et al. 2024	Grey mould infection using botrytis cineria and grape bunch morphology	Photogrammetry technique PCA and multiple linear regression	138 fully ripened grapes were analyzed at different angles. Recognizes descriptors that were closely related to grey mould
Gao et al. 2024	Cloud-edge and DT forming flexible job shop scheduling network	A genetic algorithm is used	Cloud-edge allocates resources for scheduling using automated guided vehicles and real-time responses.
Hu et al. 2024	Energy management and crop production in plant factories.	Physics-informed deep learning (PIDL)	Studying microclimate and crop conditions improves decision-making and accuracy.
Mueller et al. 2024	AI digital twin as clone of human expert	Delphi method	AI system augments the Delphi panel by adding a new perspective
Hanxiao et al. 2024	Mixed reality in developing human-robot interactions	Multiple effective motion control methods	Implemented two patterns of motion control for robotic arms with control by unity and control by ROS
Li et al. 2024	Analyze performance degradation in lithium-ion batteries of electric vehicles	Backpropagation neural network Long short-term memory model	Achieves an accuracy of 996% during partial discharge completion
Hussain et al. 2024	Analyzing the risk of failures in aging cranes	Digital-twin driven layout Aging-induced lifting capacity, ML algorithms	Improves prediction of degradation factors due to aging.

in plant factories. This enhances the system's accuracy and decision-making. By getting further assistance from AI-infused knowledge, the evolution of ChatGPT and how it improves the Delphi method's performance is examined (Mueller et al. 2024), creating an AI digital twin that is a human expert clone. An AI system can enhance a Delphi panel by contributing fresh viewpoints without taking the place of any experts. A rapidly expanding method for creating sophisticated human-robot interaction in several industries is mixed reality (MR), which can identify various patterns (Hanxiao et al. 2024). Multiple effective motion control methods are applied to different robotic arms for the development of MR applications. Control methods applied are of two types: control by Unity and control by Robot Operating System (ROS). Lithium-ion batteries (LIB) are widely used in electric vehicles (EVs) due to their long cycle life, energy density, and eco-friendliness (Li et al. 2024). Performance deteriorates as a result of many cycles of charging and draining. Half of the battery life cycle can be predicted and completed with the aid of a backpropagation neural network. To predict the available LIBs and their states, a long short-term memory model is employed. This design has demonstrated an accuracy of 99.6% during partial discharge completion in experiments. Creation of a digital twin-driven (DTD) arrangement to investigate the deterioration of aging-induced lifting capacity (LC) and assess the risk of failures in aging crane towers is also explored (Hussain et al. 2024). By experimenting with real-time data, this layout combines theoretical and numerical understanding of crane deterioration behavior. Table 4 shows the role of machine learning-based digital twins in industries.

3. Conclusion and Future Scope

In this chapter, DT technology used in various healthcare and industrial applications is discussed. The processes and techniques involved in the accurate prediction and monitoring of therapy effects are analyzed. Integrating AI, IoT, and ML algorithms with DT technology helps in data acquisition from physical objects and comparing the updates in the replica created in the digital world. This improves accuracy in prediction, diagnosis, and the therapeutic process. In the future, DT performance can be explored for various diseases using the appropriate ML algorithms to achieve early prediction of disease to save an individual's life.

References

Abilkaiyrkyzy, Akbobek, Fedwa Laamarti, Mufeed Hamdi, and Abdulmotaleb El Saddik. (2024). Dialogue System for Early Mental Illness Detection: Towards a Digital Twin Solution. *IEEE Access*.

Annamalai, Chinnappa, Chellavelu Vijayakumaran, Vijayakumar Ponnusamy, and Hyunsung Kim. (2023). Optimal ElGamal Encryption with Hybrid Deep-Learning-Based Classification on Secure Internet of Things Environment. *Sensors* 23(12): 5596.

Attaran, Mohsen, and Bilge Gokhan Celik. (2023). Digital Twin: Benefits, use cases, challenges, and opportunities. *Decision Analytics Journal*: 100165.

Avanzato, Roberta, Francesco Beritelli, Alfio Lombardo, and Carmelo Ricci. (2024). Lung-DT: An AI-Powered Digital Twin Framework for Thoracic Health Monitoring and Diagnosis.

Aziz, Shahid, Dong Won Jung, and Anas Bin Aqeel. (2024). Digital Twins in Smart Manufacturing. In *Handbook of Manufacturing Systems and Design*. Taylor & Francis.

Benson, and Mikael. (2023). Digital Twins for Predictive, Preventive Personalized, and Participatory Treatment of Immune-Mediated Diseases. *Arteriosclerosis, Thrombosis, and Vascular Biology* 43(3): 410–416.

Chai, Sheng, and Lantian Guo. (2024). Edge Computing with Fog-cloud for Heart Data Processing using Particle Swarm Optimized Deep Learning Technique. *Journal of Grid Computing* 22(1): 1–18.

Chebanov, Dmitrii K., and Vsevolod A. Misyurin. (2023). Predictive Modeling of Clinical Trial Outcomes for Novel Drugs using Digital Twin Patient Cohorts and GenerativeAI. *medRxiv*: 2023–09.

Chen, and Hanxiao. (2024). Motion Control of Interactive Robotic Arms Based on Mixed Reality Development. *arXiv preprint arXiv:2401.01644*.

Chen, Jiayuan, Changyan Yi, Hongyang Du, Dusit Niyato, Jiawen Kang et al. (2023) A revolution of personalized healthcare: enabling human digital twin with mobile AIGC." *arXiv preprint arXiv:2307.12115*.

Chen, Kaikang, Bo Zhao, Yanli Zhang, Liming Zhou, Kang Niu et al. (2023). Digital Twin-Based Vibration Monitoring of Plant Factory Transplanting Machine. *Applied Sciences* 13(22): 12162.

Cheok, E. W. W., Qian, X., Chen, C., Quek, S. T., and Si, M. B. I. (2024). A local digital twin approach for identifying, locating and sizing cracks in CHS X-joints subjected to brace axial loading. *Engineering Structures*, 299: 117085.

Chu Y, Li S, Tang J, and Wu H. (2023). The potential of the Medical Digital Twin in diabetes management: a review. Front Med (Lausanne). 2023 Jul 20;10:1178912. doi: 10.3389/fmed.1178912. PMID: 37547605; PMCID: PMC10397506.

Currie, G. M., and Rohren, E. M. (2023). May. Radiation dosimetry, artificial intelligence and digital twins: old dog, new tricks. In *Seminars in Nuclear Medicine* (Vol. 53, No. 3, pp. 457–466). WB Saunders.

Doungtap, S., Petchhan, J., Phanichraksaphong, V., and Wang, J. H. (2023). Towards Digital Twins of 3D Reconstructed Apparel Models with an End-to-End Mobile Visualization. *Applied Sciences*, 13(15): 8571.

Ekambaram, D., Ponnusamy, V., Natarajan, S.T., and Firos Khan, M. F. S. (2023). Artificial Intelligence (AI) Powered Precise Classification of Recuperation Exercises for Musculoskeletal Disorders. *Traitement du Signal*, 40(2).

Eldeeb, Hossien, Shimaa Naser, Lina Bariah, Sami Muhaidat, and Murat Uysal. (2023). Digital Twin-Assisted OWC: Towards Smart and Autonomous 6G Networks.

Falkowski, Piotr, Tomasz Osiak, Julia Wilk, Norbert Prokopiuk, Bazyli Leczkowski. et al. (2023). Study on the Applicability of Digital Twins for Home Remote Motor Rehabilitation. *Sensors*, 23(2): 911.

Gao, Qianfa, G. U., Fu, Linli Li, and Jianfeng Guo. (2024). A framework of cloud-edge collaborated digital twin for flexible job shop scheduling with conflict-free routing. *Robotics and Computer-Integrated Manufacturing* 86: 102672.

Hu, Guoqing, and Fengqi You. (2024). AI-enabled cyber-physical-biological systems for smart energy management and sustainable food production in a plant factory. *Applied Energy*, 356: 122334.

Hussain, Mudasir, Zhongnan Ye, Hung-Lin Chi, and Shu-Chien Hsu. (2024). Predicting degraded lifting capacity of aging tower cranes: A digital twin-driven approach. *Advanced Engineering Informatics*, 59: 102310.

Jamshidi, M. B., Sargolzaei, S., Foorginezhad, S., and Moztarzadeh, O. (2023). Metaverse and microorganism digital twins: A deep transfer learning approach. *Applied Soft Computing*, 147: 110798.

Jiao, Yujun, Xukai Zhai, Luyajing Peng, Junkai Liu, Yang Liang. et al. (2024). A digital twin-based motion forecasting framework for preemptive risk monitoring. *Advanced Engineering Informatics*, 59: 102250.

Khan, Sagheer, Aaesha Alzaabi, Tharmalingam Ratnarajah, and Tughrul Arslan. (2024). Novel statistical time series data augmentation and machine learning based classification of unobtrusive respiration data for respiration Digital Twin model. *Computers in Biology and Medicine,* 168: 107825.

Kobayashi, Kazuma, and Syed Bahauddin Alam. (2024). Explainable, interpretable, and trustworthy AI for an intelligent digital twin: A case study on remaining useful life. *Engineering Applications of Artificial Intelligence,* 129: 107620.

Kreuzer, Tim, Panagiotis Papapetrou, and Jelena Zdravkovic. Artificial Intelligence in Digital Twins-a Systematic Literature Review. *Available at SSRN 4681915.*

Li, Wei, Yongsheng Li, Akhil Garg, and Liang Gao. (2024). Enhancing real-time degradation prediction of lithium-ion battery: A digital twin framework with CNN-LSTM-attention model. *Energy,* 286: 129681.

Liang, Liang, Deng Ruofan, and Hao Jia. (2024). Construction of a Digital Twin Based Monitoring and Early Warning System for Ammonia Equipment. *Journal of The Institution of Engineers (India): Series C*: 1–12.

Liu, Wentao, Sudao He, Jianpeng Mou, Ting Xue, Hongtian Chen. et al. Digital twins-based process monitoring for wastewater treatment processes." *Reliability Engineering & System Safety*: 109416.

Lombardo, Gianfranco, Marco Picone, Marco Mamei, Monica Mordonini, and Agostino Poggi. (2024). Digital Twin for Continual Learning in Location Based Services. *Engineering Applications of Artificial Intelligence,* 127: 107203.

Lu, Chengjie, Qinghua Xu, Tao Yue, Shaukat Ali, Thomas Schwitalla, and Jan F. Nygård. (2023). EvoCLINICAL: Evolving Cyber-Cyber Digital Twin with Active Transfer Learning for Automated Cancer Registry System. *arXiv preprint arXiv:2309.03246.*

Ma, X., Qi, Q., and Tao, F. (2024). A Digital Twin–Based Environment-Adaptive Assignment Method for Human–Robot Collaboration. *Journal of Manufacturing Science and Engineering*, 146(3).

Macías, Aurora, David Muñoz, Elena Navarro, and Pascual González. (2024). Data fabric and digital twins: An integrated approach for data fusion design and evaluation of pervasive systems. *Information Fusion* 103: 102139.

Maheshwari, P., Kamble, S., Kumar, S., Belhadi, A., and Gupta, S. (2023). Digital twin-based warehouse management system: a theoretical toolbox for future research and applications. *The International Journal of Logistics Management.*

Menon, Devika, Bharath Anand, and Chiranji Lal Chowdhary. (2023). Digital Twin: Exploring the Intersection of Virtual and Physical Worlds. *IEEE Access.*

Mohamed, Nader, Jameela Al-Jaroodi, Imad Jawhar, and Nader Kesserwan. (2023). How Healthcare Systems Engineering can Benefit from Digital Twins? In *2023 IEEE International Systems Conference (SysCon)*, pp. 1–6. IEEE.

Mohamed, Nader, Jameela Al-Jaroodi, Imad Jawhar, and Nader Kesserwan. (2023). Leveraging Digital Twins for Healthcare Systems Engineering. *IEEE Access.*

Mohammed, M. A., Lakhan, A., Zebari, D. A., Abd Ghani, M. K., Marhoon, H. A., Abdulkareem, K. H. et al. (2024). Securing healthcare data in industrial cyber-physical systems using combining deep learning and blockchain technology. *Engineering Applications of Artificial Intelligence*, 129: 107612.

Mousavi, Z., Varahram, S., Ettefagh, M. M., Sadeghi, M. H., Feng, W. Q., and Bayat, M. (2024). A digital twin-based framework for damage detection of a floating wind turbine structure under various loading conditions based on deep learning approach. *Ocean Engineering*, 292: 116563.

Moztarzadeh, O., Jamshidi, M., Sargolzaei, S., Jamshidi, A., Baghalipour, N., Malekzadeh Moghani, M. et al. (2023). Metaverse and Healthcare: Machine Learning-Enabled Digital Twins of Cancer. *Bioengineering*, 10(4): 455.

Moztarzadeh, O., Liska, J., Liskova, V., Skalova, A., Topolcan, O., Jamshidi, A. et al. 2023. Predicting Chronic Hyperplastic Candidiasis Retro-Angular Mucosa Using Machine Learning. *Clinics and Practice*, 13(6): 1335–1351.

Mueller, Roland M., Katja Thoring, Hermann W. Klöckner, and Kai Larsen. (2024). Crafting Future Scenarios with the Help of AI: Potentials of a Hybrid Delphi Expert Panel.

Nagy, Marek, George Lăzăroiu, and Katarina Valaskova. (2023). Machine Intelligence and Autonomous Robotic Technologies in the Corporate Context of SMEs: Deep Learning and Virtual Simulation Algorithms, Cyber-Physical Production Networks, and Industry 4.0-Based Manufacturing Systems." *Applied Sciences,* 13(3): 1681.

Nair, Midhun, and Renny Edwin Fernandez. (2024). Advancing Mixed Reality Digital Twins through 3D Reconstruction of Fresh Produce. *IEEE Access.*

Opoku, De-Graft Joe, Srinath Perera, Robert Osei-Kyei, Maria Rashidi, Keivan Bamdad et al. (2024). Digital twin for indoor condition monitoring in living labs: University library case study. *Automation in Construction,* 157: 105188.

Ponnusamy, Vijayakumar, A., Vasuki, J., Christopher Clement, and Eswaran, P. (2022). AI-Driven Information and Communication Technologies, Services, and Applications for Next-Generation Healthcare System. *Smart Systems for Industrial Applications*: 1–32.

Qu, Qian, Mohsen Hatami, Ronghua Xu, Deeraj Nagothu, Yu Chen et al. (2024). Microverse: A Task-Oriented Edge-Scale Metaverse.

Rouhollahi, Amir, James Noel Willi, Sandra Haltmeier, Alireza Mehrtash, Ross Straughan et al. (2023). Cardiovision: a fully automated deep learning package for medical image segmentation and reconstruction generating digital twins for patients with aortic stenosis. *Computerized Medical Imaging and Graphics,* 109: 102289.

Rubeis, and Giovanni. (2023). "Hyperreal Patients. Digital Twins as Simulacra and their impact on clinical heuristics." In *Medizin–Technik–Ethik: Spannungsfelder zwischen Theorie und Praxis*, pp. 193–207. Berlin, Heidelberg: Springer Berlin Heidelberg.

Sasikaladevi, N., and Revathi, A. (2024). Digital twin of renal system with CT-radiography for the early diagnosis of chronic kidney diseases. *Biomedical Signal Processing and Control,* 88: 105632.

Satuluri, V. K. R Rajeswari, and Vijayakumar Ponnusamy. (2022). Diagnosis of Non-invasive Glucose Monitoring by Integrating IoT and Machine Learning." *IEIE Transactions on Smart Processing & Computing,* 11(6): 435–443.

Sheng, B., Wang, Z., Qiao, Y., Xie, S.Q., Tao, J., and Duan, C. (2023). Detecting latent topics and trends of digital twins in healthcare: A structural topic model-based systematic review. *Digital Health,* 9: p. 20552076231203672.

Spitzer, M., Dattner, I., and Zilcha-Mano, S. (2023). Digital twins and the future of precision mental health. *Frontiers in Psychiatry,* 14: 1082598.

Stergiou, C. L., Koidou, M. P., and Psannis, K. E. (2023). IoT-Based Big Data Secure Transmission and Management over Cloud System: A Healthcare Digital Twin Scenario. *Applied Sciences,* 13(16): 9165.

Subasi, Abdulhamit, and Muhammed Enes Subasi. (2024). Digital twins in healthcare and biomedicine. pp. 365–401. *In: Artificial Intelligence, Big Data, Blockchain and 5G for the Digital Transformation of the Healthcare Industry,* Academic Press.

Sun, T., Wang, J., Suo, M., Liu, X., Huang, H., Zhang, J. et al. (2023). The Digital Twin: A Potential Solution for the Personalized Diagnosis and Treatment of Musculoskeletal System Diseases. *Bioengineering,* 10(6): 627.

Szabó, Géza, and József Pető. (2024). Intelligent wireless resource management in industrial camera systems: Reinforcement Learning-based AI-extension for efficient network utilization. *Computer Communications.*

Taherkhani, Roohollah, Mohammad Amin Ashtari, and Mohamadmahdi Aziminezhad. (2024). Digital Twin-Enabled Infrastructures: A Bibliometric Analysis-Based Review. *Journal of Infrastructure Systems* 30(1): 03123001.

Trappey, A. J., Li, F., Lee, C. H., Mo, J. P., Stjepandic, J., and Jiao, R. (2024). How to manage and balance uncertainty by transdisciplinary engineering methods focusing on digital transformations of complex systems. *Advanced Engineering Informatics,* 59: 102330.

Turab, Muhammad, and Sonain Jamil. (2023). A Comprehensive Survey of Digital Twins in Healthcare in the Era of Metaverse. *BioMedInformatics* 3(3): 563-584.

Turcanu, Ion, German Castignani, and Sébastien Faye. (2024). On the Integration of Digital Twin Networks into City Digital Twins: Benefits and Challenges.

ÜNAL, Aliye Kübra, and Aynur TORAMAN. (2023). Evaluation of Digital Twin Technology. *SDÜ Sağlık Yönetimi Dergisi,* 5(1): 1–25.

Upadrista, Venkatesh, Sajid Nazir, and Huaglory Tianfield. (2024). Blockchain-Based Digital Twin to Predict Heart Attacks. In *Blockchain for Healthcare 4.0,* pp. 291–306. CRC Press.

Vallée, Alexandre. (2023). Digital twin for healthcare systems. *Frontiers in Digital Health* 5: 1253050.

Vodyaho, Alexander, Nataly Zhukova, Radhakrishnan Delhibabu, and Alexey Subbotin. (2024). Continuous agile cyber–physical systems architectures based on digital twins. *Future Generation Computer Systems,* 153: 350–359.

Wang, Baicun, Huiying Zhou, Xingyu Li, Geng Yang, Pai Zheng. et al. (2024). Human Digital Twin in the context of Industry 5.0. *Robotics and Computer-Integrated Manufacturing,* 85: 102626.

Wang, S., Zhang, J., Wang, P., Law, J., Calinescu, R., and Mihaylova, L. (2024). A deep learning-enhanced Digital Twin framework for improving safety and reliability in human–robot collaborative manufacturing. *Robotics and Computer-Integrated Manufacturing,* 85: 102608.

Wen, Jiaqi, Bogdan Gabrys, and Katarzyna Musial. (2024). Towards Digital Twin-Oriented Complex Networked Systems: Introducing heterogeneous node features and interaction rules. *Plos one* 19: no. 1: e0296426.

Yankeelov, Thomas E., David A., Hormuth, Ernesto ABF Lima, Guillermo Lorenzo et al. (2024). Designing clinical trials for patients who are not average. *Iscience* 27: no. 1.

Zanchin, A., Sozzi, M., Giora, D., Kalantari, M., Belfiore, N., Terleth, J. et al. (2024). Digital Twins analysis as a tool to find new descriptors for grapevine bunch morphology categorisation and grey mould infection risk evaluation. *Biosystems Engineering,* 237: 71–82.

14

Next-Gen Financial Management

Smart Systems and Industry 5.0 Integration for a Sustainable World

Kiran Kumar M.,[1,*] *Shilpa A. C.*,[2] *Shreelatha H. R.*[3] and *Chidambaram N.*[4]

1. Introduction

Financial management is undergoing a colossal shift, propelled by rapidly advancing technologies converging into an Industry 5.0 future. Industry 4.0 served a purpose around automation, the use of cyber-physical systems, and big data but primarily involved a larger focus on efficiency, productivity, and decision-making through machines. Industry 5.0 emphasizes humans and smart systems working together in collaborative and responsible ways and prompts us to incubate value creation where the importance of inclusivity, resilience, and societal impact is captured in how we perform our financial and accounting systems.

Smart systems such as AI, ML, blockchain, big data analytics, and RPA have entered into our financial accounting system to execute every sort of function in our organizations, even beyond traditional accounting and reporting, to the most advanced investment decision-making, risk management, fraud detection, and customer service (Nahavandi 2019). FinTech and decentralized finance (DeFi) services are demonstrating how smart systems can democratize access to financial services, increase transparency, and reduce costs. Yet, integrating such technologies within an Industry 5.0 context is vital to support sustainable, agency-based, and ethically aligned choices. As the global economy continues to grapple with important issues like climate change, inequality, and resource scarcity, financial institutions are under increasing scrutiny to integrate ESG perspectives

[1] CMS Business School, Jain (Deemed-to-be University), Bangalore, India.

[2] BGS Institute of Technology, Adichunchanagiri University, Karnatka, India.
Email: shilpaac@bgsit.ac.in, 0009-0000-1037-8478

[3] S-Vyasa Deemed to be University, Benguluru, India, Email: shreelatha.hr@gmail.com, 0000-0003-2570-5827

[4] Department of Management Studies, KGiSL Institute of Technology Thudiyalur Road, Saravanampatti Coimbatore – 641 035.
Email: chidambaram.n@kgkite.ac.in, https://orcid.org/0009-0007-1767-385X

* Corresponding author: kiranm287@gmail.com, 0000-0003-2084-9418

DOI: 10.1201/9781003655190-17

into their long-term strategies. While smart systems have the capacity to track sustainability metrics in real time, optimize impact investing, and improve green financing channels through responsible usage, their potential to reinvent financial systems through co-creation aligned with the UN Sustainable Development Goals (SDGs) becomes even greater through the lens of Industry 5.0, allowing for personalization, trust, and accountability through the collective knowledge of financial ecosystems (Coelho et al. 2023).

This chapter reviews a project looking at the strategic integration of smart technologies within the human-centric principles of Industry 5.0 with the ultimate aim to create sustainable, resilient, and inclusive financial systems. The chapter aims to examine the critical success factors, the technological enablers, and organizational model structures for next-gen financial management—a shift from efficiency to long-lasting societal value. More specifically, the chapter considers how smart systems can aid ESG pursuits, the factors by which Industry 5.0 can expand human–machine relationships within financial services, and what might inhibit progress while creating opportunities. This research is relevant and important given the current shocks global financial markets face, the emergence of rapid digitalization, and the growing demand for sustainable finance. By positioning smart financial systems within the context of Industry 5.0, this project introduces a new perspective that ties together technology, human values, and sustainability in a meaningful way. The research should have implications for not just financial companies and technology developers but also for policymakers and other stakeholders invested in building a more equitable and resilient financial future. The remainder of this chapter can be organized as follows: Section 2 is an extensive literature review identifying the transition from Industry 4.0 to Industry 5.0 and the role of smart systems in relation to finance, sustainability, and governance. Section 3 details our methodological approach. Section 4 takes a closer look at smart systems in the context of functional application within the next generation of financial management, which could include automation, blockchain, big data, and cybersecurity. Section 5 assesses the implementation of Industry 5.0 principles in respect to finance, mainly with the notion of humans and machines working together—in a personalized way. Section 6 presents trends of sustainable finance models, Section 7 examines the phenomenon, challenges, and future directions in research and innovation. Section 8 concludes the chapter identifying our overall findings and competitive vision for a more human, resilient, and ethically responsible financial future.

2. Literature Review

Industry 5.0 represents a major change from the machine-centric automation of Industry 4.0 to a human-centered system where advanced technologies are structured to augment human intelligence and creativity rather than override it. According to the European Commission, value attributes of Industry 5.0, such as "sustainability," "resilience," and "inclusiveness," would expect to see "technological advancement" with "societal wellbeing" (Dalenogare et al. 2018). This reorientation introduces new conceptual and practical models around innovation; this includes the use of AI, robotics, and IoT; importantly, it also includes the ethics framework, alongside collaborative human interaction with technology. In terms of the finance sector, this means different activities from simply automating "efficiently," but toward financial ecosystems that aim for genuine engagement with stakeholders and personalized, environmentally responsible outcomes. In the literature, there is increasing recognition that Industry 5.0 is not a negation of Industry 4.0, but a parallel evolution. Industry 4.0 is characterized by the emphasis on the optimization of supply and production processes, while Industry 5.0 would engage humans as central to the process of technological innovation. This cognitive shift has direct implications for financial management. Human supervision of applied AI and ethical and purposeful decision-making are becoming not optional requirements (Arner et al. 2015).

Smart systems include digital technologies that enable intelligent automation, real-time data processing, and predictive analytics. In the financial management space, smart systems are being used to increase the efficiency of operations in risk analysis, fraud detection, compliance, and financial forecasting, among others, with improvements to accuracy, transparency, and speed. Studies have demonstrated the role of FinTech in enabling smarter, more careful, and data-driven financial decisions (Gomber et al. 2018). Algorithms using artificial intelligence and machine learning are being employed to model market behaviors and corporate diurnal life cycles and to optimize investment portfolios and fulfill complex financial transactions, on occasion at human-less stages of decision-making. In accounting and auditing domains, robotic process automation and blockchain technologies are disrupting traditional practices and roles by establishing immutable record-keeping and automating monotonous tasks. Smart contracts are disrupting transactional scenes; sentiment analysis and natural language processing tools are being deployed for real-time market analyses and assessments. All of these smart systems are not only leading to increased operational efficiencies but play a role in influencing strategic decision-making with incremental insights into customer behaviors, market volatility, and systemic risks (Tyagi 2023).

As ESG factors become more commonplace in financial discussions, it is clear that digital finance is playing a role as a facilitator of sustainability. The existing literature indicates an upward trend in research that examines the various ways in which digital platforms can foster green finance, climate-related disclosures, and impact investing. For example, research highlighted how using digital tools allows for a more precise analysis of ESG risks and opportunities by investors, enabling smart systems to keep track of carbon footprints, fulfillment of supply chain compliance, and assessment of social outcomes in real time when considering financial decision-making (Schuetz and Venkatesh 2019). The literature noted that blockchain is also examined for its potential to improve transparency as well as traceability in sustainable finance. In this context, distributed ledger technologies offer a credible platform for verifying ESG claims and reducing greenwashing with more responsible sustainability-linked financial products, akin to a highly controlled and accountable supply chain aligned with the UN SDGs.

While smart systems and Industry 5.0 have transformative potential in financial management, integration continues to be difficult. A common theme in the literature is the technological fragmentation and legacy technologies in financial organizations. Many organizations struggle with integrating new digital tools into established digital stacks due to the cost of implementation, lack of interoperability, and uncertainty because of regulations. Even further, ethical issues and implementation governance are a significant barrier to adoption in business. Data privacy, algorithmic bias, and accountability will need to be addressed through appropriate frameworks and transparent practices. That said, integration of Industry 5.0 principles provides tremendous opportunities. As the authors have demonstrated, the relationship between smart technologies and human creativity can increase resilience in financial organizations, foster workers, and improve the potential for a more inclusive model of the economy (Singh et al. 2024).

There are multiple theoretical frameworks underpinning the application of smart systems/ Industry 5.0 in financial management. One relevant framework is socio-technical systems theory, which proposes that the best performance is achieved through the joint optimization of social and technical systems (Hong et al. 2018). With respect to finance, this would entail the development of digital infrastructures geared toward the technical efficiency of financial procedures that also accentuate human values, accountability, and trust. Another relevant framework is stakeholder theory, which argues that organizations must take into account the needs of many different stakeholders in their decision-making processes. Stakeholder theory applied to smart financial systems would support the idea of technologies that balance profitability and global responsibility. Similarly, sustainable finance theory highlights long-term values generated through environmental responsibility and ethical governance, and it has conceptual links to digital advances bringing together innovation and sustainability (Hasan and Salah 2018).

3. Research Methodology

The methodology of this study is to employ a qualitative dominant mixed-methods research design to analyze the interaction of smart systems and Industry 5.0 in the context of financial management, specifically around sustainability. Based on the complex, interdisciplinary nature of the subject—interfacing technology, finance, a human-centric design approach, as well as sustainability—a multi-dimensional methodological approach is required for both theoretical applications and empirical understanding. The methodology is a blend of methods aimed to understand specific trends, contextualize real-life applications, and reveal compelling challenges and new opportunities for realizing the digital transformation of finance under Dimension 5.0. This study follows an exploratory and descriptive research design. The exploratory aspect is needed to explore a relatively new meaning of Industry 5.0 applied to finance, and the descriptive parts capture what is presently occurring in terms of practice and strategy in this environment. This dual approach further allows the exploration of emerging patterns that can help establish a conceptual framework that draws from theoretical literature and practical case-based evidence. This study method also employs a comparative case analysis research method, analyzing selected organizations or platforms, recognizing these as exemplary organizations or platforms that embrace Industry 5.0 and smart financial practice.

Data collection is undertaken via secondary data sources and primary qualitative approaches. Secondary data sources included academic journal articles, industry reports, white chapters, and policies published between 2015–2025. These secondary sources provided the contextual foundations and theoretical rationale and empirical examples of digital finance, smart systems, and sustainability frameworks. Primary data was collected through semi-structured interviews with financial analysts, technology experts, FinTech executives, and sustainability officers. Interviews were conducted in a way to elicit expert views on real-world challenges, integration processes, and future outlooks regarding smart systems and human-centered innovation in finance. Twelve interviews were conducted that involved 45–60-minute conversations, with selection of interview participants involving purposive sampling; selection of interviewees was based on their professional experience and relevance to the topic. The study utilized qualitative analysis software, for example NVivo, for coding and conducting thematic analysis of the interview transcripts. A framework analysis approach helped to structure the data into thematic domains that included digital innovation, human–machine collaboration, ESG integration, and organizational change. When possible, descriptive analytics were used to present the structured data from industry reports regarding green finance investment volume, the adoption of AI in finance, and the rates of regulatory acceptance.

Participants were selected using a purposive sampling technique so that all participants had direct knowledge of smart financial technologies (FinTech), sustainability reporting, or Industry 5.0–aligned initiatives. The participants came from a diversity of financial sectors represented by traditional banks, FinTech firms, regulators, and consultants. The geographical spread includes Europe, Asia, and North America to provide the opportunity for a comparative view of the regional contexts driving the uptake of smart and sustainable finance. While the number of participants is small, the diversity of industries and roles represented ensures a good-quality qualitative data set that can portray different perspectives on the research topic. The small sample group is allowable in the exploratory nature of the study and reflects the richness of information gained from expert interviews. All research activities are conducted in accordance with ethical research standards. Informed consent was obtained from each participant before conducting each individual interview, where consent was informed through written information about the purpose of the research, the use of the data, and that there was no obligation for the participant to be involved in the interviews and data collection process. Furthermore, the study took specific steps to remain anonymous and confidential so the participants can be assured that their identity and/or organization could not be recognized. In addition, the study was conducted in accordance with data protection laws, such as the General Data Protection Regulation (GDPR) as related to European participants. Each step of the research process, including the interviewee selection and interpretation of their statements and comments, was done to avoid bias. Finally, triangulation, where secondary sources were reviewed to further enhance the validity of this research study, was also undertaken.

4. Smart Systems in Next-Gen Financial Management

The advancement of financial systems as a function of Industry 5.0 is characterized by the introduction of intelligent digital technologies intended to not only improve efficiencies but also enable humans to make decisions around human-centric, sustainable, and resilient functions. Smart systems, which are defined by the use of AI, machine learning, blockchain, big data, and/or automation, are forming the basis for how we manage finances for profit and non-profit organizations; it impacts how we generate financial data, how we analyze that data, and how we respond based on the information we have inferred from it. These technologies are becoming increasingly important for achieving agility, scalability, and accountability in financial services, as well as restoring transparency and ethical practice.

4.1 Automation and AI in Accounting, Auditing, and Forecasting

The implementation of automation and AI in financial administration has disrupted conventional accounting, auditing, and forecasting. Most accounting, auditing, and forecasting functions are being transformed from manual, repetitive, and time-consuming procedures and processes to sophisticated and autonomous processes that increase accuracy, lower labor costs, and help achieve operational goals—all within the context of planning the finances of an organization. In reality, within Industry 5.0 none of these applications should be considered solely effective organizational tools or methods for improving efficiency, but rather as co-equal intelligent partners with human professionals. These systems add cognitive computing to the judgment of the profession and human interaction to real-time data analysis. RPA is the background technology that has introduced foundational automation of rule-based and transactional processing functions in finance applications. Many of the accounting functions that relied heavily on human interaction and labor have now transitioned to RPA workflows; for example: data entry, invoice processing, payroll processing, account reconciliations, etc. (Tyagi 2024a, Tyagi 2024b, and Tyagi 2024c). When working autonomously, RPA utilizes bots enabling RPA functions built into usable financial platforms. For example, Xero, QuickBooks, Oracle Financials, etc., have the framework of AI and RPA to categorize costs, match payments, verify transactions, and reconcile accounts with almost perfect accuracy and speed. In fact, various financial platforms are not limited to robotic processes that follow a set of programmed and decision-oriented rules but are capable of learning from historical observations to provide dynamic learning, renegotiation of rules, and adjustment to expected transaction processes to achieve goals with recommendations consistent with the financial policies or goals of an organization. AI has transformed auditing from a periodic, sample-based process to one of continuous and full-population audits (Xu et al. 2024). Traditional auditing involved human auditors manually reviewing a limited number of transactions in a limited time window, which allowed for delays and the potential for incidental error. An AI auditor can ingest, parse, and analyze a massive volume of transactional data—including current and historical data—simultaneously in real time and check for inconsistencies, anomalies, or breaches of compliance across an entire dataset. Cognitive audit tools like MindBridge AI and Deloitte's Argus utilize a series of machine learning models, trained on their own historical assertions and non-bias data, to identify patterns that indicate fraud, accounting errors, or a regulatory breach. Cognitive audit tools not only identify issues but also contextualize anomalies by assigning a measure of risk and suggesting context-specific measures for the auditor to follow if the auditor is compelled to follow up with an investigation on the area in question.

AI systems are also being used to monitor compliance in several jurisdictions with varying regulations. By scanning financial document sources, communication sources, and log files of transactions and datasets, the AI knows its requirements and can find breaches or other general compliance issues, create reports when breaches require regulatory reporting, and ensure compliance with anti-money laundering (AML) and Know Your Customer (KYC) policies, etc. AI in auditing is an evolving practice, and like ML, AI is moving toward robustness and explainability in AI (XAI), which emphasizes transparency and explainability of algorithmic outcomes—a significant requirement in various regulated industries.

One of the most revolutionary use cases for AI in finance-related decision making is forecasting and strategic planning. Financial forecasting has always relied on static models built on frameworks using historical data finishes. Financial forecasting often failed to adequately reflect market dynamics or shocks from external sources. The good news is that the technological abilities of the algorithms have advanced to a point where they are able to produce forecasts in real time and incorporate a variety of financial and non-financial variables. AI can estimate in a real-time environment and include financial metrics with structured data combined with unstructured data. The models dynamically incorporate data that comes from both structured and unstructured information. For example, NLP is a powerful tool that allows the AI system to consume news feeds, keywords, and other factors from across the globe and assess the macro and regional impact on currency swings or commodity prices. Sentiment analysis works similarly, finding trends from inquiries on social media or community reviews. An understanding of these causes and events becomes virtually impossible in our current era of increased uncertainty, where prediction tools and models struggle to deal with complexity and multi-layered systemic risks. AI is the perfect tool to test these inferences, conduct what-if analysis, better allocate capital, and develop formations against risk mitigation. AI contextualizes rapid information and produces valuable scenarios quickly, thereby removing "human bias" dramatically. In addition, these forecasting systems enable automated decision making and support prescriptive analytics, where future trends and outcomes are forecasted along with prescribed financial decisions. For example, a financial planning system may prescribe a transfer of funds from high-volatility, high-bias investments to ESG-compliant bonds because a future tightening of regulations or a shift in investor attitudes is expected.

While automating and using AI, we need to be aware of the challenges. Managing data quality, identifying algorithmic bias, and providing model transparency and governance present challenges (Ridzuan et al. 2024). Although AI is a strong tool for finding correlations, AI is limited in its contextual judgment and ethical consideration, which requires us to maintain a Human-in-the-Loop framework, one of the main tenets of Industry 5.0. With the hybridization of human judgment with AI-supported automation, as we advance, financial functions will be effectively resilient, adaptive, and ethically based through human and AI partnerships.

4.2 Blockchain and Secure Transactions

The blockchain represents a fundamental shift in financial record keeping through the use of a decentralized, distributed ledger. In contrast to records that are owned or controlled by one organization or another, the blockchain allows multiple participants to share a single source of information related to transaction history. This creates an immutable and verifiable set of transaction records that eliminates the need for centralized intermediaries and minimizes the risk of fraud, data manipulation, or unauthorized access. This capability to enhance financial transparency is appealing to financial management applications that must consider trust, compliance, and accuracy.

The ability of blockchain to eliminate trusted centralized parties in finance ultimately relies on the concept of smart contracts. Smart contracts are code stored on the blockchain that will execute itself at a designated time and execute specified parts of a contract with respect to the agreement that it has been programmed to enforce automatically if conditions have been met. Smart contracts are designed to minimize dependency on third-party brokers, legal infrastructure, and verification through simple and nimble means, minimizing transaction costs. Smart contracts for corporate finance can be leveraged to facilitate real-time payment settlement, automated invoicing, escrow, and syndicated lending arrangements. The need for recording and reconciling service transactions and representing any risk of counterparty default can potentially be eliminated by the ability to effect transactions immediately.

Blockchain technology is especially disruptive in industry-wide use cases, and one of the most notable examples is in supply chain finance where there are multiple parties needing transparency in the entire process from raw material procurement to sale. Blockchain solutions allow financial institutions and suppliers to follow the character and movement of goods, provenance, and financial status in real time and therefore create incredible visibility into goods; this means that the supply chain will operate more efficiently and, for smaller suppliers, have more credit available to reduce information asymmetry. Asset tokenization is another excellent use case. Tokenization enables traditional assets, including real estate, commodities, and equity, to have a digital representation on a blockchain. These assets can also provide for fractional ownership and be traded in a peer-to-peer, non-digital exchange with instant settlement, supporting liquidity and, in some cases, enhancing stakeholder participation in previously unpredicted markets. In the realm of financing, international cross-border payments are a major industry, and blockchain has reduced the settlement period from several days to seconds and removed the reliance on expensive correspondent banking and settlement institutions. RippleNet, Stellar, and Corda are all developing blockchain solutions that will allow banks and other financial institutions to perform international payments in real time with a low fee for settlement and securely; this will become increasingly important for small and medium-sized enterprises (SMEs) as well as migrants sending remittances back home who rely upon fast, affordable international exchange services.

In addition to transaction efficiency, blockchain further enhances regulatory compliance, auditability, and ESG accountability. Blockchain creates immutable records, deterring retrospective modification, and it is fundamentally aware of trust in transaction links, making it ideal for identifying and tracking financial disclosures, or not, like carbon footprint, resource consumption, or ethical labor-certified sourcing. For example, ESG-linked assets—from green bonds to impact investment funds—can embrace blockchain to create investor visibility into an overt and traceable data trail, creating solid validation of sustainability claims and reducing the risk of greenwashing (Mulligan et al. 2023). The additional advantage of blockchain is that it will allow for automated compliance monitoring by embedding rules and reporting logic into smart contracts directly, providing a visible audit trail on the terms in the issuance of securities ahead of a permanent record. The benefit to regulators would be the opportunity to be closer to market development, providing them with real-time access to transactional data to enable more proactive supervision as opposed to an auditor's delayed and retrospective audits. The transition from static to real-time compliance under the corrective aspect of regulation is material to market integrity, cuts regulatory costs, and places optimal governance structures in place at minimal cost. Although blockchain technology has the potential to revolutionize financial management, it is not without its challenges. Scalability, energy consumption, interoperability between blockchains, and regulatory uncertainty are all persistent concerns. Nevertheless, new capabilities driven by solutions such as proof-of-stake consensus algorithms, layer-2 scaling capabilities, and enterprise blockchains such as Hyperledger Fabric will ultimately remove some of the obstacles to blockchain adoption in enterprise finance. Figure 1 provides an integration map of smart technologies and their application in key financial domains.

4.3 Big Data Analytics and Real-Time Decision-Making

Big data analytics is leading the way in financial innovation. It gives institutions the ability to capture and analyze massive quantities of structured and unstructured data. For next-generation financial management, big data is not just about financial key performance indicators (KPIs); it is about the inclusion of non-financial indicators such as customer sentiment, online activity, environmental trends, and political incidents. Capturing multiple data streams from many sources and feeding them

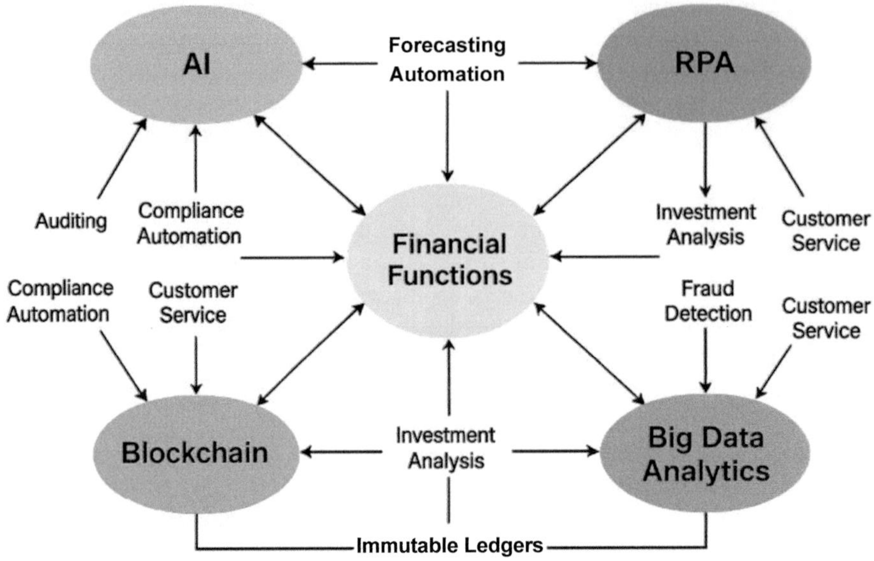

Figure 1: Integration Model of Smart Systems in Financial Functions.

through complex analytical models in real time provides invaluable knowledge that can drastically enhance the speed and quality of financial decision making.

Modern smart financial systems can easily ingest and analyze data from multiple internal and external sources - transactional databases, real-time and creative messaging from social media, CRM systems, IoT sensors, economic tickers, mission-control world displays, etc. All these data sources are continually ETL (Extract, Transform, Load) into centralized or lake/hybrid data model storage options, where next-generation analytical engines (AI, ML, NLP) are regularly used to identify correlations, trends, anomalies, and risk patterns. Financial institutions can use these capabilities to move from a lagging and stationary "fixed view" of performance to a proactive and predictive performance view. In the area of risk management, for instance, a great deal of big data analytics now allows predictive risk models to be created and changed with real-time adjustments (Ferrer 2018). Static historical credit scores are moving toward ongoing behavioral models, using spending behavior, location, and even mobile behavior for periods of time. These behavioral models are quite useful for advancing financial inclusion as they allow institutions to identify creditworthiness among populations that do not have a formal credit history. In investment and asset management, big data supports quantitative trading algorithms, sentiment analysis, and automated portfolio rebalancing decisions. Hedge funds and trading platforms use real time data to assess high-frequency trading strategies and assess volatility.

In the banking industry, real-time analytics can unlock hyper-personalisation of products and services. Financial institutions can now look at transaction histories, social media engagement, lifestyle data, and the nature of location data when considering the provision of services. The applications of this level of targeted service can be incredibly sophisticated, from relevant loan offerings to contextual marketing at the point of sale. For example, a customer may receive investment options aligned with sustainable finance principles if they are regularly shopping at eco-friendly retailers (Pan and Zhang 2020). The customer's transaction history profile now reflects not only behavioral tendencies but also ethical and moral values. This layered level of personalisation can generate greater customer engagement while creating financial services represented by the human-centricity ethos of Industry 5.0. Big Data has transformed the mechanisms to detect fraud. Previous rule-based methodologies in fraud detection traditionally struggled to keep pace with the sheer volume of specialized fraud schemes. Contrast this with using machine learning

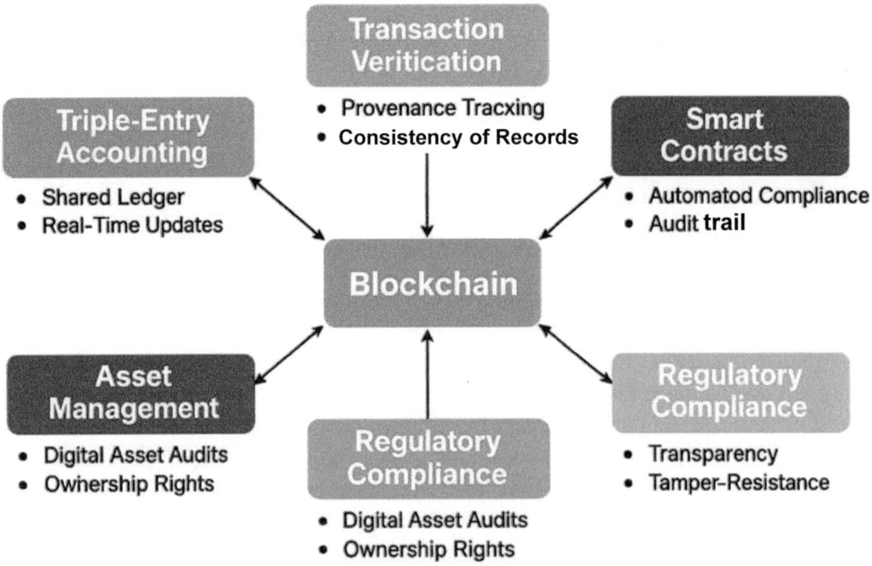

Figure 2: Big Data Architecture in Smart Financial Services.

models built off datasets of extraordinary size that only take milliseconds to detect subtle anomalies across thousands of unique transaction attributes. Fraudulent behaviors can be flagged or stopped completely within milliseconds of occurring in real time. Insurance companies are employing a very similar methodology, with real-time analytics driving a dynamic risk-profiling methodology rather than static demographic references for premium calculations. Insurance companies are now also interested in behavioral measures focused on real-time actions like driving patterns using telematics devices. Big data also assists with regulatory compliance and AML measures. RegTech powered by big data can scan tens of thousands of transactions for suspicious activity and patterns, provide real-time alerts on transactions, and automate reporting to a regulatory agency. These systems help reduce regulatory compliance requirements for financial services organizations while improving the transparency and speed of reporting requirements when regulations change. While big data analytics has many advantages, it also presents issues with data privacy, bias, and governance. Given the reliance on extensive personal datasets, appropriate data protection measures and ethical frameworks should be put in place. The introduction of big data into financial services must include explicit consent practices, transparency in algorithms, and full accountability for future decisions made using data models. Figure 2 provides a layered view of data ingestion, transformation, AI/ML engines, dashboards, and ESG output metrics.

4.4 Cybersecurity and Risk Management

The fast track of digital conversion of financial systems, accelerated by cloud computing, real-time offering of data, and interrelated systems, has not only made cybersecurity and risk more complicated but has elevated their importance as well. Financial systems, records, and decision-making tools are becoming increasingly reliant on digital infrastructure, and accordingly the vulnerability to cybercrimes, systemic breaches, and corrupted data has risen to match. This dynamic has led to the emergence of smart systems utilizing AI to defend financial services and improve their capability to manage complex systems of risk. AI-enabled cybersecurity tools have begun to be integrated into the core architecture of new financial services platforms. These tools provide ongoing activity with security, learning from patterns in user behaviors, historical transaction records, and ongoing network activity to identify, assess, and mitigate threats in almost real-time contexts. Actionable machine learning algorithms can detect new patterns of unexpected activity that indicate fraudulent event behavior, possible insider threat, malware attack, or data exfiltration attempts and do so prior to any extent of damage. These systems are utilizing problem-solving capabilities beyond rule-based defenses and adopt the adaptive approach of evolving in tandem with the ever-changing world of threats, offering proactive, rather than reactive, security.

AI-enabled real-time online intrusion detection systems (IDS) analyze millions of network interactions per second for patterns associated with suspicious activity. AI-enabled intrusion detection systems rely on behavioral authentication technologies and multi-factor identity verification to protect access to sensitive financial data. Data is also protected as it is transmitted and stored through end-to-end encryption technology—whether implemented via blockchain or cryptographic protocols—to ensure the risk of interception and modification is greatly reduced. Furthermore, as a result of having the mechanics of automated incident responses on AI-enabled cybersecurity platforms, once a threat is identified, predetermined mitigation protocols are activated automatically—it does not need manual action; it just happens. Given the losses that occurred from cyberattacks during the global pandemic, and the likelihood of cyberattacks continuing to negatively impact the bottom line for organizations, rapid mitigation activated without human intervention has the potential to lower any financial loss and, yes, also the reputational risk, which is worrisome in trust-based industries (Rejeb et al. 2020). In addition to cybersecurity, risk management is seriously changing its landscape through smart technologies, integrating AI and big data analytics to develop and use algorithmic risk models, which follow credit, financial and liquidity risk, and operational risk in perpetuity, and with real-time variables and scenarios for risk assessments of a broader focus such as identifying government-induced geopolitical risks, financial market volatility indices, social sentiments, climate risk, and other risks beyond those produced from periods once or twice a day, week, month, or year.

Integrated risk management dashboards are being implemented across financial institutions, enabling executives to have a one-stop-shop view of key risk indicators (KRIs) in a dashboard format that can graphically present the data in real time. Risk dashboards allow company leaders to identify emerging risks quickly, stress-test their financial positions to see how they will react under multiple scenarios, and use this data to make informed and strategic interventions. As regulatory guidance becomes more complex and granular, there is value in having visibility to meet regulatory needs and minimize an organization's exposure. In the compliance arena, inventory and deploy Regulatory Technology (RegTech) that streamlines and automates manual, time-consuming, and error-prone processes. RegTech platforms leverage AI and natural language processing to read regulatory texts, turn them into rules, and automatically implement them in your business. RegTech solutions will not only automate document retrieval but will consistently track all relevant transactions and users to identify unusual activity by consistently monitoring for any breaches of AML, KYC (Know Your Customer), and other regulatory requirements. RegTech tools streamline compliance costs and help mitigate both penalties and reputational risk.

One instance of this evolution can be seen in the use of cloud-based RegTech solutions, which are capable of providing scalable and sprawling compliance frameworks capable of allowing instantaneous jurisdictional change. This is particularly relevant for global financial institutions across different regulatory jurisdictions, within which manual compliance checks would be impossible to sustain at scale (Zaidi et al. 2023). However, despite the noted technological advancement, the parameters of the management of cybersecurity and algorithmic risk do not come without their separate set of challenges; e.g., algorithmic opacity, bias in threat detection models, false positives, and over-reliance on automation could all expose contemporary regulated firms to potential new vulnerabilities. Moreover, sophisticated cybercriminals are routinely using AI themselves to allow for more targeted and adaptive attack paths, all of which reduces the distance in the cyber arms race between cybercriminals and their targets (Lee et al. 2017). The factors outlined above highlight the need for a human-in-the-loop approach. While digital systems can analyze and process unprecedented volumes of data at incredible speeds, it is still the human who ultimately forms the context, makes ethical decisions, and interprets strategically. The human component

Table 1: Smart Systems and Their Financial Functions in Industry 5.0.

Smart System	Core Functions in Finance	Industry 5.0 Impact
Artificial Intelligence (AI)	Forecasting, anomaly detection, compliance automation	Augments human judgment and decision support
Machine Learning (ML)	Pattern recognition, fraud detection, risk scoring	Learns and evolves with human feedback loops
Robotic Process Automation (RPA)	Automated transactions, invoice processing, payroll	Reduces repetitive work while enabling human oversight
Blockchain	Secure transactions, smart contracts, immutable ledgers	Fosters trust, transparency, and traceability
Big Data Analytics	Predictive analytics, hyper-personalization, ESG tracking	Real-time insights for adaptive financial services

assertion resonates with the first principles of Industry 5.0, which prioritize the role of human values within a more interactive future with intelligent automation systems, especially in stakeholder-trust-critical sectors like financial security. Table 1 provides a classification of smart systems mapped to key financial functions and their value contribution under Industry 5.0 paradigms.

5. Industry 5.0 Integration Finance

Utilizing Industry 5.0 in financial management involves a strategic pivot away from pure technological efficacy to a more holistic approach that places human values, social purpose, and sustainability at the center of decision-making. Where its predecessor focused largely on full automation and cyber-physical systems, Industry 5.0 promotes human-technology collaboration, customization, and inclusion. In finance, this approach means creating systems in which technological intelligence complements the decision-making of the human user; where ethical considerations are preprogrammed into innovation; and where profit is only one measure of long-term and sustainable value to society.

5.1 Human-Centric Finance: Collaboration Between Humans and Machines

The core of Industry 5.0 is that many key technologies emerging will be human-centered, reframing technology's role as a partner to human intelligence instead of a substitute. In financial services, we are encouraged to work together with technology, empowered to impart higher-quality decisions to our clients using intelligent systems to handle complexity and scale. Take, for example, financial analysts using AI-powered tools to generate investment scenarios; AI can highlight different results, however, it is the human who applies their judgment to contextualize and assess ethical considerations of the outcomes. Working together builds creativity, empathy, and ethical foresight, all important human-only capabilities. Digital advisors, or robo-advisors, are working toward increased autonomy but are increasingly integrating human-in-the-loop systems that allow clients to override algorithmic recommendations or dictate them according to their personal preferences or ideals. Blending technical certainty with human instinct is a signature of the ethos of Industry 5.0 and will lead to diverse, resilient, and trusted financial services (Zairis et al. 2024). Figure 3 provides a spectrum from fully automated to fully human decisions, with Industry 5.0 promoting a balanced hybrid in the center.

5.2 Customization and Hyper-Personalization in Financial Services

Industry 5.0 provides the ability to achieve unparalleled personalization in financial management through the use of real-time analytics and deep learning models. Hyper-personalization allows

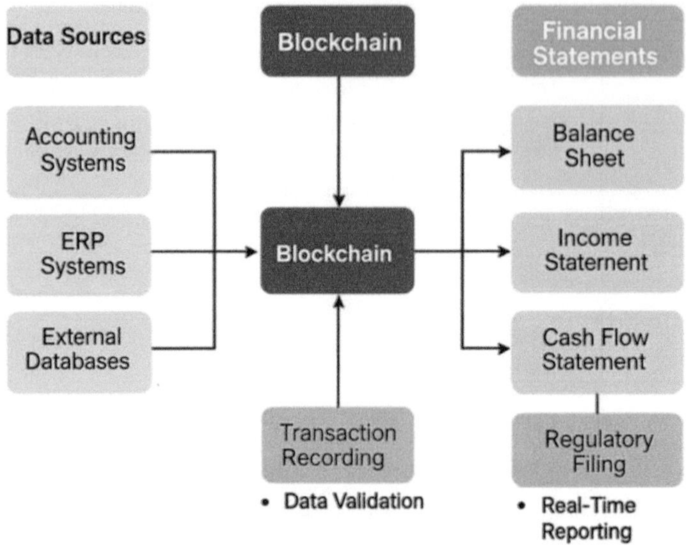

Figure 3: Human-AI Partnership Spectrum in Finance.

financial institutions to craft products and services tailored to people based on individuals' behaviors, preferences, and values—not only based upon demographic segmentation. For example, AI systems can make personalized investment recommendations crowdsourced from a client's environmental, social, and governance values when selecting a personalized portfolio to match existing wealth. Financial planning tools now offer adaptive platforms that can react to life event changes, lifestyle stage, goals, aspirations, and risk aversion or appetite. The shift from standardized products to bespoke financial ecosystems represents a radical paradigm shift from a one-size-fits-all mentality to a model that is systemic with Industry 5.0's motivation, which is not just about creating value, but also about allowing the user to be empowered via intuitive, responsive, and ethical design.

5.3 Intelligent Automation and Human Oversight

Automation remains important, but Industry 5.0 refers to intelligent automation—a type of automation enhanced by human oversight and ethical governance. In finance, for example, this means incorporating explainability, accountability, and fairness within automation. Biases in credit scoring algorithms are now audited for bias and accountability to ensure that automated decisions are not discriminatory (Marrone and Hazelton 2019). Human oversight matters for other reasons, such as detecting potential risks that are more subtle than the machine was designed to catch or data that humans understand in context but which, because of ambiguity, the machine cannot easily interpret (Tyagi 2024b). For example, compliance officers now use RegTech tools to almost instantly skim through thousands of transactions, but still find a way to make contextually appropriate judgments with the human touch. This could involve a more layered and robust decision-making process and ensure technology supports human outcomes rather than replaces them.

5.4 Real-World Examples of Industry 5.0 in Finance

Although instances of Industry 5.0 in finance are beginning to appear from both incumbent organizations and FinTech startup innovation, this process is just beginning. In Europe, the BBVA bank has rolled out an AI-human hybrid system that helps customers with even more robust customer engagement while maintaining actual human advisors that can deal with more complex queries. The AI is designed to learn from and with customers and can determine when to use a human adviser. More recently, OCBC Bank in Singapore is using a human-in-the-loop AI system

Table 2: Comparative Features of Financial Automation: Industry 4.0 vs. Industry 5.0.

Feature	Industry 4.0	Industry 5.0
Automation Focus	Full automation, efficiency-centric	Human-AI collaboration, purpose-driven
Decision-Making	Data-driven, algorithm-based	Augmented by human values and oversight
Personalization	Limited, demographic-based segmentation	Hyper-personalization, value-aligned offerings
Sustainability Integration	Add-on, often externally enforced	Core design component
Governance Model	Technical governance	Ethical and transparent co-governance

in a wealth management context. The AI provides algorithmic advice that is still supplemented by relationship managers with personal interactions. The keyword again is "relationship," which is not limited to efficient transactions—it incorporates trust, core human values we apply to financial services. Table 2 provides a comparison of financial automation principles across Industry 4.0 and Industry 5.0 frameworks.

6. Sustainable Financial Management Models

Sustainability is no longer a secondary issue in financial management; it is emerging as primary for sustainable long-term value creation opportunities, as well as risk mitigation opportunities. Environmental degradation, social inequality and governance failures can't be ignored as they exert both direct and indirect pressures on financial markets. As more firms incorporate sustainability into financial expression strategies, this also remains the experience in corporate finance and business models. The blending of Industry 5.0 and smart systems highlights the mindset and technology platforms we must invest in to develop financial models that represent not only economic performance, but also environmental stewardship and social equity as a third pillar in what might be characterized as a potential paradigm shift in financial performance normative. If we don't act, the evidence is building that sustainability will sooner rather than later replace the current existing financial performance model.

6.1 Sustainability in Finance: ESG, Green Finance, and Ethical Investing

ESG factors are changing the ways ESG are being included into both investment and corporate finance decision criteria very rapidly. ESG probably will sell out a place in portfolio screening criteria, credit ratings, credit issuer disclosure, credit issuer analysis, and company disclosures about ESG metrics. Institutional investors, sovereign wealth funds, and pension funds are likely allocating more resources based on ESG ratings because they understand the financial risks of climate change, social unrest, and bad governance. Ethical investing and green financing frameworks appear to be on an upward trajectory, arguably promoted by a growing cohort of impact-oriented investors. Financial instruments such as green bonds, social bonds, and loans at public or wholesale market rates designed with sustainability goals are intended to deliver money to environmentally responsible and socially responsible projects (Corbet et al. 2018). Financial institutions may soon be held accountable for not only reporting ESG metrics, but will also be expected to produce plans on how to improve ESG metrics. Smart systems including AI and big data analytics will support plans to increase ESG performance by improving the identification, measurement, and management of ESG risk and opportunity. Figure 4 illustrates a circular model showing AI-assisted ESG scoring, blockchain for traceability, big data for trend analysis, and human oversight.

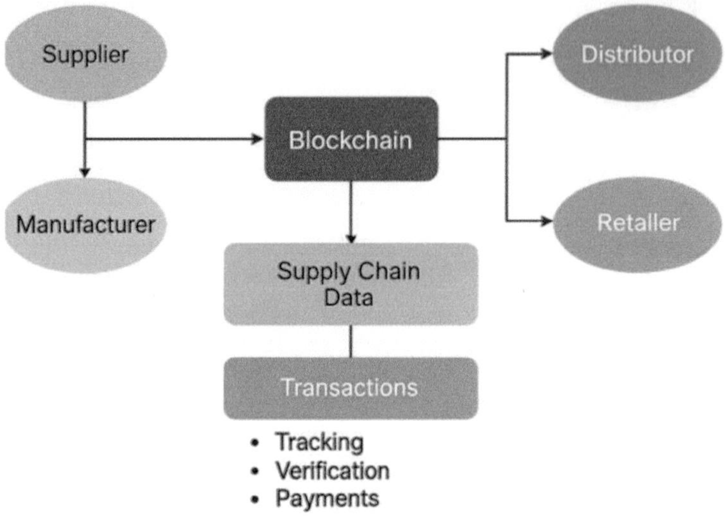

Figure 4: ESG-Aligned Smart Financial Decision Framework.

6.2 Smart Systems for Environmental and Social Monitoring

Smart financial systems offer the necessary technical foundation in order to measure, monitor, and manage sustainability-related data in real time. Satellite imaging and IoT sensors can be used to monitor carbon emissions, water usage, and land degradation; the result can be measured in financial models that account for relevant hazards to your investment risk profile. Financial products can be designed to include timely data, such as ESG scores derived from AI algorithms that combine all of the unstructured data that can be collected. Blockchain also becomes useful in realizing verifiable and unalterable sustainability verification record keeping. Tracking supply chain data, carbon credits, and compliance documentation represented on blockchain/ledger is designed for sustainability and societal reporting purposes, but even more especially to protect investor confidence from liabilities related to unscrupulous adventurism from limited financial oversight reporting obligations. Leveraging all of this technology means sustainable financial decision making becomes credible.

6.3 Financial Models Promoting Circular Economy

One area of emerging opportunities in sustainable finance is the opportunity to incorporate circular economy concepts, through financial products and services that promote reuse, recycling, and best use of resources over time. Previous straight linear financial models, dependent on consumption and disposability, are increasingly not seen as feasible to achieve environmental sustainability efforts. Utilizing intelligent and digital financial platforms can promote financing for the circular economy by creating pay-per-use plans or asset leasing and performance contracting. For example, Internet-of-Things (IoT) financed solutions can offer just-in-time financing based on machinery or vehicle real-time reporting that tracks usage, which allows lenders to extend usage-based lending. AI digital platforms can evaluate the project length and expenditure of circular economy efforts, ultimately integrating the middle ground of profitability and viability into credit risk assessments or investment approaches (Samadi-Parviznejad 2022). Thus, while these options reinforce a commitment to environmental sustainability, they create different revenue streams and business models for financial participants. Just as with performance-based financial instruments, including sustainability-linked loans today where interest rates are conditional on achieving ESG goals, financing contemporaneously with the circular economy can be structured to encourage desired practices.

6.4 Role of Policy, Regulation, and Governance

Regulators and governance are crucial to bringing sustainable financial models to fruition. Policymakers are instrumental in establishing standards around ESG disclosure, green taxonomy, and risk transparency. As the EU Sustainable Finance Disclosure Regulation (SFDR) and Task Force on Climate-related Financial Disclosures (TCFD) evolve into mandatory reporting templates, they raise good practice from merely positive intent to the expectation of sustainability by the financial system. Governance arrangements must also adapt to meet a new decision-making census to assure that algorithms and smart systems used in sustainable finance decision-making support transparency, influenceability, and accountability. We need to apply ethics-based governance to prevent greenwashing in addition to the assurance that the algorithms and automated digital tools that inform our decision-making do not create additional biases, preserve existing inequities, or varnish the attention of accountability (Charoenwong et al. 2024). A series of multi-stakeholder discussions, actions, etc., with regulators, industry and peak bodies, and technology providers is pivotal to creating trusted, sustainable financial systems of high integrity.

7. Challenges and Future Directions

Although the engagement of smart systems coupled with Industry 5.0 could revolutionize financial management, the transition may prove challenging. There are many technological, organizational, regulatory, and ethical challenges that might present barriers to the realization of potential benefits from next-generation financial ecosystems. Furthermore, there are many barriers, but with challenges arise opportunities for innovation and societal change. This chapter identifies key integration barriers and discusses opportunities for future research, policy, and practice.

The use of sophisticated smart systems across a legacy financial system represents a major obstacle toward transformation. Many traditional financial institutions use legacy technologies resulting in interoperability challenges and high integration costs. Along with the investments of cloud computing, data warehousing, and threat management for AI, blockchain, and real-time analytics, the overhead costs can be excessive for any organization, particularly smaller business models. Furthermore, data fragmentation and data quality problems severely hinder the ability of smart systems to produce useful outputs. In order for intelligent decision-making tools to perform properly, they need to be trained on diverse, accurate, and contextual data. Each institution may have varied data meanings, a misalignment of standards within the data, or use siloed databases, all of which threaten how comprehensive the AI model might be. Likewise, the emergence of Industry 5.0 is touted to be delayed as a result of how impoverished or abused the data may increasingly become across more institutions (Lim 2024).

One significant non-technical barrier is the lack of professionals with the interdisciplinary expertise that allows them to position themselves at the interface between finance, technology, and sustainability. There is recognition that we need skills like data scientists and digital finance specialists but there is still a gap in our workforce when it comes to skills such as ethical AI, ESG modeling, and circular economy finance. We need upskilling and reskilling programs to ensure that the existing financial workforce is able to adopt smart systems and operationalize sustainability. There is also a cultural issue within organizations. Outdated expectations of work, lack of digital vocations among decision-makers (Tyagi 2024a), and an inclination to avoid risk and make decisions based on historical performance, understandably create barriers to acceptance of smart systems, and ultimately benefit to an organization. In order to migrate to an Industry 5.0 mindset, it is necessary to apply technological investments; along with this comes the necessity of cultural changes and leadership that is committed and energized by innovation and sustainability.

The evolution of smart financial systems has advanced at a rapid rate and surpassed regulatory constraints in many areas. Key issues concerning data privacy, algorithmic accountability, and the regulation of cross-border finance remain unresolved. Regulators are struggling with innovations,

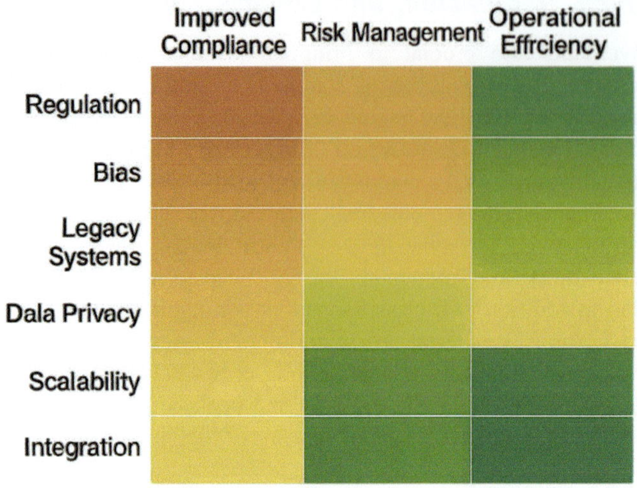

Figure 5: Challenges vs. Opportunities Heatmap for Industry 5.0 Finance.

protecting consumers, and systemic stability. Ethical questions are at the forefront. In particular, algorithmic bias, surveillance capitalism, and automated "black-box" decision-making can undermine fairness in AI-driven decision processes—a great concern in the regulation of algorithmic trading and investment. However, the emergence of ESG products has also raised issues related to a general concern of "greenwashing" where investment products and funds are mischaracterized as sustainable. The importance of this research reveals the necessity for stronger governance frameworks, third-party verification processes, and global seamless consistency and/or standards.

The future of financial management will be characterized by the tension between automation and empathy, intelligence and ethics, and novelty and inclusion. The financial systems of the future will be able to create efficiencies in capital flows while also achieving broader social and environmental outcomes. As DeFi, asset tokenization, and AI-enabled ESG platforms gain traction by being open and applied in a values-based marketplace, the new financial products and services will benefit a more aware and informed clientele. Meanwhile, Central Bank Digital Currencies (CBDCs) may also catalyze a change in monetary policy, payment systems, and financial inclusion if introduced and implemented under the principles of Industry 5.0. It will be important that all the stakeholders that make up the financial ecosystem—financial institutions, technology suppliers, governments, and academia—work together to develop the infrastructure and regulatory regimes to deal with this evolution. It is also important that our research around responsible innovation, ethical AI governance, and sustainable finance metrics keeps pace with the developments in technology so as not to slip behind in being aligned with the broader societal objectives (Adhikari and Ramkumar 2023). Figure 5 presents a heatmap which identifies major challenges presented in rows such as regulation, bias, legacy systems, and their potential innovation leverage points in columns such as improved compliance, risk management, and operational efficiency. The color gradient from red to green severity to opportunity respectively.

To tackle these issues and usher in a sustainable, human-centered future for financial systems, multiple significant recommendations can be made (Ren et al. 2023). Financial institutions should be more concentrated on interoperable infrastructure development; build agent intelligence and ESG skill development pipelines; utilize transparent and auditable algorithms to earn public trust; Regulators should begin to move toward a proactive and technology-neutral policy development process that promotes responsible innovation without impeding innovation; Academia and think tanks should keep on bringing research development at the intersection of finance. Civil society organizations monitor and uphold ethical standards and chart the public interest; above all stakeholders

Table 3: Key Integration Challenges in Smart Financial Ecosystems.

Challenge Area	Description	Impact on Adoption
Legacy Systems	Incompatible infrastructure	Slows down transformation
Data Quality and Silos	Fragmented, inconsistent data structures	Low model performance and reliability
Regulatory Ambiguity	Lack of clear compliance frameworks for emerging tech	Creates hesitation and risk-averse behavior
Skills Gap	Shortage of interdisciplinary professionals	Limits effective use of smart systems
Ethical and Bias Concerns	Algorithmic bias, lack of explainability	Impacts trust and fairness

need to adopt a systems thinking orientation; financial transformation cannot occur independent of social, technological, and ecological transformations. Table 3 presents the challenges that hinder the integration of smart systems in sustainable and human-centric financial environments.

8. Conclusion

The integration of intelligent systems and Industry 5.0 signals a defining turning point for financial management, moving away from efficiency and towards human values, sustainability, and the long-term impact on society. This research has shown that AI, blockchain, big data, RPA, etc., can aid modernization of finance by not only introducing technology but also reconstructing finance towards a more ethical, inclusive, socially responsible, and resilient framework when seen in the context of Industry 5.0. There will however need to be a commitment to ensure technology is integrated in consideration of human oversight, ethical governance, and throughout stakeholder engagement processes. We must also overcome barriers to the diffusion of Industry 5.0 financial models, and challenges listed previously such as legacy systems, regulatory uncertainty, and data privacy, and build the workforce of the future with sustainable skills and training through a pooling of forces, and innovation and policy. In addition, as the possibilities for green finance, ESG, and personalizable financial services can potentially display the importance of financial institutions in the sustainability agenda. Although amidst profound contextual change, the outcomes from related practices remain relevant if human agency and systemic accountability are the foundation for finance, we can create the future of financial management development beyond profit maximization to promote equitable economic development. Financial management in a time of rapidly changing geographies and uncertainty fits well with diversification offered by Industry 5.0 and enables better understanding of both a smart, sustainable, and inclusive future of finance.

References

Adhikari, N., and Ramkumar, M. (2023). IoT and Blockchain Integration: Applications, Opportunities, and Challenges. *Network*, 3(1): 115–141. https://doi.org/10.3390/network3010006.

Arner, D. W., Barberis, J. N., and Buckley, R. P. (2015). The evolution of Fintech: a new Post-Crisis paradigm? *SSRN Electronic Journal*. https://doi.org/10.2139/ssrn.2676553.

Charoenwong, B., Kowaleski, Z. T., Kwan, A., and Sutherland, A. G. (2024). RegTech: Technology-driven compliance and its effects on profitability, operations, and market structure. *Journal of Financial Economics*, 154: 103792. https://doi.org/10.1016/j.jfineco.2024.103792.

Coelho, P., Bessa, C., Landeck, J., and Silva, C. (2023). Industry 5.0: The arising of a concept. *Procedia Computer Science*, 217: 1137–1144. https://doi.org/10.1016/j.procs.2022.12.312.

Corbet, S., Lucey, B., Urquhart, A., and Yarovaya, L. (2018). Cryptocurrencies as a financial asset: A systematic analysis. *International Review of Financial Analysis*, 62: 182–199. https://doi.org/10.1016/j.irfa.2018.09.003.

Dalenogare, L. S., Benitez, G. B., Ayala, N. F., and Frank, A. G. (2018). The expected contribution of Industry 4.0 technologies for industrial performance. *International Journal of Production Economics*, 204: 383–394. https://doi.org/10.1016/j.ijpe.2018.08.019.

Ferrer, E. C. (2018). The Blockchain: a new framework for robotic swarm systems. In *Advances in intelligent systems and computing* (pp. 1037–1058). https://doi.org/10.1007/978-3-030-02683-7_77.

Gomber, P., Kauffman, R. J., Parker, C., and Weber, B. W. (2018). On the Fintech Revolution: Interpreting the Forces of Innovation, Disruption, and Transformation in Financial Services. *Journal of Management Information Systems*, 35(1): 220–265. https://doi.org/10.1080/07421222.2018.1440766.

Hasan, H. R., and Salah, K. (2018). Blockchain-Based proof of delivery of physical assets with single and multiple transporters. *IEEE Access*, 6: 46781–46793. https://doi.org/10.1109/access.2018.2866512.

Hong, L., Luo, M., Wang, R., Lu, P., Lu, W., and Lu, L. (2018). Big data in health care: applications and challenges. *Data and Information Management*, 2(3): 175–197. https://doi.org/10.2478/dim-2018-0014.

Lee, D., Chuen, D. L. K., and Deng, R. H. (2017). *Handbook of Blockchain, Digital Finance and Inclusion: Cryptocurrency, Fintech, Insurtech, and Regulation*. Academic Press.

Lim, T. (2024). Environmental, social, and governance (ESG) and artificial intelligence in finance: State-of-the-art and research takeaways. *Artificial Intelligence Review*, 57(4). https://doi.org/10.1007/s10462-024-10708-3.

Marrone, M., and Hazelton, J. (2019). The disruptive and transformative potential of new technologies for accounting, accountants and accountability. *Meditari Accountancy Research*, 27(5): 677–694. https://doi.org/10.1108/medar-06-2019-0508.

Mulligan, C., Morsfield, S., and Cheikosman, E. (2023). Blockchain for sustainability: A systematic literature review for policy impact. *Telecommunications Policy*, 48(2): 102676. https://doi.org/10.1016/j.telpol.2023.102676.

Nahavandi, S. (2019). Industry 5.0—A Human-Centric solution. *Sustainability*, 11(16): 4371. https://doi.org/10.3390/su11164371.

Pan, Y., and Zhang, L. (2020). Roles of artificial intelligence in construction engineering and management: A critical review and future trends. *Automation in Construction*, 122: 103517. https://doi.org/10.1016/j.autcon.2020.103517.

R, T. B., R, J., and Shanavas, A. (2024). Sustainable Finance And Use Of Artificial Intelligence In Investment Decision Making. *International Journal of Advanced Research*, 12(09): 1212–1218. https://doi.org/10.21474/ijar01/19554.

Rejeb, A., Keogh, J. G., and Treiblmaier, H. (2020). How Blockchain Technology can Benefit Marketing: Six pending research areas. *Frontiers in Blockchain*, 3. https://doi.org/10.3389/fbloc.2020.00003.

Ren, Y., Ma, C., Chen, X., Lei, Y., and Wang, Y. (2023). Sustainable finance and blockchain: A systematic review and research agenda. *Research in International Business and Finance*, 64: 101871. https://doi.org/10.1016/j.ribaf.2022.101871.

Ridzuan, N. N., Masri, M., Anshari, M., Fitriyani, N. L., and Syafrudin, M. (2024). AI in the Financial Sector: The Line between Innovation, Regulation and Ethical Responsibility. *Information*, 15(8): 432. https://doi.org/10.3390/info15080432.

Samadi-Parviznejad, P. (2022). The role of big data in digital transformation. *Journal of Data Analytics*, 1(1): 42–47. https://doi.org/10.59615/jda.1.1.42.

Schuetz, S., and Venkatesh, V. (2019). Blockchain, adoption, and financial inclusion in India: Research opportunities. *International Journal of Information Management*, 52: 101936. https://doi.org/10.1016/j.ijinfomgt.2019.04.009.

Singh, R., Gupta, A., and Mittal, P. (2024). A Systematic literature review on blockchain-based smart contracts: platforms, applications, and challenges. *Deleted Journal*. https://doi.org/10.1145/3704741.

Tyagi, A. K. (2023). Transformative effects of CHATGPT on the modern era of education and Society. In *Advances in systems analysis, software engineering, and high performance computing book series* (pp. 374–387). https://doi.org/10.4018/978-1-6684-8531-6.ch019.

Tyagi, A. K. (2024a). *Blockchain–Artificial Intelligence-Based Secured Solutions for Smart Environment* (pp. 547–577). https://doi.org/10.1002/9781394303564.ch23.

Tyagi, A. K. (2024b). *Blockchain-Enabled Internet of Things (IoTs) Platforms for IoT-Based Healthcare and Biomedical Sectors* (pp. 201–217). https://doi.org/10.1002/9781394287420.ch10.

Tyagi, A. K. (2024c). *Sensors and Digital Twin Application in Healthcare Facilities Management* (pp. 369–390). https://doi.org/10.1002/9781394287420.ch19.

Xu, R., Zhu, J., Yang, L., Lu, Y., and Da Xu, L. (2024). Decentralized finance (DeFi): a paradigm shift in the Fintech. *Enterprise Information Systems*, 18(9). https://doi.org/10.1080/17517575.2024.2397630.

Zaidi, N., Maurya, M., Grima, S., and Tyagi, P. (2023). Exploration of Artificial intelligence (AI) in Banking sector: A bibliometric analysis. pp. 205–233. *In: Apress eBooks*. https://doi.org/10.1007/978-1-4842-9810-7_13.

Zairis, G., Liargovas, P., and Apostolopoulos, N. (2024). Sustainable Finance and ESG Importance: A systematic literature review and research agenda. *Sustainability*, 16(7): 2878. https://doi.org/10.3390/su16072878.

Section 4

Open Issues and Future Opportunities

15

Industrial IoT Transformation in Aircraft Manufacturing and Maintenance, Repair, and Overhaul Operations

Vineet Bhatia,[1,*] *Sumati Sidharth*,[1] *Varsha Bhatia*,[2] *Bhavesh Bhatia*[3] and *Vivek Jaglan*[4]

1. Introduction

The convergence of Information Technology (IT) (data, business systems like ERP) and Operational Technology (OT) (physical control systems like SCADA, PLCs in factories) has ushered in an era of new industrial revolution—transforming the way manufacturing systems operate, communicate and deliver value. The IIoT itself is what helps make that revolution possible, and it is a technological ecosystem of interconnected things (devices and sensors), and the ports they use to connect and communicate, composed of machines, computers, people, things, intelligent devices and objects—all of which collect and share data. Contrasting with consumer IoT applications that mainly focus on convenience and lifestyle enhancement, IIoT applications link the physical world with digital infrastructures to build Cyber-Physical Systems (CPS) with the ability to sense, analyze, and optimize industrial processes in real time. This convergence enables manufacturers greater visibility into their operation than ever before, optimization of processes without human intervention, and autonomous decision-making. IIoT in manufacturing is more than just another technology drive—it heralds a shift to data-driven manufacturing.

Conventional manufacturing systems followed pre-defined time schedules and reaction-based maintenance strategies, which used to culminate in inefficiencies, unplanned downtime, and non-optimal utilization of resources (Soori et al. 2023b). Today's manufacturing environments are challenged with a host of issues IIoT technologies have the best chance to resolve. Increased

[1] Department of Technology Management, Defence Institute of Advanced Technology Pune India.
Email: sumatisidharth@gmail.com, https://orcid.org/0000-0002-1879-596X
[2] Sunstone Pune India.
Email: varshabhatia.in@gmail.com, https://orcid.org/0009-0003-5717-7616
[3] Department of Computer Science Western University Londin Onterio Canda.
Email: bhavesh.b@protonmail.com, https://orcid.org/0009-0007-9680-3842
[4] City Institute of Higher Learning (C-IHL) Moncton, New Brunswick Canada.
Email: jaglanvivek@gmail.com, https://orcid.org/0000-0003-3532-6353
* Corresponding author: bhatiavineet.in@gmail.com, https://orcid.org/0009-0006-1133-8914

DOI: 10.1201/9781003655190-19

global competition has put more pressure on manufacturers to lower costs while enhancing quality and delivery performance. IIoT offers the technological basis to meet these issues via increased connectivity, intelligence, and automation.

More than 70% of manufacturing companies have started implementing IIoT in some capacity, from pilot projects to full-scale deployments, according to recent industry surveys (Subramanyan 2025). Yet the maturity levels differ considerably between industries, company sizes, and regions. The IIoT applications in manufacturing cut across the entire value chain, from production planning and supply chain management to quality control and maintenance optimization. Predictive maintenance has been one of the most important use cases, allowing manufacturers to forecast machine breakdowns prior to their occurrence, minimizing unplanned downtime and associated maintenance expenses (Carvalho et al. 2019).

Quality control applications use IIoT sensors and analytics to find defects in real time, which reduces waste and boosts customer satisfaction. Also, supply chain visibility applications offer end-to-end tracking and optimization, enabling more responsive and efficient logistics operations.

Despite the clear benefits, IIoT implementation in manufacturing faces several key challenges. Cybersecurity concerns are a top issue. Increased connectivity creates new attack paths and vulnerabilities that need careful management (Sisinni et al. 2018). Legacy system integration is another major challenge, as many manufacturing facilities still use equipment installed decades ago that lacks modern connectivity. Skills gaps and resistance to change in organizations also slow down IIoT adoption, requiring significant investments in training and change management. The economic advantages of IIoT in manufacturing are considerable but require careful justification and measurement. Recent studies have indicated that successful IIoT implementations can cut operational costs by ten to twenty percent (10–20%) and enhance productivity by fifteen to twenty-five percent (15–25%) (Kang et al. 2016). However, achieving these advantages requires a large initial investment and may take several years to break even.

Technological advancement in IIoT development, which includes latest technologies like 5G communications, edge computing, and artificial intelligence (AI), is opening up fresh opportunities for manufacturing applications. 5G networks aim to provide the ultra-low latency and high reliability needed for critical manufacturing applications, such as autonomous mobile robots and real-time quality control systems (Cheng et al. 2018). Edge computing allows data processing and decision-making right where data is generated, cutting down on latency and bandwidth use while improving system responsiveness.

2. IIoT Architecture and Enabling Technologies

The Industrial Internet of Things represents a complex ecosystem of interconnected technologies, protocols, and systems designed to operate reliably in demanding industrial environments. Understanding the architectural foundations and enabling technologies is crucial for successful IIoT implementation and optimization in manufacturing contexts. This section provides a comprehensive examination of IIoT system architecture, communication protocols, and the key technologies that enable industrial-grade IoT deployments.

2.1 IIoT System Architecture

Industrial IoT systems employ a structured five-layer framework that builds upon traditional hierarchies while integrating networking and computational technologies. This architecture delivers operational flexibility and intelligent decision-making capabilities across various manufacturing environments (Wan et al. 2016). The summary of five layers is presented in Table 1 below:

Table 1: Five-Layer IIoT Architecture Framework.

Layer	Primary Functions	Key Components	Technical Capabilities	Manufacturing Applications
Layer 1: Device Layer	Physical data generation and consumption	• Programmable Logic Controllers (PLCs) • Distributed Control Systems (DCS) • SCADA systems • Industrial sensors • Actuators and controllers	• Temperature, pressure, vibration monitoring • Flow rate and position tracking • Embedded computing capabilities • Direct network participation • Local processing for immediate response	• Real-time equipment monitoring • Process parameter control • Safety system integration • Quality measurement at source
Layer 2: Edge Computing Layer	Local processing and real-time analytics	• Industrial gateways • Edge servers • Fog computing nodes • Protocol translators	• Data aggregation from multiple sources • Minimal latency control algorithms • Bandwidth optimization through filtering • Autonomous operation during outages	• Immediate safety responses • Quality control decisions • Local optimization routines • Continuous operation assurance
Layer 3: Network Communication	Data transmission and protocol management	• Industrial Ethernet infrastructure • EtherNet/IP, PROFINET, EtherCAT • Wi-Fi and cellular networks • LPWAN technologies • 5G URLLC capabilities	• Deterministic communication timing • High-speed, low-latency connections • Mobile equipment connectivity • Wide-area sensor networks • Ultra-reliable transmission	• Motion control systems • Mobile robot coordination • Remote sensor integration • Critical timing applications
Layer 4: Cloud Platform (Wang et al. 2018)	Scalable computing and enterprise integration	• AWS, Microsoft Azure, Google Cloud • Data storage repositories • Analytics engines • Machine learning frameworks • Database management systems	• Historical data warehousing • Complex predictive modeling • AI algorithm execution • Enterprise system integration • Computational resource scaling	• Long-term trend analysis • Predictive maintenance algorithms • Supply chain optimization • Enterprise-wide reporting
Layer 5: Application Layer	User interaction and business intelligence	• Manufacturing Execution Systems (MES). • Enterprise Resource Planning (ERP) • Human-Machine Interfaces (HMI) • Specialized analytics applications	• Process optimization tools • Predictive maintenance dashboards • Quality control interfaces • Business intelligence reporting • Real-time monitoring displays	• Production planning optimization • Maintenance scheduling • Quality assurance workflows • Performance analytics

Figure 1 illustrates the **IIoT System Architecture**.

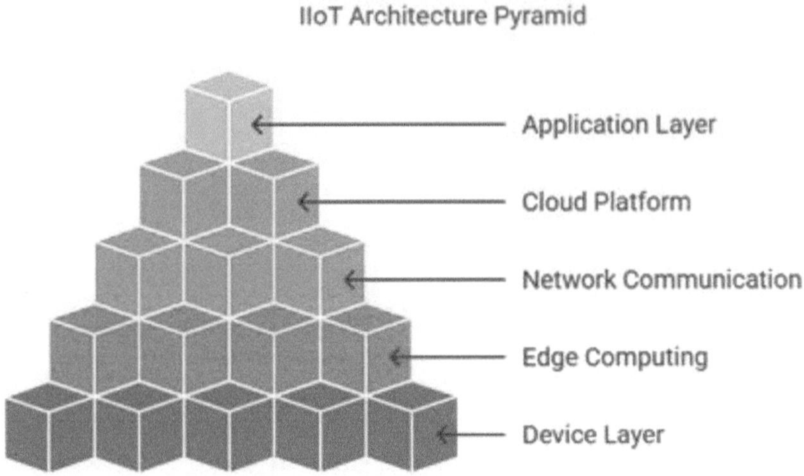

Figure 1: IIoT System Architecture.

2.2 Architecture Evolution and Modern Capabilities

2.2.1 Traditional vs. Contemporary Systems

The transformation from legacy (old) industrial systems to modern IIoT architectures represents a fundamental shift in manufacturing technology paradigms. This evolution encompasses changes in communication protocols, processing capabilities, network dependencies, and data management approaches. The details are given in Table 2.

Table 2: Traditional vs. Contemporary Systems Comparison.

Aspect	Legacy Systems	Modern IIoT Architecture
Communication Protocols	• Proprietary protocols • Specialized hardware interfaces • Vendor-specific solutions	• Standardized protocols (MQTT, CoAP, HTTP/HTTPS) • Universal connectivity standards • Cross-vendor interoperability
Processing Capabilities	• Centralized processing • Limited local intelligence • Reactive responses only	• Distributed edge computing • Local decision-making • Proactive and predictive responses
Network Dependencies	• Complete reliance on central systems • Single points of failure • Limited offline capabilities	• Autonomous edge operation • Resilient distributed architecture • Continuous operation assurance
Data Management	• Local data storage only • Limited analytics capabilities • Manual reporting processes	• Cloud-scale data repositories • Advanced analytics and AI • Automated intelligence generation

2.2.2 Layer Integration and Data Flow

2.2.2.1 Horizontal Integration Capabilities

Effective IIoT implementation requires seamless integration across all architectural layers to ensure optimal data flow and system performance. The horizontal integration capabilities enable real-time communication, data sharing, and coordinated decision-making between different system components. The details are given in Table 3.

Table 3: Horizontal Integration Capabilities.

Integration Type	Technologies Used	Business Benefits
Device-to-Device	• Direct sensor networks • Local communication protocols • Peer-to-peer connections	• Reduced latency for critical responses • Enhanced system reliability • Improved operational efficiency
Edge-to-Cloud	• Secure data pipelines • Selective data transmission • Bandwidth optimization	• Comprehensive analytics capabilities • Long-term data retention • Enterprise-wide visibility
Enterprise Systems	• ERP integration APIs • MES connectivity • Business process automation	• Seamless workflow integration • Automated decision-making •Strategic planning support

2.2.3 Technical Implementation Considerations

2.2.3.1 Communication Protocol Selection Criteria

Selecting the appropriate communication protocol is crucial for IIoT system success, as different protocols serve specific use cases and operational requirements. The protocol selection directly impacts system performance, security, scalability, and integration capabilities with existing manufacturing infrastructure. The details are given in Table 4.

Table 4: Communication Protocol Selection Criteria.

Protocol	Optimal Use Cases	Key Advantages	Implementation Requirements
MQTT	• High-volume sensor data • Publish-subscribe scenarios • Quality-of-service requirements	• Lightweight messaging • Broker-based architecture • Multiple QoS levels	• MQTT broker infrastructure • Network security protocols • Message routing configuration
CoAP	• Resource-constrained devices • RESTful web integration • Request-response patterns	• Web protocol compatibility • Confirmable message delivery • Automatic retransmission	• CoAP server deployment • Resource discovery setup • Security credential management
OPC UA	• Industrial equipment integration • Complex data modeling • Security-critical applications	• Platform independence • Rich data type support • Built-in security features	• OPC UA server configuration • Certificate management • Information model design

Figure 2 illustrates various communication protocols employed in IIoT implementations.

2.2.3.2 Edge Computing Deployment Factors

Edge computing deployment requires important consideration of hardware, software, and integration factors to ensure reliable industrial-grade operation. The deployment strategy must address environmental challenges, real-time processing requirements, and seamless integration with existing manufacturing systems. The details are given in Table 5.

2.3 Cloud Computing Integration

Cloud computing has revolutionized how manufacturers approach Industrial Internet of Things deployments. Modern factories generate massive amounts of data that traditional on-site systems simply can't handle effectively. The cloud offers the perfect solution—unlimited storage, powerful processing capabilities, and advanced analytics tools that turn raw sensor data into meaningful business insights. This shift isn't just about technology; it's about competitive advantage. Companies that successfully integrate cloud solutions with their IIoT systems can respond faster to market

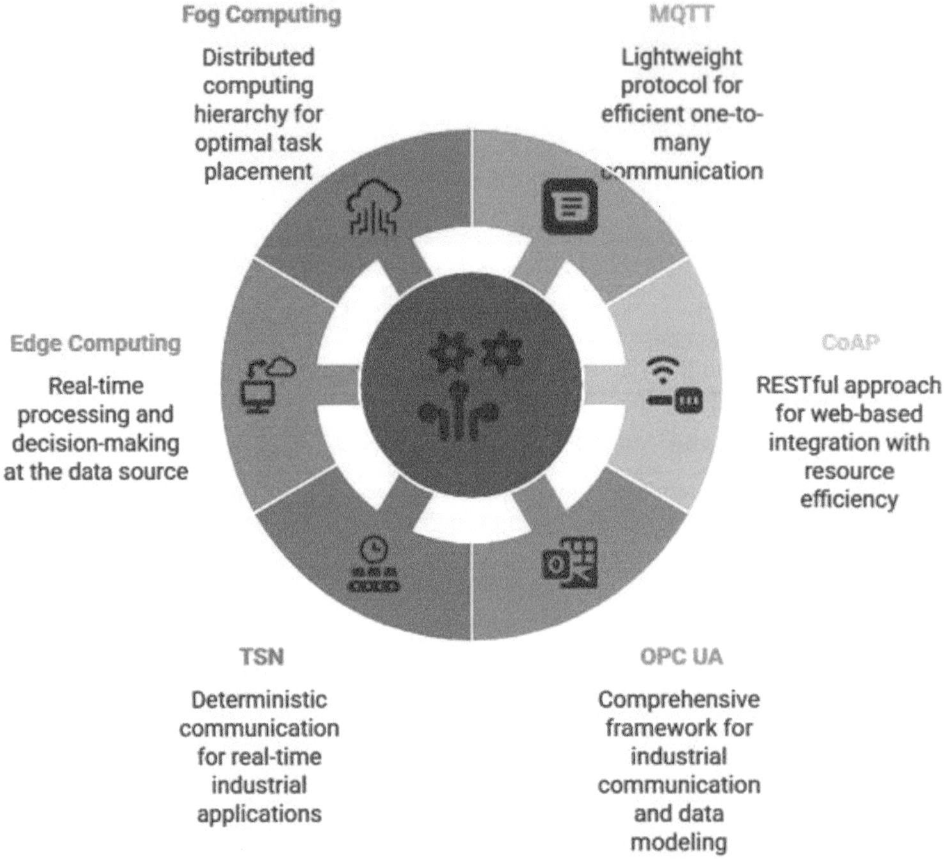

Figure 2: IIoT Communication Protocols.

Table 5: Edge Computing Deployment Strategy.

Deployment Factor	Requirements	Implementation Approach
Hardware Selection	• Industrial-grade ruggedization • Wide temperature range operation • Vibration and EMI resistance	• Certified industrial computing platforms • Environmental testing validation • Redundant system design
Software Architecture	• Real-time operating systems • Deterministic processing • Fault tolerance mechanisms	• RTOS implementation • Priority-based task scheduling • Automatic failover capabilities
Integration Strategy	• Legacy system compatibility • Protocol translation • Gradual migration support	• Gateway-based integration • Phased deployment approach • Comprehensive testing protocols

changes, reduce operational costs, and improve product quality significantly. However, the journey requires careful planning around data security, regulatory compliance, and system performance requirements. The details are given in Table 6.

Table 6: Cloud Integration Strategies for IIoT Systems.

Feature	Cloud-Based Solutions	Hybrid Cloud Architecture	On-Premises Only	Key Benefits
Scalability	✓ Unlimited expansion	✓ Flexible scaling options	✗ Hardware limitations	Grow resources as needed
Data Storage	✓ Massive capacity	✓ Distributed storage	✗ Physical constraints	Handle IoT data volumes
Real-time Processing	✗ Latency issues	✓ Edge + cloud combo	✓ Minimal latency	Critical for time-sensitive operations
Advanced Analytics	✓ AI/ML capabilities	✓ Best of both worlds	✗ Limited processing power	Predictive maintenance insights
Cost Efficiency	✓ Pay-as-you-use	✓ Optimized spending	✗ High upfront costs	Reduce capital expenditure
Data Security	~ Vendor dependent	✓ Enhanced control	✓ Full control	Maintain sensitive data protection
Maintenance	✓ Provider managed	✓ Shared responsibility	✗ Full internal burden	Focus on core business
Integration Speed	✓ Rapid deployment	✓ Phased implementation	✗ Complex setup	Faster time to market
Regulatory Compliance	~ Varies by provider	✓ Customizable approach	✓ Direct oversight	Meet industry requirements
Business Continuity	✓ Built-in redundancy	✓ Multiple backup layers	✗ Single point failure	Ensure operational resilience

2.4 Artificial Intelligence (AI) and Machine Learning (ML)

AI and ML technologies are transforming IIoT applications by enabling systems to learn from data, recognize patterns, and make autonomous decisions. Manufacturing applications of AI and ML span from predictive maintenance and quality control (Frankó et al. 2022) to process optimization and supply chain management (Sharp et al. 2018, Choudhry et al. 2022).

Predictive Analytics applications use historical and real-time data to predict future equipment behavior, quality outcomes, and process performance. ML algorithms analyze sensor data, maintenance records, and operating parameters to identify patterns that precede equipment failures or quality issues. These predictions enable proactive maintenance scheduling and process adjustments that minimize downtime and improve product quality (Milić et al. 2023).

Computer Vision applications leverage image processing and deep learning algorithms to automate quality inspection, part identification, and assembly verification tasks. Modern computer vision systems can detect defects smaller than human visual capabilities and operate continuously without fatigue. Integration with IIoT systems enables real-time quality feedback and process adjustments based on visual inspection results.

Autonomous Systems represent the convergence of AI, IIoT, and robotics technologies to create self-governing manufacturing systems. Autonomous mobile robots use IIoT connectivity to navigate manufacturing facilities, coordinate with other systems, and adapt to changing conditions.

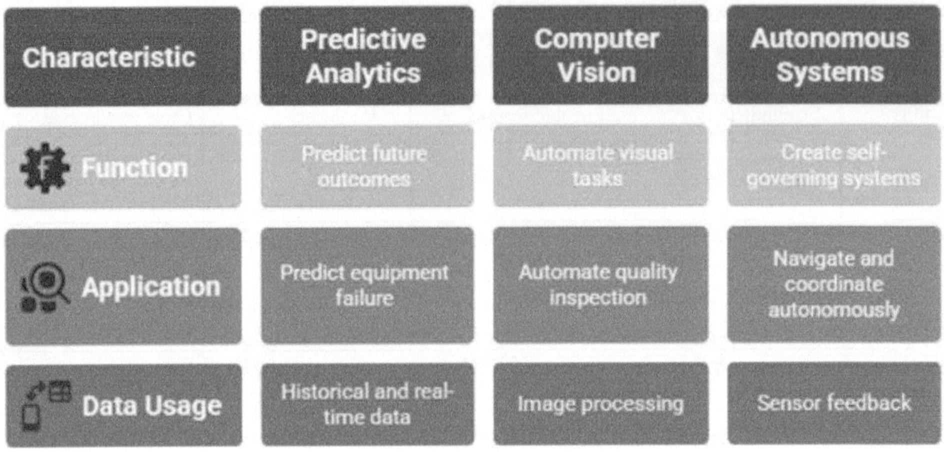

Figure 3: AI and ML in IIoT Application.

Autonomous process control systems adjust operating parameters in real time based on sensor feedback and learned optimization strategies.

Figure 3 illustrates the AI and ML applications in IIoT.

2.5 Cybersecurity: A Core Concern for IIoT in Aerospace

Cybersecurity represents one of the most critical aspects of IIoT implementation, as increased connectivity creates new attack vectors and vulnerabilities that can impact manufacturing operations, safety, and intellectual property (Sadeghi et al. 2015).

- **Growing Digital Risks:** As IIoT connects more of the factory floor and maintenance systems, the risks multiply: manufacturing, safety, and even valuable intellectual property could be at stake through new digital vulnerabilities.
- **Built-In Security:** Effective cybersecurity for IIoT isn't just a one-off project—it needs to be woven into every phase, from how devices are first designed and installed, right through to daily operations and regular maintenance.
- **Defense in Depth:** A smart way to protect IIoT setups is by stacking several security layers at once—a strategy called Defense in Depth. Instead of leaning on just one defensive measure, this approach combines things like network segmentation, strict access rules, strong data encryption, round-the-clock monitoring, and rapid-response incident plans. The big win here? Even if a hacker gets past one layer, the rest stay intact, so the whole system doesn't go down with a single breach (Mosteiro-Sanchez et al. 2020)called Industrial IoT (IIoT.
- **Who Gets In, Who Doesn't:** Managing who can do what is just as vital as keeping intruders out. That's where Identity and Access Management (IAM) (Vallois et al. 2021) comes in—IAM systems make sure only the right people and machines can access IIoT resources. These systems handle three things in particular: verifying identities, checking permissions, and keeping detailed logs of activity. In today's IIoT world, reliable IAM means supporting everything from passwords and certificates to tokens and even biometric scans, while giving administrators fine-tuned control over who can access each part of the system.
- **Guarding the Data:** Locking down sensitive information is a high priority. Encryption—scrambling data so it can't be read if intercepted—keeps data safe whether it's traveling between devices, sitting in storage, or being used actively. Effective encryption depends on robust key management, which handles secure key creation, distribution, and regular updates, supporting a wide range of encryption standards.

- **Stay Alert, React Fast:** You can't always stop a cyberattack, but you can catch it early and limit the damage. Security Information and Event Management (SIEM) platforms pull together security alerts and logs from all over the IIoT environment, acting as a sort of central security hub. These systems analyze data from every device and sensor, spotting potential problems in real time and giving teams the chance to jump on issues before they spiral out of control (Saheed et al. 2025).

Figure 4 illustrates basic to advance cyber security strategies.

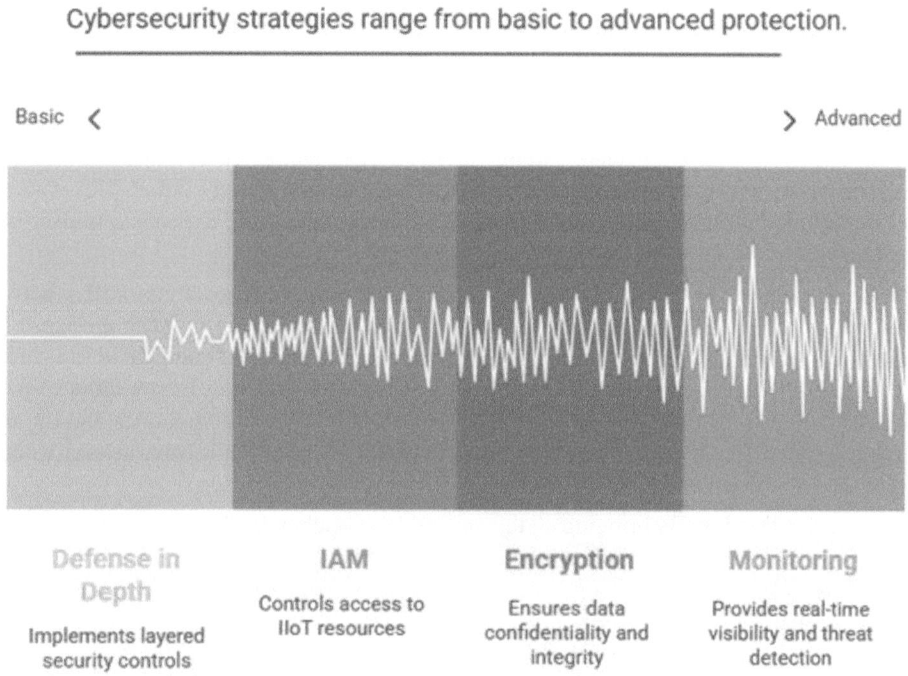

Figure 4: Cybersecurity Framework (Basic to Advanced Protection).

3. Applications of IIoT in Manufacturing

- **Game-changing Connections:** The Industrial Internet of Things (IIoT) has completely changed the game in manufacturing, bringing together machines, sensors, and software in networks that give companies a clearer view of their operations than ever before.
- **From Visibility to Action:** With IIoT, factories get real-time insights, granular control, and the ability to fine-tune every part of the production line. This isn't just about spotting problems in the moment—it's about continuously improving everything from supply chain to shop floor.
- **Real-world Impact:** In this section, we'll break down how IIoT is actually being put into action across various manufacturing settings. We'll look at exactly how companies are rolling out these technologies, the clear gains they've seen in efficiency and cost, the real-world challenges they've faced, and how they've gotten results.
- **Looking Back, Looking Ahead:** Some uses of IIoT are already tried-and-trusted, with proven payoffs; others are just coming into their own, promising even bigger returns as they mature.
- **No Silver Bullet:** We'll also touch on the bumps in the road—integration headaches, cybersecurity worries, and the scramble for new skills—that can slow down or complicate adoption, even as the technology continues to mature.

This section explores the diverse applications of IIoT technologies in manufacturing environments, examining their implementation approaches, benefits, challenges, and real-world outcomes.

3.1 Predictive Maintenance

Predictive maintenance is one of the most widely adopted applications of IIoT in manufacturing, transforming traditional reactive and preventive maintenance approaches into data-driven, condition-based strategies (Aboshosha et al. 2023, Elkateb et al. 2024, Nangia et al. 2020).

- **Beyond the Repair Shop:** Predictive maintenance has become one of the real success stories for IIoT in manufacturing. It has moved us far beyond just fixing broken machines or following a rigid maintenance schedule—in its place, we get smart, data-powered upkeep that addresses issues before they cause trouble.
- **Spotting Trouble Early:** The big idea in predictive maintenance is constantly keeping an eye on important equipment stats—things like vibration, temperature, and noise—so you can catch problems as they start to develop, not after something's already failed. This way, repairs and replacements can be planned for natural downtime, avoiding expensive mid-production breakdowns.
- **The Tools Behind the Tech:** Making this work means equipping machines with a network of sensors that act as extra senses for maintenance teams. Vibration sensors pick up signs of wear, imbalance, or bearing issues in moving parts. Temperature sensors sound the alarm if electrics overheat or if lubrication problems crop up. Current sensors track how hard motors are working and can flag unusual electrical patterns that spell trouble. Even microphones and oil analysis sensors get involved, listening for odd noises or checking whether lubricants are still doing their job.
- **Smarter, Not Harder:** By bringing all this sensor data together, predictive maintenance systems let teams move away from guesswork and time-based routines, zeroing in on equipment that truly needs attention—driving up uptime, trimming waste, and slashing costs along the way (Mourtzis et al. 2021).

The data collected from these sensors is transmitted through IIoT networks to analytics platforms that apply machine learning algorithms to identify patterns associated with equipment degradation. These algorithms learn from past equipment failures and normal performance data to create predictive models tailored for each individual machine. More sophisticated systems also integrate various data points like weather conditions, production schedules, and maintenance logs to make their predictions even more accurate (Schleichert and Bringmann 2017). Figure 5 illustrates the IoT-based Predictive Maintenance System.

Implementation Benefits: Organizations implementing predictive maintenance typically achieve significant improvements in equipment reliability and maintenance efficiency. Studies indicate that predictive maintenance can reduce unplanned downtime by 30–50% while decreasing maintenance costs by 20–25% (Mobley 2002). The ability to predict failures allows maintenance teams to prepare necessary parts and resources in advance, reducing repair time and improving first-time fix rates.

Beyond direct cost savings, predictive maintenance provides strategic benefits including improved production planning reliability, enhanced safety through early detection of dangerous conditions, and extended equipment lifespan through optimal maintenance timing. The data collected for predictive maintenance also provides insights into equipment performance trends and operational efficiency opportunities.

Challenges and Solutions: Implementing predictive maintenance faces several technical and organizational challenges. Data quality issues arise from sensor calibration drift, communication failures, and environmental interference. False positives can erode user confidence and increase maintenance costs, while false negatives can result in unexpected failures. Organizations address these challenges through comprehensive sensor validation programs, advanced data-cleansing algorithms, and continuous model refinement processes (Compare et al. 2020, Haque et al. 2024).

Organizational challenges include resistance from maintenance personnel who may view predictive systems as threats to their expertise and changes to established maintenance procedures. Successful implementations address these challenges through comprehensive training programs, change management initiatives, and demonstration of system value through pilot projects.

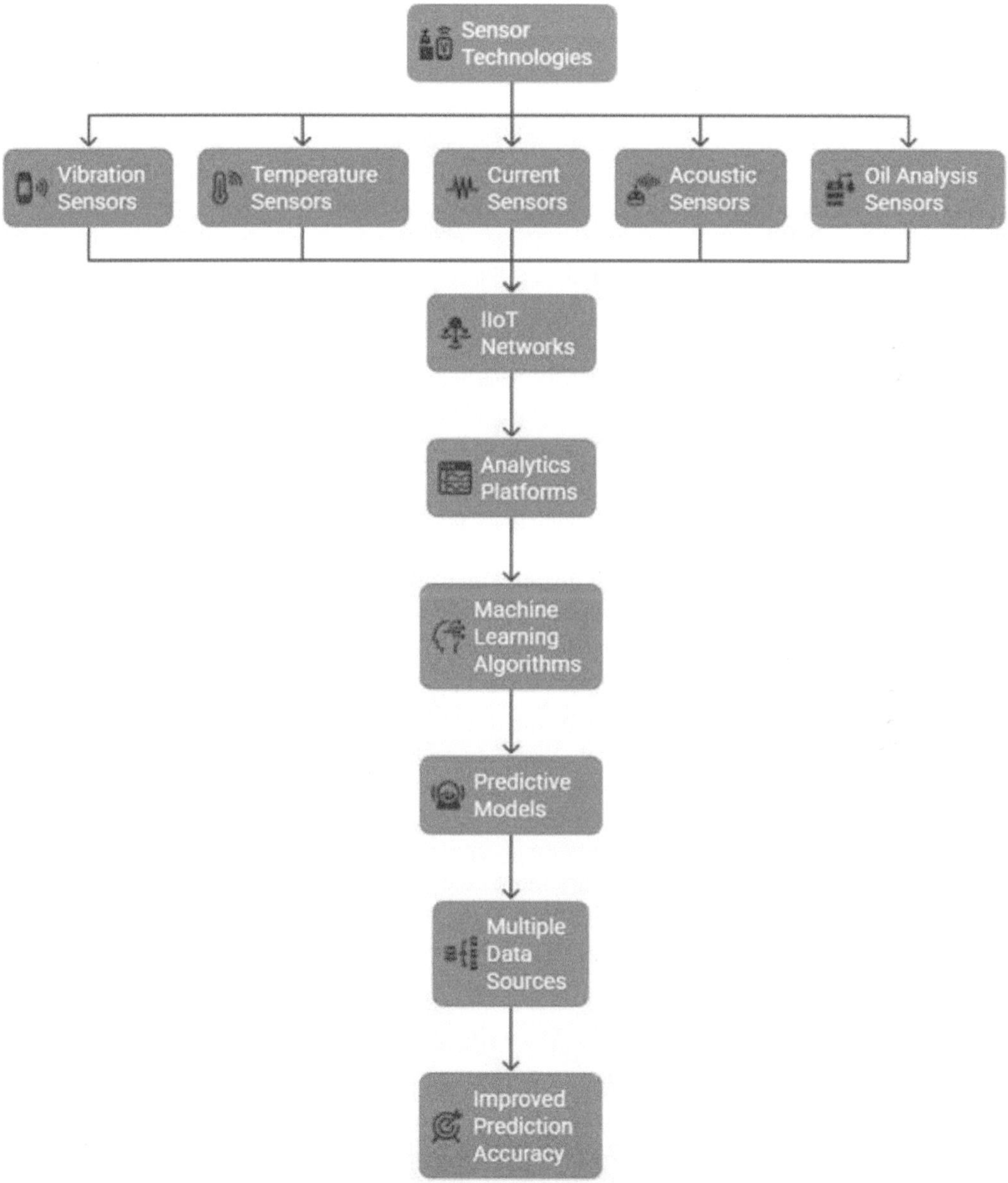

Figure 5: IoT-based Predictive Maintenance System.

3.2 Quality Control and Inspection

IIoT-enabled quality control systems represent a paradigm shift from traditional sampling-based quality approaches to comprehensive, real-time quality monitoring and control. These systems integrate sensors, vision systems, and analytics platforms to monitor product quality throughout the manufacturing process, enabling immediate corrective actions and continuous process improvement (Sen et al. 2023, Zhan et al. 2021).

Real-Time Quality Monitoring: Modern IIoT quality systems deploy sensors at critical points throughout the manufacturing process to monitor parameters that affect product quality. Dimensional sensors measure part geometries and tolerances in real time, providing immediate feedback on process variations. Force and torque sensors monitor assembly operations to ensure proper component installation and prevent damage. Environmental sensors track conditions such as temperature, humidity, and cleanliness that can impact product quality (Vermesan et al. 2022).

Vision systems integrated with IIoT networks provide automated visual inspection capabilities that can detect defects invisible to human inspectors or traditional measurement systems. Deep learning algorithms trained on historical defect data can identify subtle quality issues and classify defect types automatically (Saif et al. 2022). These systems operate continuously without fatigue and can inspect 100% of production rather than statistical samples (Hussain et al. 2020).

Statistical Process Control Integration: IIoT quality systems enhance traditional statistical process control (SPC) methods by providing real-time data collection and analysis capabilities. Control charts are automatically updated with each measurement, enabling immediate detection of process shifts and trends. Advanced analytics identify correlation patterns between process parameters and quality outcomes, supporting root cause analysis and process optimization initiatives (Goecks et al. 2024).

Automated SPC systems can trigger immediate process adjustments when control limits are approached, preventing the production of defective parts. Integration with manufacturing execution systems enables automatic notification of quality issues to appropriate personnel and initiation of corrective action procedures (Elkateb et al. 2024). Figure 6 illustrates the IIoT-based quality control systems.

Benefits and Outcomes: Organizations implementing IIoT quality systems typically achieve significant improvements in product quality metrics. First-pass yield rates improve by 15–30% as real-time monitoring enables immediate detection and correction of quality issues. Customer complaint rates decrease substantially as quality problems are identified and corrected before products reach customers. The comprehensive data collection enabled by IIoT quality systems provides unprecedented insights into quality trends and improvement opportunities. Manufacturers can identify the root causes of quality issues more quickly and implement targeted improvements that prevent recurrence. The integration of quality data with other manufacturing systems enables optimization of the entire production process for quality outcomes. Figure 7 illustrates the IIoT role in quality control.

3.3 Supply Chain Visibility and Management

IIoT technologies are revolutionizing supply chain management (SCM) by providing end-to-end visibility, real-time tracking, and intelligent coordination across complex global supply networks. These applications address critical challenges including inventory optimization, logistics efficiency, and supply chain resilience (Büyüközkan and Göçer 2018).

Asset Tracking and Visibility: IIoT-enabled asset tracking systems deploy sensors and communication devices on products, components, and packaging materials to provide real-time location and condition information throughout the supply chain (Taj et al. 2023). GPS tracking devices monitor the location of shipments in transit, while environmental sensors track temperature, humidity, and shock exposure that could affect product quality.

Characteristic	Real-Time Monitoring	Statistical Process Control
Data Acquisition	Continuous sensor data streams	Real-time data updates control charts
Defect Detection	Vision systems identify subtle defects	Identifies process shifts and trends
Process Adjustment	Immediate adjustments prevent defective parts	Automated adjustments when limits are approached
System Response	Immediate feedback on process variations	Notifies personnel, initiates corrective actions

Figure 6: IIoT based Quality control systems.

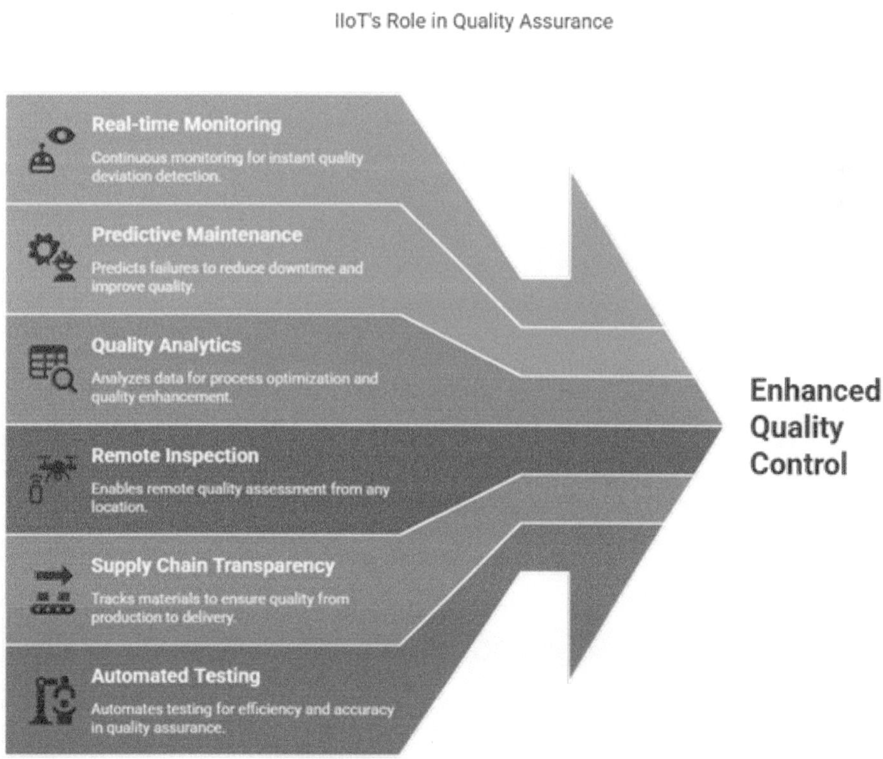

Figure 7: IIoT in Quality Assurance.

"Radio frequency identification (RFID)" and "near-field communication (NFC)" technologies enable automatic identification and data capture at key supply chain nodes including receiving, storage, and shipping operations. These technologies eliminate manual data entry errors and provide complete traceability of materials and products throughout the supply chain.

Inventory Optimization: IIoT sensors deployed in warehouses and storage facilities provide real-time inventory visibility that enables more accurate demand forecasting and inventory optimization (Jarašūnienė et al. 2023). Weight sensors on storage shelves automatically track inventory levels, while motion sensors monitor material movement patterns. This data feeds into inventory management systems that optimize reorder points, safety stock levels, and replenishment timing (Ugbebor et al. 2024).

Advanced analytics applied to IIoT supply chain data identify patterns and trends that improve demand forecasting accuracy (Sallam et al. 2023). Machine learning algorithms analyze historical consumption patterns, seasonal variations, and external factors such as weather and economic conditions to predict future demand more accurately. These improved forecasts enable organizations to reduce inventory carrying costs while maintaining service levels.

Supplier Performance Management: IIoT systems provide comprehensive visibility into supplier performance across multiple dimensions including delivery performance, quality metrics, and service levels. Real-time tracking of supplier deliveries enables accurate measurement of on-time delivery rates and identification of chronic performance issues. Quality sensors at receiving operations automatically assess incoming material quality and provide feedback to suppliers (Behrendt et al. 2021). The integration of supplier performance data with procurement and sourcing systems enables data-driven supplier selection and management decisions. Organizations can identify the best-performing suppliers for different categories and negotiate performance-based contracts that align supplier incentives with organizational objectives. Figure 8 illustrates the IIoT supply chain management.

Figure 8 illustrates the role of IIoT in supply chain management.

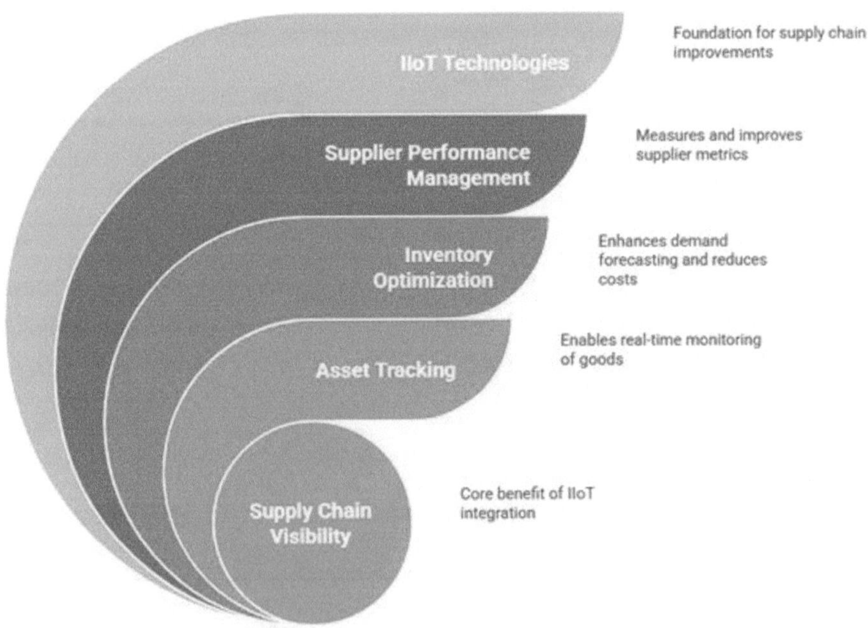

Figure 8: IIoT in Supply Chain Management.

3.4 Smart Factory and Production Optimization

Smart factory implementations represent the most comprehensive application of IIoT technologies, integrating sensors, analytics, and automation systems to create intelligent, adaptive manufacturing environments. These systems optimize production processes in real-time, coordinate activities across multiple production lines, and enable flexible manufacturing approaches that can respond rapidly to changing market demands (Zhong et al. 2017).

Production Line Integration: Smart factory systems integrate IIoT sensors and controllers across all production equipment to create synchronized, intelligent production lines (Soori et al. 2023c). Sensors monitor machine performance, product quality, and material flow, while controllers coordinate activities between different production stages. This integration enables optimal production scheduling, minimizes work-in-process inventory, and maximizes equipment utilization.

Advanced scheduling algorithms use real-time production data to optimize production sequences, minimize changeover times, and balance workload across available resources (Quijano et al. 2025). These systems can automatically adjust production schedules in response to equipment breakdowns, material shortages, or rush orders while maintaining overall production efficiency.

Energy Management and Sustainability: IIoT energy management systems monitor and optimize energy consumption across manufacturing facilities ("Industrial IoT-Based Energy Monitoring System: Using Data Processing at Edge" n.d.). Smart meters and sensors track energy usage at the equipment level, identifying opportunities for energy savings and cost reduction. Integration with utility demand response programs enables facilities to adjust energy consumption during peak pricing periods. Advanced analytics identify energy consumption patterns and correlate energy usage with production activities to optimize energy efficiency. Machine learning algorithms predict energy demand and automatically adjust HVAC, lighting, and equipment operation to minimize energy costs while maintaining production requirements.

Flexible Manufacturing Systems: IIoT enables the implementation of flexible manufacturing systems that can rapidly reconfigure to produce different products or adapt to changing production requirements (Javaid et al. 2022). Modular production equipment connected through IIoT networks can be quickly reprogrammed and reconfigured for different products. Mobile robots equipped with IIoT connectivity can be reassigned to different tasks based on production needs. Digital twin technologies create virtual representations of manufacturing systems that enable simulation and optimization of production scenarios before implementation. These digital models use real-time data from IIoT sensors to maintain accuracy and enable continuous optimization of manufacturing operations (Soori et al. 2023a).

Figure 9 illustrates the IIoT-based Smart Factory and Production Optimization.

3.5 Worker Safety and Environmental Monitoring

IIoT applications in worker safety and environmental monitoring address critical requirements for maintaining safe working conditions, ensuring regulatory compliance, and protecting both workers and the environment. These systems provide continuous monitoring capabilities that exceed traditional safety approaches and enable proactive intervention before hazardous conditions develop (Misra et al. 2022).

Personal Safety Monitoring: Wearable IIoT devices monitor worker vital signs, location, and environmental exposure to ensure worker safety in hazardous environments. Heart rate monitors detect signs of heat stress or physical overexertion, while GPS tracking devices ensure workers can be located quickly in emergency situations (Moon and Ju 2024). Gas detection sensors worn by workers provide early warning of exposure to hazardous chemicals or oxygen-deficient atmospheres. Integration with facility safety systems enables automatic emergency response

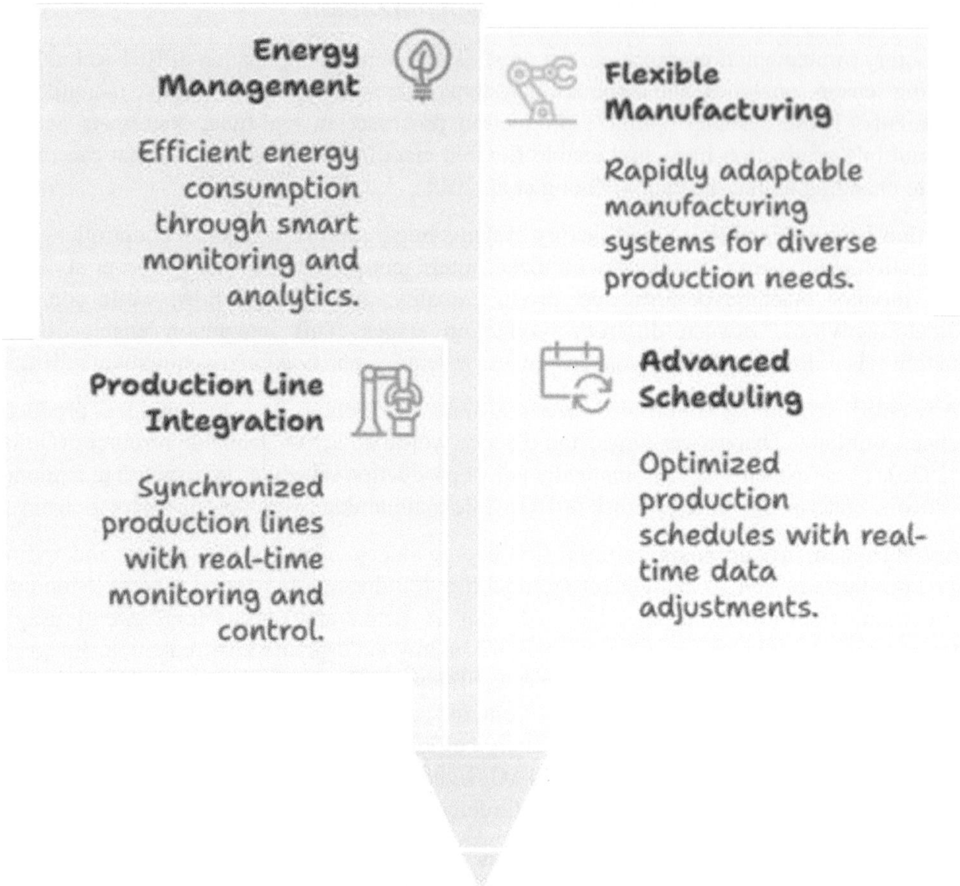

Figure 9: IIoT-based Smart Factory and Production Optimization.

when safety thresholds are exceeded. Emergency notification systems can automatically contact emergency responders and provide real-time location information for affected workers. Evacuation systems can guide workers to safety through optimal routes based on current facility conditions.

Environmental Compliance: IIoT environmental monitoring systems track emissions, waste generation, and resource consumption to ensure regulatory compliance and support sustainability initiatives. Air quality sensors monitor particulate levels, chemical concentrations, and other pollutants in real time. Water quality sensors track discharge parameters and identify potential contamination events (Lou et al. 2022). Integration with regulatory reporting systems automates compliance documentation and ensures timely submission of required reports. Predictive analytics identify trends that could lead to compliance violations and enable proactive corrective actions. The comprehensive data collection provided by IIoT systems also supports environmental management system certification and sustainability reporting requirements.

Facility Safety Management: Building-wide IIoT safety systems monitor conditions throughout manufacturing facilities to identify and prevent safety hazards. Fire detection systems use advanced sensors to identify potential fire conditions before flames are visible. Gas leak detection systems monitor for hazardous chemical releases and automatically initiate containment procedures. Integration with facility management systems enables coordinated response to safety incidents and optimization of safety system performance. Predictive analytics applied to safety system data identify patterns that could indicate developing hazards or equipment malfunctions that could compromise safety (Behrendt et al. 2021).

3.6 Why Companies Are Turning to IIoT—And What Gets in the Way

The implementation of IIoT technologies in manufacturing environments delivers substantial benefits but also presents substantial challenges that must be carefully managed to ensure positive outcomes. Let's break down what IIoT brings to the table, from daily operations to long-term strategy, and why some organizations still struggle to make the most of it.

3.6.1 Operational Gains from IIoT

- **Boosting Production:** By connecting machines and tracking performance in real time, companies often see a 10–20% jump in efficiency. Less downtime, smarter equipment use, and fine-tuning processes all add up (Manyika et al. 2015).
- **Higher Quality:** Catching glitches early—sometimes before anyone notices—keeps quality high. Real-time sensors and automated controls help nip problems in the bud.
- **Saving Money:** Lower maintenance bills, less energy used, less waste, better inventory control—these add up to big savings. In fact, some companies cut total costs by 15–25% after rolling out IIoT across their operations.
- **Empowering People:** Routine, boring tasks get automated, freeing up staff for more valuable work. Plus, workers get better information to make decisions, which makes everyone more productive.

3.6.2 Strategic Wins with IIoT

- **Staying in the Game:** It's not just about running better—it's about competing better. IIoT can help companies react faster to customer needs, offer more consistent products, and even develop new ways of doing business.
- **Smarter Choices:** With data pouring in from everywhere, managers can spot hidden inefficiencies and make decisions based on hard evidence, not just hunches.
- **Innovation Opportunities:** The flood of data from IIoT systems can spark new ideas, leading to fresh products and services.

3.6.3 Roadblocks Along the Way

- **Tech Hurdles:** Juggling new and old systems is tricky. Getting everything to talk to each other takes both elbow grease and a lot of know-how.
- **Security Concerns:** More connections mean more ways for hackers to sneak in. Protecting data and devices is a top priority—and a constant challenge.
- **Data Dilemmas:** If the data coming in is messy or unreliable, the whole system suffers. Cleaning and organizing it is half the battle.
- **Old Equipment Blues:** Fitting IIoT into existing, sometimes outdated, systems can be a headache, often requiring custom work.
- **People and Skills:** Not every company has a team of IIoT experts on staff. Training current employees or hiring outside help is often a must.
- **Resistance to Change:** New tools mean new routines, and not everyone is eager for a shake-up. Overcoming skepticism and getting buy-in takes patience and persistence.

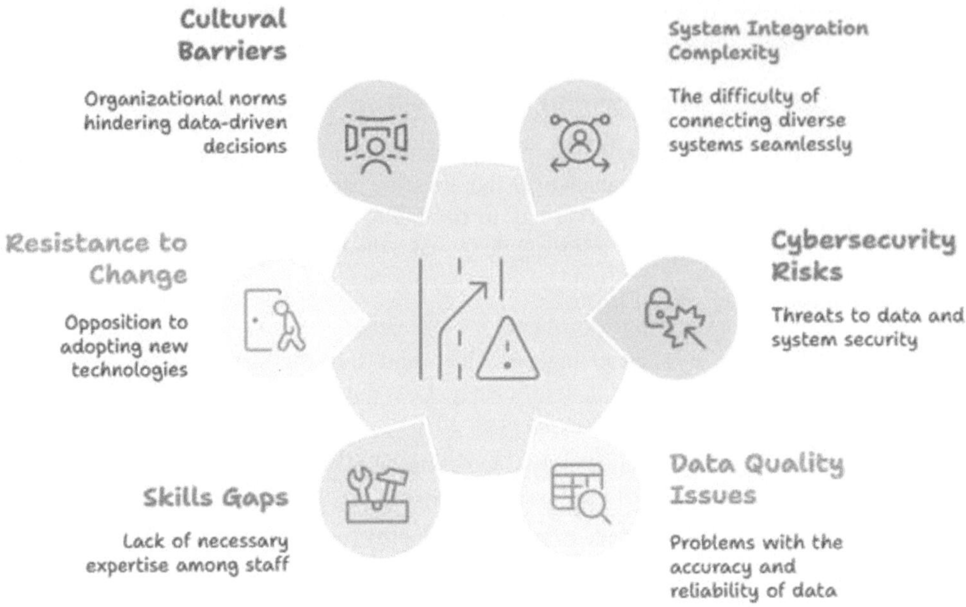

Figure 10: Challenges in IIoT Implementation.

4. IIoT Applications in Aircraft Manufacturing

The application of IIoT in aircraft manufacturing is transforming traditional factories into smart, connected environments, leading to significant advancements in production processes, quality assurance, and supply chain management.

4.1 Production Line Optimization

IIoT solutions enable manufacturers to run more intelligent operations by linking assets, providing real-time analytics, and continuous monitoring. This connectivity allows for greater visibility into production-line performance, helping plant managers identify inefficient processes, such as bottlenecks, and make immediate changes to improve overall equipment efficiency (OEE) (Javaid et al. 2021). Automation driven by IIoT reduces reliance on manual labor, minimizing human error and accelerating manufacturing processes. Automated systems, like those used for drilling and fastening, can precisely drill holes and place rivets. This consistent accuracy significantly reduces errors in assembly and lessens worker fatigue. As a result, assembly times for crucial aircraft components can be up to 50% faster.

4.2 Quality Control

IIoT plays a crucial role in enhancing product quality and consistency in aircraft manufacturing. Real-time data capture from IIoT-connected machines provides more accurate and consistent information than traditional methods, enabling manufacturers to identify defects and issues proactively (Zhang et al. 2024). Advanced analytics, powered by machine learning and AI, offer profound insights into industrial operations.

- **A Smart Step Forward:** IIoT is now a cornerstone for boosting quality and uniformity in aircraft manufacturing, moving way past the limits of old-school methods.
- **Data You Can Trust:** Unlike manual inspections, IIoT-connected machines deliver real-time, accurate data that's also consistent—no more guesswork or spot checks. This constant flow of

information lets teams spot potential defects almost as soon as they appear, often before they turn into major headaches.

- **From Analysis to Action:** Advanced analytics, supercharged by AI and machine learning, take things even further. These tools don't just report issues—they predict them, sometimes automating quality checks and flagging small problems that might slip by human inspectors.
- **Seeing the Unseen:** Machine vision cameras paired with AI can catch tiny flaws in real time, especially during critical steps like fuselage assembly (Association for Advancing Automation and 900 2025). Deep learning technology can spot issues invisible to the naked eye, cutting down on expensive rework and costly delays.
- **The Big Picture:** By combining live monitoring with smart analytics, aircraft makers end up with fewer defects, more reliable products, and a manufacturing process that runs like a well-oiled machine.

4.3 Supply Chain Management and Asset Tracking

- **No More Needle-in-a-Haystack Searches:** IIoT is changing the game for aerospace supply chains by giving everyone in the loop—suppliers, makers, and even customers—a real-time view of where components are, what shape they're in, and how they're moving.
- **Get an Instant Heads-Up:** If something goes wrong in transit—say, a part is at risk or gets damaged—the system can alert the right people immediately so they can step in fast.
- **Crystal-Clear Visibility:** Real-time tracking of parts and shipments helps companies stay on top of inventory, avoid bottlenecks, and keep production lines humming. This visibility is a lifesaver when facing tricky issues like fluctuating demand or unexpected shipping delays (Pop et al. 2023).
- **Smarter Inventory Decisions:** When machines share live production data, procurement teams get a clear picture of what's needed and when, making it easier to avoid overstocking or running out of critical parts. This leads to leaner inventory, less waste, and a supply chain that's both nimble and cost-effective.

4.4 Use Cases

4.4.1 Fujitsu's COLMINA Manufacturing Platform

Fujitsu enhances Mitsubishi Heavy Industries' manufacturing by

- **Collaborative Innovation:** Fujitsu partnered with Mitsubishi Heavy Industries' Commercial Aircraft Division to develop a tailored IoT solution aimed at boosting manufacturing efficiency (Bhatia et al. 2024).
- **COLMINA Platform in Action:** The solution runs on Fujitsu's COLMINA platform, which harnesses the cloud to optimize processes across the entire supply chain.
- **User-Centered Design:** The system features dashboards built with user experience in mind, making it easy for teams to check real-time production stats, progress, and alerts.
- **Faster Problem Resolution:** Maintenance crews can quickly spot issues on these dashboards and take immediate action, minimizing downtime and keeping operations flowing smoothly.

4.4.2 Airbus Digital Shadow

- **Real-Time Virtual Twin:** Airbus has introduced a "digital shadow" at its Saint Eloi plant—this is essentially a live, virtual model of the actual assembly line, mirroring real-world activities.
- **Data-Driven Insights:** The system continuously processes data from machinery and conveyors, generating simulations that help identify where and how assembly operations can be improved.

- **Smart Tool Integration:** IoT sensors have been fitted onto assembly tools to monitor how much torque is applied during critical fastening tasks.
- **Instant Quality Control:** If a tool detects torque outside the safe range, it automatically stops and sends an alert to technicians, preventing potential defects before they happen.

4.4.3 Airbus Mixed Model Line

- **Flexible Production:** Airbus now operates a "mixed model line" capable of building wings for both the A–330 and A–350 aircraft on a single production line.
- **Connected Systems:** This setup relies on advanced digital controls that use IoT technology to coordinate machinery, tools, materials, and work-in-progress parts in real time.
- **Seamless Coordination:** The interconnected system ensures all elements communicate smoothly, allowing for quick switches between different wing models without causing disruptions.

5. IIoT Solutions For MRO Providers

The aerospace Maintenance, Repair, and Overhaul (MRO) sector faces mounting pressures to keep aircraft operational, uphold the highest quality standards, and comply with strict regulations without compromise, and the Industrial Internet of Things (IIoT) is driving a digital transformation by replacing outdated, paper-heavy processes with smart, data-driven systems that streamline every aspect of maintenance; by shifting from fragmented spreadsheets to connected, real-time analytics, IIoT empowers maintenance teams to make faster, smarter decisions, ultimately reducing aircraft turnaround times and enhancing overall efficiency.

5.1 Real-Time Asset Monitoring

- **Constant Health Checks:** Thanks to IIoT sensors, airlines and MRO providers can now keep tabs on every critical part of an aircraft, with live data flowing in from thousands of sensors.
- **Data Overload:** A single flight can generate terabytes of sensor data—just imagine one engine alone being monitored at 5,000 different points every second!
- **Immediate Insights:** This flood of real-time information makes it easier to spot brewing issues early, fix them before they escalate, and keep aircraft in peak condition.
- **Track Everything, Everywhere:** IIoT-powered real-time location systems (RTLS) let teams instantly locate key components—like landing gear, engine turbine disks, or fan blades—inside sprawling MRO hangars, slashing the time lost to searching.
- **Avoid Costly Delays:** Quick detection of potential bottlenecks means faster fixes. That's key in aerospace, where even a two-day delay in engine maintenance can cost millions in lost revenue.

5.2 Condition-Based and Predictive Maintenance

- **Moving Beyond Calendars:** The industry is moving away from rigid, calendar-based maintenance (which can lead to unnecessary work and higher costs) toward smarter, data-driven models.
- **Predict, Don't React:** With IIoT, maintenance teams can crunch sensor data to forecast when parts are likely to fail, then schedule repairs or replacements just in time—before anything actually breaks down.

- **Cut Costs and Downtime:** This isn't just about keeping planes in the air—it's about saving money. Predictive maintenance can slash overall maintenance costs by 12–18% and unplanned downtime by 15–20%.
- **Planned, Not Panic:** Being able to plan for repairs instead of reacting to emergencies means both lower costs and more reliable aircraft, keeping operators and passengers happy.

5.3 Inventory and Logistics Optimization

- **Always Know Where Parts Are:** IIoT gives MRO teams a clear, real-time picture of every part and material's location, condition, and status—making sure the right piece is always on hand when it's needed.
- **No More Waiting Around:** Precise tracking helps reduce the time aircraft spend grounded, which is critical in a business where every minute on the ground hits the bottom line.
- **Smooth Logistics:** By tracking everything—from turbine disks to fan blades—IIoT systems ensure that missing parts never hold up a repair job.
- **Workflows that Work:** Streamlined digital inventory and logistics processes get rid of wasteful search times, speed up order fulfillment, and keep inventory costs down by making sure the right parts are always available at the right time.

5.4 Digital Workforce Productivity

IIoT contributes to digital workforce productivity by automating data collection and inspection tasks, thereby reducing the risk of human error. It provides real-time production information, such as consignment bills and production orders, and enables instant monitoring of production status, communication of requirements, and feedback to streamline workflows. This digitization of manufacturing processes helps reduce errors and accelerate operations. Furthermore, IIoT systems can collect and analyze data from multiple sources, allowing organizations to identify inefficiencies and take immediate steps to improve them, empowering employees with accurate, up-to-date information for data-driven decisions. This augmentation of human capabilities through intelligent systems is a key aspect of the evolving industrial revolution in aviation. Figure 11 illustrates the IIoT technologies in aircraft MRO.

Figure 11: IIoT Application in Aircraft MRO.

Table 7: Application Area in Aircraft Base Maintenance.

Application Area in Aircraft Base Maintenance	IIoT Features	Real-World Benefits
Hangar Operations	Sensors monitor temperature, humidity, and air quality; smart lighting and energy systems respond to usage.	Creates safer, more efficient workspaces; cuts energy waste and keeps conditions ideal for delicate tasks.
Equipment & Tool Tracking	RFID tags and IoT sensors on tools and gear provide live updates on location and calibration status.	No more hunting for missing tools; ensures only properly calibrated equipment is used, preventing delays.
Work Progress Tracking	Digital systems and sensors track task completion, technician hours, and flag delays automatically.	Gives managers clear visibility into job status, helps catch slowdowns early, and keeps projects on track.
Heavy Maintenance Checks (C/D)	Automated inspection systems (cameras, ultrasonic sensors) feed data to central monitors.	Catches hidden flaws like corrosion or wear that manual checks might miss, boosting safety and reliability.
Parts & Inventory Management	IIoT tracks high-value parts; smart shelves log removals/installations; sensors monitor storage conditions.	Cuts errors from wrong parts, ensures components are stored correctly, and keeps inventory accurate.
Quality Control & Compliance	Sensors and cameras automatically document procedures and verify inspections are complete.	Easier compliance with strict regulations; creates a clear, digital paper trail for audits and records.
Resource Allocation	IIoT systems track staff locations, skills, and workloads in real time.	Smarter assignment of teams, faster response to complex tasks, and better use of specialized expertise.
Workflow Optimization	Live sensor data highlights bottlenecks (e.g., inspection bays with consistent delays).	Identifies inefficiencies for process improvements, reducing holdups and speeding up overall maintenance.
Predictive Scheduling	Combines past maintenance data with live aircraft monitoring to plan future checks.	Helps airlines schedule downtime more effectively, minimizing disruptions to flight operations.
Structural Inspections	IIoT-enabled tools (ultrasonic gauges, boroscopes) automatically detect and record structural issues.	Comprehensive digital records, early detection of problems, and fewer chances for human oversight.
Engine & Component Overhauls	Sensors monitor engines and parts during disassembly, inspection, and reassembly.	Data guides repair/replace decisions, cutting costs and extending component life where possible.
Paint & Coating	Environmental sensors control and monitor conditions during painting and curing.	Ensures perfect finish quality by maintaining ideal temperature and humidity, with alerts if things go off track.

5.5 IIoT In Aircraft Base Maintenance At MRO

IIoT applications in airline base maintenance operations represent a significant evolution from traditional scheduled maintenance approaches. Base maintenance, which involves major inspections, overhauls, and modifications during extended aircraft downtime, benefits enormously from IIoT integration. Table 7 presents application areas of IIOT in aircraft base maintenance.

Figure 12 illustrates the **IIoT in Base Maintenance at MRO**.

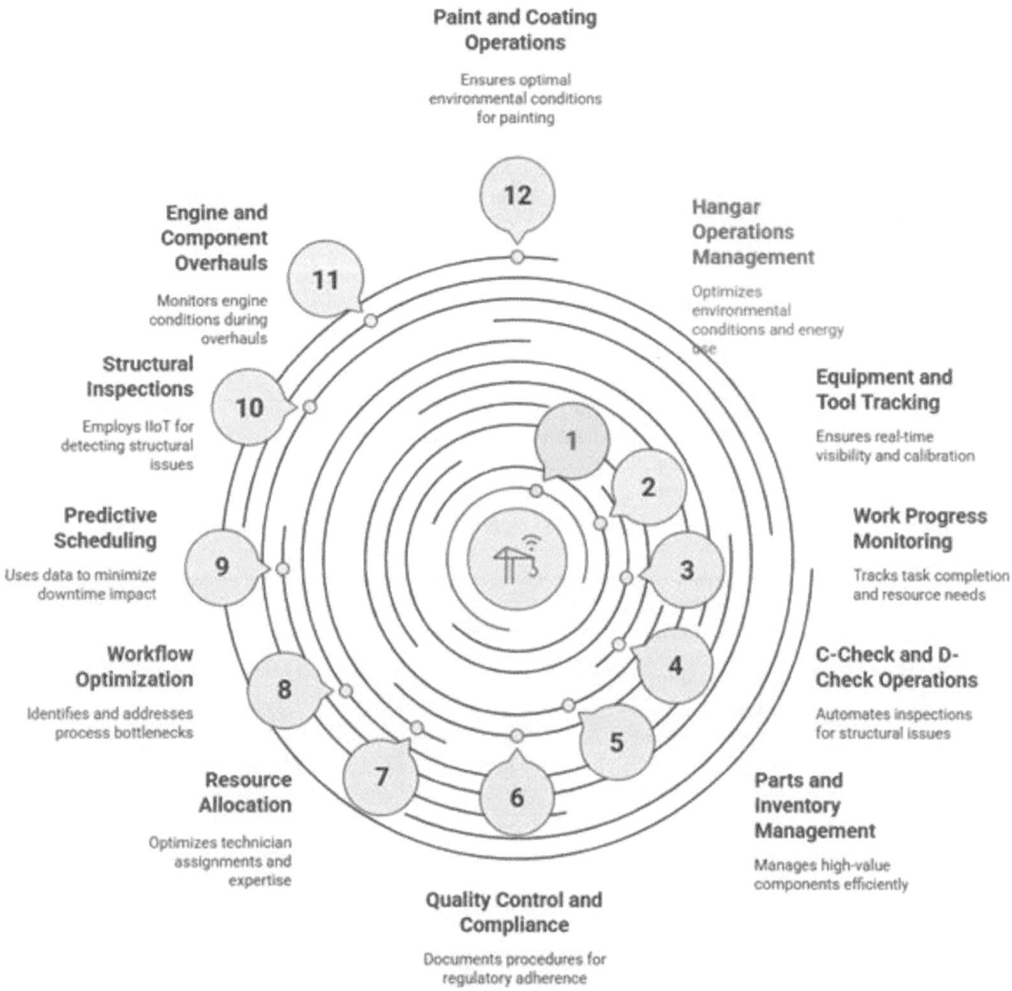

Figure 12: IIoT in Base Maintenance at MRO.

5.6 Technology IIoT in Aircraft MRO

The world of aerospace manufacturing and MRO is changing fast, thanks to the Industrial Internet of Things—better known as IIoT. Think of IIoT as a digital web that connects all sorts of machines, tools, and systems using an array of sensors, smart software, and reliable networks. By pulling together live data from across the shop floor and maintenance bay, companies can spot issues as they happen, automate complex processes, and make better decisions on the fly. The result? Work gets done faster, with fewer hiccups, less downtime, and safer conditions—plus, the bottom line looks a lot healthier. A tabular overview of the impact of IIoT in aerospace manufacturing and MRO, along with technology specific to aircraft MRO, is shown in Table 8:

Table 8: IIoT Technologies for Aerospace .

Technology Category	Specific IIoT Technologies & Applications in Aircraft MRO	Description & Benefits
Sensors & Data Acquisition	- Vibration Sensors: Monitor engine and component vibrations to detect anomalies indicative of wear or impending failure. - Temperature & Pressure Sensors: Track critical operating parameters of engines, hydraulics, and other systems. - Acoustic Sensors: Detect unusual sounds that could indicate mechanical issues. - Accelerometers: Measure acceleration and orientation for structural health monitoring. - Fuel Flow Sensors: Monitor fuel consumption and identify inefficiencies. - RFID Tags/Beacons: Track tools, parts, and baggage within MRO facilities and across the supply chain.	These sensors are embedded in aircraft components and MRO equipment to collect real-time data on their condition and performance. This data forms the foundation for predictive maintenance and operational insights, allowing for early detection of issues and precise monitoring.
Connectivity	- Wireless Communication (Wi-Fi, Bluetooth, Zigbee, 5G): Transmit collected sensor data from aircraft to ground systems and within MRO facilities. - Satellite Communication: For in-flight data transmission from aircraft to ground control (e.g., via SWIM - System Wide Information Management). - Edge Computing: Process data closer to the source (on the aircraft or at the MRO facility) to reduce latency and bandwidth requirements.	Enables seamless and reliable data flow from diverse sources to centralized analysis platforms. Different connectivity options are chosen based on data volume, range, and power consumption requirements.
Data Analytics & AI/ ML	- Predictive Analytics Algorithms: Analyze historical and real-time sensor data to forecast component failures and remaining useful life. - Machine Learning (ML): Develop models that learn normal operating parameters and identify deviations, improving prediction accuracy over time. - Artificial Intelligence (AI): Power advanced decision-making, such as optimizing maintenance schedules or prescribing specific repair actions. - Big Data Platforms: Store and manage the massive amounts of data generated by connected aircraft.	Transform raw sensor data into actionable insights. AI and ML algorithms identify patterns, predict outcomes, and automate decision-making for more efficient and proactive maintenance.
Digital Twins	- Aircraft Digital Twin: A virtual replica of a physical aircraft or its components, continuously updated with real-time data from sensors. - MRO Process Digital Twin: Simulate maintenance procedures and workflows in a virtual environment to optimize efficiency and identify bottlenecks.	Allow for real-time monitoring, simulation of scenarios, and predictive analysis without impacting the physical asset. Digital twins help in troubleshooting, planning maintenance actions, and training technicians.

Table 8 contd. ...

...Table 8 contd.

Technology Category	Specific IIoT Technologies & Applications in Aircraft MRO	Description & Benefits
Augmented Reality (AR) & Virtual Reality (VR)	- AR for Maintenance: Technicians use AR overlays on real components to access digital manuals, guided repair procedures, and real-time data. - VR for Training: Simulate complex maintenance tasks in a virtual environment for technician training, reducing risks and costs.	Enhance human capabilities in MRO. AR provides interactive visual aids for complex repairs, while VR offers immersive and safe training environments.
Robotics & Automation	- Automated Guided Vehicles (AGVs): Transport parts and tools within MRO hangars. - Inspection Drones: Conduct visual inspections of aircraft exteriors, especially in hard-to-reach areas, improving safety and speed. - Robotic Arms: Perform repetitive or hazardous tasks like painting, sanding, or specific component inspections.	Automate routine, dangerous, or physically demanding tasks, improving speed, accuracy, and safety in MRO operations.
Cloud Computing	- Data Storage & Processing: Securely store and process vast amounts of aircraft and MRO data. - Scalable Infrastructure: Provide the computing power needed for complex analytics and AI models. - Remote Access: Enable remote monitoring and support for aircraft and MRO operations.	Provides the flexible and scalable infrastructure required for handling large datasets and running sophisticated analytical tools, enabling collaboration and remote operations.

6. Case Studies: Leading Aerospace Manufacturers

Leading aerospace manufacturers are embracing the Industrial Internet of Things (IIoT), using these technologies to boost efficiency, improve safety, and streamline their intricate operations.

6.1 Airbus

Airbus is a leader in the adoption of IIoT within the aerospace industry, deeply integrating Industry 4.0 principles across its entire product lifecycle.

6.1.1 Manufacturing Initiatives

- **Factory of the Future:** Airbus has implemented a digital manufacturing initiative that integrates IIoT sensors on tools and machines and equips workers with industrial smart glasses. This setup reduces errors, increases safety, and dramatically boosts productivity—one cabin-seat marking process saw a 500% improvement (Buntz n.d.).
- **Wearable Technology:** Workers use smart glasses for real-time guidance and quality checks, nearly eliminating errors in complex assembly steps (Buntz n.d.).
- **Shop Floor Integration:** IIoT sensors provide real-time data on production status, enabling dynamic adjustments and improved workflow across assembly lines (Buntz n.d.).

- **Connected Experience:** Airbus has launched initiatives like its "Connected Experience" solution, which utilizes IIoT to pull data from machines and conveyors on the assembly line. This data is analyzed and aggregated to create a "digital shadow," providing a real-time visualization of the entire assembly line and highlighting surface inefficiencies ("Connectivity Innovations | Airbus" 2024).

6.2 Boeing

As a global leader in aircraft manufacturing, Boeing has widely embraced automation and AI. They've done this to handle the growing complexity of aircraft design, cut down on manufacturing errors, and streamline their production.

6.2.1 Manufacturing Initiatives

Boeing has transitioned towards smart factories where AI and automation collaborate with human operators to create highly efficient and precise manufacturing environments.

- **Enterprise Sensor Integration (ESI):** Boeing uses its ESI platform to create an integrated IIoT ecosystem, connecting RFID, GPS, BLE, Wi-Fi, and other sensors for real-time tracking and management of tools, parts, and assets across its factories. This enables end-to-end visibility, automates tool management, optimizes inventory, and reduces costs (RFID JOURNAL n.d.).
- **Smart Factories:** Boeing's Sheffield facility uses IIoT for live machine performance data, just-in-time tool ordering, and reduced operator complexity. Hundreds of automated machines provide real-time data, informing smarter decisions and streamlining maintenance (Boeing n.d., Patel et al. n.d.).
- **Wearables and AR:** Technicians use wearable devices (e.g., smart glasses) to access instructions and drawings hands-free, reducing errors and walking time. Integrated sensors on clothing and throughout the factory track technician location and workflow, further improving efficiency.
- **AI and Automation:** Boeing combines IIoT sensor data with AI for predictive maintenance, real-time monitoring, and digital twins, enhancing quality control and reducing downtime, especially in major programs like the 737, 777X, and 787 Dreamliner (Orca Lean n.d.).

6.3 Dassault Aviation

Dassault Aviation, through Dassault Systèmes, is a significant proponent of digital transformation in aerospace, particularly through its 3DEXPERIENCE platform and virtual twin technology.

6.3.1 Manufacturing Initiatives

Dassault Aviation's IoT deployment in manufacturing include:

- **Production System Monitoring**: IoT sensors and connectivity are used to continuously monitor production lines, capturing real-time data on equipment status, process parameters, and product quality. This data is centralized and analyzed to improve decision-making and operational oversight (Research and Markets n.d.).
- **Predictive Maintenance:** By collecting and analyzing machine and process data, Dassault Aviation implements predictive maintenance strategies. This reduces downtime by enabling proactive interventions before failures occur, increasing both reliability and efficiency.
- **Digital Twin Technology:** Dassault Aviation employs digital twins—virtual replicas of aircraft, manufacturing environments, and processes. IoT data feeds these digital twins, allowing for simulation, optimization, and validation of manufacturing steps before physical implementation

(Dassult 2021). This ensures "right first time" manufacturability and reduces costly errors or rework.

- **Operator Augmentation:** IoT-enabled devices and augmented reality solutions provide operators with real-time information directly on the assembly line. This enables hands-free access to instructions, quality checks, and training, helping to minimize errors and improve productivity.
- **Process Automation:** IoT is integral to automating production lines, enabling smart, interconnected machinery that can adapt to changing requirements and optimize workflows without manual intervention.

6.4 Bombardier

Bombardier, a top global producer of business jets, is dedicated to innovating new products and mobility solutions. Their focus remains on eco-friendly design, enhanced safety, and cutting-edge technological advancements. While direct IIoT project specifics are less detailed in the provided materials compared to other OEMs, Bombardier's strategic initiatives demonstrate a clear embrace of the underlying principles and technologies.

6.4.1 Manufacturing Initiatives

- **Digital Twin and Digital Thread:** Bombardier is expanding its use of Siemens' Xcelerator portfolio, including NX™ for 3D modeling, simulation, and manufacturing, and Teamcenter® for product lifecycle management. This enables the creation of a comprehensive digital twin of aircraft and manufacturing processes, supported by a digital thread that connects data and workflows across engineering, production, and maintenance (SIEMENS n.d.).
- **Real-Time Process Visibility:** Solutions like Apriso provide real-time visibility, control, and synchronization of processes and materials on the shop floor, a hallmark of IoT-enabled manufacturing environments (Aerospace Manufacturing and Design n.d.).
- **Big Data and Connected Aircraft:** Collaborations with GE Aviation have led to the Smart Link Plus program, which equips aircraft with connectivity solutions to collect and analyze operational data. This system enables predictive maintenance, fleet-wide performance monitoring, and faster return-to-service—functions that rely on IoT-style data collection and analytics (GE Aerospace n.d.).
- **Low-Code Applications for Integration:** By adopting Siemens' Mendix™ low-code platform, Bombardier is enabling the development of custom apps that connect disparate data sources, further enhancing data accessibility and process integration across manufacturing and engineering (SIEMENS n.d.).

6.5 Leading MRO Providers Implementing IIoT Solutions

By integrating IIoT solutions into their maintenance workflows, major MRO organizations are moving beyond traditional reactive maintenance approaches toward proactive, data-driven strategies. This technological adoption enables real-time monitoring of equipment health, automated diagnostics, and predictive failure analysis, ultimately resulting in reduced operational costs and improved asset reliability. Table 9 combines data of major MROs to demonstrate how these MRO companies are implementing IIoT technologies.

Table 9: Key MRO Utilizing IoT.

MRO Name	Key IoT Technologies & Applications	Primary Advantages	References
Lufthansa Technik	AVIATAR Platform: Predictive Health Analytics, Condition Monitoring (real-time aircraft/system health), APU & Cabin Temperature Monitoring, Digital Technical Logbook.	Enhanced aircraft availability, reduced flight delays/ cancellations, lower costs, improved safety, optimized maintenance schedules, extended asset lifespans.	(Frank Martens n.d.)
ST Engineering	Smart MRO Strategy: Robotics (drones for visual inspections), Data Analytics for Predictive Maintenance (engine health, MRO scheduling), Automated Intralogistics (material transport).	Optimized resource use, reduced wastage, lower operating costs, reduced carbon footprint, real-time activity monitoring, improved turnaround times.	("Smart MRO | ST Engineering" n.d.)
Delta TechOps	Trax eMRO & eMobility Solutions: Digitizing maintenance processes, streamlining workflows, improving data integrity, cloud-based platform for scalability.	Enhanced efficiency and accuracy, improved data integrity, elevated operational performance, boosted technician efficiency, seamless scalability.	("Delta TechOps Adopts Trax eMRO System" n.d.)
Turkish Technic	Hybrid Auto-ID IoT Solution: Passive UHF RFID (tools, equipment, shelving), Active ZigBee sensors with GPS (real-time personnel/vehicle tracking), Maintag's Flytag (aircraft components), ACARS integration.	Boosted personnel and tool usage efficiency, reduced unused inventory, 20% less technician search time, real-time resource management and scheduling.	(Swedberg 2015)

7. Challenges and Limitations of IIoT Adoption in Aerospace

Several challenges must be addressed to fully realize the transformative potential of these technologies. These challenges are summarized in Table 10.

8. Conclusion

This chapter concludes with the following points:

8.1 IIoT as a Transformative Force in Aerospace

- **Beyond Incremental Change:** The Industrial Internet of Things (IIoT) represents not simply a new set of technologies but a fundamental shift in how the aerospace industry approaches manufacturing and maintenance. This paradigm is redefining efficiency, safety, and productivity across both production and MRO operations.
- **A Digital Backbone for Industry 4.0:** By establishing a robust, interconnected digital network that links machines, processes, and personnel, IIoT enables the continuous collection and real-time analysis of operational data—cornerstones of the Industry 4.0 revolution.

8.2 Operational and Strategic Benefits

- **From Reactive to Proactive:** IIoT is empowering aerospace companies to move beyond reactive practices toward proactive, predictive models. For example, advanced predictive maintenance leverages real-time data to anticipate equipment issues, thereby minimizing unplanned downtime and reducing operational expenses.

Table 10: Challenges of IIoT Adoption in Aerospace.

Challenge Category	Key Issues	Impact Areas	Specific Concerns
Data Security and Privacy Concerns	• Creation of new security risks through increased IIoT device deployment • Exposure of sensitive operational data to unauthorized access • Increasing cybercriminal targeting of IIoT systems	• Operational data vulnerability • Network security breaches • Critical aerospace operations	• Device hijacking and attacker control of IIoT endpoints • Manipulation of automated functions • Production disruptions • Theft of confidential information • Privacy concerns for sensitive operational data • Design data protection issues
Integration Complexities with Legacy Systems	• Compatibility issues between older equipment and IIoT systems • Lack of modern interfaces in existing infrastructure • Costly and complex retrofitting requirements	• Equipment modernization • Data connectivity • System integration	• Data silos from isolated legacy systems • Strategic planning requirements • Investment challenges • Need for edge computing solutions • Protocol translation requirements
Lack of Standardization and Interoperability	• Fragmented IIoT landscape with proprietary solutions • Competing standards creating connectivity challenges • Difficulty connecting devices from different vendors	• System flexibility • Scalability limitations • Vendor relationships	• Limited uniformity across platforms • Vendor lock-in situations • Ecosystem dependencies • Need for open standards • Industry collaboration requirements
High Initial Investment and ROI Uncertainty	• Substantial initial capital expenditure requirements • Complex cost structure including multiple components • Difficulty demonstrating immediate ROI	• Financial barriers • Investment decisions • Long-term planning	• Smart sensor costs • Connectivity infrastructure expenses • Data storage and analytics platform costs • Software and hardware upgrade requirements • Barriers for smaller MROs • Long-term benefit realization challenges • Complex supply chain considerations
Skill Gap and Workforce Adaptation	• Shortage of qualified professionals • Rapid technology evolution outpacing skills development • Automation-driven workforce changes	• Human resources • Training and development • Job market dynamics	• Data analytics expertise shortage • AI and cybersecurity skill gaps • Network management competency needs • Potential job displacement • Retraining and upskilling requirements • Investment in development programs
Regulatory Compliance and Certification	• Stringent regulatory frameworks and certification processes • Complex and evolving regulations • Safety and reliability requirements	• Compliance management • Certification processes • Regulatory adherence	• IIoT device compliance requirements • Data flow regulations • Automated process standards • Safety and security standard adherence • Dynamic regulatory changes • Enforcement priority variations • Continuous assessment needs • Audit trail requirements

- **Enhanced Decision-Making:** With comprehensive data analytics, organizations can make more informed decisions, allocate resources more effectively, and achieve higher overall equipment effectiveness.
- **Safety and Sustainability:** IIoT supports a safer workplace by providing instant monitoring and alerts for both equipment and personnel. Simultaneously, it promotes environmental sustainability by optimizing energy consumption and reducing material waste.

8.3 Industry Leadership and Innovation

- **Global Adoption by Major Players:** Leading manufacturers—including Airbus, Boeing, Dassault Aviation, and Bombardier—are at the forefront of IIoT integration, often combining it with artificial intelligence, machine learning, and digital twin technologies.
- **Airbus** employs its "Connected Experience" platform to visualize assembly lines and deploy robotics for scalable production.
- **Boeing** utilizes AI-driven robotics, machine vision, and digital twins to refine manufacturing and predict equipment failures before they occur.
- **Dassault Aviation's** 3DEXPERIENCE platform leverages virtual twins to unify design, production, and maintenance, with the goal of maximizing fleet availability and maintenance efficiency.
- **Bombardier** has invested in AI and big data partnerships to boost operational performance, and established an Innovation and Design Centre to accelerate the development of next-generation aircraft.
- **Complementary Technologies:** The convergence of IIoT with AI, machine learning, digital twins, and emerging 5G networks is creating new possibilities for intelligent, autonomous, and resilient aerospace operations.

8.4 Challenges and the Path Forward

- **Navigating Obstacles:** The adoption of IIoT in aerospace is not without challenges: data security concerns, integration with existing legacy systems, a lack of universal standards, and significant upfront investment requirements are notable hurdles.
- **Collaborative Solutions:** Industry leaders are addressing these issues through ongoing technological innovation and strategic partnerships, ensuring that IIoT deployments are both secure and scalable.
- **Future Outlook:** As AI, machine learning, and 5G continue to evolve, the potential for even smarter, more autonomous aerospace systems grows. For companies in this sector, continued investment in these interconnected technologies is not optional—it is essential for maintaining competitiveness, ensuring safety, and achieving sustainable growth in global aviation.

References

Aboshosha, Ashraf, Ayman Haggag, Neseem George, and Hisham A. Hamad. (2023). IoT-Based Data-Driven Predictive Maintenance Relying on Fuzzy System and Artificial Neural Networks." *Scientific Reports* 13 (1): 12186. https://doi.org/10.1038/s41598-023-38887-z.

Aerospace Manufacturing and Design. n.d. Bombardier Aerospace Selects Apriso. Aerospace Manufacturing and Design. Accessed July 13, 2025. https://www.aerospacemanufacturinganddesign.com/news/aerospace-manufacturing-design-bombardier-apriso-020211/.

Andreas Behrendt, Enno de Boer, Tarek Kasah, Bodo Koerber, Niko Mohr, and Gérard Richter. (2021). Leveraging Industrial IoT and Advanced Technologies for Digital Transformation. McKinsey & Company. https://www.mckinsey.com/~/media/mckinsey/business%20functions/mckinsey%20digital/our%20insights/a%20manufacturers%20guide%20 to%20generating%20value%20at%20scale%20with%20iiot/leveraging-industrial-iot-and-advanced-technologies-for-digital-transformation.pdf.

Association for Advancing Automation and 900. (2025). Advancing Quality Control with AI-Powered Machine Vision. Automate. 2025. https://www.automate.org/blogs/advancing-quality-control-with-ai-powered-machine-vision.

Bhatia, Vineet, Ajay Kumar, Sumati Sidharth, Sanjeev Kumar Khare, Surendra Chandrakant Ghorpade. et al. (2024). Industry 4.0 in Aircraft Manufacturing: Innovative Use Cases and Patent Landscape. In *Industry 4.0 Driven Manufacturing Technologies*, edited by Ajay Kumar, Parveen Kumar, and Yang Liu, 103–37. Cham: Springer Nature Switzerland. https://doi.org/10.1007/978-3-031-68271-1_5.

Boeing. n.d. Let's Connect: Digital Thread Advances Manufacturing. Accessed July 13: 2025. https://www.boeing.com/content/theboeingcompany/us/en/innovation/innovation-quarterly/2022/10/let-s-connect-digital-thread-advances-manufacturing.

Brian Buntz. n.d. The Top 20 Industrial IoT Applications. Accessed July 13: 2025. https://www.iotworldtoday.com/iiot/the-top-20-industrial-iot-applications.

Büyüközkan, Gülçin, and Fethullah Göçer. (2018). Digital Supply Chain: Literature Review and a Proposed Framework for Future Research. *Computers in Industry* 97 (May): 157–77. https://doi.org/10.1016/j.compind.2018.02.010.

Carvalho, Thyago P., Fabrízzio A. A. M. N., Soares, Roberto Vita, Roberto da P. et al. (2019). A Systematic Literature Review of Machine Learning Methods Applied to Predictive Maintenance. *Computers & Industrial Engineering* 137 (November): 106024. https://doi.org/10.1016/j.cie.2019.106024.

Cheng, Jiangfeng, Weihai Chen, Fei Tao, and Chun-Liang Lin. (2018). Industrial IoT in 5G Environment towards Smart Manufacturing. *Journal of Industrial Information Integration* 10 (June): 10–19. https://doi.org/10.1016/j.jii.2018.04.001.

Choudhry, Mani Deepak, Jeevanandham S, Biji Rose, and Sruthi Mol P. (2022). Machine Learning Frameworks for Industrial Internet of Things (IIoT): A Comprehensive Analysis. In *2022 First International Conference on Electrical, Electronics, Information and Communication Technologies (ICEEICT)*, 1–6. https://doi.org/10.1109/ICEEICT53079.2022.9768630.

Claire Swedberg. (2015). Turkish Technic Uses RFID, ZigBee and Other Auto-ID Technologies to Improve Efficiency. *RFID JOURNAL*, March. https://www.rfidjournal.com/news/turkish-technic-uses-rfid-zigbee-and-other-auto-id-technologies-to-improve-efficiency-2/73345/.

Compare, Michele, Piero Baraldi, and Enrico Zio. (2020). Challenges to IoT-Enabled Predictive Maintenance for Industry 4.0. *IEEE Internet of Things Journal* 7 (5): 4585–97. https://doi.org/10.1109/JIOT.2019.2957029.

Connectivity Innovations | Airbus. (2024). September 9, 2024. https://www.airbus.com/en/innovation/digital-transformation/connectivity-innovations.

Dassult. (2021). Virtual Twin Experiences. Dassault Systèmes. November 16: 2021. https://www.3ds.com/virtual-twin.

Delta TechOps Adopts Trax eMRO System. n.d. Https://Www.Eplaneai.Com. Accessed July 13: 2025. https://www.eplaneai.com/news/delta-techops-adopts-trax-emro-system.

Depti Patel, Robert J. Rencher, and Luis A. Vinuelas. n.d. SYSTEM AND METHOD FOR USING AN INTERNET OF THINGS NETWORK FOR MANAGING FACTORY PRODUCTION. US 20160202692, filed 2015.

Elkateb, Sherien, Ahmed Métwalli, Abdelrahman Shendy, and Ahmed E. B. Abu-Elanien. (2024). Machine Learning and IoT – Based Predictive Maintenance Approach for Industrial Applications. *Alexandria Engineering Journal* 88 (February):298–309. https://doi.org/10.1016/j.aej.2023.12.065.

Frank Martens. n.d. AVIATAR. Accessed July 13: 2025. https://www.lufthansa-technik.com/en/aviatar.

Frankó, Attila, Gergely Hollósi, Dániel Ficzere, and Pal Varga. (2022). Applied Machine Learning for IIoT and Smart Production—Methods to Improve Production Quality, Safety and Sustainability. *Sensors* 22 (23): 9148. https://doi.org/10.3390/s22239148.

GE Aerospace. n.d. Bombardier Connects with GE to Go Big on Big Data, Offers Free 'Smart' Box for Challenger, Global Customers | GE Aerospace News. Accessed July 13: 2025. https://www.geaerospace.com/news/press-releases/systems/bombardier-connects-ge-go-big-big-data-offers-free-smart-box-challenger.

Goecks, Lucas Schmidt, Anderson Felipe Habekost, Antonio Maria Coruzzolo, and Miguel Afonso Sellitto. 2024. Industry 4.0 and Smart Systems in Manufacturing: Guidelines for the Implementation of a Smart Statistical Process Control. *Applied System Innovation* 7 (2): 24. https://doi.org/10.3390/asi7020024.

Haque, Roksana, Ammar Bajwa, Noor Alam Siddiqui, and Ishtiaque Ahmed. (2024). Predictive Maintenance In Industrial Automation: A Systematic Review Of Iot Sensor Technologies And Ai Algorithms. *American Journal of Interdisciplinary Studies* 5 (01): 01–30. https://doi.org/10.63125/hd2ac988.

Hussain, Tanveer, Khan Muhammad, Javier Del Ser, Sung Wook Baik, and Victor Hugo C. de Albuquerque. (2020). Intelligent Embedded Vision for Summarization of Multiview Videos in IIoT. *IEEE Transactions on Industrial Informatics* 16 (4): 2592–2602. https://doi.org/10.1109/TII.2019.2937905.

Industrial IoT-Based Energy Monitoring System: Using Data Processing at Edge. n.d. Accessed July 9, 2025. https://www.mdpi.com/2624-831X/5/4/27.

James Manyika, Michael Chu, and Peter Bisson. (2015). The Internet of Things: Mapping the Value beyond the Hype. McKinsey Global Institute. https://www.mckinsey.com/~/media/mckinsey/industries/technology%20media%20and%20telecommunications/high%20tech/our%20insights/the%20internet%20of%20things%20the%20value%20of%20digitizing%20the%20physical%20world/the-internet-of-things-mapping-the-value-beyond-the-hype.pdf.

Jarašūnienė, Aldona, Kristina Čižiūnienė, and Audrius Čereška. (2023). Research on Impact of IoT on Warehouse Management. *Sensors* 23 (4): 2213. https://doi.org/10.3390/s23042213.

Javaid, Mohd, Abid Haleem, Ravi Pratap Singh, Shanay Rab, and Rajiv Suman. (2021). Upgrading the Manufacturing Sector via Applications of Industrial Internet of Things (IIoT). *Sensors International* 2 (January):100129. https://doi.org/10.1016/j.sintl.2021.100129.

Javaid, Mohd, Abid Haleem, Ravi Pratap Singh, and Rajiv Suman. (2022). Enabling Flexible Manufacturing System (FMS) through the Applications of Industry 4.0 Technologies. *Internet of Things and Cyber-Physical Systems* 2 (January):49–62. https://doi.org/10.1016/j.iotcps.2022.05.005.

Kang, Hyoung Seok, Ju Yeon Lee, SangSu Choi, Hyun Kim, Jun Hee Park. et al. (2016). Smart Manufacturing: Past Research, Present Findings, and Future Directions. *International Journal of Precision Engineering and Manufacturing-Green Technology* 3 (1): 111–28. https://doi.org/10.1007/s40684-016-0015-5.

Lou, Helen H., Rajib Mukherjee, Zhenyu Wang, Tim Olsen, Urmila Diwekar. et al. (2022). A New Area of Utilizing Industrial Internet of Things in Environmental Monitoring. *Frontiers in Chemical Engineering* 4 (March). https://doi.org/10.3389/fceng.2022.842514.

Milić, Saša D., Željko Đurović, and Mirjana D. Stojanović. (2023). Data Science and Machine Learning in the IIoT Concepts of Power Plants. *International Journal of Electrical Power & Energy Systems* 145 (February):108711. https://doi.org/10.1016/j.ijepes.2022.108711.

Misra, Sudip, Chandana Roy, Thilo Sauter, Anandarup Mukherjee, and Jhareswar Maiti. (2022). Industrial Internet of Things for Safety Management Applications: A Survey. *IEEE Access* 10:83415–39. https://doi.org/10.1109/ACCESS.2022.3194166.

Mobley, R. Keith. (2002). 4 - Benefits of Predictive Maintenance. In *An Introduction to Predictive Maintenance (Second Edition)*, edited by R. Keith Mobley, 60–73. Plant Engineering. Burlington: Butterworth-Heinemann. https://doi.org/10.1016/B978-075067531-4/50004-X.

Moon, Juhyun, and Byeong-Kwon Ju. (2024). Wearable Sensors for Healthcare of Industrial Workers: A Scoping Review. *Electronics* 13 (19): 3849. https://doi.org/10.3390/electronics13193849.

Mosteiro-Sanchez, Aintzane, Marc Barcelo, Jasone Astorga, and Aitor Urbieta. (2020). Securing IIoT Using Defence-in-Depth: Towards an End-to-End Secure Industry 4.0." *Journal of Manufacturing Systems* 57 (October):367–78. https://doi.org/10.1016/j.jmsy.2020.10.011.

Mourtzis, Dimitris, John Angelopoulos, and Nikos Panopoulos.(2021). Design and Development of an IoT Enabled Platform for Remote Monitoring and Predictive Maintenance of Industrial Equipment. *Procedia Manufacturing*, 10th CIRP Sponsored Conference on Digital Enterprise Technologies (DET 2020) – Digital Technologies as Enablers of Industrial Competitiveness and Sustainability, 54 (January):166–71. https://doi.org/10.1016/j.promfg.2021.07.025.

Nangia, Shikhil, Sandhya Makkar, and Rohail Hassan. (2020). IoT Based Predictive Maintenance in Manufacturing Sector. *SSRN Electronic Journal*. https://doi.org/10.2139/ssrn.3563559.

Olaf Peter Schleichert, and Björn Bringmann. (2017). Predictive Maintenance Taking Pro-Active Measures Based on Advanced Data Analytics to Predict and Avoid Machine Failure. Deloitte Touche Tohmatsu Limited, UK. https://www.deloitte.com/content/dam/assets-zone2/de/de/docs/about/2024/Deloitte_Predictive-Maintenance_PositionPaper.pdf.

Orca Lean. n.d. From 737 to the Future: How Boeing Leverages AI & Automation. Accessed July 13, 2025. https://www.orcalean.com/article/from-737-to-the-future-how-boeing-uses-automation-and-ai-to-optimize-manufacturing.

Pop, Gheorghe Ioan, Aurel Mihail Titu, and Alina Bianca Pop. (2023). Enhancing Aerospace Industry Efficiency and Sustainability: Process Integration and Quality Management in the Context of Industry 4.0. *Sustainability* 15 (23): 16206. https://doi.org/10.3390/su152316206.

Quijano, Jorge, Nohemi Torres Cruz, Leslie Quijano-Quian, Eduardo Rafael Poblano-Ojinaga, and Salvador Anacleto Noriega Morales. (2025). Data-Driven Scheduling Optimization for SMT Lines Using SMD Reel Commonality. *Data* 10 (2): 16. https://doi.org/10.3390/data10020016.

Ravi Subramanyan. (2025). Expanding Industrial IoT in 2025: Survey Reveals Growth. January 28, 2025. https://www.hivemq.com/blog/expanding-industrial-iot-in-2025-survey-reveals-growth/.

Research and Markets. n.d. Dassault Aviation SA - Digital Transformation Strategies. Accessed July 13, 2025. https://www.researchandmarkets.com/reports/5846398/dassault-aviation-sa-digital-transformation.

Rfid Journal. n.d. Boeing: Creating Effective IIoT Ecosystems. Accessed July 13, 2025. https://www.rfidjournal.com/video/boeing-creating-effective-iiot-ecosystems/192360/.

Sadeghi, Ahmad-Reza, Christian Wachsmann, and Michael Waidner. (2015). Security and Privacy Challenges in Industrial Internet of Things. In *Proceedings of the 52nd Annual Design Automation Conference*, 1–6. DAC '15. New York, NY, USA: Association for Computing Machinery. https://doi.org/10.1145/2744769.2747942.

Saheed, Yakub Kayode, Adekunle Isaac Omole, and Musa Odunayo Sabit. (2025). GA-mADAM-IIoT: A New Lightweight Threats Detection in the Industrial IoT via Genetic Algorithm with Attention Mechanism and LSTM on Multivariate Time Series Sensor Data. *Sensors International* 6 (January):100297. https://doi.org/10.1016/j.sintl.2024.100297.

Saif, Yazid, Yusri Yusof, Kamran Latif, Aini Zuhra Abdul Kadir, Maznah binti Iliyas Ahmad et al. 2022. Development of a Smart System Based on STEP-NC for Machine Vision Inspection with IoT Environmental. *The International Journal of Advanced Manufacturing Technology* 118 (11): 4055–72. https://doi.org/10.1007/s00170-021-08095-y.

Sallam, Karam, Mona Mohamed, and Ali Wagdy Mohamed. (2023). Internet of Things (IoT) in Supply Chain Management: Challenges, Opportunities, and Best Practices. *Sustainable Machine Intelligence Journal* 2 (March):(3):1–32. https://doi.org/10.61185/SMIJ.2023.22103.

Sen, Sachin K., Gour C. Karmakar, and Shaoning Pang. (2023). Critical Data Detection for Dynamically Adjustable Product Quality in IIoT-Enabled Manufacturing. *IEEE Access* 11:49464–80. https://doi.org/10.1109/ACCESS.2023.3276942.

Sharp, Michael, Ronay Ak, and Thomas Hedberg. (2018). A Survey of the Advancing Use and Development of Machine Learning in Smart Manufacturing. *Journal of Manufacturing Systems*, Special Issue on Smart Manufacturing, 48 (July):170–79. https://doi.org/10.1016/j.jmsy.2018.02.004.

Siemens. n.d. Bombardier to Digitally Transform the Aircraft Engineering. Siemens Digital Industries Software. Accessed July 13, 2025. https://newsroom.sw.siemens.com/en-US/bombardier-siemens-xcelerator/.

Sisinni, Emiliano, Abusayeed Saifullah, Song Han, Ulf Jennehag, and Mikael Gidlund. (2018). Industrial Internet of Things: Challenges, Opportunities, and Directions." *IEEE Transactions on Industrial Informatics* 14 (11): 4724–34. https://doi.org/10.1109/TII.2018.2852491.

Smart MRO | ST Engineering. n.d. Accessed July 13, 2025. https://www.stengg.com/en/aerospace/aerospace-mro/smart-mro.

Soori, Mohsen, Behrooz Arezoo, and Roza Dastres. (2023a). Digital Twin for Smart Manufacturing, A Review. *Sustainable Manufacturing and Service Economics* 2 (April):100017. https://doi.org/10.1016/j.smse.2023.100017.

Soori, Mohsen, Behrooz Arezoo, and Roza Dastres. (2023b). Internet of Things for Smart Factories in Industry 4.0, a Review. *Internet of Things and Cyber-Physical Systems* 3 (January):192–204. https://doi.org/10.1016/j.iotcps.2023.04.006.

Soori, Mohsen, Behrooz Arezoo, and Roza Dastres (2023c). Internet of Things for Smart Factories in Industry 4.0, a Review. *Internet of Things and Cyber-Physical Systems* 3 (January):192–204. https://doi.org/10.1016/j.iotcps.2023.04.006.

Taj, Soonh, Ali Shariq Imran, Zenun Kastrati, Sher Muhammad Daudpota, Raheel Ahmed Memon et al. (2023). IoT-Based Supply Chain Management: A Systematic Literature Review. *Internet of Things* 24 (December):100982. https://doi.org/10.1016/j.iot.2023.100982.

Ugbebor, Friday, Michael Adeteye, and John Ugbebor. (2024). Automated Inventory Management Systems with IoT Integration to Optimize Stock Levels and Reduce Carrying Costs for SMEs: A Comprehensive Review. *Journal of Artificial Intelligence General Science (JAIGS) ISSN:3006-4023* 6 (1): 306–40. https://doi.org/10.60087/jaigs.v6i1.257.

Vallois, Valentin, Ahmed Mehaoua, and Mourad Amziani. (2021). Blockchain-Based Identity and Access Management in Industrial IoT Systems. In *2021 IFIP/IEEE International Symposium on Integrated Network Management (IM)*, 623–27. https://ieeexplore.ieee.org/document/9463959.

Vermesan, Ovidiu, Marcello Coppola, Roy Bahr, Ronnie Otto Bellmann, Jøran Edell Martinsen. et al. (2022). An Intelligent Real-Time Edge Processing Maintenance System for Industrial Manufacturing, Control, and Diagnostic. *Frontiers in Chemical Engineering* 4 (July). https://doi.org/10.3389/fceng.2022.900096.

Wan, Jiafu, Shenglong Tang, Zhaogang Shu, Di Li, Shiyong Wang et al. (2016). Software-Defined Industrial Internet of Things in the Context of Industry 4.0. *IEEE Sensors Journal* 16 (October): 1–1. https://doi.org/10.1109/JSEN.2016.2565621.

Wang, Jinjiang, Yulin Ma, Laibin Zhang, Robert X. Gao, and Dazhong Wu. (2018). Deep Learning for Smart Manufacturing: Methods and Applications. *Journal of Manufacturing Systems*, Special Issue on Smart Manufacturing, 48 (July):144–56. https://doi.org/10.1016/j.jmsy.2018.01.003.

Zhan, Peng, Shaokun Wang, Jun Wang, Leigang Qu, Kun Wang et al. (2021). Temporal Anomaly Detection on IIoT-Enabled Manufacturing. *Journal of Intelligent Manufacturing* 32(6): 1669–78. https://doi.org/10.1007/s10845-021-01768-1.

Zhang, Qiang, Yifan Zhang, Qun Luo, Cijun Yu, Ningdong Yu et al. (2024). Cloud-Edge-End-Based Aircraft Assembly Production Quality Monitoring System Framework and Applications. *Journal of Manufacturing Systems* 75 (August): 116–31. https://doi.org/10.1016/j.jmsy.2024.06.002.

Zhong, Ray Y., Xun Xu, Eberhard Klotz, and Stephen T. Newman. (2017). Intelligent Manufacturing in the Context of Industry 4.0: A Review. *Engineering* 3 (5): 616–30. https://doi.org/10.1016/J.ENG.2017.05.015.

16

Digital Twins Enabled by AI and the Internet of Things for Intelligent Cities and Ecologically Designed Ecosystems

M. Ramprasath,[1,*] *Elangovan G.,*[1] *Prakash Duraisamy,*[2] *V. Kavitha*[1] and *V. Arun*[3]

1. Introduction

The construction industry has traditionally struggled with problems like inefficiency, late payments, noncompliance with regulations and interactions, distrust among participants, and information chain of custody disintegration, despite its critical role as an essential source of the global economy (Adel et al. 2022). Meanwhile, smart devices and sensors have made it practically impossible to manually coincide, annotate, and process the massive amounts of raw data (Big Data) accumulated from meticulously monitoring the built environment (Wang 2022). Integrating novel technologies such as blockchain technology, digital twins (DT), the Internet of Things (IoT), and artificial intelligence (AI) holds great promise for tackling the highlighted difficulties (Abioye 2021).

Digital twins (DTs) are becoming increasingly popular in today's constructed environments due to their capacity to improve collaboration and information flow across the asset lifespan (Zhao 2022). Additionally, they enable consumers to stay informed about the present performance of assets through simulations and suggest analytics-based decision-making for intelligent asset management (Hunhevicz 2022). This allows for the real-time monitoring of asset behaviour. Innovative built environments rely on DTs, which have undergone significant changes due to advancements in information technology (IT) and house all those technologies for connection and

[1] Department of Data Science and Business Systems, SRM Institute of Science and Technology, Tamil Nadu.
Emails: elangovg2@srmist.edu.in,http://orcid.org/0000-0003-2385-2129; kavithav2@srmist.edu.in, https://orcid.org/0000-0002-2728-5646

[2] Computer Science and Engineering, University of Wisconsin-Green Bay, Wisconsin, USA.
Email: duraisap@uwgb.edu,

[3] Department of Computing Technologies Systems, SRM Institute of Science and Technology, Tamil Nadu.
Email: arunv2@srmist.edu.in, https://orcid.org/0000-0001-6561-4044

* Corresponding author: ramprasm@srmist.edu.in, http://orcid.org/0000-0002-2667-9184

DOI: 10.1201/9781003655190-20

deployment. However, this technology is still in its early stages and is mainly used for creating virtual versions of material artefacts for prototyping, modelling, and verification (Yitmen 2022). In addition, DTs are typically used in a disjointed fashion in practice, especially during the building operation and maintenance phase.

The Centre for Digital Built Britain in the United Kingdom (CDBB) has introduced a set of principles referred to as the Gemini Principles (Yitmen 2021) to facilitate the use of digital approaches to design, construction, operation, and maintenance, as well as integrate the built environment more efficiently. These principles stress the importance of knowledge management value in a way that reduces data volume while increasing data value. Data trustworthiness, security, quality, and transparency are cornerstones of the Gemini Principles, reminiscent of the primary features of blockchain, a widely used distributed ledger technology (DLT). Blockchain is a suitable technology that meets the requirements of the Gemini Principles because of its concurrency (Teisserenc 2021). Also, in a complicated ecosystem where stakeholders engage with DTs, decentralised software programmes that use distributed ledgers and decentralised databases will offer a safe way to communicate information throughout a product's lifecycle. This method can positively solve data exchange problems while still guaranteeing the necessary availability, confidentiality, and integrity (Putz 2021). With blockchain technology, users can rest assured that their transactions are legitimate. Additionally, high-value transactions can be easily verified and tracked by utilising encryption and consensus procedures. Consequently, there is a lot of hope for solving problems with communication, trust, and accountability with blockchain technology's application to the construction sector and its combination with DTs and BIM for lifelong information management.

1.1 Sustainable Building Environment (SBE)

Sustainable building environments (SBEs) incorporate eco-friendly and resource-efficient practices into every stage of a building's lifecycle, from planning and construction to operation, maintenance, and eventual demolition (Martin 2022). By emphasising energy efficiency, reducing waste, and utilising environmentally conscious materials, SBEs aim to mitigate the environmental impact of buildings. This method considers several key issues, including biodiversity, air quality, and water conservation. The goal of sustainable building design in a sustainable built environment (SBE) is to create habitable spaces that can withstand and even thrive in harsh weather conditions while minimising the adverse environmental impacts and resource consumption required to power them (Elmousalami 2021). Sustainable construction technology, intelligent systems, and renewable energy sources are integral to SBE's plan to strike a balance between economic and social benefits and ecological concerns (Seyedrezaei 2023). Sustainable built environments (SBEs) prioritize a healthy environment while also prioritizing the health, productivity, and comfort of building occupants and the surrounding community. To minimize energy use without sacrificing enjoyment or safety, SBEs are progressively integrating technologies such as the Internet of Things (IoT) and Digital Twins (DT) to monitor and optimize performance in real-time, as shown in Figure 1, which shows the Sustainable Building Lifecycle (Debrah 2022). In addition, SBEs contribute to the shift toward greener city planning, as they are compatible with larger sustainability initiatives, such as zero-energy buildings. Consequently, SBEs make building and infrastructure management more efficient, effective, and environmentally friendly. Lending and industrialization are part of it (Piras 2024).

1.2 AI-powered digital twins

The utilisation of AI-powered digital twins (DTs) is crucial for enhancing commercial construction energy efficiency and significantly decreasing the release of carbon dioxide. These systems improve the administration of energy tactics, leading to higher environmental quality and efficiency, by merging complicated algorithms with real-time data. What follows is an outline of the primary ways in which the DTs powered by AI accomplish these modifications:

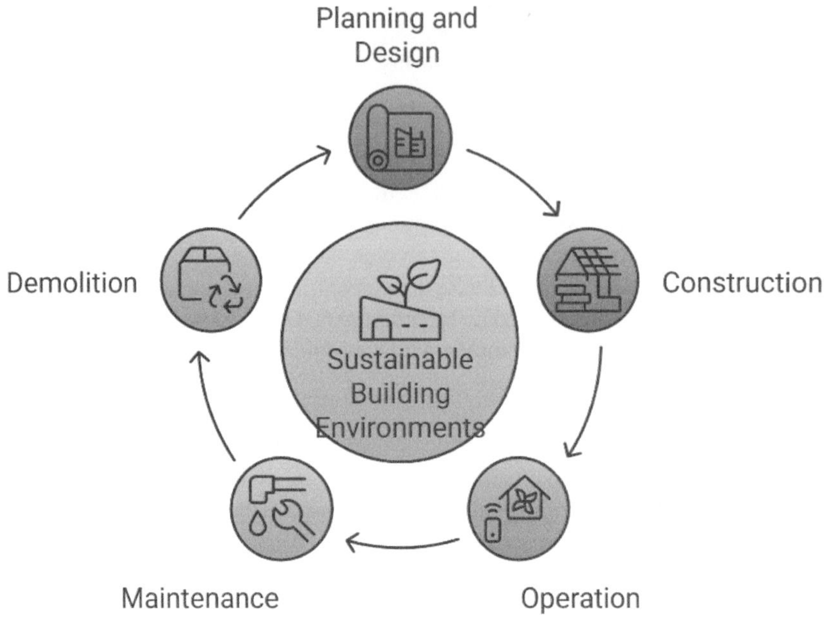

Figure 1: Sustainable Building Lifecycle.

- Integrating IoT sensors for real-time data collection on occupancy, energy use, and surroundings allows DTs to offer precise monitoring and management of electric power systems (Renganayagalu 2024). In addition, there are dynamic HVAC systems that may adjust their operations based on occupancy and user preferences, allowing human-centred energy management systems to maximise energy use without sacrificing comfort (Rafsanjani, H.N,2024).

AI technologies, such as random forest regression, optimise electricity production in microgrids for smart buildings while balancing ecological and economic factors. Preventive analytics and optimisation also include algorithmic techniques for machine learning (Chen 2024). By providing complex optimisation scenarios, DTs optimise the usage of energy from renewable sources while reducing outputs and operational expenses in multi-objective optimisation (Chen 2024).

Despite the many benefits of DTs driven by AI, there are still challenges to integrating different technologies. Several concerns need fixing before digital twins can be used to their full potential in energy management.

2. Related Work

2.1 Literature Evaluations on Digital Twins

As the field of digital twins continued to gain traction, researchers sought to categorise and define the various types of digital twins. In their 2018 article, “Digital Twin in Manufacturing: A Categorical Literature Review and Classification,” Kritzinger et al. produced one of the most frequently cited works in the field. There have been 736 citations to this publication. A classification system for digital twin research was established by the authors. More than half of the studies focused on the topic, aligning with the paper’s content. Furthermore, in other publications, the writers differentiated between digital twins. The two most common models in the literature are digital models, in which there is no automatic data flow connecting the physical and virtual versions, and digital shadows, in which an automated information flow exists but only in one direction, from the physical to the virtual.

According to the authors, a "real" digital twin features bidirectional automated data flow; however, very few studies discuss this concept. A literature review using Scopus data is suggested in this paper. Fifteen pieces in 2017 sought to pin down what exactly a digital twin is.

2.2 Literature Reviews on Digital Twins

Several recent reviews have introduced new methods of analysis. A meta-analysis of fourteen digital twin-related literature articles published between 2018 and 2021 is presented in the work of Rossman and Hertweck (2022), titled "Digital Twins: A Meta-Review on Their Conceptualization, Application, and Reference Architecture" [5]. The authors construct a nine-layer digital twin architecture based on the analysis:

1. Physical things, physical duplication.
2. Generating data.
3. Connectivity with the network.
4. Integrating and storing data.
5. Representing and preparing data.
6. Information model, procedures, virtual twin, or simulated being.
7. Mini services are those that deal with infrastructure and operations.
8. Ensuring the security of systems and protecting the privacy of data.
9. A company's model and procedures.

Kuehner, Scheer, and Strassburger presented a meta-review named "Digital Twin: Finding Common Ground—A Meta-Review" in 2021. The digital twin was the subject of twenty-four articles used for this research. The writers produced a comprehensive categorisation of the evaluated reviews. Most authors (83%) think that the idea of a digital twin is still in its infancy. The study delves into the advantages and disadvantages highlighted by the reviews. Digital twins have several applications, but some of the most important include condition tracking and monitoring, data management, system analysis, prescription, and prediction.

Table 1 shows the comparing communicated digital twin intentions according to objective and year.

Table 1: Comparison of previous papers based on the objectives.

References	Main Objective	Years
10	A look at how infrared imaging is used in structures.	2022
22	An overview of DT that makes sure circumstances are healthy	2023
23	A look at how electronic devices might help with environmentally friendly building.	2024
24	A summary of the concept of Industry 4.0 and how it can make production more sustainable	2021
25	A look at how BIM is changing into DT	2021

2.3 Bibliometric Methods for Studying Literature in Digital Twins

The use of bibliometric tools offers a straightforward solution to these issues. Analysis of citations and co-citations was the primary focus of the initial bibliometric investigations. Publications and authors can have their impacts measured using these techniques. According to (Mejia et al. 2022), development evaluation of released topics has been among the most critical areas of bibliometrics since 2013. Significant publication databases can be analysed using new machine-learning approaches and bibliometric trend analysis, enabling reliable conclusions to be drawn. To date, bibliometric methods have been employed in a limited number of digital twin evaluations. In their 2019 monograph "Digital Twins in the High-Technology Manufacturing Industry," Borovkov,

Gamzikova, Kukushkin, and Ryabov attempted to apply bibliometric approaches to digital-twin analysis. Using a bibliometric lens, the book analyses the state of digital twins.

3. Usage of Digital Twins in the Construction Industry

3.1 Digital Twins: Use Cases in Building and Maintenance Management

Digital Twins (DTs) offer numerous benefits to building projects and can significantly enhance project outcomes. By combining real-time data with simulator capabilities, digital twins (DTs) improve interaction, efficiency, and decisions across the entire endeavour lifecycle. The project's participants can evaluate progress and make informed decisions with the help of DTs since they provide a common virtual representation of the project (Owais 2024). They make it easier to organise and exchange information, which keeps everyone on the same page during the modular construction and operation phases (Yang 2024). By eliminating manual processes and enhancing effectiveness, Building Digital Twins (BDTs) improve site management on construction sites by automating data collection and information extraction. Furthermore, DTs provide continuous evaluation and feedback loops, which are crucial to lean construction methodologies (Owais 2024), which seek to improve processes further and decrease waste. By simulating various situations and allowing proactive risk management, digital twins (DTs) improve predictive capacities. They address common problems in the construction business by using cutting-edge technology, such as AI and the Internet of Things, which have the potential to increase the security and dependability of data (Zhang 2024). Problems with data integration and quality prevent DTs from being widely used in the construction industry despite their obvious benefits.

3.2 Digital Twins for Smart Cities and Sustainability

Because of their predictive capabilities and capacity to integrate data in an immediate fashion, digital twins (DTs) are crucial in smart cities for enhancing sustainability and allocating resources more effectively. With the help of Digital Twins (DTs), city planners and decision-makers may boost efficiency, green their operations, and make the most of available resources. Predictive maintenance is one application where DTs are used to save waste and extend asset life by anticipating structural demands and providing proactive interventions (Ersan 2024). Automatic resource allocation is made possible using digital twins (DTs) by simulating various scenarios to guarantee water, energy, and resource efficiency (Wang 2024). In addition, DTs help with sustainable urban planning by analysing environmental data in real time, which means they may help find suitable places to put energy sources that are renewable and cut down on carbon emissions (Wang 2024). In addition, digital twins can optimise urban agriculture, enhance sustainability, and increase food security in smart agriculture using mobile, climate-adaptive plant cultivation systems (Cardoso 2023). Policy and development decisions can be guided by digital twins (DTs), which integrate various information like traffic and emissions statistics, supporting data-driven decision-making and leading to more sustainable outcomes (Sohail 2024). Scenario modelling allows city planners to test out "what-if" scenarios and see how different policies or innovations could affect sustainability (Pankowska 2024).

4. Working Together with Digital Twins

DTs are bridging the gap between the actual and online worlds. You can use them to define the manufacturing, process, and warehousing of machines, devices, and robots in detail. Industry 4.0 benefits of DTs include enhanced data security and quality, decreased costs, and quicker decision-making. In their description of DTs, the contributors (Schluse 2018) included data, functions, and interfaces for communicating, as well as one virtual model of a machine, robot, or device.

DTs comprise both virtual and physical entities, as well as the information that connects them (Sahal 2021). Working together entails not only exchanging data but also dividing up responsibilities so that everyone can respond appropriately. In their discussions on the value of Sahal ML for robotics, the writers of (Alsamhi 2020) emphasised their significance. Enhancing service quality, connection, and dependability are the cornerstones of the partnership.

In addition, using drones in tandem with the IoT can improve public safety (Ansari 2019), the intelligence of apps for smart cities, and the quality of service. The process of multi-user collaboration and activity identification is detailed in (Alsamhi 2020). A detailed description of DTs and human partnership can be found in (Gravina 2020). Collaboration on industry platforms is not without its difficulties, though, as the writers pointed out (Andreassen 2018). When working in a group, users can complete complex tasks far more quickly and efficiently than when working alone (Wang 2020). Additionally, to counteract COVID-19, the developers of (Sahal 2021) utilised blockchain technology to enable decentralised multi-drone collaboration. Drones and the Internet of Things (IoT) collaborate to make cities more intelligent and environmentally friendly, while green IoT is a product of this partnership (Sahal 2021). Nevertheless, the research above fails to include the collaboration and uses of DTs. As a result, we go over how DTs may work together to enhance logistics and transportation apps.

Collecting data from innovative Internet of Things devices or their DTs on the assembly lines is essential for creative industries. Unreliable sensors, RFID, actuators, or associated data transducers (DTs) can introduce and generate erroneous data, affecting the analysis and decision-making process. Based on this, the contributor (Sahal 2021) has suggested an integrated approach to interaction for cooperative DTs to offer auto-detection within the cyber-physical manufacturing environment by leveraging the intelligence of operational data. By communicating with other cooperative DTs at the edge level, the suggested technique can determine if the DT is behaving erratically. Innovative factory architecture based on service-based DTs was introduced by the authors (Catarci 2019). The automated integration of related human processes and the division of parallels with internet services were demonstrated in the technology. Programmes, manufacturing, production, and operational forecasting were the primary points of the authors' introduction to DT and big data in the innovative sector (Leng 2019). Additionally, in creative industries, the technologies that enable DT were explored. He and colleagues introduced innovative industries that are sustainable and driven by DT. Additionally, DTs integrated with data analytics and Industry 4.0.

4.1 Blockchain Technology and Digital Twins

In some industries, DT technology has been integrated with blockchain technologies (Debrah 2022). The integration of blockchain technology into distributed technology (DT) has led to the proposal of Manu Chain, a continuous bi-level approach (Leng 2019). Using blockchain's decentralised properties, the Manu Chain model has established two tiers of intelligence: one for outstanding, self-organising tasks and another for iterative, coarse-grained, holistic optimisation.

The authors of (Renganayagalu 2024) have developed an algorithm to ensure the transfer of values across DTs in economic systems using DLT. To further enhance transparency for all parties involved, we offer a DLT-based architecture for secure data sharing that can track an asset's lifecycle, including its ownership details and associated events.

4.2 A Review of Intelligent Digital Twins and Predictive Data

To facilitate informed, physics-based DTs, an anticipatory DTs technique has been suggested, which makes use of comprehensible machine learning and a library of component-based reduced-order models (Kapteyn 2020). A fixed-wing UAV case study was used to accomplish the prescient DTs approach. When a UAV's structural sensors identify deterioration or breakage to one of its wings, the UAV can decide whether to execute a bold move to prevent a structural breakdown or a cautious one.

The smart city DT infrastructure proposed in (Austin 2020) leverages knowledge representation, reasoning, and ML formalisms to facilitate data processing, event detection, and robotic decision-making in a complementary and supportive manner. Service that is based on predictions is an implementation of Industry 4.0 that relies on predictive data analysis for maintenance. Given its potential influence on the industrial sectors, it is one of the more contemporary commercial uses that have been studied. To shed light on how DTs and conservation work together in the manufacturing sector, the writers of (Errandonea 2020) examined previous research on the topic. In addition, the authors outlined a list of prerequisites for utilising big data for proactive repair in Industry 4.0 applications. In addition, a three-layer DT reference model for predictive maintenance defined by ledgers has been suggested by the authors in (Altun 2019).

These layers are edge, fog, and cloud. When it came to data processing, Song et al. presented an analytical preservation model that utilised decision trees in multiple ways (Song 2016). DTs represented the numerous actual machines located in various places. Data from product nameplates, which relate to individual physical devices, are merged with additional simulation models in each DT. The database stored the duration of operation and recorded data collected from sensors linked to the physical machine. To facilitate decision-making in complex systems, DTs also employed modular-based maintenance strategies (Vathoopan 2018). Based on DT's development, the proposed model was a corrective maintenance strategy. When it comes to maintenance prediction utilising DTs, (Aivaliotis et al. 2019) suggested a physics-based solution. Minimising modelling efforts and providing a structure for modelling for various resources to enable DTs were the main points of discussion regarding the proposed technique. The machine's dynamic behaviour was modelled using grey, black, and white boxes. The use of virtual sensor modelling facilitated the data collection process in the simulation. To ensure machine DT, we also incorporated modelling characteristics that were changed often.

5. Architecture for Collaborative Digital Twins Driven by Data and Built on the Blockchain

To connect blockchain technology with an operational assessment of data, the suggested framework is considered a step above DT implementation in production systems. It is possible to build and deploy the proposed framework on top of the DT stage, which utilises blockchain technology to ensure scalability, security, trust, decentralisation, data collaboration, peer interaction, and accessibility.

5.1 Essential Elements for Collaborating on Digital Twins

Here, we outline the ground rules for how DTs must collaborate in the field with dispersed production. To meet the needs of digital twins' collaboration, the literature lists nine criteria. Data sources, or DTs, of tangible objects such as equipment, devices, and individuals, are part of every production process and are accessible across the entire network used by production units. The company's decision-making process benefits from this data because it is updated frequently and in real time (R1 & R2).

Collaborative DT data is notable since it is a data-driven learning system. When it comes to advanced risk prediction (R3), DTs serve as the data analysis engine, as they continuously update the data fed into the learning models. For the most part, simulation models help foresee the outcomes of physical asset modifications. Nevertheless, DTs aid in comprehending both the present and potential future states of the physical asset (R4). To improve product design, conduct immediate troubleshooting, and incorporate new ideas in a collaborative context, virtual visibility of potential hazards within physical assets is beneficial.

Additionally, DT networks are used to enhance the interconnection of entities involved in the network's operations, including devices, machines, individuals, departments, and organisations.

Participants' interactions through the DTs network enable trustworthy data exchange, which in turn allows the sharing of data both internally and externally (R5). Nevertheless, to maintain confidence between network peers, the connected DTs network must undergo authentication (R6). To keep safe cooperation and interactions within the DTs network, registration establishes a level of trust. As an undetectable representation of physical objects and their functions within the collaborative environment, DT serves as a virtual counterpart to the real thing. For instance, precise tracking throughout the DT network is made possible by the items' transparent visibility through the DT model (R7).

Each node in a centrally managed network reports to a single administrator. All nodes in a decentralised network are autonomous entities rather than being controlled by a central server. Importantly, decision-making is necessary for DTs' partnership because centralised control has the problem of lacking global data, and decentralisation has the problem of lacking data on individual failures (R8 & R9).

5.2 Elements of the Suggested Architecture

Emerging blockchain technology has the potential to circumvent security concerns that have halted DT projects. The permanent nature and diversity of blockchain technology allow DT efforts to adapt more rapidly and efficiently to their surroundings. To better equip DTs with digital identities and data-tracking capabilities, a decentralised blockchain network can be utilised. Additionally, authenticating DTs in industry contexts nowadays is more closely tied to decentralised, secure, and safe data transfer. Lacking information accessibility, traceability, or permanence, the authors of earlier work presented cryptocurrency for DTs in the process, which ensures safe, secure, and dependable transactions. The created DTs use smart contracts to monitor and control the transactions. To facilitate the independent sharing of knowledge across DTs, blockchain technology is utilised for data storage and exchange. Accordingly, the previously mentioned high-level needs could be met by combining DTs, blockchain, and artificial intelligence to execute DTs collaboration in distributed manufacturing. Considering these needs, the conceptual framework for DT coordination based on blockchain technology needs two primary components. Table 2 presents the need for cooperation between digital twins.

Table 2: Need for cooperation between digital twins.

Paper	Area of Investigation	Outcome
R1	Gathering information	Facilitating the use of data in decision-making
R2	Frequency of data updates	Offering up-to-the-minute information on the physical twin
R3	Analysing data	Being able to foresee possible hazards in advance
R4	Skills for modeling	Making it possible to view the things virtually
R5	Data sharing	Permits the exchange of data, both internally and externally
R6	Data sharing verification	Upholding confidence between network peers
R7	Authorization openness	Making it possible to track data across the enterprise
R8	Capabilities for distributed decision-making	Offering a thoughtful method of reaching decisions through consensus
R9	Independence from a central authority	Providing dependable and effective solutions

5.3 Collaborative DTs for Predictive Analytics Driven by Data on a Ledger

Building a framework for making and revising data-driven collaborating DTs based on ledger technology makes use of this aspect. It shows how to use predictive analytics by creating web-based and offline models with data from historical DTs stored in a ledger and data from live-streaming

DTs. The following are the two primary parts needed to prepare the data-driven, collaborative DTs based on ledgers for predictive analytics:

5.3.1 Blockchain-Based DTs Framework

Secure distributed operational data management and analytics involving many parties is possible with the use of DLT and a blockchain network connecting several DTs. Figure 2 illustrates the foundation DT framework, which is based on ledgers. Here are the five components that must be defined at the technical level for the ledger-based Digital Twin model: (1) identified DT holder, (2) DT grade, (3) timestamp, (4) contract, and (5) archive database [10,41]. The ledger contains all the necessary information to keep track of the requirements for material objects, such as the owner of DT. The process of creating a transaction involves using data that is deemed the DT condition within a date and time stamp, which is generated by the communication mechanism that allows unidirectional data transfers across DTs and their physical counterparts. Data about DTs, transactions, and activities are recorded on DLT. Additionally, for collaborating DTs based on DLT, the transfer of value among DTs is guaranteed. This improves the standard of making choices by increasing the accuracy of real-time predictions, such as the status of updated DTs within production systems, thanks to the ledger-based running database syncing this data in real time. The collaborative DTs-based systems will determine the DTs' data structure and collaboration needs, including the sharing of information and analytics. Therefore, it is necessary to specify which categories of participants are highlighted by DTs. From the get-go, the collaboration scenario should also dictate the nature of the participants' communicative activities.

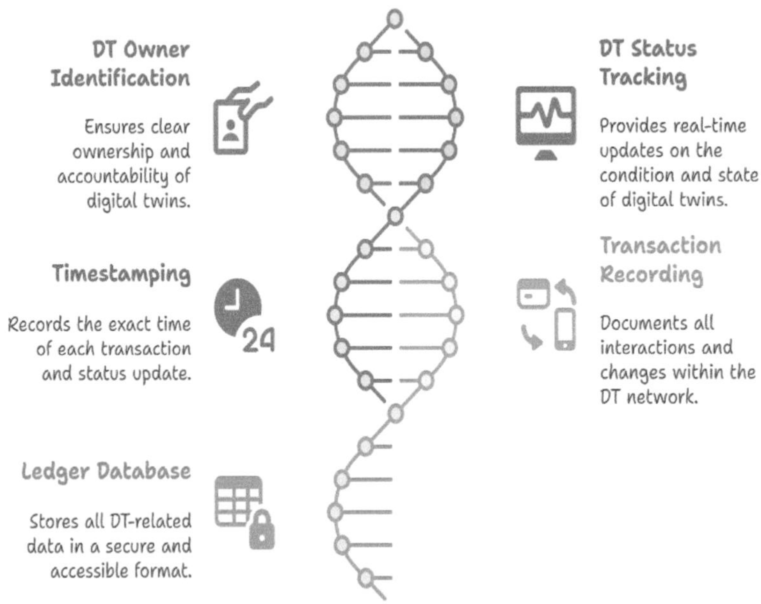

Figure 2: Foundations of Ledger-Based DT Model.

5.3.2 Applying data-driven digital twins to predictive analytics

Understanding the DT position, interacting with other DTs at the edge level, learning from different DTs, and sharing shared semantic information inside manufacturing processes are all advantages of the data-driven DT cooperation idea within the suggested framework. For data-driven collaborative DTs based on ledgers, the process of creating a predictive model is discussed. There are primarily two steps to the process: making an offline prediction model and implementing it online. Heterogeneous machine-learning frameworks, such as Apache Spark MLlib, will be utilised to create the offline prediction model, which is a classifier. Inexpensive machine-learning (both traditional and deep) techniques are all part of Apache Spark MLlib's toolbox. We will build and train an offline predictive model using DT operational data from the past, which is stored on a ledger. The smart contracts could be used to assess the created predictive model for the internet-based forecasting model. Therefore, the suggested framework can assist decision-makers in taking the necessary steps to reduce production system downtime and in dynamically replanning a set of safety measures.

5.3.3 Choice-Making through Cooperation

To enhance the DTs' cooperation, this part is utilised in the creation of the distributed agreement method. To ensure the safety of the manufacturing process, a suggested conceptual framework is to develop a working DT system that facilitates distributed decision-making. Thus, a decentralised algorithm for decision-making will be created by combining the features of the consensus mechanism with an evolving prediction system that utilises operational data driven by DT in real time.

6. Healthcare Internet of Things (H-IoT)

The door to groundbreaking digital healthcare and business studies has been opened by current trends and breakthroughs in information technology (Dhanvijay 2019), and connected sensing and communicating devices make up what is known as the Internet of Things (IoT). Because they can look back at all the problems and difficulties, IoT devices have caused chaos in the medical diagnostic space. The proliferation of biomedical data generated by these sensors has the potential to revolutionise the creation of automated systems for collecting medical records. Big data is crucial in enhancing these medical systems' ability to diagnose, treat, and make decisions when Internet of Things (IoT) sensors are integrated with sophisticated ML-based algorithms. Research interest in the Internet of Everything (IoE) has grown due to the Internet of Things (IoT) in biomedical applications. This involves somatic therapies and patient tracking and surveillance (Khan 2020). Miniature sensors also improve healthcare by constantly monitoring patients' vital signs; this bodes well for early detection and management and strengthens the human wellness system. Figure 3 shows internet of things (IoT) and its revolutionary impact on healthcare.

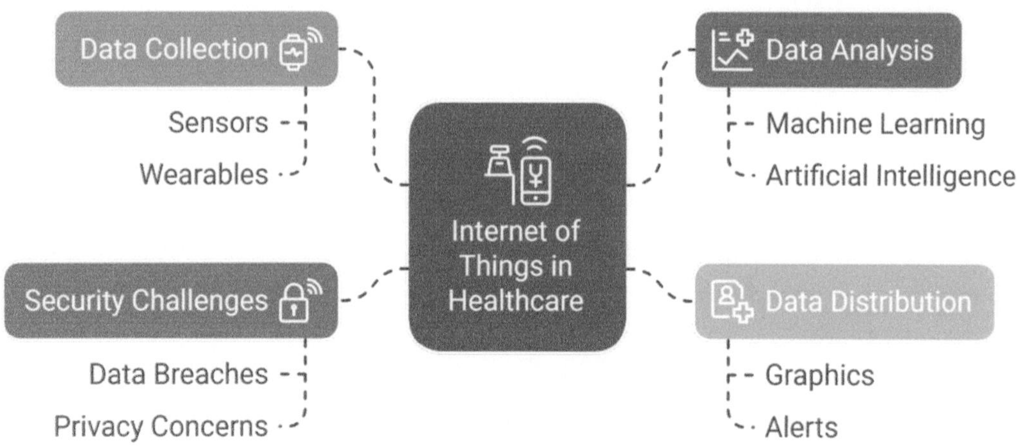

Figure 3: Internet of Things in Healthcare: Revolutionising Medical Practices.

As more sensors, wearables, and other linked devices become widely available, medical experts will have access to a wealth of client health data that can enhance the effectiveness of therapies and outcomes. Healthcare practitioners, however, will need efficient systems for handling information if they are to make full use of this data. Table 3 shows the problem in the IoT healthcare section and their solutions.

Data created by IoT devices is gathered, stored, analysed, and then disseminated as part of connected health data cleansing. Patient vitals, exercise, sleep, and prescription compliance are all examples of information that can fall into this category. Using this information, doctors may keep tabs on their patients from afar, tailor their care to everyone's needs, and spot the first symptoms of illness. Data gathering is the initial stage of H-IoT information processing. Linked devices, such as sensors and wearables, gather data. Next, the information is sent to a central database or storage system on the cloud. Healthcare practitioners from different places should be able to access the information in the file, and it should be able to handle enormous quantities of data. Data analysis follows data storage and collection. Finding trends and patterns in data is the goal of data analysis, which uses sophisticated algorithms. Some examples of this include being able to spot modifications to a patient's vital signs, abnormalities in their behaviour, and potential health problems down the road. Another application of machine learning and artificial intelligence is analysing data for possible health concerns. Data distribution is the last stage of information processing for H-IoT. Healthcare professionals need data that is both readily available and comprehensible. Graphics and charts showing important subdivisions and patterns in the data are examples of what can be included. Healthcare practitioners can also be notified of significant changes in patient health status through alerts and notifications.

The mind-sponge theory is one solution to the problems caused by the proliferation of H-IoT devices—which produce massive amounts of information that healthcare practitioners may find challenging to handle. Theoretically, healthcare practitioners should be able to "sponge up" and analyse this data to improve patient care through well-informed decision-making. To do this, we may leverage strategies for machine learning and sophisticated statistical analysis to find identifiable patterns and trends in the data and to give doctors the knowledge that is most important to them. Medical professionals may enhance customer satisfaction and overall care quality by using the power of IoT data in this way (Vuong 2022).

Due to the fast development and implementation of IoT systems, the security vulnerabilities in IoT devices will inevitably grow over time. Because data breaches might result in the loss of life, the security challenges around H-IoT are highly concerning. In a healthcare setting, Figure 3 showcases

Table 3: An analysis of the current problems in the healthcare sector and their potential Internet of Things (IoT) solutions.

S.No	List of Problems	IoT based Solution
1	Documentation of patient information (COVID-19)	i) Raises consciousness about what triggered the COVID-19 pandemic. ii) It can be useful for monitoring and analyzing patients› progress toward recovery. iii) The Internet of Things gadget efficiently sends, saves, and analyzes patient data to provide future therapy that is better.
2	Handling cases of COVID-19	i) IoT gadgets can also keep an eye on and control things like temperature, humidity, and other external variables. ii) IoT gadgets have services for location that work in real time. These are used to help treat people with COVID-19.
3	Intelligent hospital shortage	i) Smart hospitals are now possible thanks to the Internet of Things (IoT), which enables the use of automation and linked, dedicated networks. ii) Internet of Things (IoT) technology enables automated, interconnected healthcare facilities.
4	Relying on faulty judgment	i) With their ability to forecast and consistently monitor patient data, IoT devices greatly assist in making precise decisions. ii) It takes effective communication and IoT technology to perform surgeries accurately. iii) Writes down every detail and comes up with sound judgments.

the innovative characteristics of H-IoT. Patients with diabetes will be given identifications that can be scanned and linked to a forum in the cloud. This platform will house patients' EHRs, prescriptions, vital lab results, and medical history information. This would facilitate easier access to these documents for doctors and nurses through desktops, laptops, and tablets. The varied security rules and access control structures made it challenging to meet the requirements of making patients feel comfortable and secure while providing their private medical data.

6.1 IoT in Big Data

Since big data enables the collection and evaluation of massive amounts of health-related data, it is crucial to H-IoT. With the use of big data in H-IoT, statistical models have been created. These models can help find potential health risks and anticipate disease progression. AI algorithms allow doctors to sift through mountains of patient information in search of patterns to guide their treatment plans. All things taken together, H-IoT analytics using big data can better understand developments and trends, which can improve patients' health, reduce healthcare costs, and simplify healthcare delivery. The personalised healthcare system's e-health services can meet older people's clinical needs. In their comprehensive study, (Jagadeeswari et al. 2018) examined the medical industry's upcoming use of big data analytics, mobile apps, the Internet of Things (IoT), fog computing, and cloud computing. In addition, the disadvantages of creating an improvised healthcare system for detecting early illnesses were examined in this study, and practical solutions to these constraints were suggested. Our evaluation will provide the groundwork for a more robust medical industry. In their study (Yang et al. 2019), proposed an intelligent, extensive data storage system driven by the H-IoT and prioritises data security. The Figure 4 shows the big data application components. Additionally, this proposed system would have adaptive access control, safeguard customers' healthcare data at a high level, distinguish between emergencies and routine access control, and facilitate smart data duplication to reduce significant data storage needs. Encryption and storage in a system allowed for the secure sharing of medical files produced by H-IoT with healthcare professionals from different medical areas. Moreover, these access control methods allow authorised users to decode sensitive medical records and aid in administering lifesaving first aid to patients. An innovative, secure, and self-adaptive two-factor gate control system was proposed in this study. It can handle both everyday and emergency scenarios with ease. Patients' medical records can be

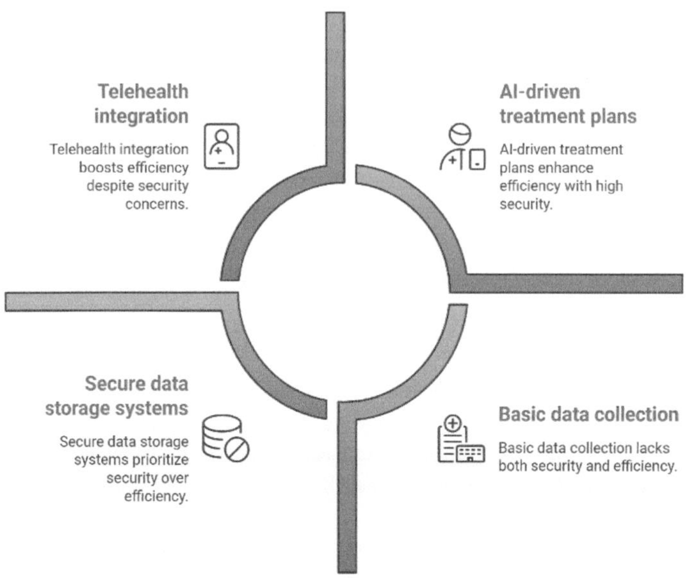

Figure 4: Big Data Applications in Healthcare.

recovered in severe emergencies using a shatter-glass access method. Lastly, the study concluded that the proposed framework is safe. Many specialised applications, such as H-IoT systems and robotics, rely on big data to amass massive amounts of data. Regarding big data analytics, systems driven by the H-IoT are crucial. On the other hand, sensing might make it harder to make reliable predictions. Consequently, a health-related Internet of Things (IoT) system that leverages AI to treat Parkinson's disease.

In the past decade, new findings and approaches have been published, and the use of big data in healthcare is expanding at a rapid pace. Big data approaches have been efficiently applied in healthcare research and biomedical information technology. Clients, investigators, hospitals, gadgets, and mobile phones have recently produced a deluge of healthcare data. Clinical organisations are continuously making this data to find treatments for new ailments. Scientists have developed a wide range of new devices to gather this data. Similarly, patients are beginning to manage their treatment needs with the help of these instruments, which come in the form of smartphone applications. Further, these gadgets and applications can be integrated with telehealth and telemedicine through the Internet of Things. The benefits and difficulties of big data in the biological field were covered by (Mishra et al. 2024).

6.2 H-IoT in Edge Computing

As demonstrated in (Rani 2022), intelligent systems that constantly track human actions have been developed thanks to enhancements in sensing modalities. Frameworks that use wearable sensors to forecast an individual's activity level using a Recurrent Neural Network (RNN) on devices like desktops and laptops have recently gained popularity. Various wearable healthcare instruments provided the input data, including gyroscope, accelerometer, magnetometer, and electrocardiography (ECG) sensors. We contrasted the proposed strategy with more conventional approaches using existing datasets. The study concluded that the proposed plan is superior to more traditional methods based on the results of the experiments.

New developments in edge computing have facilitated the creation of useful healthcare apps, such as e-clouds for tracking patients from a distance. For this, you'll need improved quality of service, lots of sensors, and easy visualisation, which should take precedence. Regarding optimising technologies like telesurgery, the study's experimentation showed that the proposed strategy performed better than other popular methods.

Intelligent systems have become more popular due to the explosion of data and the Internet of Things. In addition, smart city and cloud computing are crucial in intelligent healthcare systems. When maximising system resources and service flow, emergency departments (EDs) need real-time systems and customised methodologies for modelling and simulation. So, to simulate the non-consumable resources, (Oueida 2018) proposed a system and provided theoretical validation and description. Additionally, the RPN might be used in real-time situations by considering metrics for performance, like the mean time spent waiting for patients, the percentage of resource utilisation, and the length of stay (LOS) for patients. The proposed system was modelled in the study to enhance the LoS, decrease patient waiting time, and optimise resource use.

6.3 Challenges of H-IOT

A significant obstacle to the growth of H-IoT is the several difficulties it faces with performance. These include protection, delays, dependability, and efficiency. The ideal characteristics of an enhanced H-IoT system are minimal latency, high reliability, low power consumption, and robust security. Several writers' struggles with Internet of Things (IoT) technology are examined in depth below:

Regarding health-related Internet of Things (H-IoT) applications, low latency is key for things like telemedicine, distant patient tracking, and rapid real-time response handling of information. Because of low latency, medical professionals can respond quickly to situations and make choices.

Nevertheless, optimising network components and using specific software and hardware can make achieving low latency challenging. One major obstacle to H-IoT is immediate data communication and processing; otherwise, healthcare professionals may be unable to make well-informed decisions. Because of the importance of timing in H-IoT systems, a very brief latency is required to handle information from the beginning to the conclusion of transmission. While deep learning approaches improve the big data analytical system's ability to process enormous data sets, current obstacles prevent us from reliably extracting all the data from the given dataset. As a result, robust algorithms that can theoretically explore and fill up the dataset's collection of data are urgently needed.

6.4 Difficulties with Protection

Because patient data is sensitive, security is a top priority in the healthcare IoT. The increased connectivity and accessibility of H-IoT devices and systems increase their vulnerability to assaults and security compromises. Health records might be a gold mine for scammers looking to commit identity theft, fraud, or other crimes.

The biggest issue is keeping energy consumption low. To add features intended to reduce security risks, an advanced neural network or supervised learning method that is battery-efficient and capable of handling light encryption is required. The system mentioned before does, however, involve computational complexity. A fog paradigm may be used with a federal system close to the security equipment.

6.5 Obstacles to Scalability

Overcoming scalability issues is crucial for facilitating the widespread use of H-IoT systems. Also, a correct access method should ensure no packets are lost during data transfer in H-IoT systems. Beyond this, the inherent lack of stability in most current routing protocols in H-IoT systems is a major drawback. Furthermore, there are several significant concerns regarding penetration and localisation. For seamless connectivity, implantable sensors and wearable devices must be standardised. There cannot be any computational delays or extra expenses incurred when processing the generated data flow.

6.6 Importance of Health-Related Internet of Things

Modern healthcare practitioners must overcome many obstacles. By delivering relevant data at the correct time, H-IoT raises medical IQ. At the outset, it simplifies processes by doing away with superfluous steps, dramatically lowering service costs and freeing up more quality time for doctors and patients. An increase in third-party payments could generate funding if surveys like HCAHPS show greater patient satisfaction and patient outcome scores. Better patient and business results are additional benefits of H-IoT. In 2012, the medical sector generated approximately $200 billion, according to the IMS Institute. A higher IQ in the medical area significantly decreases the likelihood of making clinical errors. Thanks to new biometric equipment and gadgets, patients can get instantaneous input from their doctors. To avoid having to wait fourteen days for a response from a patient, the IoE allows for immediate actions to be performed once obtained. Better patient care is one of the many advantages of integrating big data. The most critical thing is that the IoE gives medical professionals the data they need to keep people healthy (El-Sappagh 2020).

7. Future direction

7.1 Protecting Personal Information

The vast amounts of data generated by digital transport networks, along with the potential risks to personal information, make DT security and privacy issues complex in the innovative transportation sector. Hence, rather than transmitting the raw data, connected devices should perform local data analysis via federated learning and subsequently communicate solely the model to the blockchain. Thus, blockchain technology can address the problem of security, while centralised education can address the issue of privacy. When used together, these methods can significantly improve the safety and confidentiality of DTs in transportation networks.

7.2 Connection

As the number of intelligent gadgets integrated into smart transportation systems continues to increase, ensuring that all these devices can connect in real time remains a significant challenge. Thanks to the proliferation of connected devices in intelligent transportation, next-generation networks for communication, such as Beyond Fifth Generation (B5G) or Sixth Generation (6G), are essential. To keep the transportation system running smoothly if smart devices become detached, blockchain technology could allow devices to borrow data from nearby devices. It is possible to avoid missing data, guarantee high accuracy, and guarantee full connectivity by running machine learning at the edge.

7.3 International Logistics Systems

Logistics networks rely on DTs to improve transportation on things like roads, trains, streets, oceans, and smart automobiles, which can gather data from their environment and use it to generate a map. To reduce traffic, they can subsequently share their data with others. In a decentralised network, blockchain technology works in tandem with shared data, which may include information such as parking availability, road closures, traffic speeds, and more.

7.4 When, How Fast, and How Responsive

Problems with the timeliness and velocity of DTs extend beyond transportation and logistics. Several shifts in logistics and transportation might result from changes in timing and speed. For starters, meeting the demands of customer service—which necessitates pinpoint precision and lightning-fast reactions—requires better decision-making and reaction times. Transportation and logistics companies aren't interested in data, but they do need real-time visibility.

7.5 Transport Sector Packaging and Containers

Most of the logistics-related transportation occurs within packaging. Numerous obstacles must be overcome in the logistics sector to create, administer, and oversee packaging and containers effectively. As a result of COVID-19, the demand for containers and packaging is on the rise, driven by the expansion of online shopping. Additionally, it is essential to consider various types of packing. As a result of these findings, operational efficiency and waste caused by inefficient use are greatly diminished. By combining DTs with computer vision, a container model can be built, and these issues can be identified automatically. Decisions regarding the state of containers can be influenced by previous data stored in the blockchain at the moment when DT was started. As a defect in the container, the decision-making process needs fixing, including using and maintaining it. Combining blockchain technology with DTs allows for the creation of a more efficient, environmentally friendly, and lightweight system for container management and product packaging.

7.6 The Method for Deciding

Other configurations of systems could be derived by extending the proposed framework to incorporate two layers of the decision-making process. The methodology for making choices consists of two levels: (1) avoiding risks and (2) reconfiguring the system in the event of unexpected events through dynamic planning. In the train transportation sector, for instance, the system can utilise IoT to predict failures, identify their causes, and initiate maintenance. In a different context, the system can foretell the hazards associated with the product's condition within a truck in the transportation logistics sector. These forecasts will allow for a consensus-based decision-making process to guide the vehicle to the most risk-averse factory.

8. Conclusion

This theoretical model explains how innovative distributed production can benefit from DTs' cooperation with the technology of blockchain. Cooperation among DTs facilitates the interaction mechanism that allows for an understanding of DT status, sharing of goals, exchange of information, interaction, reciprocal learning, and adaptation. We outline the problems with DTs' collaboration that have been reported in the literature and demonstrate how blockchain technology can address them. The suggested architecture can enhance real-time analysis of information and DT collaboration. Also included are smart logistics and railway predictive maintenance, two areas that could benefit from the implementation of the conceptual framework in intelligent transportation. In conclusion, we emphasised the way forward for scholars interested in this fascinating field.

References

Abioye, S. O., Oyedele, L. O., Akanbi, L., Ajayi, A., Delgado, J. M. D., Bilal, M. et al. (2021). Artificial intelligence in the construction industry: A review of present status, opportunities and future challenges. *Journal of Building Engineering*, 44: 103299.

Adel, K., Elhakeem, A., and Marzouk, M. (2022). Decentralizing construction AI applications using blockchain technology. *Expert Systems with Applications*, 194: 116548. https://doi.org/.

Aivaliotis, P., Georgoulias, K., Arkouli, Z., and Makris, S. (2019). Methodology for enabling digital twin using advanced physics-based modelling in predictive maintenance. *Procedia CIRP*, 81: 417–422.

Alsamhi, S. H., and Lee, B. (2020). Blockchain for multi-robot collaboration to combat COVID-19 and future pandemics. arXiv:2010.02137.

Alsamhi, S. H., Lee, B., Guizani, M., Kumar, N., Qiao, Y., and Liu, X. (2021). Blockchain for decentralized multi-drone to combat COVID-19 and future pandemics: Framework and proposed solutions. Transactions on Emerging Telecommunications Technologies, e4255. https://doi.org/.

Alsamhi, S. H., Ma, O., and Ansari, M. S. (2020). Convergence of machine learning and robotics communication in collaborative assembly: Mobility, connectivity and future perspectives. *Journal of Intelligent & Robotic Systems*, 98: 541–566.

Alsamhi, S. H., Ma, O., Ansari, M. S., and Gupta, S. K. (2019). Collaboration of drone and Internet of Public Safety Things in smart cities: An overview of QoS and network performance optimization. *Drones*, 3 13.

Alsamhi, S. H., Sahal, R., Hawbani, A., Al-qaness, A. A., Lee, B., and Guizani, M. (2021). Green Internet of Things using UAVs in B5G networks: A review of applications and strategies. *Ad Hoc Networks*, 117: 102505.

Altun, C., and Tavli, B. (2019). Social Internet of digital twins via distributed ledger technologies: Application of predictive maintenance. pp. 1–4. *In:* 2019 27th Telecommunications Forum (TELFOR). Belgrade, Serbia: IEEE.

Andreassen, T. W., Lervik-Olsen, L., Snyder, H., Van Riel, A. C. R., Sweeney, J. C., and Van Vaerenbergh, Y. (2018). Business model innovation and value-creation: The triadic way. *Journal of Service Management*, 29: 883–906.

Austin, M., Delgoshaei, P., Coelho, M., and Heidarinejad, M. (2020). Architecting smart city digital twins: Combined semantic model and machine learning approach. *Journal of Management in Engineering*, 36(4): 04020026.

Cardoso, J. L. F. P., and Rhodes, D. H. (2023). Digital twin use case for smart, sustainable cities. pp. 99–115. *In:* Proceedings of the 2023 Conference on Systems Engineering Research. Hoboken, NJ, USA.

Catarci, T., Firmani, D., Leotta, F., Mandreoli, F., Mecella, M., and Sapio, F. (2019). A conceptual architecture and model for smart manufacturing relying on service-based digital twins. pp. 229–236. *In: 2019 IEEE International Conference on Web Services (ICWS)*. Milan, Italy: IEEE.

Debrah, C., Chan, A. P., and Darko, A. (2022). Artificial intelligence in green building. *Automation in Construction*, 137: 104192.

Dhanvijay, M. M., and Patil, S. C. (2019). Internet of Things: A survey of enabling technologies in healthcare and its applications. *Computer Networks*, 153: 113–131.

Elmousalami, H. H., Darwish, A., and Hassanien, A. E. (2021). The truth about 5G and COVID-19: basics, analysis, and opportunities. pp. 249–259. *In:* Hassanien, A. E., and Darwish, A. (Eds.). *Digital transformation and emerging technologies for fighting COVID-19 pandemic: Innovative approaches*. Springer International Publishing. https://doi.org/.

El-Sappagh, S., Abuhmed, T., Islam, S. R., and Kwak, K. S. (2020). Multimodal multitask deep learning model for Alzheimer's disease progression detection based on time series data. *Neurocomputing*, 412: 197–215.

Errandonea, I., Beltrán, S., and Arrizabalaga, S. (2020). Digital twin for maintenance: A literature review. *Computers in Industry*, 123: 103316.

Ersan, M., Irmak, E., and Colak, A. M. (2024). Applications, insights and implications of digital twins in smart city management. pp. 378–383. *In:* 2024 12th International Conference on Smart Grid (icSmartGrid). Setubal, Portugal: IEEE.

Heaton, J., and Parlikad, A. K. (2020). Asset information model to support the adoption of a digital twin: West Cambridge case study. *IFAC-PapersOnLine*, 53: 366–371.

Hunhevicz, J. J., Motie, M., and Hall, D. M. (2022). Digital building twins and blockchain for performance-based (smart) contracts. *Automation in Construction*, 133: 103981.

Jagadeeswari, V., Vairavasundaram, S., Logesh, R., and Vijayakumar, V. (2018). A study on medical Internet of Things and big data in personalized healthcare system. *Health Information Science and Systems*, 6: 14.

Kapteyn, M. G., Knezevic, D. J., and Willcox, K. (2020). Toward predictive digital twins via component-based reduced-order models and interpretable machine learning. In Proceedings of the AIAA Scitech 2020 Forum (p. 0418). Orlando, FL, USA.

Khan, M. A., Quasim, M. T., Alghamdi, N. S., and Khan, M. Y. (2020). A secure framework for authentication and encryption using improved ECC for IoT-based medical sensor data. *IEEE Access*, 8: 52018–52027.

Kuo, T. T. (2020). The anatomy of a distributed predictive modeling framework: Online learning, blockchain network, and consensus algorithm." *JAMIA Open*, 3: 201–208.

Leng, J., Jiang, P., Xu, K., Liu, Q., Zhao, J. L., Bian, Y. et al. (2019). Makerchain: A blockchain with chemical signature for self-organizing process in social manufacturing. *Journal of Cleaner Production*, 234: 767–778.

Leng, J., Yan, D., Liu, Q., Xu, K., Zhao, J. L., Shi, R., et al. (2019). ManuChain: Combining permissioned blockchain with a holistic optimization model as bi-level intelligence for smart manufacturing. *IEEE Transactions on Systems, Man, and Cybernetics: Systems*, 50: 182–192.

Li, Q., Gravina, R., Li, Y., Alsamhi, S. H., Sun, F., and Fortino, G. (2020). Multi-user activity recognition: Challenges and opportunities. *Information Fusion*, 63: 121–135.

Martin, M., Chong, A., Biljecki, F., and Miller, C. (2022). Infrared thermography in the built environment: A multi-scale review. *Renewable and Sustainable Energy Reviews*, 165: 112540.

Obushnyi, S., Kravchenko, R., and Babichenko, Y. (2019). Blockchain as a transaction protocol for guaranteed transfer of values in cluster economic systems with digital twins. pp. 241–245. *In:* 2019 IEEE International Scientific-Practical Conference Problems of Infocommunications, Science and Technology (PIC S&T). Kyiv, Ukraine: IEEE.

Oueida, S., Kotb, Y., Aloqaily, M., Jararweh, Y., and Baker, T. (2018). An edge computing based smart healthcare framework for resource management. *Sensors*, 18: 4307.

Owais, O. A., Poshdar, M., Bidhendi, A., Jaafar, K., and Sarhan, S. (2024). From concept to concrete: Digital twins enabling different levels of lean construction. pp. 767–779. *In: Proceedings of the 32nd Annual Conference of the International Group for Lean Construction (IGLC 32)*. Auckland, New Zealand.

Pankowska, M., and Żytniewski, M. (2024). Digital twins for smart city. pp. 269–286. *In: Smart Spaces. Elsevier*.

Piras, G., Agostinelli, S., and Muzi, F. (2024). Digital twin framework for built environment: A review of key enablers. *Energies*, 17: 436.

Putz, B., Dietz, M., Empl, P., and Pernul, G. (2021). Ethertwin: Blockchain-based secure digital twin information management. *Information Processing & Management*, 58: 102425.

Rafsanjani, H. N., Nabizadeh, A. H., and Momeni, M. (2024). Digital twin energy management system for human-centered HVAC and MELs optimization in commercial buildings. [Conference/journal details if available].

Rani, P., Kavita, Verma, S., Rawat, D. B., and Dash, S. (2022). Mitigation of black hole attacks using firefly and artificial neural network. *Neural Computing and Applications*, 34: 15101–15111.

Renganayagalu, S. K., Bodal, T., Bryntesen, T.-R., and Kvalvik, P. (2024). Optimising energy performance of buildings through digital twins and machine learning: Lessons learnt and future directions. pp. 1–6. *In: Proceedings of the 2024 4th International Conference on Applied Artificial Intelligence (ICAPAI)*. Halden, Norway: IEEE.

Sahal, R., Alsamhi, S. H., Breslin, J. G., Brown, K. N., and Ali, M. I. (2021). Digital twins collaboration for automatic erratic operational data detection in Industry 4.0. *Applied Sciences*, 11: 3186.

Schluse, M., Priggemeyer, M., Atorf, L., and Rossmann, J. (2018). Experimentable digital twins—Streamlining simulation-based systems engineering for Industry 4.0. *IEEE Transactions on Industrial Informatics*, 14: 1722–1731.

Seyedrezaei, M., Becerik-Gerber, B., Awada, M., Contreras, S., and Boeing, G. (2023). Equity in the built environment: A systematic review. *Building and Environment*, 245: 110827.

Sohail, A., Shen, B., Cheema, M. A., Ali, M. E., Ulhaq, A., Babar, M. A. et al. (2024). Beyond data, towards sustainability: A Sydney case study on urban digital twins. arXiv:2406.04902.

Song, Z., and Canedo, A. M. (2016). Digital twins for energy efficient asset maintenance. U.S. Patent Application 15/052,992, filed February 25, 2016.

Teisserenc, B., and Sepasgozar, S. (2021). Adoption of blockchain technology through digital twins in the construction industry 4.0: A PESTELS approach." *Buildings*, 11: 670.

Vathoopan, M., Johny, M., Zoitl, A., and Knoll, A. (2018). Modular fault ascription and corrective maintenance using a digital twin. *IFAC-PapersOnLine*, 51(1): 1041–1046.

Vuong, Q. H., Nguyen, M. H., and La, V. P. (Eds.) (2022). The mindsponge and BMF analytics for innovative thinking in social sciences and humanities. *Walter de Gruyter GmbH.*

Wang, H., and Wang, Y. (2024). Smart cities net zero planning considering renewable energy landscape design in digital twin. *Sustainable Energy Technologies and Assessments*, 63: 103629.

Wang, Q., Yin, Y., Chen, Y., and Liu, Y. (2024). Carbon peak management strategies for achieving net-zero emissions in smart buildings: Advances and modeling in digital twin. *Sustainable Energy Technologies and Assessments*, 64: 103661.

Wang, T., Li, J., Kong, Z., Liu, X., Snoussi, H., and Lv, H. (2020). Digital twin improved via visual question answering for vision-language interactive mode in human–machine collaboration. *Journal of Manufacturing Systems*, 58: 261–269.

Wang, W., Guo, H., Li, X., Tang, S., Xia, J., and Lv, Z. (2022). Deep learning for assessment of environmental satisfaction using BIM big data in energy efficient building digital twins. *Sustainable Energy Technologies and Assessments*, 50: 101897.

Yang, G., Jan, M. A., Rehman, A. U., Babar, M., Aimal, M. M., and Verma, S. (2020). Interoperability and data storage in Internet of multimedia things: Investigating current trends, research challenges and future directions. *IEEE Access*, 8: 124382–124401.

Yang, J., and Ng, S. T. (2024). Prospects for digital twin technology in the building modular construction and operation phases: A game theory-based analysis. *Journal of Cleaner Production*, 470: 143344.

Yang, Y., Zheng, X., Guo, W., Liu, X., and Chang, V. (2019). Privacy-preserving smart IoT-based healthcare big data storage and self-adaptive access control system. *Information Sciences*, 479: 567–592.

Yang, Z., Tang, C., Zhang, T., Zhang, Z., and Doan, D. T. (2024). Digital twins in construction: Architecture, applications, trends and challenges [Journal/Conference details if available].

Yitmen, I., Alizadehsalehi, S., Akıner, I., and Akıner, M. E. (2021). An adapted model of cognitive digital twins for building lifecycle management. *Applied Sciences*, 11: 4276.

Zhang, C., Zhou, G., Li, H., and Cao, Y. (2020). Manufacturing blockchain of things for the configuration of a data- and knowledge-driven digital twin manufacturing cell. *IEEE Internet of Things Journal*, 7: 11884–11894.

Zhao, J., Feng, H., Chen, Q., and Garcia de Soto, B. (2022). Developing a conceptual framework for the application of digital twin technologies to revamp building operation and maintenance processes. *Journal of Building Engineering*, 49: 104028.

17

Ethical and Legal Considerations for Artificial Intelligence, Blockchain, and the Internet of Things

Mohan Singh,[1,*] *Seema Singh,*[2] *Kapil Dev Tyagi*[3]
and *Vaibhav Bhushan Tyagi*[4]

1. Introduction

The intense fusion of AI, blockchain, and IoT is creating unprecedented opportunities and challenging ethical and legal issues. These are the technologies that are changing businesses and industries, and even challenging our sense of what is possible. So-called apps like this do, however, also provoke important questions about data privacy, security, fairness, and responsibility. This chapter discusses the complex ethical and legal issues that relate to using AI, blockchain, and IoT, stressing the need for proactive management of ethical risk.

- **The Movement of Technologies:** IoT, AI, and Blockchain in Business. Internet of Things, Artificial Intelligence, and Blockchain are considered ideal technologies that could lead to the digital disruption of industries. Each of these technologies offers specific functionalities, but their appropriate combination can open up new possibilities for systemic efficiency and innovation. IoT-related The Internet of Things (IoT) is a system of interrelated computing devices, mechanical and digital machines, objects, etc., provided with unique identifiers and the ability to transfer data over a network without requiring human-to-human or human-to-computer interaction (Kamilaris 2021). AI provides new possibilities to effectively analyze a lot of data, learn patterns, and make smart decisions. A decentralized, immutable, and transparent

[1] Department of ECE, G.L. Bajaj Institute of Technology and Management, Greater Noida-201308 (India).
[2] Department of Education, Lloyd Institute of Management & Technology, Greater Noida-201306 (India).
Email: seemasingh1613@gmail.com, https://orcid.org/0009-0009-4776-224
[3] Department of ECE, Jaypee Institute of Information Technology, Noida-201309 (India).
Email: kapiltyagiitd@gmail.com, https://orcid.org/0000-0002-4596-2321
[4] Department of ECE, ISBAT University, Kampala-8383, Uganda.
Email: tyagi.fict@isbatuniversity.com https://orcid.org/0000-0001-8153-3607
* Corresponding author: mohansingh571@gmail.com, https://orcid.org/0000-0002-2973-3676

DOI: 10.1201/9781003655190-21

ledger for recording transactions, which is enabled by blockchain technology, makes it more secure and reliable (Zheng 2018).

- **Statement of Ethical and Legal Issues:** With the pervasion of these technologies into everyday life, it is important to discuss the ethical and legal implications of the creation and use of these technologies. People who interact with imperfect AI know all too well how these systems can make mistakes—and how those errors can materialize in the world. Both ethical frameworks and legal regulations should be refined to reflect technological progress and direct the development and deployment of AI, blockchain, and IoT systems so as to align with the values of society and serve the welfare of humans or the community. The intersection of AI, IoT, and blockchain poses unprecedented ethical and legal issues that can only be addressed effectively with the active involvement of multiple skills and stakeholders. Neglecting these challenges risks falling prey to unintended negative effects, losing public trust, and preventing these revolutionary technologies from being increasingly embraced.
- **Purpose of the Chapter:** The purpose of this chapter is to present a detailed overview of the ethics and legal issues with regard to the convergence of AI, blockchain, and IoT. This chapter aims to inform policymakers, industry, and academia by shedding light on ethical dilemmas and the legal context that should be considered in order to assess risk analysis for responsibility toward innovation in these advancing areas. The chapter will also offer some guidelines and best practice recommendations for compliance with ethical and legal frameworks, in order to gain assurance that these technologies are employed for the good of society.
- **Structure Overview:** The structure of this chapter is planned so that it will extensively address the ethical and legal aspects of AI, blockchain, and IoT technologies. Here, it is going to be introduced and clarified what AI is, and an overview of the USA and EU trends and strategy on ethics and law of AI in healthcare. The chapter will then discuss the ethical issues surrounding these technologies, including privacy, bias, transparency, accountability, and security. Also, it will analyze the regulatory environment in the US and Europe in connection with the use of these technologies, encompassing data protection, cybersecurity, intellectual property, and liability rules. Practical case studies and examples in the ethical and legal implications and opportunities of AI, blockchain, and IoT applications will be presented. The chapter will conclude by providing suggestions and directions for facilitating adherence to ethical and legal requirements, encouraging responsible innovation, and ensuring that the full potential of these disruptive technologies is realized.

2. Foundational Concepts

Figure 1 provides a systematic view of the ethical and legal issues in three fast-growing technologies: AI, Blockchain, and the IoT.

- **Artificial Intelligence (AI):** AI is an offshoot of computer science and represents a machine's ability to exhibit intelligent human behavior. Such systems can accomplish things like learning, problem solving, decision making, and language understanding, leveraging methods such as machine learning, deep learning, and natural language processing. Its applications span many industries: healthcare, finance, transportation, entertainment, and it's changing the way we engage with technology and the way that we solve hard problems.
- **Blockchain:** The concept of using a distributed, immutable ledger to record transactions across a network of computers. It's an inalterable and secure infrastructure that eliminates the centralized guardian from the picture without compromising on trust, transparency, and security by using cryptography and consensus mechanisms. You can use it for much more than just cryptocurrency, like supply chain monitoring, healthcare, voting systems, and digital identity claiming, adding a layer of transparency and security.

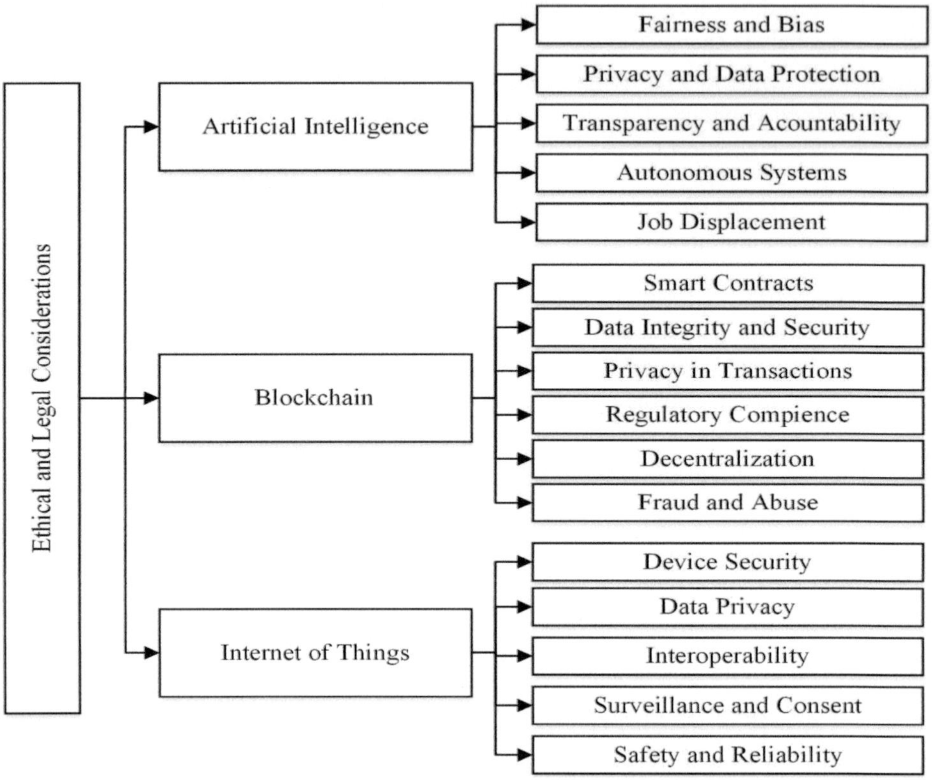

Figure 1: Ethical & Legal Considerations.

- **Internet of Things (IoT):** The IoT is a system of interrelated computing devices, mechanical and digital machines, objects, animals, or people that are provided with unique identifiers and the ability to transfer data over a network without requiring human-to-human or human-to-computer interaction. They can interconnect and communicate with each other and with central control systems to enable automation, monitoring, and control in different application fields. IoT is revolutionizing sectors such as manufacturing, agriculture, healthcare, and smart cities, leading to enhanced efficiency, improved productivity, and better data-based decision making. The intersection of AI, blockchain, and IoT is opening new doors for creative products and solutions in various sectors when it comes to efficacy, safety, and decision-making.

Table 1 presents a systematic introduction to the major points in the ethical and legal aspects of these technologies (AI, Blockchain, and the IoT).

2.1 Ethics and Law in Technology

- **Ethics:** Ethics is the study of moral principles and values that determine how humans act and make decisions. Technology and ethical values. In technology, ethics can be defined as the study of the moral aspects of new technologies and their consequences for human life and the world. Ethical approaches offer guidance on how to innovate responsibly and to develop and implement technologies in ways that respect human rights, promote fairness, and limit harm.
- **Law:** The system of rules which a particular country or community recognizes as regulating the actions of its members and which it may enforce by the imposition of penalties. This is the nature of technology law—technology, broadly speaking, is about how technology is made, distributed, and used, including not just data privacy and security, IP, and liability. Law is supposed to strike a balance between innovation and society's rights and other social aspects.

Table 1: Structured introduction of key ethical and legal issues.

Technology	Ethical Considerations	Legal Considerations
Artificial Intelligence (AI)	Bias & fairness in decision making, Privacy concerns, Accountability & transparency. It is request that the following be added to allow them to easily distribute their planned Bible studies via given platform, Jobs Displacement Risks.	Privacy laws (GDPR, CCPA), Responsibility of AI-based decisions, Intellectual property & proprietary rights, Law of robotics for decision-making machines
Blockchain	Clash between transparency and privacy, Decentralization & ethical governance, Fraud prevention & trust issues, Environmental monitoring on mining casualties	Compliance issues in the sphere of smart contracts, legality & taxation of cryptocurrency, the laws on combating money laundering (AML), Data privacy & cross-border transfers
Internet of Things (IoT)	Security & ethical hacking issues, User data privacy & surveillance risk, Consent & informed use of IOT devices, Ethical systems by design	Rules governing the security of IoT devices, Adhering to the laws on consumer protection, Data ownership & responsibility for breaches, International regulation of connected devices

- **Differences and Interplay**: Ethics and law are two separate but related notions that act in concert in the regulation of technology. Morals are frequently the foundation on which laws and protocols are built, whereas legal structures serve as a system of established legislation by which morality can be maintained and power exercised. This nuance in ethics often captures not the legal but a socially and morally enforced obligation that may or may not find an equivalent in the law. The relationship between ethics and law is critical to the responsible development and use of technology, public trust, and societal benefits.

3. Ethical Implications of AI and Digital Twins

The fast-paced development and deployment of AI and DT in numerous industries introduce a tangled and multi-faceted web of ethical conundrums that need to be analyzed thoroughly. These technologies present enormous opportunities for innovation and advancement, and at the same time, substantial threats to personal rights, societal values, and the foundations of our communities, as shown in Figure 2. One of the main ethical issues is related to privacy and security of data, since AI systems actually rely on large datasets to learn and to make decisions (Binns 2018). The collection, storage, and processing of personal data have ethical implications for informed consent and transparency, as well as the potential for misuse or unauthorized access (Bakiner 2023). Algorithmic fairness is a third profound moral imperative, as machine learning algorithms trained on biased data

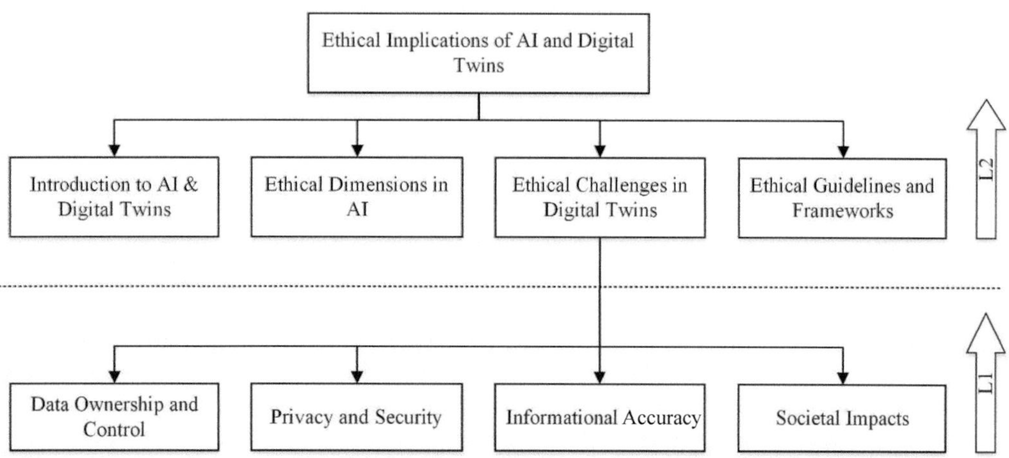

Figure 2: Ethical Implications of AI and Digital Twins.

Table 2: Ethical Implications in AI and Digital Twins.

Aspect	Ethical Implications in AI	Ethical Implications in Digital Twins
Privacy & Data Security	Threat of unauthorized access to highly sensitive personal information, Privacy issues with surveillance & tracking, Matters of data ownership and informed consent	Real-world data collection/simulation may breach users' privacy, Threat to digital twin models from cyber-attacks, Ambiguity about who owns the data and is accountable for it.
Bias & Fairness	AI models can embed bias into hiring, health care, and law enforcement, ethical issues in the fairness of decision-making, Open issues in systems automation.	Digital twins can mimic biased representations in the real world, and this has the potential to result in unfair simulations or predictions. Danger of unethical utilization in predictive modeling (e.g., perpetuation of disparities)
Autonomy & Accountability	Ambiguity in responsibility for AI decisions, morals of AI taking control from Humans, ethical questions related to AI-led warfare or policing.	Accountability problems when digital twins come up with bad predictions, Ethical implications of substituting real-world decision-making with simulated knowledge
Societal Impact & Employment	Automation, which accompanies AI, results in job displacement. Ethical issues related to AI replacing human creativity & intuition, the possibility of misuse in social trickery & disinformation	Digital twins may widen economic inequalities if availability is restricted by industry or group. Ethical implications when employing digital twins for modeling human behavior & surveillance
Environmental & Sustainability Concerns	Huge energy use in training AI models, AI becomes a force of nature in crisis response, mechanism for wasteful automation of no value.	Environmental implications due to data storage and processing requirements, Moral dilemmas of weighing the advantages of digital twin against ecological footprint

can perpetuate and even amplify existing social inequalities, and lead to discriminatory outcomes in hiring, lending, and the criminal justice system. Table 2 provides a vast overview of ethical difficulties in AI and Digital Twins.

3.1 Introduction to AI and Digital Twins

3.1.1 The Game Changer

AI has turned out to be the most powerful technology, driving radical changes in all aspects of human life and businesses in the world. More recently, digital twins—the virtual replicas of physical things, processes, or systems created to simulate, optimize, or predict their physical counterparts—have become powerful methods that offer an unprecedented comprehension of real-world behavior and operations. As AI makes more progress and acts more autonomously, ethics has a central part to play in the forward march of responsible development and application. Increasing the autonomy of AI to make decisions implicitly requires reconsideration of the model of responsibility (Dignum 2018).

- ➢ **AI Definition**: AI is the theory and development of computer systems that can perform tasks that typically require human intelligence, such as visual perception, speech recognition, decision-making, and translation between languages.
- ➢ **Digital Twin Definition**: A digital twin is a digital replica of a physical asset, system, or process that centrally contains a highly accurate and complete model at its core, including all the data driving the dynamic behaviors and characteristics of how it operates and interacts with its environment to enrich and optimize the performance and value of the asset. The digital twin concept is used in a number of industries such as manufacturing, healthcare, infrastructure, and energy, enabling functions such as predictive maintenance, performance investigation and optimization, and real-time monitoring. Digital twin usage provides automation and insight value due to enhanced operations, less downtime in machinery, better decision-making, known indicators of when equipment needs servicing, and innovation occurring in a virtual environment without risk, and a 24/7 online digital doppelganger world of scenario evaluations. In addition,

live vehicle data can be provided to AI algorithms for the prediction of part performance (Fuller 2020).

- **Use Cases**: AI-based applications span across various domains, including healthcare, finance, transportation, and education, and provide applications such as better diagnosis, personalized services, and self-driving cars. In healthcare, for example, AI systems contribute to the early detection and treatment of diseases, as well as to the discovery of new drugs and treatments; and in finance, AI solutions are used for fraud detection and risk assessment, as well as algorithmic trading.

3.2 Ethical Dimensions in AI

AI gives rise to many ethical considerations regarding bias, fairness, accountability, and transparency, which require careful thought to minimize any negative impacts and to ensure fair treatment. Algorithmic bias, in favor of biased training data or mathematically challenged algorithms, can reinforce and even worsen systemic societal biases, which can result in unfair outcomes in hiring, lending, and the criminal justice system. The risk of machine AI displacing human workers and further widening economic inequalities calls for pre-emptive actions to mitigate job losses and promote inclusive opportunities.

- **Autonomy vs. Control:** As AI systems are granted more autonomy, it becomes a topic of discussion as to how much influence and control humans should have in decision-making, especially critical ones. Ensuring the right level of autonomy while maintaining control is vital to avoid undesirable consequences and to guarantee that AI systems serve human values and societal objectives.
- **Bias and Discrimination:** AI platforms trained on data with inherent biases can perpetuate and exacerbate existing patterns of bias within society in areas such as hiring, credit, and criminal sentencing. Addressing the problem of algorithmic bias requires a delicate balance between collecting unbiased data and designing an algorithm that can also meet equity and fairness at run time. The use of AI in public administration affects biases of the algorithms, which can result in suspicious or unjust distribution of services to the general public (Vatamanu 2025).
- **Moral agency:** A second central ethical issue regards attributing culpability and responsibility for the actions and decisions of AI, particularly in the case of harm caused by AI or in the context of its failures. Transparency, explainability, and accountability are central in ensuring that AI systems are used responsibly and ethically. The development of morality-capable agents, in particular AI systems capable of making ethical decisions, raises profound questions regarding the nature of morality and the role of machines in ethical decision-making.
- **Ethics by Design:** Just as with safety, where comprehensive procedures for embedding safety into systems have occurred, embedding ethics in the design of AI systems is vital to foster responsible AI-driven innovation and help prevent the creation or multiplication of potential harms. Value-aligned AI design, guided by the aspiration to promote human well-being, autonomy, and dignity, would ensure that AI systems are compatible with human values and human society. Methods for value elicitation and the representation of human values, the translation of human values into technical requirements, and the handling of moral overload need to be developed to demonstrate that designs are getting close to achieving the values of interest (Dignum 2017).

3.3 Ethical Challenges in Digital Twins

There are ethical issues such as data privacy, security, and ownership involved in digital twins, and tenacious protection mechanisms need to be established to ensure the privacy of sensitive information from being violated and used. Data confidentiality plays an important role in digital twin applications, many of which store and process sensitive information on people, objects, and activities, leading to strict requirements for data protection and privacy compliance. Surveillance and manipulation: the shared concept of digital twins for surveillance and manipulation may cause privacy and power issues to merge and may necessitate an analysis of the potential consequences of such in terms of individual autonomy and control. Real-time encryption and data security are crucial to avoid unauthorized access, tampering, or sabotage of digital twins, and to avoid potential cascading effects on the physical assets or systems that they simulate. Digital twins often involve proprietary data systems and algorithms, so the issue of data ownership, IP ownership, and access to data comes into play.

- **Data Integrity and Consent**: The integrity of data feeding the digital twin is fundamental to valid simulations and predictions. When data of people is used in digital twins, collecting informed consent is necessary to preserve their privacy and autonomy. The absence of human control over DT applications can lead to a decrease in human autonomy, deskilling, and isolation. Although the potential of digital twins for forecasting, monitoring, control, and optimization of complex systems is recognized by experts (Kapteyn 2022).
- **Prediction and Preemption**: Computer systems being able to predict possible futures and intercede proactively leads to questions about autonomy and the unintended. It is of great significance for the reliability and credibility of digital twin predictions to maintain transparency, explainability, and validation.
- **Simulated Realities and Identity**: The collapse and erasure of the boundary between Reality, Physical, and Virtual by digital twins raises deeper philosophical questions around reality, identity, and experience in a human-made universe. The dependence of digital twins on data collected from IoT sensors and other input devices introduces opportunities for issues related to data quality, accuracy, and completeness. The quality of data is crucial for the reliability and validity of simulations and the prediction of the digital twin.

3.4 Dual-Use and Misuse Potential

Dual-use technologies, such as AI, blockchain, and IoT, introduce ethical issues because they can be used for good and ill and require scrutiny of security risks and misuse protection. The built-in properties of blockchains, such as immutability and decentralization, create challenges from a regulatory and law enforcement perspective. The intermingling of the technologies and services may provide opportunities for innovation, but also leads to complicated ethical and legal issues, which need to be meticulously considered (Yang 2022).

- **Surveillance applications:** AI to living room, living room to spying room—AI surveillance raises enormous questions about privacy, freedom, and possible misuse; human rights and freedoms need to be considered carefully in order not to trample civil liberties. For IoT, the proliferation of internet-connected devices brings to the fore notions of surveillance, data integrity, and fears of being targeted for cyberattacks.
- **Military applications of AI-based twins:** AI in autonomous weapons systems poses moral questions on responsibility, human oversight, and the risk of inadvertent escalation, requiring international cooperation and regulation. Ethical implications arise from the weaponization of technology and the evisceration of human control in warfare. If digital twins are used in military applications, they could bring with them ethical issues related to the weaponization of technology and a lack of human control in warfare.
- **Falsification or manipulation of digital twin models:** Attackers could directly impact the fidelity of digital twin models and, as a result, cause faulty decisions, system malfunctions, or harm in the physical world. DT models and the data those models depend on can be manipulated to make flawed decisions, malfunction, or even result in physical harm. Digital twins can be vulnerable to different security and privacy issues, including data breaches, unauthorized access, and cyberattacks (Wang 2023).

3.5 Ethical Guidelines and Frameworks

Ethics for AI, blockchain, and IoT provide a foundation for responsible development, deployment, and use, promoting transparency, accountability, and respect for human values. As AI, blockchain, and IoT evolve, we need to develop and enforce sound ethical guidelines and policies to safeguard the responsible deployment and utilization of these technologies to yield a beneficial impact.

- **IEEE, OECD, UNESCO, and regional principles:** Institutions such as IEEE, OECD, and UNESCO are issuing ethical principles and guidelines for AI, offering principles for responsible AI innovation and use. AI governance aims to mitigate risks by preventing incidents of privacy breaches, misuse of personal information, bias, and discrimination. Numerous authorities commit to taking into account the ethical principles, although the interpretation and application of these principles in practice differ widely between recommendation documents (Dignum 2023). Recent research has focused mainly on the legislation of North American and European regulators. While international public organizations have also released visions for the development of ethical AI, none of these instruments have received a comprehensive literature review reflecting multiple perspectives. AI ethics guidelines stress the need to combine guidance with substantive ethical analysis and proper strategies of implementation.
- **Concepts:** Beneficence, Non-Maleficence, Justice, Autonomy, Explainability, Transparency: Ethical frameworks usually highlight principles like beneficence (doing good), non-maleficence (not causing harm), justice (fairness), autonomy, explainability, and transparency to drive the design and deployment of various AI, blockchain, and IoT technologies.

4. Legal Frameworks and Compliance Requirements

4.1 Overview of Regulatory Landscape

Regulation around AI, blockchain, and IoT is changing, with governments and regulatory bodies globally struggling to comprehend existing laws and apply them to the new features offered by these technologies. The lack of established regulation of blockchain technology is worrisome with regard to privacy, confidentiality, data security, and governance, as shown in Figure 3. The current laws and regulations, originally designed for a centralized governance mechanism, usually cannot adapt easily to the decentralized nature of the blockchain, making enforcement an issue and creating challenges for regulators (Menon 2021).

Table 3 provides a comprehensive overview of legal frameworks and compliance requirements across various domains.

- **Evolving nature of technology law:** The constant and fast development of new technologies means that the law needs to evolve nimbly and responsively to ensure that its regulatory frameworks continue to be fit for purpose to deal with evolving risks and opportunities. In the AI, blockchain, and IoT world, developments change at lightning speed, and the law needs to adapt fast to respond to new risks and opportunities, as well as to support creativity.
- **Fragmented global legal frameworks:** Legal systems vary by jurisdiction, meaning that an organization's ability to innovate and collaborate across borders can be limited by uncertainty and complexity. The piecemeal nature of global legal systems about AI, blockchain, and IoT is complex for global businesses to navigate, and organizations have to think carefully in terms of the ability to negotiate different regulatory landscapes. The absence of harmonized legal definitions across borders creates burdens for international businesses and requires careful consideration of multiple regulatory and compliance landscapes. Uneven regulation within a region/country can hinder the application of IoT in use cases like tracking and agri-food supply (Elijah 2018).

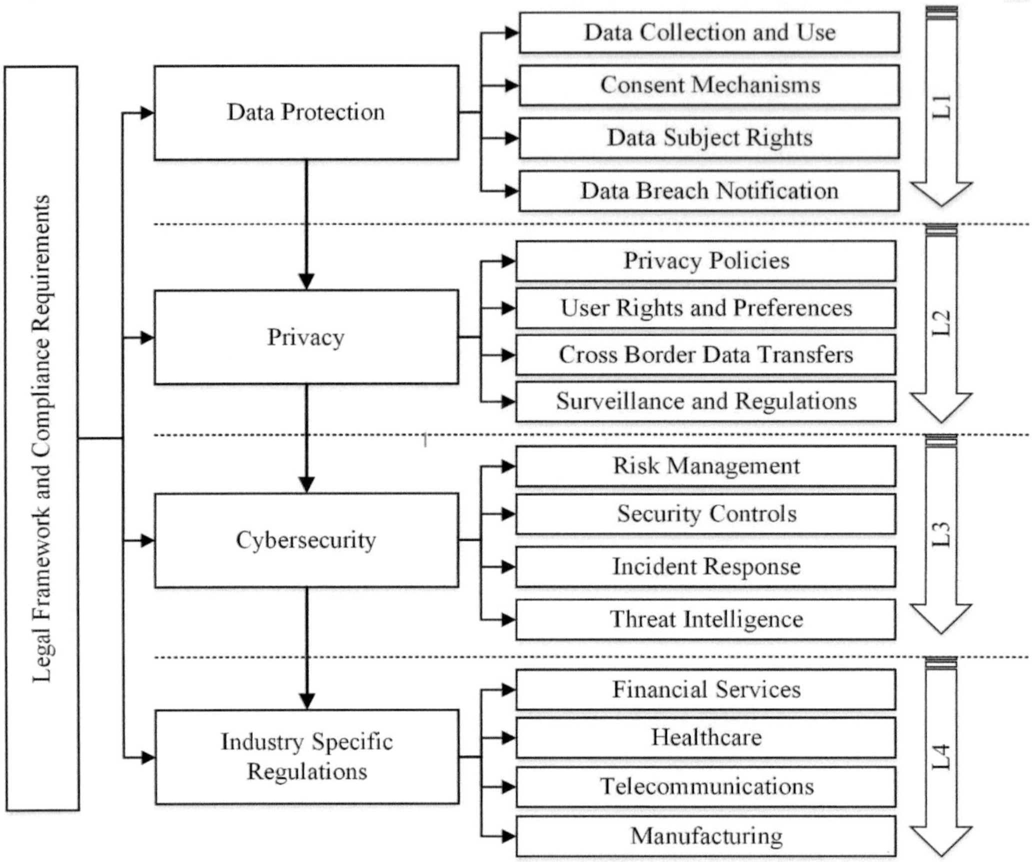

Figure 3: Legal frameworks.

4.2 Data Protection and Privacy Laws

The data protection and privacy laws are essential to dictate how personal data should be collected, processed, and used within the context of AI, blockchain, and IoT, to respect individuals' rights and to maintain responsible and fair data handling.

4.2.1 General Data Protection Regulation (GDPR)

GDPR has strict rules on the processing of personal data and addresses consent, data minimization, transparency, and security, which apply to AI, blockchain, and IoT applications that handle the personal data of individuals of the EU/EEA. The GDPR is designed to standardize data privacy laws across the EU and afford better protections and rights to individuals to access and control their personal information, establishing a myriad of requirements on companies related to safeguarding that data, rights of the data subject under the GDPR The GDPR confers rights on individuals (for example, the right of access to their data, the right to rectify or erase it, the right to restrict its processing, the right to data portability, and the right to object).

- ➢ **Legal bases for processing, consent and data subjects' rights:** The GDPR provides that organizations must have a legal basis for processing personal data (e.g., consent, performance of a contract, legitimate interests) and is designed to safeguard data subjects' rights, such as the right to access, rectify, and erase their data.
- ➢ **Challenges of real-time AI/IoT systems:** Real-time AI/IoT systems confront challenges for satisfying the data protection principles (i.e., data minimization, purpose limitation, and

Table 3: Legal Frameworks and Compliance Requirements

Aspect	Key Legal Frameworks	Compliance Requirements
Data Protection & Privacy	General Data Protection Regulation (GDPR), California Consumer Privacy Act (CCPA), Bill to protect individual privacy (India), Health Insurance Portability and Accountability Act (HIPAA)	The system should be secure by design and filled with user data securely collected, stored & processed.
Cybersecurity Regulations	NIST CSF (US), The ISO/IEC 27001 Standard for Information Security, EU Cybersecurity Act	Compulsory risk assessments & incident reporting, companies need to have good encryption practices, scheduled reporting, and compliance monitoring.
AI Governance & Ethics	EU AI Act (Europe), AI Bill of Rights (US), OECD AI Principles, National AI Strategy (multiple countries)	Transparency in the decision-making of AI, Responsible use of AI: fairness & accountability, Oversight & auditing of AI models for bias
Blockchain & Cryptocurrency Regulations	Promote the FATF guidelines on Proliferation Financing, S.E.C. (security exchange commission) laws, Laws relating to anti-money laundering (AML), Regulation (EU) on Markets in Crypto-Assets (MiCA) for exchanges	KYC (Know Your Customer) compliance, legal registration & taxation for crypto assets, Fraud protection & risk management
Internet of Things (IoT) Regulations	Internet of Things Cybersecurity Improvement Act (US), European IoT security regulations, IoT laws for various industries, Consumer Protection Laws for IoT Devices	Device login and encryption in an isolated environment, Consumer data privacy protections, Industry adherence to global safety standards
Intellectual Property & Copyright Laws	Berne Convention Relating to Copyrights and Neighboring Rights, The Digital Millennium Copyright Act (DMCA), Agreement on the Trade-Related Aspects of Intellectual Property Rights (TRIPS), Treaties of the World Intellectual Property Organization (WIPO)	The safety of patents, copyrights & trademarks, licensing & fair use policies, and the enforcement of intellectual property rights

transparency) and need to develop new approaches related to data governance and privacy-enhancing technologies. AI compliance with regulations, such as the AI Act of the EU, is difficult to implement, especially for edge, decentralized, resource-constrained devices in terms of processing power.

4.2.2 Challenges for real-time AI/IoT

systems: Real-time AI/IoT systems face challenges for compliance with data protection principles (especially data minimization, purpose limitation, and transparency), as well as the requirement to develop new solutions at the level of data governance and privacy-protecting technologies. Implementation of AI compliances (after regulations, e.g., AI Act of the EU) in machine systems on the edge—near resource accuracy constraints, decentralized, resource-limited nodes—would not be an easy exercise (in terms of compute). It is impractical to seek explicit authorization from data generators at this order of magnitude, and it is unclear whether there are other legal bases that enable processing of such personal data (Li 2023).

4.2.3 Other Regional Regulations

Data protection and privacy are also becoming areas in which regional regulations are developed, translating into a complex array of laws and requirements for compliant behavior by global companies. Liability is another major issue, especially in terms of AI and the responsibility for decisions made by AI that turn out to be harmful to people or organizations. For an organization operating the future system, the most important legal evaluation use case is its application in securing evidence for subsequent criminal prosecution. These regulations differ from those in the GDPR, and organizations operating in those areas should remember them. The Digital Personal Data Protection Act in India, the Lei Geral de Proteção de Dados in Brazil, and the Personal Information Protection Law in China all form comprehensive regulations regarding the processing of personal data, provide particular rights to individuals, and establish specific regulatory and other measures for data controllers and processors. Responsible AI, blockchain, and IoT developed by organizations—covering the intersection of data protection regulation with the latest technology trends—require comprehension of legal, ethical, and technical aspects. The legal and regulatory landscape for AI and robotics discusses issues such as safety standards, autonomous system liability, data protection, and intellectual property rights.

4.3 Legal Issues in Blockchain

Blockchain is deeply coupled with innovative legal issues of smart contracts, data persistence, and decentralized governance that need to be carefully analyzed from the perspectives of contract law, IPR, and compliance.

- **Smart legal systems:** Legal systems that can integrate smart contracts have to deal with theoretical gaps and conflicts with traditional legal principles, as well as with issues of interpretability, enforceability, and liability, regarding questions ranging from error management to dispute resolution and the legal recognition of code. Smart contracts are functions enabling automation of transactions in a transparent and immutable way. But the enforceability and interpretation of these smart contracts are legal challenges. Smart contracts are extremely popular because once they are fulfilled, they get executed automatically. Smart contracts facilitate business automation that takes place when contractual terms are satisfied. The condition and legal recognition of the smart contract are related to applicable law, which sometimes requires adjustments to traditional contract law principles to fit into the conditions. The immutable nature of blockchain raises questions around amending or removing data under Data Protection Laws. This is also a major challenge for the legal liability of a faulty smart contract to be executed. Legal systems addressing issues of liability and dispute resolution are required (Velasco 2022).
- **Immutability vs. Legal Rights:** The immutability of blockchain can conflict with obligations to comply with various laws, including data protection laws (e.g., the right to be forgotten under the GDPR), which require new solutions for data modifications and deletions. The non-modifiable nature of blockchain is a difficult scenario for adherence to data protection laws like the right to be forgotten under GDPR; therefore, there is a need to design systems for data modification or deletion without compromising blockchain integrity. The immutability of blockchain provides great transparency; however, it is a significant obstacle for data protection laws, such as the right to delete or change personal data. Compliance with privacy legislation and financial regulations (such as Know Your Customer requirements for financial transactions) poses a unique problem for integrating blockchain-based systems, for which there exists no optimal solution in the consequent selection of cryptographic operations, access control schemes, and governance strategies. In a blockchain system, the challenge of ensuring trustworthiness and selective data modification requires the application of advanced cryptographic methodologies, such as redactable blockchains or zero-knowledge proofs, which can provide data privacy and compliance with regulatory requirements. The GDPR also provides individuals with the right to request deletion of their data in some circumstances, which directly goes against one of the core features of blockchain: immutability. This tension could be solved with more developed solutions, for example, personal data stored off-chain with only hashes on the blockchain, or capabilities for data pseudonymization or encryption.

➢ **Decentralization and Jurisdiction:** The borderless nature of the blockchain ecosystem poses challenging issues of jurisdiction, governance, and regulatory supervision that can only be resolved meaningfully through international cooperation and the harmonization of legal standards. The dispersed and decentralized nature of blockchain technology complicates both the definition of jurisdiction and the identification of authorities responsible for overseeing it, as transactions and data may apply across multiple legal jurisdictions, hence the need for international cooperation and harmonization of legal norms. The decentralized nature of blockchain technology makes it difficult to pin down jurisdiction, as users (nodes, participants, etc.) could be distributed in different legal jurisdictions, making it a challenge for law enforcement and regulatory requirements. International cooperation and the establishment of shared standards are necessary to create jurisdictional regimes to regulate blockchain-based activities in jurisdictions worldwide in order to solve the legal issues of cross-border transactions and to achieve unified enforcement. Legal frameworks need to deal with issues of jurisdiction, responsibility of network actors, and dispute resolution in a decentralized world order. Lack of central authority in blockchain-based systems does not facilitate the judicial process for fraud or theft, or other legal aspects, and therefore alternative methods for dispute resolution and legal schemes for addressing unlawful behavior have to be provided. Legal systems should consider new issues such as the legal nature of crypto assets, anti-money laundering (AML), and investor protection in decentralized financial systems (Masluk 2018). One of the key challenges with blockchain is that it's hard to track nefarious transactions and who is behind them. Blockchain has enabled transactions to be documented in a peer-to-peer and less opaque way.

4.4 Legal Challenges in AI and IoT

Privacy and security issues are crucial in AI and IoT applications since the collection of massive personal data, processing, and analysis are involved, which are based on stringent laws and regulations to control information protection; otherwise, the risk will be huge for those technology deployments, and the data plan will be ignored. Due to the skyrocketing gathering, treatment, and utilization of personal data, privacy and security have become major concerns for the technology of AI and IoT (Alenizi 2024). Privacy and security are again at the forefront in the era of AI and IoT applications, and privacy should be preserved for various reasons (heavy collection and processing of personal data, monitoring of private spaces).

➢ **Autonomous Systems Liability:** Establishing liability for harm caused by autonomous AI systems or connected IoT devices raises novel legal questions, requiring us to rethink how law—including product liability, negligence, or accountability law—could apply to autonomous systems. Attribution of liability for robots requires an in-depth analysis of legal concepts and new rules designed to take into account the nature of AI and the IoT. Determining who to hold responsible for damage caused by autonomous AI systems or interlinked IoT devices becomes a very difficult legal problem, which includes questions of product liability, negligence, and responsibility, and for which we need new legal frameworks because the existing ones simply would not fit. The growing autonomy of AI systems gives rise to a range of difficult liability and fault questions in cases where the relevant human input is reduced or absent. Autonomous AI systems from a legal perspective raise doubts about the relevance of traditional concepts of liability and require a new framework for determining responsibility for damage caused by AI agents; moreover, considerations include the level of autonomy, predictability, and human intervention (Bashayreh 2021).

➢ **Intellectual Property:** The fact that AI systems—in theory—can produce original works implies that issues related to intellectual property rights, authorship, and ownership in connection with AI-generated output arise, which require clarification regarding the applicable legal frameworks designed to accommodate IP rights such as copyrights, patents, and trade secrets arising from AI-generated content. The IP implications of AI-created content (including copyright, patents, and trade secrets) will need to be carefully considered, and the legal landscape defined to manage questions of ownership, authorship, and infringement. Defining relevant legal frameworks to manage cases concerning copyright, patents, and trade secrets in AI-generated content is crucial for driving innovation and promoting equity in the distribution of related economic benefits (Guadamuz 2016). As AI systems have grown more capable of producing independently generated work, in the form of creative content and inventions, the question of authorship and ownership has become complex, forcing the current legal structure to be revisited.

- **Compliance in regulated sectors:** Adoption of AI, blockchain, or IoT in regulated sectors such as healthcare and finance mandates diligent adherence to sector-specific regulations, data protection laws, and ethical standards, demanding strong compliance and risk management programs. Use of AI, blockchain, and IoT in regulated areas (e.g., health and finance) requires industry-specific standards and compliance with data protection legal norms and ethics codes for responsible innovation and risk mitigation. AI, blockchain, and IoT adoption in such regulated industries as finance and healthcare depends on adherence to industry-specific regulations, data protection laws, and ethical codes. Thanks to blockchain, the supply chain can be traced, monitored, and controlled at all levels: from the origin of the resources to the final use by the end-user. Blockchain technology is handy when the supply chain needs to be tracked end-to-end.

4.5 Compliance and Risk Mitigation Strategies

- **Privacy by Design and Default:** Applying privacy-by-design principles when developing AI, blockchain, and IoT leads to the incorporation of privacy features in the design and development stage and helps reduce data collection, increase data security, and encourage transparency in the use of data. Privacy by design and by default should be applied to AI, blockchain, and IoT systems at the design and development stage to minimize data collection, improve data security, enhance transparency of data processing activities, and reduce privacy risks while ensuring compliance with data protection rules. The practical use of privacy by design in the context of AI-, blockchain-, and IoT-based systems serves to embed privacy protection into the early stages of product development, reduce data collection, increase security by default, and foster transparency regarding data processing operations. To minimize data collection, improve data security, and instill transparency into data processing procedures to mitigate privacy risks and meet requirements imposed by privacy laws, considering privacy during the design and development of AI-, blockchain-, and IoT-based systems is essential.
- **Impact Assessments (Data Protection Impact Assessments/DPIA, Algorithmic Impact Assessments/AIA):** Carrying out comprehensive impact assessments in advance of implementing AI, blockchain, and IoT systems makes it possible to anticipate and manage ethical, legal, and social risks, minimizing these risks and protecting fundamental rights. Carrying out impact assessments with the deployment of AI, blockchain, and IoT is needed in order to define potential ethical, legal, and societal implications, which would, in turn, allow organizations to take appropriate mitigation measures to ensure the protection of fundamental rights. This will enable organizations to prevent potential ethical, legal, and societal harm by ensuring any risks are identified and addressed early, and mitigation measures are properly put into place to protect fundamental rights when implementing AI, blockchain, and IoT systems. Impact assessments are necessary to detect potential ethical, legal, and social risks in AI, blockchain, and IoT systems, so that organizations can apply remedial actions and protect fundamental rights.
- **Certification and Standardization:** Embracing industry standards, certifications, and best practice methodologies in AI, blockchain, and IoT development and deployment demonstrates

a firm commitment to quality, security, and ethical behavior, which are all foundations of trust and responsibility in these technologies. Complying with industry norms, certifications, and best practices for AI, blockchain, and IoT development and deployment means a commitment to ensuring quality, security, and ethical behavior, bolstering trust and responsibility with these technologies. Trust and accountability in AI, blockchain, and IoT are improved by following best practices, industry standards, and certifications in the development and deployment of these technologies, demonstrating dedication to quality, security, and ethical behavior.

- **Regulatory sandboxes for innovation testing**: In the area of testing innovations, setting up regulatory sandboxes that allow innovators to test AI, blockchain, and IoT-based technologies in a controlled space encourages testing and learning for policy makers, reduces risks, and ensures responsible innovation. Sandboxing can provide a platform for innovators to test new AI, blockchain, and IoT solutions in a controlled environment, thereby promoting experimentation and regulatory learning that can help limit the pace of potential risks and support responsible innovation.

5. Transparency and Accountability in Blockchain-IoT

The intersection of blockchain with IoT raises novel transparency and accountability issues, which both present challenges and offer solutions, requiring new tools and techniques to ensure responsible and ethical use. The combination of blockchain and IoT can, however, be associated with specific challenges and opportunities in terms of transparency, accountability, and the need for novel mechanisms and guidelines to drive their responsible and ethical use, as shown in Figure 4. When blockchain is used in conjunction with IoT, it generates distinct transparency and accountability challenges, creating an urgent need to develop new approaches and frameworks to ensure ethical and socially responsible use. Table 4 demonstrates how blockchain increases visibility and transparency while introducing serious accountability issues in Blockchain–IoT ecosystem integration.

5.1 Convergence of Blockchain and IoT

5.1.1 Blockchain and IoT- The symbiotic technology move. Introduction about Blockchain and IoT

Together, these are two technologies that, when combined, have the power to revolutionize many industries. On its part, blockchain provides a real-time, secure, transparent, and immutable public record for transactions and data transfer. Battle et al. have secured this with robust security and improved data integrity, ensuring that data logging is tamper-proof and IoT devices are communicating more securely, which can mitigate cyber-attacks such as tampering and unauthorized intrusion. The application of blockchain to IoT in supply chain management can lead to greater transparency as not only does it provide an immutable history for the product origin, and the product location, but also the product status, and this allows accountability and trace of the product during the supply chain.

- **Use-cases:** Smart supply chains, smart homes/cities, health monitoring, digital identity: Blockchain–IoT convergence unlocks these and many other use cases, such as smart supply chains, smart homes, smart cities, health monitoring, and digital identity management. The convergence of blockchain and IoT enables novel applications including smart supply chains, smart homes/cities, health-care monitoring, and digital identity management.
- **Decentralized Automation:** The combination of IoT sensors with smart contract–based blockchain infrastructure leads to autonomous systems with no human in the loop, introducing new forms of automation and enabling new levels of efficiency. The use of smart contracts on the blockchain, powered by IoT sensor readings, permits a self-regulating decision-making process without human input, ensuring optimized operational efficiency.

Table 4: Transparency and Accountability in Blockchain-IoT.

Aspect	Transparency Considerations	Accountability Considerations
Data Integrity & Verification	IoT data with tamper-proof logs using blockchain, Clear documentation of data origins, Public vs. private blockchain for secure data sharing	Who will be held accountable for incorrect IoT data stored on the blockchain? Assigning blame for data manipulation or misuse requires verifying that your stakeholders are adhering to data validation protocols.
Privacy & Security	Transparent encryption for communication between IoT devices. Access controls can be secured by blockchain smart contracts, Transparent policy concerning private vs. public data use.	Liability for breaches caused by insecure IoT devices: Who is responsible for adhering to privacy laws (GDPR, CCPA)? Ethical issues of aggregated and wide-scale data collection and monitoring
Smart Contracts & Automation	Transparent implementation of pre-defined IoT rules through blockchain smart contracts, Auditable contract history, preventing fraud and misuse, eliminates reliance on 3rd parties with decentralized automation.	Who is responsible if an automated smart contract doesn't work or performs incorrectly? The legal status of smart contracts in various jurisdictions, attributing responsibility when effects are unintended
Supply Chain & Asset Tracking	Real-time blockchain-based monitoring of IoT assets, Clear records eliminate supply chain fraud, and Traceability provides authenticity and ethics.	Product liability doctrine for defects in Internet of Things-enabled devices, IoT-driven proof for fair labor and ethical sourcing, and the enforcement of compliance by global trade regimes
Regulatory & Compliance Requirements	Regulatory reporting in IoT transparency by blockchain, Open-access device & network compliance tracking, and traceable audit logs for regulators in the industry	Who is responsible for noncompliance in IoT devices? Legal exposure in unauthorized adoption of IoT devices, and sanctions for not respecting the security and transparency rules

5.2 Transparency Challenges

While the combination of blockchain for IoT is undoubtedly promising, it also presents significant transparency challenges that need to be considered in order to ensure the responsible and ethical use of this technology. The fact that blockchain is immutable, although an advantage for security, can cause problems when trying to cope with IoT data sources, as biased or incorrect information can allow false statements to permeate the network. Another important issue lies in data privacy when communicating between IoT devices and blockchain, since the data collected by IoT devices can be recorded on the blockchain, which leads to risks regarding privacy and confidentiality.

- **Technical Complexity:** The private sector's use of cryptography and blockchain as its underlying architecture renders it difficult for the public to understand. Because of the technical complexity involved in blockchain and cryptography, the average person does not understand it, which can lead to public distrust despite the technology's potential to promote transparency and accountability. The public's limited comprehension of blockchain technology and cryptography can make it harder to create an atmosphere of transparency and accountability.
- **Black-box Algorithms**: Decisions in AI-based IoT systems are frequently opaque and inscrutable, which complicates the ability to interpret and verify the justification for autonomous actions and recommendations. The transparency of blockchain can be compromised by the low explainability of AI-driven IoT decisions, which might make it hard to understand and verify why automated actions or recommendations are made.
- **Data Provenance:** It is difficult to maintain data provenance on IoT networks due to the distributed nodes of sensors and the possibility of malicious data manipulation, which can make it hard to trace the origin and veracity of sensor data. The decentralized nature of devices, along with data tampering possibilities, makes it challenging to maintain data provenance in IoT networks, rendering it difficult to trace the source or originality of sensor data.

5.3 Accountability Dilemmas

Accountability in Blockchain-IoT environments introduces difficult trade-offs, particularly regarding liability for autonomous actions and data breaches, and consequently requires the development of coherent legal and ethical frameworks. It is difficult to establish which autonomous device actor is liable for damages caused by autonomous IoT devices in blockchain networks, as devices can make decisions autonomously based on smart contracts.

- **Decentralized architecture:** This issue relates to the decentralized nature of blockchain, which means that no one can be held responsible for system failures or security breaches, as there is no single central authority. The distributed aspect of blockchain means no one entity is responsible for system failures or security issues, simply because no single organization runs the system. The attribution of blame is also difficult; the blockchain is so disintermediated that there is no central body that can be held responsible if the system fails or an exploit occurs. In multi-stakeholder, peer-to-peer settings, blame attribution is hard because multiple parties are involved in a transaction or decision process.
- **Exploitability and Legal Risks:** The reliance on smart contracts for automated decisions in blockchain-IoT systems introduces risks related to the legal enforceability of the code, and unforeseen consequences, exposing developers and users to failure and liability for bugs in the smart contracts. The use of smart contracts in Blockchain-IoT is one way such decision that can drive the efficiency of systems, and it raises questions about the legal status of code and the liability of developers and users for the consequences of smart contracts' errors. In the context of blockchain and the Internet of Things, this raises questions about the automation of processes through smart contracts, the enforceability of code, and the liability of developers and users for potential errors or unforeseen side effects resulting from smart contracts. When there is an error in a smart contract code or when the code is hacked, standard legal remedies may not be applicable, creating problems for dispute resolution and accountability.

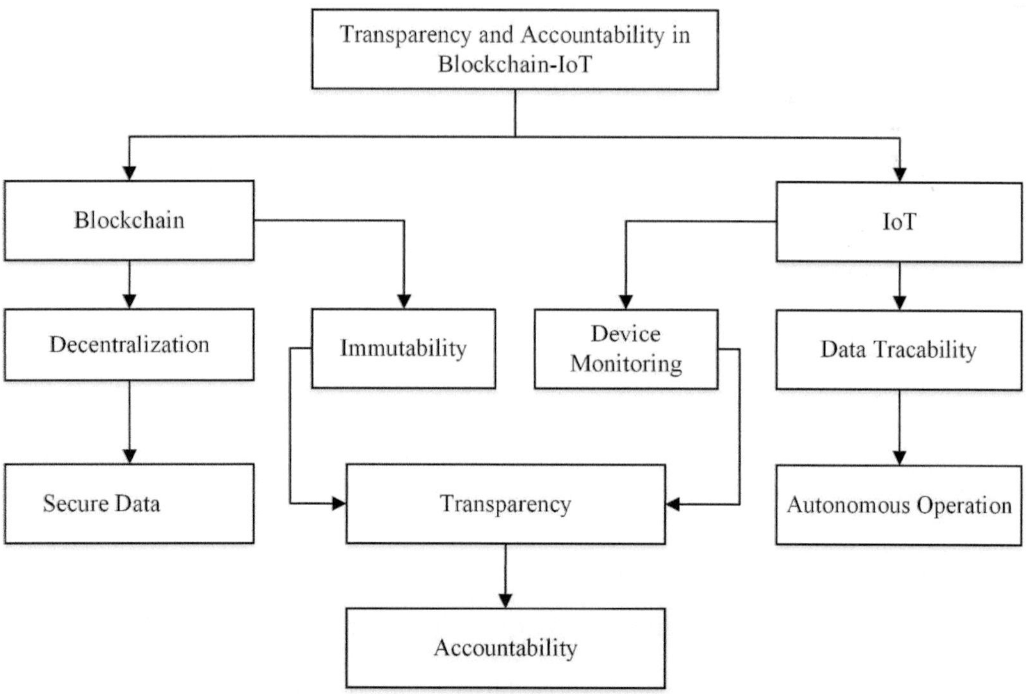

Figure 4: Transparency and Accountability emerge in Blockchain-IoT Systems.

- **Systemic Failure:** The latter difficulty, that of assigning liability in complex systems, the intended behavior of which is not necessarily able to be fully specified and where events in the operating environment are unpredictable, will create responsibility/liability gaps where the systems are concerned and robust human control and foresight are restricted. Within complex AI systems, the difficulty of responsibility attribution can lead to responsibility and, therefore, liability gaps when the AI's functioning does not lend itself to full specification and when the unpredictability of events in the operating environment may limit robust human control and foresight.

5.4 Strategies for Enhancing Trust

Reliable and predictable solutions to stimulate ethical and responsible AI, blockchain, and IoT must be holistic and transformative, with trust, transparency, and accountability as integrated components. In order to ensure the ethical development and responsible use of AI, blockchain, and IoT systems, strategies need to be put in place to build trust, transparency, and accountability. Complicating trust-building in automation involves more than simply producing reliable models. To maintain ethically responsible AI, blockchain, and IoT system development, a new proactive approach is needed to bring about accountability, transparency, and trust.

- **Transparent Protocols:** Using an open-source development process and auditable logic is crucial in building trust for AI algorithms and systems. It is of utmost importance to make AI robust and transparent to enable clear and understandable explanations for high-risk AI applications. To develop reliable AI, we need high-risk AI applications to be accurate, robust, transparent, and explainable. We need to use open-source, traceable reasoning at the core of AI algorithms and systems.
- **Audit-Trail/Logging:** Full-fledged log and audit-trail mechanisms support forensic investigations and accountability in Blockchain–IoT environments. The total system is auditable by any authorized user, maintaining a chain of custody. The use of extensive audit trails and logging functionalities can enable forensic investigation and accountability in Blockchain–IoT systems. To support forensic capabilities and accountability in Blockchain–IoT systems, detailed audit trails and logs can be utilized.
- **Governance Models:** Formulate clear governance structures and regulatory sandboxes for AI, blockchain, and IoT that offer compliance guidance and ethical standards. Creating clear governance structures and regulatory sandboxes in the AI, blockchain, and IoT arenas can inform rules of ethical and lawful behavior. For AI, blockchain, and IoT, regulatory sandboxes and governance frameworks can be developed to shape ethical behavior and compliance (Díaz-Rodríguez 2023).

5.5 Case Studies

Examining live examples that show ethical and legal problems encountered in AI, blockchain, and IoT will be useful for stakeholders. Real-world case studies that illustrate the ethical and legal implications of AI, blockchain, and IoT in practice can provide helpful insights for all stakeholders. By considering use cases demonstrating ethical and legal concerns in AI, blockchain, and IoT, we hope to provide valuable guidance to interested parties.

- **Blockchain-IoT-based Food Traceability:** Several companies have started using IoT with blockchain for food traceability, such as tracing whether a food item is coming from a nearby farm, and so on. The application of the whole agricultural supply chain onto the base system—covering product registration, transactions, and transport—allows the consumer to verify the accuracy and efficacy of their purchases (Jothikumar 2021). Furthermore, combining IoT and blockchain improves the performance of food traceability and information transparency.

Table 5: Ethical challenges in data-driven systems along with key risks and mitigation strategies.

Ethical Challenge	Description	Key Issues	Potential Impacts/ Risks	Mitigation Strategies
Privacy and Data Protection	The gathering, storing, and, to some degree, processing of mountains of personal and sensitive data.	Minimizing Data: Collecting too much data, Anonymization that doesn't work, and re-identification, Security breaches that leave sensitive data and their owners' privacy in public photos or videos, Stealth data collecting without the knowledge of users, Misuse of biometric data	Distrust and Reputational harm, Identity theft, and theft of financial information., Discrimination and abuses undermine the autonomy of individual personal data. Legal and regulatory sanctions, for example, GDPR fines	Adopt Privacy by Design strategies, Strong encryption and user access controls, Regular security audits and vulnerability assessments, no data retention policy, Privacy-enhancing technologies (PETs) such as differential privacy
Bias and Discrimination	A major challenge is that data-driven systems often learn and replicate hidden, biased knowledge, which can eventually lead to unfair results.	Biases from the past due to social inequality, and the bias of sampling due to no representative data (collecting bias). Measurement bias based on skewed methods of collection, Algorithmic bias in decision making, and Cognitive bias by developers impacting model design	Discrimination in hiring, credit, health care, policing, Stereotyping at its best, Increasing social disparities, Erosion of public trust, Legal threats, and reputational harm	Data sources and developer teams should be diverse. Model bias audits regularly, Fair input constraints in algorithms, Use bias impacts assessments, and advance the development of AI with inclusivity.
Transparency and Explainability	Complex models also have the effect of making decision-making opaque, weakening transparency and trust.	Opaque algorithms are hard to explain, Dedication in the AI's decisions, Low transparency for users in explaining AI-generated decisions, Issues with compliance (i.e., GDPR's right to explanation)	Culture of not being held accountable for mistakes and bias, Overcoming mistakes is a journey in itself, Disparagement of trust and acceptance of AI, Risks related to non-compliance with regulation.	Create an Explainable AI (XAI) methodology. If possible, rely on interpretable models. Have good documentation and audit trails, and have humans in the loop for important choices.
Accountability and Responsibility	Ambiguity about who is responsible for mistakes or damage due to AI-directed decisions.	Share of responsibility between more than one player, Pick an object independently without human help. Negative outcomes from automation by parties outside the system	No compensation to injured persons, Difficult to pin down blame, Difficulties in corrective action enforcement, Risk of moral disengaging	Set up precise legal scenarios regarding the issue of AI liability, put in place rigorous and comprehensive oversight provisions, require algorithmic impact assessments, and clearly define stakeholder roles and responsibilities

Table 5 cont. ...

...Table 5 cont.

Ethical Challenge	Description	Key Issues	Potential Impacts/ Risks	Mitigation Strategies
Informed Consent	Making sure users are fully aware of and consent to the ways their data is utilized.	Complex terms of service no one reads, Scope creep: Data collected for one purpose and reused elsewhere, Unpredictable AI and what future uses look like, Withdrawal of consent is hard	Extent of individual freedom, Legal non-compliance risks, Public mistrust and backlash, Using the vulnerable population as a means to an end	Write consent forms in clear, easy-to-understand language, Granular consenting options, Easy withdrawal possibilities to be made available, Promptly & regularly inform their users about data-sharing settings.
Data Quality and Integrity	Data quality, such as accuracy, completeness, and consistency, influences AI outcomes.	Erroneous or outdated data, Missing values that give the wrong impression, Compiling of different sources, Bad input making bad predictions	Wrong predictions and decisions, Discriminatory outputs on inconsistent data, Squandered resources and inefficiencies, Uncertainty about the reliability of AI	Have robust data quality checks in place, Frequent updates and validation of the dataset, transparently present limitations of data, and Model-based governance for precision
Unintended Consequences and Societal Impact	How AI will impact people, communities, and society at large.	Automation job displacement, Echo chambers and filter bubbles, Surveillance and privacy decay, The digital divide is reinforcing inequalities, Dual-use dilemmas (use/misuse in warfare)	Social, financial & political instability, Polarization and democracy decline, Loss of individual freedom, growing socio-economic gap, AI applications with detrimental effects such as autonomous weapons	Pre-deployment impact assessments: Assist in the development of pre-deployment impact assessments, ensure the participation of multiple sections of the community in the construction of AI, include ethical considerations in system design decisions, and shape policies to regulate societal impact
Data Ownership	The struggle for ownership and control of the data that has been gathered.	The theoretical framework for data ownership is not legally clear. There is a lack of intellectual property protection for user-generated data.	Loss of user control over data, the use of data for commercial purposes without recompense, Litigation around data sharing and monetization	Establish clear legal regimes for who owns data, give people control of and monetization for their data, and promote social/governance models of collaboration (e.g., data trusts, etc.).

- **Power grid optimization:** Power grid operators can automate energy delivery and maximize resource utilization by using smart contracts in conjunction with IoT sensors. Still, the issue of data privacy and the risk of cybersecurity breaches need to be resolved to ensure the trustworthiness and reliability of grid assets.
- **Contact tracing in outbreak situations:** AI-supported contact tracing apps can quickly identify and notify users who may have come into contact with contagious diseases. But gathering and analyzing private health information is concerning, and the risk of surveillance and discrimination must be weighed carefully. The privacy implications of the collection and use of personal health information should be considered, and the risk of surveillance and discrimination should be carefully examined.

The intersection of Artificial Intelligence, Blockchain, and the Internet of Things provides new possibilities as well as new challenges regarding ethical and legal issues. It is necessary for the parties involved to join forces and create inclusive enablers and approaches for responsible innovation, protection of citizens' rights, and trust in these promising technologies.

6. Ethical Challenges in Data-Driven Systems

Data-intensive systems can bring enormous benefits in terms of innovation and efficiency but also pose a range of ethical challenges. These are problems that arise out of the nature of gathering data, processing and analyzing it, and deploying systems that base decisions on that data, as shown in Figure 5. Table 4 summarizes the ethical challenges in data-driven systems and the corresponding high-level risks and mitigations.

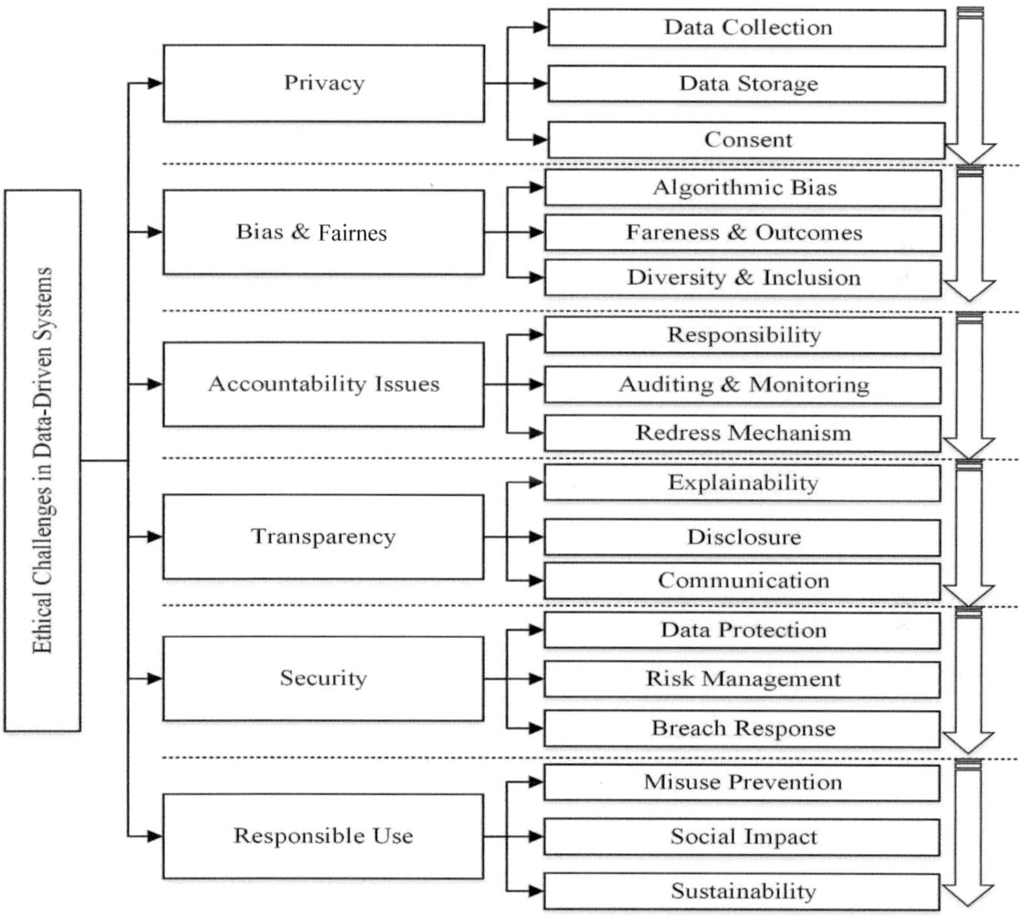

Figure 5: Ethical Challenges in Data-driven Systems.

6.1 Introduction to Data-Driven Systems

The rise of data-driven systems, powered by the growing presence of large datasets and powerful computing resources, is reshaping many domains, such as healthcare, finance, transportation, and urban planning. Data-driven systems, driven by the proliferation of large datasets and high-powered computing, have profoundly transformed many fields, including healthcare, finance, transportation, and urban planning. Data-driven systems, enabled by advances in the availability of huge datasets and powerful computation, have led to transformative innovations across applications, including health, finance, transportation, and urban planning. Machine learning techniques, such as neural networks, support vector machines, and decision trees, are important for data processing and analysis, enabling intelligent decisions and automation (Tzachor 2020). These techniques are applied to predictive modeling, pattern recognition, and anomaly detection, thus improving productivity and efficiency in many industries. There are machine learning approaches—e.g., neural networks, support vector machines, or decision trees—which are essential for the processing and analysis of data, expanding automation and sophisticated decision-making. But the growing reliance on data-based systems also raises significant ethical issues that must be carefully considered. Yet computational systems that rely on data also carry significant ethical implications that need to be carefully handled.

6.2 Consent, Control, and Autonomy

Ethical challenges of consent, control, and autonomy are even more relevant in this type of data-reliant system, thus warranting close examination to ensure the rights and interests of individuals are safeguarded. Obtaining valid informed consent depends on open data collection practices, easily comprehensible privacy notices, and user-friendly devices that permit individuals to give explicit permission for the use of their data. This includes stating the purpose of the data, data sharing, and data retention times.

- **Informed Consent:** Explicit consent is required for data capture using IoT devices, which often capture data in the background. Meaningful consent is difficult to achieve because, among other reasons, privacy policies are complex, there is a risk of coercion or duress, and there are issues associated with granting and withdrawing consent. AI advancement continually challenges privacy, especially when entities still hold substantial rights in the data collected, even if individuals consider themselves the owners of that data. Moreover, blockchain as a decentralized ledger can be used for self-sovereign identity (SSI), which grants users the right to manage their identities and selectively share data with third parties, making blockchain a possible solution for some of these issues.
- **Persistent Data Harvesting:** The continual collection of data on individuals (often with or without their knowledge or consent) poses moral and privacy issues regarding control, autonomy, and personal data. Users should be offered choices that enable them to control their privacy preferences and decide whether they want to be part of data collection. Establishing mechanisms that give individuals greater control over data should be a priority in fostering an atmosphere of transparency and accountability.
- **Behavioral Influence**: Algorithms of digital platforms are capable of influencing users' decisions by tailoring information, interactions, and choices. Algorithmic openness and accountability are crucial to preventing such distortions. Transparency is also key in building trust and confidence among users in how algorithms affect their online activity.

6.3 Data Ownership and Rights

Ownership and rights of data are ethical and legal issues that are relevant in the era of AI, blockchain, and IoT and require a deep study of the various aspects involved. The concept of data ownership is often ambiguous, especially when it comes to personal data generated through complex relationships between people, devices, and online services. Even where companies offer people tools that allow them to download their data, the process of data gathering, sharing, and usage is so complex that most people are not equipped to do so.

- **Who Owns the Data?** Defining frameworks for who owns or has rights to data, in clear and fair ways, is critical in the increasingly complicated data landscape, ensuring that people maintain control over their personal information. The ownership of data has been a widely discussed issue, specifically within domains like digital agriculture, where divergent perspectives on who is entitled to govern agricultural data can be identified. The return on investment is very good.

The ability to reconcile the private and public sides of data is key, as is repairing farmers' trust relationships.

- **Ethical Data Stewardship** (Responsible storage, sharing, and use of data): Ethical data stewardship is critical to maintaining the public's trust and acting with integrity in managing data. This also includes developing and enforcing stringent protocols and standards in terms of how data is ethically collected, stored, shared, and utilized. The embedding of traceability mechanisms can contribute to an enhancement of transparency and accountability in data-driven practices.
- **Self-Sovereign Identity (SSI):** The utilization of blockchain could potentially enable self-sovereign identity systems, which allow individuals to own and control their digital identities and share or selectively disclose personal data to third parties. Such systems allow users to take control of their digital identities without relying on any central authority, providing better privacy and security. By distributing data control among a broad group of people, blockchain provides a transparent and credible ecosystem where it is tough to manipulate data.

6.4 Inequity and Discrimination

One of the biggest ethical challenges in AI, blockchain, and IoT is how to avoid continuing and expanding inequity and the perpetuation of discrimination, and support fairness and systematic strategies for mitigating bias must be developed. Algorithmic bias, which stems from bias or skewed representativeness in the data used to teach an algorithm, can create disparities across a range of sectors such as healthcare, finance, and criminal justice. This is because AI systems are trained on historical data, which is already biased, and so these become biased models and recreate biased or unfair results.

- **Algorithmic Bias:** Algorithmic bias is a significant risk as algorithms can inadvertently favor and discriminate against some demographic groups if the training data lacks diversity or harbors historical prejudices. The pressing issue of AI recruitment tools, as an example, is of extreme concern when it comes to technical and managerial solutions and fair algorithms for healthcare decisions.
- **Access Divide:** The access divide—that is, unequal access to technology and digital resources—raises significant ethical concerns, most especially for minorities. It is important to tackle the root causes of digital exclusion and ensure that everyone has fair access to the advantages of AI, blockchain, and IoT. The digital divide between developed and developing countries underscores the need for comprehensive approaches in terms of infrastructure, affordability, and digital literacy.
- **Inequalities and the Challenges of the Future of Work:** he transformative role of artificial intelligence in automation may increase economic inequalities by displacing workers in specific sectors. There is a requirement to proactively find ways to mitigate job losses in affected sectors and to provide opportunities for retraining. The negative impact on society from AI-driven automation needs to be addressed through assisting displaced workers, ensuring equal access to training, and through regulation that mitigates the consequences of technological progress.

6.5 Ethical Risk Mitigation

Ultimately, the responsible use of AI, blockchain, and IoT technologies requires effective ethical risk mitigation, including strong governance models, codes of ethics, and ongoing oversight. Establishing ethics for mitigating bias in AI systems is crucial, and this extends to how well those systems also learn patterns about people from data.

- **Boards and Panels:** Companies can also set up boards and panels within their organization to oversee and provide direction on ethical issues associated with AI, blockchain, and IoT initiatives to ensure that ethics is incorporated into the development and release processes. Third-party ethics commissions provide a structured approach for a deep dive into the moral implications of AI efforts, driving adherence to ethical norms and best practices.

➢ **Algorithmic Audits and Impact Assessments:** Periodic algorithmic audits and impact assessments can detect and address biases and unintended consequences in AI systems to ensure alignment with ethical principles and societal norms. By analyzing algorithms and measuring their social impact, potential bias and unforeseen consequences are revealed, thereby holding AI development accountable and encouraging transparency.

➢ **Stakeholder Engagement in Design and Deployment:** Including ethicists, policymakers, and community representatives in the design and deployment of AI, blockchain, and IoT projects can ensure that diverse viewpoints are taken into account and ethical concerns are addressed proactively.

The **perceived severity or impact** on a 0–100 scale is given in Table 5a and shown in Figure 6 also.

Table 5a: Ethical Challenges in Data-Driven Systems.

Ethical Challenge	Perceived Impact (0–100)
Privacy Violations	95
Bias and Discrimination	90
Lack of Transparency	85
Inadequate Consent Mechanisms	80
Data Security and Breaches	85
Accountability Gaps	75
Data Ownership Ambiguity	70
Surveillance and Misuse	80
Inequitable Access to Data-Driven Benefits	65

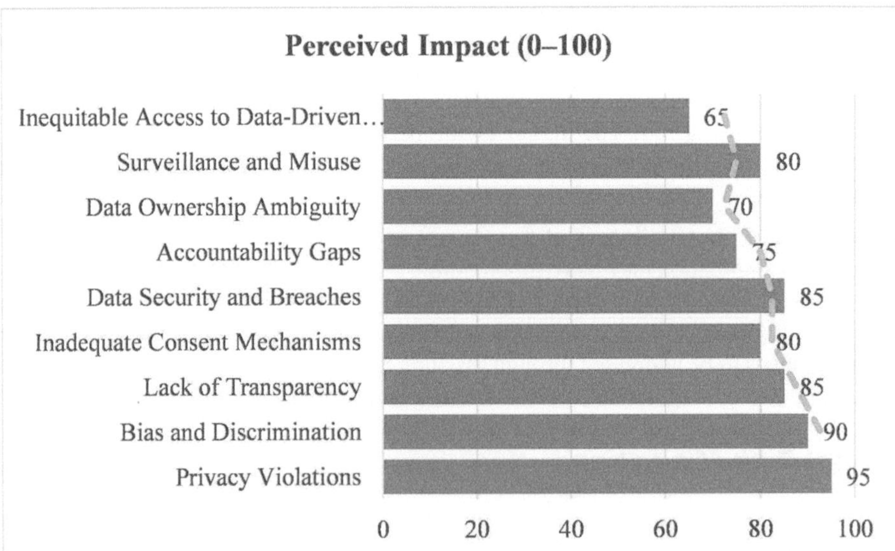

Figure 6: Perceived Severity or Impact on a 0–100 Scale.

7. Conclusion

This chapter has investigated the complex world of ethical and legal issues that accompany the implementation of AI, blockchain, and IoT technologies, highlighting the need to adopt risk management on ethical aspects in the first place. The heart of the chapter's ethical and legal discussion is its close connection, reflecting the importance of an integrated approach to governance. Ethical and legal challenges that are addressed in isolation from one another can result in governance that is both fragmented and ineffective, highlighting the necessity for integrated approaches that are sensitive to the deep connections between these challenges. Developments in AI, blockchain, and IoT are boosting opportunities and risks: these simultaneous advances in AI, blockchain, and IoT are gradually increasing opportunities and risks and therefore require holistic governance frameworks that consider the specific threats presented by such a merging of technologies. Strong policy and governance frameworks are needed to shape the responsible use and deployment of AI, blockchain, and IoT, calling for governments, industry, and civil society to work together. What the government can do: the government can take a serious look at the policy and regulation of AI, blockchain, and IoT to facilitate ethical and responsible use of these technologies. It requires interdisciplinary collaboration between industry, government, and civil society to design practical and efficient government strategies. For the future, research in this area must strive to develop flexible, collaborative, and inclusive regulatory frameworks that will follow the rapid developments of AI, blockchain, and IoT. Interdisciplinary collaboration, direction, and continued review, along with flexible regulation, are essential in addressing AI ethical challenges. It is important to continuously assess and monitor AI systems to identify and mitigate evolving ethical issues and social impacts.

References

Alenizi, A., Mishra, S., and Baihan, A. (2024). Enhancing secure financial transactions through the synergy of blockchain and artificial intelligence. *Ain Shams Engineering Journal*, 15(6): 102733. doi: 10.1016/j.asej.2024.102733.

Bakiner, O. (2023). The promises and challenges of addressing artificial intelligence with human rights. *Big Data & Society*, 10(2): 20539517231205476. doi: 10.1177/20539517231205476.

Bashayreh, M., Sibai, F. N., and Tabbara, A. (2021). Artificial intelligence and legal liability: towards an international approach of proportional liability based on risk sharing. *Information & Communications Technology Law*, 30(2): 169–192. doi: 10.1080/13600834.2020.1856025.

Binns, R., Van Kleek, M., Veale, M., Lyngs, U., Zhao, J., and Shadbolt, N. (2018, April). 'It's Reducing a Human Being to a Percentage' Perceptions of Justice in Algorithmic Decisions. pp. 1-14. *In: Proceedings of the 2018 Chi conference on human factors in computing systems*. doi: 10.1145/3173574.3173951.

Díaz-Rodríguez, N., Del Ser, J., Coeckelbergh, M., de Prado, M. L., Herrera-Viedma, E., and Herrera, F. (2023). Connecting the dots in trustworthy Artificial Intelligence: From AI principles, ethics, and key requirements to responsible AI systems and regulation. *Information Fusion*, 99: 101896. doi: 10.1016/j.inffus.2023.101896.

Dignum, V. (2017). Responsible autonomy. *arXiv preprint arXiv:1706.02513*. doi: 10.48550/arxiv.1706.02513.

Dignum, V. (2018). Ethics in artificial intelligence: introduction to the special issue. *Ethics and Information Technology*, 20(1): 1–3. doi: 10.1007/s10676-018-9450-z.

Dignum, V. (2023). Responsible artificial intelligence: Recommendations and lessons learned. pp. 195–214. *In: Responsible AI in Africa: Challenges and opportunities*. Cham: Springer International Publishing. doi: 10.1007/978-3-031-08215-3_9. doi: 10.1007/978-3-031-08215-3_9.

Elijah, O., Rahman, T. A., Orikumhi, I., Leow, C. Y., and Hindia, M. N. (2018). An overview of Internet of Things (IoT) and data analytics in agriculture: Benefits and challenges. *IEEE Internet of things Journal*, 5(5): 3758–3773. doi: 10.1109/jiot.2018.2844296.

Fuller, A., Fan, Z., Day, C., and Barlow, C. (2020). Digital twin: enabling technologies, challenges and open research. *IEEE access*, *8*, 108952-108971. doi: 10.1109/access.2020.2998358.

Guadamuz, A. (2016). The monkey selfie: copyright lessons for originality in photographs and internet jurisdiction. *Internet Policy Review*, 5(1): 1–12. doi: 10.14763/2016.1.398.

Jothikumar, R. (2021). Applying blockchain in agriculture: A study on blockchain technology, benefits, and challenges. pp. 167–181. *In: Deep learning and edge computing solutions for high performance computing*. Springer, Cham. doi: 10.1007/978-3-030-60265-9_11.

Kamilaris, A., Cole, I. R., and Prenafeta-Boldú, F. X. (2021). Blockchain in agriculture. pp. 247–284. *In:* Galanakis, C. M., (Ed.), *Food Technology Disruptions*. Elsevier Doyma. https://doi.org/10.1016/B978-0-12-821470-1.00003-3.

Kapteyn, M. G., Knezevic, D. J., Huynh, D. B. P., Tran, M., and Willcox, K. E. (2022). Data-driven physics-based digital twins via a library of component-based reduced-order models. *International Journal for Numerical Methods in Engineering*, 123(13): 2986–3003. doi: 10.1002/nme.6423.

Li, R., Allal, L. B., Zi, Y., Muennighoff, N., Kocetkov, D., Mou, C. et al. (2023). Starcoder: may the source be with you!. *arXiv preprint arXiv:2305.06161*. doi: 10.48550/arxiv.2305.06161.

Masluk, A., and Gofman, M. (2018). Protecting personal data with blockchain technology. pp. 119–125. *In: Information Technology-New Generations: 15th International Conference on Information Technology*. Springer International Publishing. doi: 10.1007/978-3-319-77028-4_19.

Menon, S., and Jain, K. (2021). Blockchain technology for transparency in agri-food supply chain: Use cases, limitations, and future directions. *IEEE Transactions on Engineering Management*, 71: 106–120. doi: 10.1109/TEM.2021.3110903.

Tzachor, A., Whittlestone, J., Sundaram, L., and Heigeartaigh, S. O. (2020). Artificial intelligence in a crisis needs ethics with urgency. *Nature Machine Intelligence*, 2(7): 365–366. doi: 10.1038/s42256-020-0195-0.

Vatamanu, A. F., and Tofan, M. (2025). Integrating Artificial Intelligence into Public Administration: Challenges and Vulnerabilities. *Administrative Sciences*, 15(4): 149. doi: 10.3390/admsci15040149.

Velasco, C. (2022, May). Cybercrime and Artificial Intelligence. An overview of the work of international organizations on criminal justice and the international applicable instruments. pp. 109–126. *In: ERA Forum* (Vol. 23, No. 1,). Berlin/Heidelberg: Springer Berlin Heidelberg. doi: 10.1007/s12027-022-00702-z.

Wang, Y., Su, Z., Guo, S., Dai, M., Luan, T. H., and Liu, Y. (2023). A survey on digital twins: Architecture, enabling technologies, security and privacy, and future prospects. *IEEE Internet of Things Journal*, 10(17): 14965–14987. doi: 10.48550/arxiv.2301.13350.

Yang, W., Wang, S., Yin, X., Wang, X., and Hu, J. (2022). A review on security issues and solutions of the internet of drones. *IEEE Open Journal of the Computer Society*, 3: 96–110. doi: 10.1109/ojcs.2022.3183003.

Zheng, Z., Xie, S., Dai, H. N., Chen, X., and Wang, H. (2018). Blockchain challenges and opportunities: A survey. *International Journal of Web and Grid Services*, 14(4): 352–375. doi: 10.1504/ijwgs.2018.10016848.

18

Future of AI-based Applications in Education Sector

Challenges and Scope

*Perarasi T.** and *Manoj R.*[1]

1. Introduction

The integration of Artificial Intelligence (AI) in education has developed considerably in the last few decades, which marks a change in data-operated models, refined by rules-based systems. The first notable application of AI in education emerged as intelligent tuition systems (ITS) in the 1970s and 1980s. These systems were designed to imitate the behavior of a human tutor by delivering personalized instruction, assessing the student's understanding, and providing feedback. Classic examples such as scholars, Guiden and Sukratic tutors used symbolic AI methods to map directed by direct logic and learner reactions (Wolf 2010).

In the 1990s and early 2000s, AI applications began taking advantage of Bayesian models, decision trees, and case-based reasoning to improve the adaptability and accuracy of these systems. Systems such as Cognitive Tutors (developed at Carnegie Mellon University) could now learn from interactions and provide dynamic problem-solving assistance.

By the 2010s, Machine Learning (ML) and Natural Language Processing (NLP) brought about major progress in learning analytics, adaptive assessment, and automatic grading. Online learning platforms such as Coursera and edX began embedding ML models to analyze learner behavior, recommend personalized content, and predict learning outcomes.

The most recent success in the 2020s is the rise of generic AIs run by large language models (LLM) like GPT-3, GPT-4 and Cloud. These models can now:

- Generate human-like text and material
- Provide real -time reaction and aid as virtual teaching assistants
- Summarize textbooks or create quizzes automatically
- Enable conjunct agents to learn self-book

This development indicates a paradigm change: AI is no longer limited to tutoring.

Department of Electronics and Communication Engineering, Bannari Amman Institute of Technology, Erode, Tamil Nadu, 638 401, India. Email: manojramesh2005@gmail.com

* Corresponding author: pera.sweety@gmail.com

DOI: 10.1201/9781003655190-22

1.1 Significance of AI in Modern Pedagogy and Institutional Transformation

AI technologies are re-defining key educational processes including teaching, learning, administration, and policy-making. Personalized learning, shifting away from "one-size-fits-all" models, is one of the most transformative contributions of AI. By continuously analyzing learner data, AI systems can tailor content, suggest remedial strategies, and adjust learning pace and complexity. For teachers, AI provides tools for automated grading, student progress monitoring, risk flagging, and predictive analytics. In higher education, AI supports research productivity, resource allocation, and admissions through data-powered insights. Globally, the institutes are taking advantage of AI to achieve the following: apply adaptive course delivery, support inclusive education through multilingual NLP tools and accessibility features, enhance safety, scheduling and resource management, enable smart complaint, introduce the ecosystem of lifelong learning using AI for upscaling and grouping, Educationally, AI strengthens creative and inquiry-based learning, where learners construct knowledge with AI agents through exploration, simulation, and feedback loops. This is particularly powerful when AI is integrated with technologies such as augmented/ virtual reality (AR/VR), learning analytics dashboards, and blockchain for credentials.

1.2 Objectives of the Chapter

The purpose of this chapter is to serve as a comprehensive guide to understanding the future direction of AI-based applications in education, while critically analyzing their technical, institutional, and moral dimensions as shown in Figure 1. The major objectives are:

- To map their role in changing the developed landscape and education systems of AI technologies—from material construction and distribution to evaluation and administration.
- To present technical system models describing the architectural and functioning of major AI applications such as intelligent tuition systems, learning analytics engines, virtual assistants and generative content platforms.
- To examine emerging trends and innovations, including interactive AI, affective-computing systems, decentralized credentials, and AI-AR/VR fusion enabling immersive education.
- To analyze the major challenges obstructing large-scale AI adoption in education—such as data privacy, algorithmic bias, infrastructure limitations, and regulatory issues.
- To propose a strategic roadmap for future AI integration, emphasizing innovation, ethical governance, scalability, and responsible inclusion in educational policy and practice.

Through this lens, the chapter will not only reveal what AI can do now but also provoke discussion on what AI should do to support just, high-quality, and meaningful education for all in the future.

2. Literature Review and Technological Background

This section synthesizes the current research and technological progress related to AI applications in education. It highlights both foundational and state-of-the-art developments, establishing the phase to understand current capabilities and imagine directions for the future.

The key takeaways from the literature are depicted in Table 1 as:

- AI in education is developing from solid automation gear to adaptive, clever and famous structures.
- Early studies focus on cognitive modeling and professional systems, while recent studies prefer to get individuals to know human-AI cooperation, real-time response and personal direction.

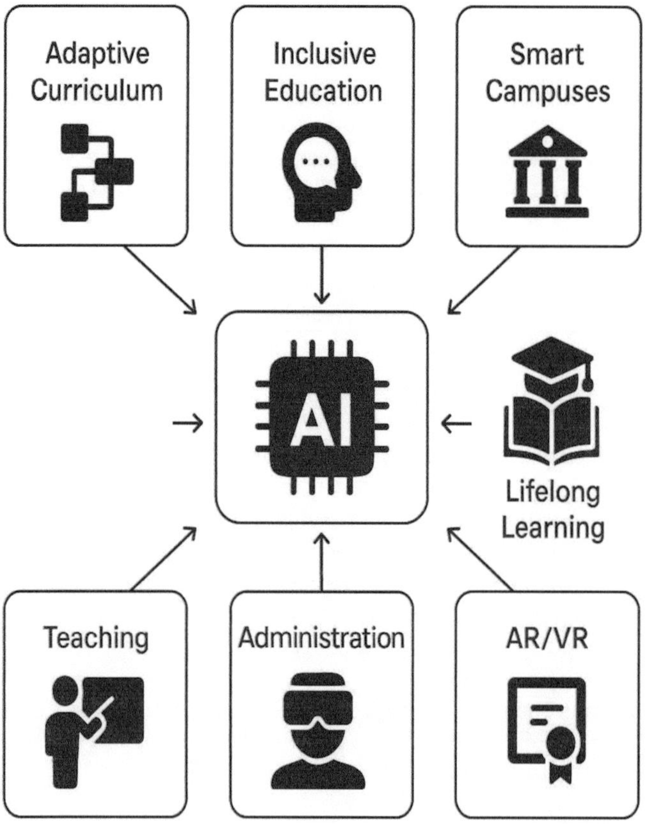

Figure 1: Future Directions of AI

- Ethics with moral design, learner-driven fashion, and an equity framework integrate morality and moral AI.

3. System Models for AI-based Educational Applications

The deployment of AI in education is fundamentally rooted in carefully designed system models that integrate various technologies to meet pedagogical goals. These models outline the architectural blueprint, data flow, and AI algorithms that collectively power personalized learning, intelligent tutoring, analytics, and content generation. Understanding these system models is vital for both researchers and practitioners aiming to develop or implement effective AI-driven educational solutions.

3.1 Personalized Learning Model

Personalized education has emerged as the cornerstone of AI applications in education, adapting instructional material and pacing to the unique requirements, preferences, and abilities of each learner. This model relies on central learner profiling, including the learner's knowledge state, learning style, and the continuous collection and analysis of data about engagement patterns and preferences. This profile forms the basis for adaptive content delivery, where learning materials are dynamically tailored to maximize understanding and retention.

The key to successful privatization lies in recognizing learning styles or preferences—such as visual, auditory, kinesthetic, or multimodal—though contemporary research tends to support a more

Table 1: Literature on Related Topics.

Topic	Focus	Key Points	Tools/Examples
ITS History	Early AI tutors	Mimicked human tutors using rule-based logic	SCHOLAR, GUIDON, Cognitive Tutors, Symbolic AI
Learning Analytics	Data-driven insights	Predicted performance, dropout risk, learning paths	OLA, LAK, ML, visualization, decision trees
NLP in Education	Language-based tools	Essay scoring, feedback, comprehension	E-rater, Grammarly, Turnitin, Biber, Jurafsky
Generative AI	LLMs in learning	Content creation, virtual tutors, grading	ChatGPT, Khanmigo, CodeGrade, GPT-4
AI-Pedagogy Models	Framework alignment	Maps AI to teaching strategies	Bloom's, SAMR, TPACK

data-driven approach, focusing on the specific needs of individual learners rather than rigid style classifications.

Technically, personalized learning systems often employ reinforcement learning algorithms that optimize content-delivery strategies based on learner responses and performance outcomes. For example, a system may experiment with different problem types or difficulty levels and observe which sequence best improves engagement and mastery. Additionally, collaborative filtering methods, borrowed from recommender systems, suggest learning resources by analyzing patterns in similar learners, enricting privatization.

Commercial platforms such as DreamBox Learning and D2L have successfully implemented these approaches. For example, DreamBox uses real-time data to customize mathematics instruction at a granular level, while D2L's adaptive learning platform individualizes content across subjects by analyzing continuous learner interactions.

3.2 Intelligent Tutoring System (ITS) Architecture

Intelligent tutoring systems represent one of the most mature and widely studied AI applications in education. Their architecture is traditionally composed of four integral components: the domain model, learner model, pedagogical model, and user interface (UI).

- The domain model specifies the subject-matter knowledge the system uses to generate instructional materials and assess learner responses.
- The learner model provides a dynamic representation of the current knowledge, misconceptions and skills of the learner, updated through ongoing interactions.
- The academic model determines how to teach, deciding on instructional strategies, feedback mechanisms, and scaffolding to maximize learning.
- The UI provides an interactive environment through which learners connect with the system.

AI techniques power ITS functionality, including Bayesian networks for probabilistic knowledge tracking and uncertainty management, decision trees for adaptive question selection, and neural networks for pattern recognition in learner behavior and response analysis. Systems such as AutoTutor leverage natural language processing to simulate conversational dialogue, creating a more engaging, human-like tutoring experience. Similarly, Carnegie Learning's Cognitive Tutor exemplifies ITS use in mathematics education, dynamically adjusting problem difficulty and hints based on learner progress.

3.3 Learning Analytics and Early Warning Systems

Learning analytics integrates AI with large-scale academic data to provide actionable insight into overall performance and institutional effectiveness, as in Figure 2. The system model normally

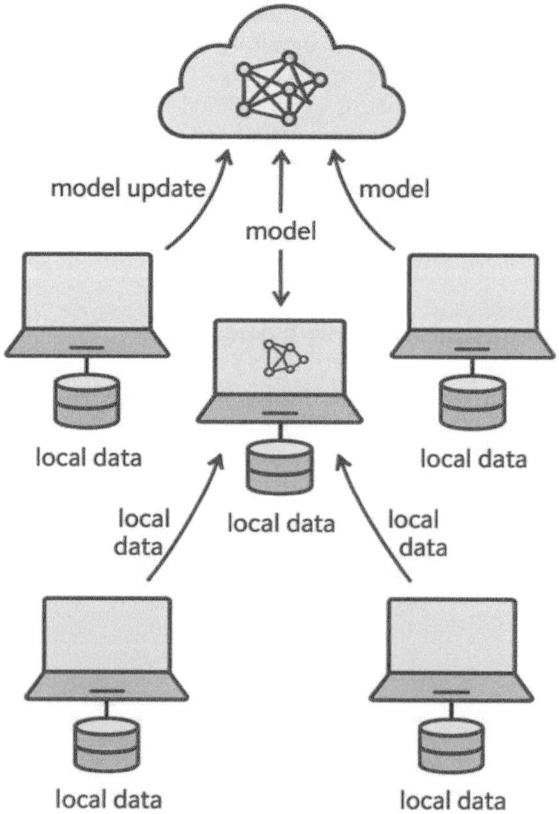

Figure 2: Learning Analytics Model

includes three principal components: data ingestion, in which data is collected from diverse resources such as LMS logs, assessment results, and engagement metrics; an analytics engine, which applies AI algorithms to discover trends, patterns, and predictions; and visualization tools, which present data through dashboards and reports accessible to teachers and administrators.

A critical application of this model is the early warning system, which uses predictive modeling techniques—regularly using machine learning algorithms—to identify learners at risk of failure or dropout. These systems enable timely interventions, including personalized tutoring or academic support.

With open-learning analytics (OLAs) and systems such as Moodle analytics plugins, institutions gain data-powered decision-making capabilities. AI models analyze behavior and performance indicators to predict outcomes, promoting proactive support strategies that can significantly improve retention and achievement rates.

3.4 AI for Content Generation and Assessment

The latest developments in large language models (LLMs) together with computer vision have transformed educational content creation and automated assessment processes. AI systems produce individualized learning materials, practice questions, and summaries with explanatory text that matches both learner requirements and course standards. AI-generated content enables real-time adaptation, making it more relevant and engaging than traditional static materials. AI-operated system applies natural language processing to grade essays and open-ended responses, giving detailed, consistent feedback at scale. Computer vision tools automatically evaluate handwritten work and diagrams, supporting mathematics and science subjects. The grading workflow in Gradescope uses

AI technology to reduce teacher workload while maintaining accuracy and fairness standards. GPT-based question generation (QG) tools provide the ability to generate assessment banks, quizzes, and personalized exercises aligned with learning objectives.

The systems mentioned here represent various dimensions of AI applications in educational models—from personalized learning pathways and human-like tutoring to data-driven insight and automated content workflows. Each model reflects a combination of pedagogical principles and state-of-the-art AI technologies, demonstrating the interdisciplinary nature of advancing educational innovation. Understanding these models equips stakeholders with the knowledge to design, implement, and evaluate systems that enhance teaching effectiveness and learner outcomes.

4. Emerging AI Applications in Education

A new educational frontier has emerged because of the fast advancement of artificial intelligence technology. The introduction of new AI applications transforms the way students learn materials, the methods teachers employ in instruction, and how educational organizations verify learner accomplishments, as depicted in Figure 3. The educational system has evolved through these innovations, which combine condensed AI technology with immersive technologies, affectionate computing, and blockchain integration to deliver personalized, engaging, and trustworthy learning environments.

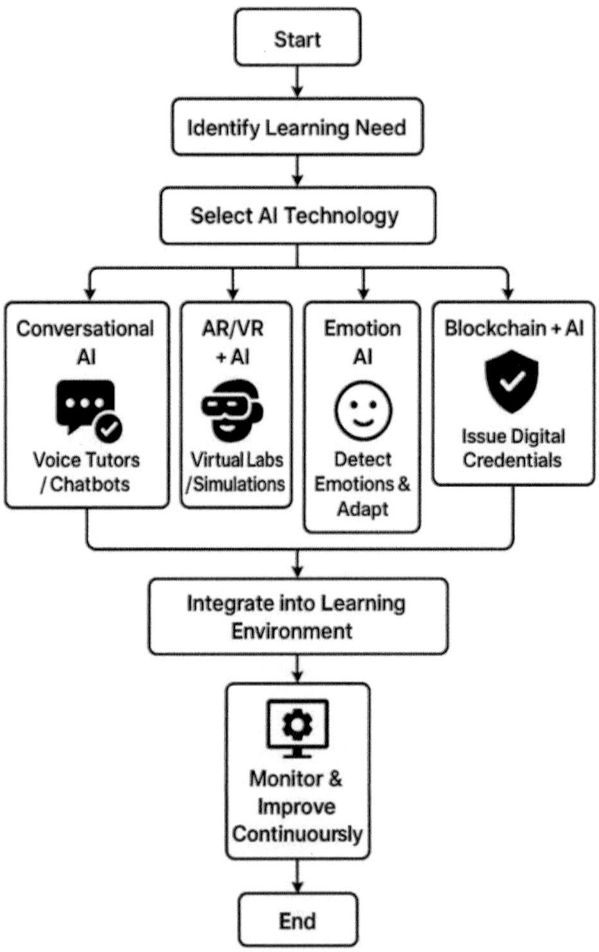

Figure 3: Emerging AI Fields.

4.1 Conversational AI and Virtual Tutors

Conversational AI has become an important tool in democratizing access to personalized learning support. Voice assistants like Amazon Alexa have expanded into education through features such as "Alexa Edu Kaushal," which helps students with complex concept clarification, homework, language practice, and on-demand support. These voice-activated tutors serve as always-available partners for learners, able to answer questions through natural language interactions, provide clarification, and guide practice sessions. In addition, the rise of multilingual conversational agents is greatly expanding access to education. Tools such as Google Dialogflow and Microsoft Azure Bot Service allow virtual tutors to communicate in multiple languages, supporting learners across linguistic backgrounds. This ensures that language barriers do not hinder access to quality education, especially in global and multicultural classrooms. Such AI-driven agents are being integrated into platforms like Duolingo and language learning apps to provide real-time conversation practice, simulating immersive learning experiences.

4.2 AR/VR + AI Integration for Immersive Learning

The combination of Artificial Intelligence (AI) with augmented reality (AR) and virtual reality (VR) creates opportunities for new interactive learning that suit various educational approaches. AI algorithms run adaptive virtual labs to enable students to conduct advanced scientific experiments within a safe virtual environment. The AI system behind Labster adapts virtual experiment complexity based on the student's performance to enhance both engagement and conceptual mastery.

AI-based computer vision provides real-time performance feedback by tracking student actions and offering customized guidance within these immersive environments. Medical training simulations use AI to evaluate both hand movements and procedural accuracy, guiding students through surgical training. The immediate correction system shortens the learning process while helping learners retain their acquired skills. AI-powered equipment offers immersive learning experiences that traditional methods—such as chemical testing, engineering prototyping, and historical simulations—cannot provide, especially in terms of cost, risk, or logical limitations.

4.3 Emotion AI and Affective Computing

The educational environment benefits from emotional computing as it detects emotions and reacts properly. AI systems, through facial identification technology and EEG sensors with their emotional states, receive real time monitoring of learner engagement. Reyleyes and Affectiva educational tools use micro-extraction analysis to monitor learners' attention during the delivery of educational content, as well as track their levels of frustration or boredom.

The emotional data obtained through AI systems enables dynamic adjustment of instructional strategies by teaching platforms. The system responds to learner confusion or disengagement by applying alternative presentation methods, including slower explanations, additional examples, or interactive activities. Such responsiveness mimics empathetic human tutoring and promotes a more supportive learning environment.

The technology also enables early detection of mental health concerns by identifying emotional distress patterns, allowing timely support intervention for student well-being.

4.4 Blockchain + AI for Educational Credentials

The combination of blockchain technology with AI brings revolutionary changes to how academic credentials are issued, verified, and used. Digital credentials built on blockchain become tamper-resistant and fraud-proof, solving problems of diploma forgery and lack of transparency. MIT and other institutions have begun issuing blockchain-based digital diplomas, enabling graduates to share secure certificates with employers globally. AI systems work with these platforms, which combine educational credentials with current job market requirements, to evaluate comprehensive learning data with skills and course results. Results show that learning machines utilize AI-driven platforms that leverage smart contracts and machine learning to propose personalized APASCILING tracks based on current labor market trends. Through this combination, education-to-employment connection becomes stronger by constantly learning and supporting the workforce.

Blockchain technology integrated with artificial intelligence establishes a reliable and efficient educational system that validates achievements through secure processes and accurately matches them with professional opportunities.

The following sections explain how new AI applications improve learning outcomes by addressing important issues related to emotional support, credential integrity, accessibility, and engagement. Joint innovation leads to an educational future that provides adaptive experiences, immersive learning, equitable access, and strong real-world outcome alignment.

5. Challenges in Adopting AI in Education

Although the transformative potential of AI in education is widely recognized, the path to its widespread and effective adoption is filled with significant challenges. These challenges span ethical, infrastructural, pedagogical, and governance domains, requiring collaboration among multiple stakeholders. Understanding these barriers is essential to designing responsible, inclusive, and sustainable AI interventions.

5.1 Ethical and Privacy Concerns

One of the most important concerns in AI-driven education systems relates to data ownership and privacy. Learners generate vast amounts of sensitive data—including personal information, learning behavior, and even emotional states—raising critical questions about who owns this data and how it is protected. Compliance with regulations such as the General Data Protection Regulation (GDPR) in Europe poses both technical and legal challenges for AI systems, requiring clear consent, transparency, and strict control over data use.

In addition, AI algorithms can inadvertently perpetuate bias and discrimination, reflecting bounds present in training datasets or design assumptions. For example, if an AI tutoring system is trained mainly on data from a certain demographic group, it may provide recommendations below the ideal for students outside this group, thus reinforcing educational inequalities (BAKER et al. 2020). Attenuation of algorithmic bias requires careful data curation, continuous auditing, and the inclusion of diverse stakeholders in AI development.

The working model for addressing Ethical AI is the applicable framework of responsibility and transparency (FAT), used in certain academic AI deployments (Salbust et al. 2019). This framework emphasizes transparency to reduce AI's accountability, ensure responsibility for outcomes, and actively combat bias to foster trust and equity.

5.2 Infrastructure and Accessibility

The digital divide is still a significant obstacle, especially in developing regions where internet connectivity, hardware accessibility, and technical literacy may be limited. Without adequate infrastructure, AI's promise to improve education risks aggravating existing inequalities instead of reducing them. In addition, the cost of deploying and maintaining AI systems—including cloud services, data storage, and specialized hardware—may be insurmountable for many educational institutions, especially public schools and under-financed universities. Models such as AI-as-a-Service (AIaaS), provided by companies such as Google Cloud and Microsoft Azure, offer scalable AI capabilities to lower up-front costs, but still require technical expertise and reliable internet.

An illustrative use case is the OneBillion Foundation in East Africa, which applies AI-operated learning tablets in low-resource settings. Their system adapts to the learner's motion with minimal teacher intervention, but this initiative highlights structural facility gaps and challenges related to sustained support.

5.3 Pedagogical Limitations

Effective AI integration depends not only on technology but also on human factors. A central challenge is the lack of sufficient teacher training to understand and benefit from AI tools. Without professional development and support, teachers may misuse or underuse AI systems, limiting their potential impact. There is also concern about excessive dependence on automation, where AI could overshadow crucial human elements such as empathy, creativity, and moral judgment in education. Excessive reliance on AI can reduce learner motivation and diminish the role of teachers as mentors and facilitators. Research suggests a mixed model, where AI augments rather than replaces teachers. For example, AI-driven tutoring systems require teacher supervision, allowing teachers to interpret AI-generated learner data and make instructional decisions.

5.4 Policy and Governance Issues

Currently, many countries lack clear regulatory structures for AI in education, creating uncertainty around liability, data protection, and moral standards. This policy vacuum slows adoption and risks inconsistent or unsafe use of AI tools, as shown in Figure 4. Moreover, intellectual property (IP) concerns arise around AI-generated educational material. Questions about who owns AI-made or AI-assisted materials and how copyright applies remain unresolved, complicating material distribution and reuse. A promising governance approach is the establishment of AI ethics committees in educational institutions to monitor deployment and ensure compliance, alongside collaborative policy efforts at national and international levels (UNESCO 2021).

Considering these challenges, consider the "AI in School" pilot program (Nemi et al. 2020) in Finland. The initiative aimed to integrate AI tutoring and analytics but faced issues including teacher readiness, data privacy negotiations, and ensuring equal access across urban and rural schools.

Figure 4: Policy Issues.

Through iterative feedback and policy adjustments, the program offers an example of a working model that balances AI innovation with ethical, infrastructural, and pedagogical constraints.

6. Future Scope and Roadmap

Artificial intelligence (AI) is quickly reshaping the educational landscape by enabling a smarter, more personalized, and integrated education environment. As this field continues to evolve, the future scope of AI applications presents a promising path to revolutionize teaching methods, curriculum design, learner engagement, and institutional management. This section highlights future-oriented subtopics—hybrid human-AI classrooms, lifelong learning, decentralized AI ecosystems, and current research opportunities—their potential applications, benefits, and challenges.

6.1 Towards Hybrid Human-AI Classrooms

The future classroom will be a dynamic fusion of human intervals and AI-operated assistance. AI systems are being designed to work as co-pilots for teachers, augmenting their capabilities rather than replacing them. Previously used experimental applications include AI-powered intelligent tutoring systems that provide personalized feedback and adapt to student learning styles. For example, a platform such as Carnegie Learning's "Mathia" uses AI to personalize mathematics tutoring, identify misconceptions, and adjust problem difficulty accordingly. In addition, AI-assisted planning tools are gaining traction, helping teachers create flexible and adaptive courses by leveraging data analytics that respond to students' learning patterns and needs in real time. Content Technologies, Inc. uses AI to create customized textbooks and learning materials based on the unique requirements of student groups. This dynamic adaptation helps eliminate the limitations of traditional "one-size-fits-all" education, improving learning outcomes and engagement.

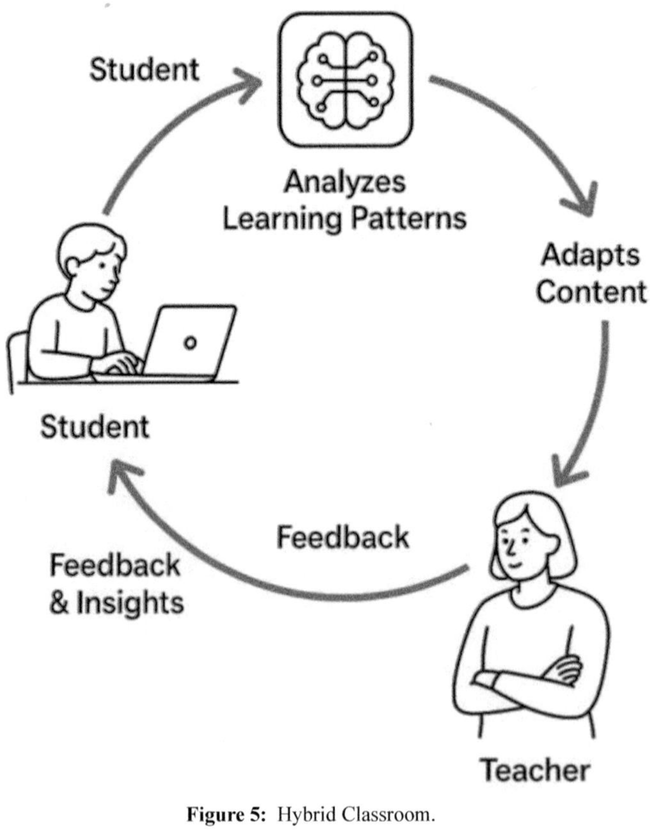

Figure 5: Hybrid Classroom.

Moreover, by automating routine administrative tasks such as grading and attendance, AI classrooms can enable teachers to invest more time in direct interaction, as Figure 5 portrays. AI-run sentiment analysis tools can also gauge students' engagement and emotional states through video or audio analysis, alerting teachers to students who may need additional support. Such applications exemplify a hybrid classroom model where AI increases teacher effectiveness and promotes a more interactive, personalized learning environment.

6.2 Lifelong Learning and Skill-based Education

In times of rapid technological progress and changing labor markets, lifelong learning and continuous skill development have become essential. AI-driven personalized learning platforms are at the forefront of this shift, enabling students to acquire new skills effectively throughout their careers. Platforms such as Coursera and LinkedIn Learning use AI algorithms to recommend courses based on students' previous activity, career goals, and emerging industry requirements. A practical application involves real-time course adjustment aligned with evolving labor market trends. AI analyzes large-scale labor market data to identify in-demand skills and certifications, dynamically updating course offerings. For example, IBM's AI-powered Skills Gateway assesses labor market gaps and proposes tailored learning routes, ensuring learners stay adaptive to future work.

Moreover, AI-enabled career coaching tools offer personalized guidance, assisting learners in making informed decisions about upskilling or reskilling in response to automation and globalization. These platforms empower learners to take charge of their education and remain competitive in an evolving job market. In addition to personal learning, AI supports corporate training by providing customized learning experiences aligned with organizational goals. For instance, AI-based learning management systems (LMSs) use machine learning to curate Docebo content and Optim for personnel, aiming to improve OPTIM, agility, and staff productivity.

6.3 Decentralized AI Ecosystems

Data privacy, security, and ethical concerns remain major challenges in educational AI adoption. The emergence of decentralized AI ecosystems offers promising solutions through the use of federated learning and community-powered model training. Federated learning enables AI models to learn across multiple decentralized devices or servers without centralizing data, thereby preserving student and institutional privacy. Applications for federated learning are especially relevant in large education consortia and networks where data-sharing restrictions exist. For example, several universities can collaborate on a predictive model for student dropout risk without exposing sensitive student data externally. This approach not only addresses privacy concerns but also ensures that AI models benefit from diverse, heterogeneous data, improving robustness and fairness.

Community-driven model training fosters inclusivity by enabling educators and researchers worldwide to contribute to and refine AI models. Open-source initiatives and AI marketplaces facilitate this collaborative environment, accelerating innovation and democratizing access to high-quality educational AI systems.

6.4 Research Opportunities

Significant research opportunities remain to realize the full potential of AI in education. One major area is the development of zero-shot and few-shot learning techniques, which allow AI models to learn new tasks with minimal labeled data. This is especially valuable for low-resource educational environments—such as rural schools or underfunded institutions—where large annotated datasets are unavailable. AI models capable of generalizing with limited examples can help reduce educational inequalities by providing intelligent tutoring and assessments in these contexts. Explainable AI (XAI) is another critical research domain. AI decisions influencing education must be transparent and interpretable to teachers, students, and parents. XAI methods help clarify how AI systems generate recommendations or predictions, enabling stakeholders to understand, validate, and challenge this output. For example, in automated grading, XAI may explain why certain answers receive specific scores, helping teachers verify fairness and accuracy.

Further research focuses on ethical AI deployment, dealing with training data, ensuring the availability of diverse student populations, and developing robust data security and consent frameworks. The intersection of AI with cognitive science and pedagogy also invites research on how to optimize AI interventions to complement human teaching styles and promote deeper learning.

The future of AI in schooling is an exciting convergence of progressive technology and pedagogical ideas that aim to create more personalized, equitable, and efficient learning experiences. Human–AI co-teaching will allow educators to be supported with intelligent instructional and management tools. Lifelong learning systems will enable the development of adaptive skills aligned with the demands of the real world. The decentralized ecosystems of AI will ensure the training of AI models that preserve privacy, while research in low-resource learning and explainability will address moral and practical challenges. Together, these developments set a path toward a smarter and more responsive educational system, capable of meeting the diverse needs of students around the world. However, successful adoption will require continuous collaboration between educators, technologists, policymakers, and researchers to navigate challenges related to fairness, privacy, transparency, and infrastructure.

7. Working Model: AI-Powered Personalized Learning Ecosystem

A widely adopted working model in AI-based education is an ecosystem of personalized learning operated by AI algorithms, as has been implemented by platforms such as DreamBox Learning and Knewton. The purpose of these systems is to provide tailored learning experiences by adapting content and feedback to the needs of individual learners.

7.1 Model Components and Architecture

1. Learner Profiling Module: Sensors, digital footprints, and interaction logs collect and update detailed data at the level of learning behavior, preferences, prior knowledge, learning pace, and engagement.
2. Adaptive Content Delivery Engine: Uses AI techniques such as reinforcement learning and collaborative filtering to recommend and dynamically adjust educational materials. For example, if a student struggles with a concept, the system adjusts difficulty or provides alternative explanations.
3. Learning Style Identification: The learning methods integrate the biological and behavioral analysis to identify the preferred methods (visual, audible, kinesthetic) and accept the instructional strategy accordingly.
4. Feedback and Assessment Module: Automated grading, generative feedback, and emotion AI use natural language processing (NLP) and computer vision to detect frustration or confusion.
5. Teacher Dashboard and Analytics: Provides teachers with real-time insights into performance, potential learning gaps, and engagement metrics, enabling targeted interventions.

7.2 AI Techniques Utilized

- Reinforcement Education: J Knowledge customizes the content index by learning from the learner's responses to optimize retention and motivation.

- Collaborative Filtering: Based on the behavior of the same learners, it suggests learning resources to increase discovery and engagement.
- Natural Language Processing: Automatic question needed generation, grading and summary, reduces teacher's workload.
- Emotion AI: Uses facial validity and sensor data to find the learner's feelings, corresponding to instruction to maintain engagement.

7.3 Real-World Application: DreamBox Learning

DreamBox is an adaptive AI-powered learning platform for mathematics instruction, mainly used in K–8 classrooms worldwide. It continuously collects data on how students interact with math problems, including time spent, mistakes made, and strategies used.

Using this data, DreamBox's AI engine adjusts lesson plans in real time, scaffolds complex concepts, and revises foundational skills as needed. Teachers gain access to dashboards that highlight individual and class trends, helping them tailor instruction in the classroom.

Studies report that Dreambox users show significant gains in mathematical skills compared to traditional methods, demonstrating the effectiveness of the model in Fletcher & Nissenbaum Personalized Learning 2012).

7.4 Significance of the Model

This customized learning ecosystem exemplifies how AI can transform education from a one-size-fits-all model to a learner-centered approach. It balances automation with human oversight, ensures inclusivity by adapting to diverse learner needs, and supports data-driven pedagogy.

The modularity of the model allows integration with emerging AI technologies like AR/VR for immersive experiences or blockchain for secure credentialing, thus serving as a foundational structure for next-generation AI educational programs.

8. Case Studies and Global Initiatives

The AI landscape in education is not only shaped by innovative technologies but also by the strategic efforts of international organizations and national authorities. These initiatives act as guiding models to scale AI applications responsibly and effectively. This section investigates major global and national case studies that illustrate diverse approaches, challenges, and successes in AI-enabled education.

8.1 UNESCO's AI in Education Framework

Globally, UNESCO has significant influence in shaping ethical, inclusive, and human-centered AI in education. The UNESCO *AI in Education Report* (2021) outlines a comprehensive framework that emphasizes transparency, fairness, and the importance of learner agency in deploying AI tools. It highlights the ethical AI toolkit for education, a practical resource designed to assess AI risks, protect learner data privacy, and help educational institutions promote inclusion.

This framework has been tested through UNESCO's collaboration with multiple countries, especially in the Global South, where infrastructure challenges and linguistic diversity are major barriers. For example, UNESCO has partnered with the African Union to support AI educational systems that provide content in several local languages and adapt to varying levels of connectivity. These systems use AI models trained on diverse datasets to reduce bias and ensure equitable learning outcomes (UNESCO 2021).

UNESCO's working model is characterized by its multi-stakeholder governance approach, involving governments, educators, technology developers, and students to ensure that AI aligns with educational goals and cultural contexts. This approach offers a replicable method for balancing innovation with ethical responsibility globally.

8.2 India's National Education Policy (NEP) 2020 and AI Integration

At the national level, India's NEP 2020 marks a strategic commitment to AI as a transformative force in education reform. The policy envisions a future where AI enables personalized, merit-based learning accessible to all socio-economic groups. The NEP proposes establishing the National Educational Technology Forum (NETF) to guide AI integration and create a regulatory ecosystem that promotes innovation while protecting learner rights.

Under this initiative, a major working model is the AI-powered DIKSHA platform, which serves millions of learners by providing personalized content, continuous assessment, and real-time feedback. The system uses machine learning algorithms to analyze learner progress and adapt instructional materials accordingly. Additionally, natural language processing modules enable content delivery in multiple regional languages, addressing India's linguistic diversity.

Pilot programs introducing AI-enabled adaptive learning in rural regions have shown promising outcomes, with AI systems dynamically adjusting problem levels and offering remedial support. These models use reinforcement learning to optimize lesson sequencing based on individual student responses. Importantly, teacher training programs built into NEP ensure that educators can interpret AI insights and integrate the technology with traditional pedagogy (Mishra and Singh 2022).

India's case highlights how a massive, diverse nation can use AI to bridge education, although challenges such as infrastructure deficits and digital literacy remain.

8.3 United States: AI4K12 Initiative

In the USA, AI4K12 initiatives represents a forward-looking approach that integrates AI literacy into a unified standards-based curriculum in elementary and secondary education. Unlike AI-driven instructional tools, AI4K12 emphasizes equipping students with an understanding of AI concepts, ethics, and social impacts.

The working model of the initiative includes three pillars: AI curriculum recommendations, AI course guidelines, and collaboration with groups of educational technology providers to create experimental units of AI learning. For example, students are exposed to learning basic machine learning concepts and ethical case studies, promoting critical engagement with AI techniques that are increasingly embedded in daily life.

This approach acknowledges that future citizens and workers will not only need to interact with AI but also understand its methods and consequences, thereby democratizing AI education. Programs like MIT's AI+ course further support this model by offering interactive modules that simulate AI decision-making and expose students to concepts such as bias and fairness (AI4K12 2023).

8.4 Finland's Elements of AI Program

Finland's Elements of AI initiative represents a successful model for broad-based AI literacy that extends beyond traditional educational environments. Launched as a free online course aimed at the general public, it combines foundational AI theory with practical examples and ethical considerations.

The program's working model is defined by accessibility—designed for learners without prior technical backgrounds—and by its integration with formal education systems, where it supplements secondary and tertiary curricula. The Elements of AI course uses interactive quizzes, real-world case studies, and discussion forums to encourage a community-based learning experience.

Table 2: Summary.

Initiative	**Region**	**Focus**	**Key Features**
UNESCO AI Framework	Global	Ethical AI use	Ethical toolkit, inclusivity, multilingual AI, global collaboration
NEP 2020 & DIKSHA	India	AI for education reform	Adaptive learning, regional languages, teacher training, NETF
AI4K12	USA	AI literacy in schools	AI curriculum, ethics, ML activities, K–12 focus
Elements of AI	Finland	Public AI literacy	Free online course, no coding required, ethics + real cases
UNESCO–AU Partnership	Africa	Inclusive access	Local language AI, offline content support, bias reduction

This initiative has been recognized for promoting national AI readiness, fostering social dialogue on AI's role, and inspiring similar programs internationally (Hagström et al. 2019). Its success shows how public education campaigns can complement formal schooling in building a digitally literate society.

Together, these global and national initiatives illustrate a rich ecosystem of AI adoption in education, ranging from infrastructure and policy reform (UNESCO) to foundational literacy (AI4K12 and Elements of AI) to grassroots implementation models (AI4K12 and Finland). Their various working models highlight the importance of: contextualizing AI tools for cultural, linguistic, and infrastructural realities; prioritizing ethics, transparency, and human oversight; equipping educators and students with skills to use and understand AI effectively; and collaborating across sectors to ensure sustainable implementation, as shown in Table 2.

These case studies collectively point to a future in which AI-enhanced education can be innovative and inclusive.

9. Conclusion

This chapter presented a comprehensive exploration of the future of AI-based applications in education, emphasizing the delicate balance between technological potential and the complex challenges inherent in real-world implementation.

The scope of AI in education is vast—from personalized learning, intelligent tutoring, and AR/VR immersive experiences, to emotion recognition and secure credentialing through blockchain. Each innovation promises to reshape how knowledge is delivered, assessed, and experienced. However, leveraging these benefits requires a grounded understanding of ethical dilemmas, infrastructural disparities, pedagogical readiness, and regulatory frameworks.

International and national initiatives—such as UNESCO's Ethical AI Framework, NEP 2020, India's AI4K12 Courses, and Finland's Elements of AI—demonstrate how visionary policies, integrated governance, and targeted educational programs can translate the promise of AI into tangible outcomes. Their working models highlight human agency, data privacy, cultural adaptation, and interdisciplinary collaboration.

Ultimately, the path forward calls for human-centered, equitable, and sustainable AI integration in education. AI should empower teachers rather than replace them, amplify learner voices rather than silence them, and democratize opportunity rather than widen divides. By fostering ethical AI ecosystems, advancing digital literacy at all levels, and ensuring transparent governance, the education sector can use AI to develop a generation ready to thrive in an AI-driven world.

As AI continues to grow, it is also aimed at enhancing human potential and promoting lifelong education - structures, tools, and mentality that guides its application in education.

References

AI4K12 Initiative. (2023). *AI4K12: Guiding AI Education in K-12*. https://ai4k12.org.

Baker, R. S. J. d., Smith, J., Johnson, E. R., and Lee, M.T. (2020). Algorithmic Bias in Education: The Case of Personalized Learning Systems. Journal of Educational Data Mining, 12(3): 1–25.

DreamBox Learning. (2023). *Personalized Math Learning*. https://www.dreambox.com.

Fletcher, J. D., and Nissenbaum, M. (2012). Effectiveness of Intelligent Tutoring Systems for K-12 Mathematics. *Journal of Educational Technology*, 49(3): 195–205.

Hagström, B., Tukiainen, M., and Rautiainen, M. (2019). Elements of AI: Democratizing Artificial Intelligence. *Finnish Ministry of Education and Culture Report*.

Holstein, K., McLaren, B. M., and Aleven, V. (2019). Designing for Complementarity: Teacher and Student Needs for Orchestration Support in AI-Enhanced Classrooms. *CHI Conference on Human Factors in Computing Systems*.

Knewton. (2023). *Adaptive Learning Platform Overview*. https://www.knewton.com.

Mishra, P., and Singh, R. (2022). AI-powered Personalized Learning in Rural India: A Case Study. *International Journal of Educational Technology*, 9(1): 45–62.

Niemi, H., Virtanen, P., and Koskinen, J. (2020). AI in Finnish Schools: Opportunities and Challenges. European Journal of Education, 55(3): 297–311.

One Billion Foundation. (2021). Scaling AI-powered Tablet Learning in East Africa. [Online] Available: https://onebillion.org/.

Selbst, A. D., Andrew, M., Karahalios, K., and Narayanan, A. (2019). Fairness and Abstraction in Sociotechnical Systems. Proceedings of the Conference on Fairness, Accountability, and Transparency.

UNESCO. (2021). *Recommendation on the Ethics of Artificial Intelligence*. United Nations Educational, Scientific and Cultural Organization.

UNESCO. (2021). *Artificial Intelligence in Education: Challenges and Opportunities*. United Nations Educational, Scientific and Cultural Organization. https://unesdoc.unesco.org/ark:/48223/pf0000376705.

19

Theoretical Foundations of Digital Twins for Predictive Intelligence in IoT-Driven Healthcare Ecosystems

Sonam Gupta,[1] *Pradeep Gupta*,[1] *Malik Shahzad Ahmed Iqbal*[2] and *Vikash Yadav*[3,*]

1. Introduction

The coming together of Internet of Things (IoT) technologies, artificial intelligence (AI), and digital twin paradigms is leading the medical sector toward a technological breakthrough. The convergence of these forces is going to fundamentally transform how healthcare is provided, monitored, and improved through the development of smart and predictive systems that can anticipate patient needs, prevent adverse incidents, and maximize treatment outcomes (Rasheed et al. 2020, Kharroubi et al. 2024). Initially, digital twins were applied in manufacturing and the aerospace industry, but due to their ability to map dynamic, real-time virtualizations of physical entities, processes, and systems, digital twins represent a paradigm-shifting concept for medical facilities.

With the advent of digital twins in healthcare, the entire industry is undergoing a revolutionary shift in the way healthcare professionals view patient treatment, diagnosis methods, and future treatment planning. As a revolutionary technology, it bridges the physical and virtual worlds, giving healthcare providers the ability to simulate, analyze, and predict patient outcomes with a level of accuracy and precision never seen before. The integration of IoT sensors, wearable devices, and advanced analytics platforms has also established an environment in which constant surveillance and real-time data analysis can generate insights that could not previously be obtained through conventional healthcare practices.

Otherwise known as predictive intelligence, the idea central to predictive intelligence in healthcare represents a shift toward more proactive rather than reactive treatment procedures. In the

[1] Department of Computer Science and Engineering, Ajay Kumar Garg Engineering College, Ghaziabad, India. Emails: guptasonam6@gmail.com, https://orcid.org/0000-0002-4323-6991; gupta.pradeep85@gmail.com, https://orcid.org/0000-0003-3142-6373

[2] Department of Computer Science & Engineering, Acharya University, Karakul, Uzbekistan. Email: malik22@acharya.ac.uz, https://orcid.org/0009-0005-5157-5570

[3] Government Polytechnic Bighapur Unnao, Department of Technical Education, Uttar Pradesh, India.

* Corresponding author: vikas.yadav.cs@gmail.com, https://orcid.org/0000-0003-1348-1379

DOI: 10.1201/9781003655190-23

past, the reactive nature of healthcare systems was marked by interventions after the manifestation of symptoms and the progression of conditions. However, as digital twin technology is combined with the IoT-driven healthcare ecosystem, it triggers a paradigm shift toward predictive and preventive care models. This transformation is especially significant in a world where healthcare expenses continue to rise, and the need for personalized, efficient, and effective medical procedures keeps increasing exponentially.

Digital twins applied to the healthcare industry rely on the theoretical frames covering a wide range of disciplinary fields, such as computer science, biomedical engineering, clinical informatics, and systems biology. This convergence has created a rich theoretical environment capable of supporting the development of complex healthcare digital twin systems that can simulate biological processes, predict disease progression, and optimize treatment protocols. The interdisciplinary character of this field requires a deep understanding of both technical capabilities and clinical needs, ensuring that digital twin implementations are not only viable but also clinically relevant and applicable.

Probably the greatest strength of the digital twin technology in healthcare is the fact that by combining multiple data sources, it can continuously and comprehensively monitor the health condition of the patient. Facilitated by the wearable gadgets and implantable medical machines, IoT sensors gather immense quantities of real-time data that can be processed and analyzed in order to create detailed virtual models of patient health. The resultant effect of this round the clock monitoring capacity is that healthcare professionals are able to identify early indicators of health decline before further complications occur and discover tendencies that could lead to the development of chronic diseases, and establish preventive measures before the latter takes place.

The predictive intelligence functionality of healthcare digital twins is especially useful for managing chronic diseases, where long-term observation and continuous adjustments to treatment regimens are essential for achieving optimal clinical outcomes. Diseases such as diabetes, heart disease, and pulmonary disorders require constant monitoring and regular changes to medications, lifestyle patterns, and clinical examination schedules. Digital twins can model the outcomes of different treatment strategies, estimate patient responses to various interventions, and recommend the most appropriate treatment plans based on patient attributes and historical data trends.

The use of artificial intelligence and machine learning algorithms in combination with digital twin technology has significantly expanded the accuracy of predictive healthcare systems. The application of advanced analytics tools—such as deep learning, reinforcement learning, and ensemble methods—enables digital twins to process multidimensional healthcare datasets and extract actionable insights that support clinical decision-making. Such AI-based strategies can detect hidden trends in patient data that are not visible to the human eye, estimate the probability of adverse events, and propose patient-specific treatment strategies based on individual profiles as well as population-level analyses.

The digital twin of healthcare would extend the IoT ecosystem to include a variety of connected devices and sensors, ranging from consumer-grade to highly specialized medical monitoring devices. This distributed system connects all these devices together, forming a fully integrated data collection network capable of capturing physiological parameters, environmental conditions, behavioral patterns, and clinical measurements in real time. The theoretical frameworks for designing and implementing such IoT-enabled healthcare systems should consider issues of data quality, interoperability, security, and scalability, but what matters most is that such digital twin systems can offer reliable and accurate predictive information.

Deployment of digital twins in healthcare also raises concerns regarding data privacy, data security, and ethical use of patient data. In healthcare, the handling of data is highly sensitive and requires a strong theoretical framework that supports privacy preservation, secure data transfer, and ethical decision-making in AI-driven healthcare systems. These concerns are especially acute as digital twin systems become more advanced and the autonomous decisions they generate begin to influence patient care and treatment recommendations.

The potential uses of digital twins in healthcare extend far beyond the close monitoring of individual patients to include hospital operations, public health, and population-level management. Digital twin technology can help healthcare facilities optimize resource distribution, anticipate equipment maintenance needs, and enhance operational efficiency. At a system level, digital twins can support epidemiological research, disease surveillance, and population health initiatives by providing detailed representations of outbreaks, identifying risk factors, and demonstrating the effectiveness of interventions.

The economic consequences of adopting digital twins in healthcare are significant, with potential advantages such as lowering the cost of healthcare delivery, improving patient outcomes, and increasing operational efficiency. Improved ability to foresee and mitigate adverse incidents, streamline treatment protocols, and reduce unnecessary medical interventions may lead to major savings for healthcare systems while improving the quality of care delivered to patients. The investment costs associated with digital twin implementation—such as infrastructure development, technology acquisition, and staff training—should also be carefully structured with long-term returns in mind.

Digital twins-based solutions within the healthcare ecosystem are comprised of several related disciplines supporting the theoretical backgrounds: cyber-physical systems theory, complex adaptive systems approaches, machine learning, and healthcare informatics. These premises collectively construct the theoretical basis for understanding how digital twins can be implemented to generate predictive intelligence in healthcare. When combined with IoT technologies, these capabilities go even further, enabling continuous data flows needed to maintain synchronized virtual representations of patients, medical devices, and healthcare facilities themselves.

The theory of digital twins as related to predictive intelligence in the IoT-driven healthcare ecosystem was discussed in this chapter to examine the main concepts, philosophy, architecture, and theoretical structures that enable digital twin implementation. To initiate the discussion, the concept of digital twins was examined, the process of their gradual transformation into healthcare practice was evaluated, and the theoretical frameworks enabling predictive intelligence were explored. This capacity is further enhanced with the integration of IoT technologies, which provide the data streams required to maintain synchronization of the virtual models of patients, medical equipment, and facilities (Liu et al. 2021, Qian et al. 2022).

Not only relevant to academia but also practical application, this theoretical exploration provides foundations that can support improved patient outcomes, reduced healthcare costs, and overall enhancements in care quality. A solid grasp of these theoretical underpinnings aids medical professionals, researchers, and technologists in the design, implementation, and optimization of digital twin systems that can meaningfully transform healthcare delivery through the power of predictive intelligence.

2. Theorical Basics of Digital Twins in Healthcare

2.1 What is It and Its Fundamental Principles

A digital twin in healthcare settings is a complex technical upgrade to common health information systems due to its bidirectional mobility features that create live interrelations among physical and virtual components (Abouella et al. 2024). These components form three core parts: the physical element, the virtual element, and the bidirectional data exchange connection, which allows real-time information synchronization between both elements. In healthcare applications, the corresponding virtualization grants the representation of patients, medical equipment, care procedure, and the whole healthcare facility, which keeps on developing with regard to general information feeds.

The theoretical concept of the digital twin in healthcare is based on several important principles that differentiate digital twins from traditional approaches such as static models or databases. First, the dynamic synchronization principle ensures that the virtual representation remains up to date with the actual state of the physical entity. This is especially important in healthcare, as patient conditions may change rapidly and timely intervention can save lives (Nadeem et al. 2025). Second, the concept of bidirectional interaction enables a digital twin not only to reflect the physical entity but also to influence it through suggestions, warnings, and machine-supported interventions.

The third principle is predictive capability, in which the digital twin is used to determine prospective states or occurrences based on prior information, real-time conditions, and advanced analytics. This predictive aspect is central to the success of intelligence-driven healthcare digital twins because reactive healthcare delivery is no longer adequate. The fourth principle relates to personalization, where each digital twin is customized to reflect individual patients, taking into account their unique physiological features, medical history, genetic tendencies, and lifestyle choices. Table 1 shows the core principles of digital twins in healthcare.

Table 1: Core Principles of Digital Twins in Healthcare.

Principle	Description	Significance in Healthcare
Dynamic Synchronization	The virtual representation continuously updates to reflect the physical entity's actual state.	Crucial in healthcare where patient conditions can change rapidly, and timely interventions can be life-saving.
Bidirectional Interaction	The digital twin not only reflects the physical entity but can also influence it through suggestions, warnings, and machine interventions.	Allows for proactive care and automated adjustments based on the virtual model's insights.
Predictive Abilities	The digital twin uses historical data, real-time conditions, and advanced analytics to forecast future states or occurrences.	Essential for moving from reactive to proactive and preventive healthcare, anticipating patient needs, and preventing adverse incidents.
Personalization	Each digital twin is tailored to an individual patient, considering their unique physiological features, medical history, genetic predispositions, and lifestyle choices.	Enables highly customized and effective treatment plans that optimize patient outcomes based on individual characteristics.

2.2 Theorized Models of Healthcare Digital Twins

The theoretical framework of healthcare digital twins is based on various models and theories. The Model-Based Systems Engineering (MBSE) approach offers a methodology for creating and sustaining digital twins by focusing on the development of complete system schemes that document functional and non-functional requirements. Within healthcare, MBSE enables the creation of digital twinning of healthcare systems that accurately model numerous physiological systems as well as medical equipment and procedures (Alshammari 2024).

Another important theoretical framework is that of Complex Systems Theory, given that healthcare systems resemble complex adaptive systems characterized by nonlinear relationships and emergent properties and behaviors of component parts. Healthcare digital twins must be capable of capturing these complex system properties in order to model and accurately predict health outcomes. Complex adaptive systems theory helps explain why individual patient digital twins may exhibit emergent properties resulting from the interplay between various physiological systems and external circumstances.

Information Systems Theory is also relevant because digital twins function as complex information-processing systems that gather, store, process, and distribute health-related information. The theory emphasizes information quality, system integration, and user interaction to achieve effective system performance. As applied to healthcare digital twins, it underscores the importance of high-quality data sources, compatibility with other healthcare information systems, and system

Table 2: Comparative Overview of Theoretical Models in Healthcare Digital Twins.

Theoretical Model	Core Concept	Application in Healthcare Digital Twins	Significance
Cyber-Physical Systems (CPS)	Integration of physical processes with computational feedback loops	Enables real-time monitoring of patient vitals and device control	Ensures safety-critical operations with real-time responsiveness
Complex Adaptive Systems	Non-linear, emergent behaviors of interrelated systems	Models dynamic interactions among physiological systems and treatments	Captures real-world complexity and unpredictability in patient outcomes
Machine Learning & AI Theory	Predictive modeling and data-driven decision-making	Enables pattern recognition, anomaly detection, and personalized treatment recommendations	Enhances accuracy of diagnostics and forecasting
Information Systems Theory	Focus on data integration, quality, and usability	Manages data flow from IoT, EHRs, and clinical interfaces	Improves interoperability, user experience, and system reliability
Real-Time Systems Theory	Guarantees timely system response under deadlines	Ensures fast reaction to critical events (e.g., arrhythmia detection, insulin regulation)	Vital for life-saving operations with temporal constraints

usability for medical workers. A comparative overview of these foundational models is summarized in Table 2, which highlights their core principles, applications, and relevance to predictive healthcare.

2.3 Architectural Frameworks

The theoretical design of a healthcare digital twin system includes several layers and elements that cooperate in generating the most comprehensive virtual models. The first layer is the physical healthcare environment, which refers to patients, medical devices, healthcare facilities, and care processes. The data provided by this layer is processed into the system of the digital twin using diverse sensing technologies and data acquisition methods. Figure 1 shows the conceptual framework of the same.

The data layer is involved in the coherent collection and preprocessing of health data (collected by various data sources such as IoT sensors, electronic health records, medical devices, external data sources like environmental sensors, and social determinants of health databases). The theoretical implementation of this layer is based on database theory, data management concepts, and information architecture to achieve data quality, integrity, and accessibility.

The modeling layer is the data center of the digital twin, which constructs and verifies virtual models. In this layer, different modeling methods such as mathematical models, statistical models, machine learning models, and simulation models are used to build faithful representations of physical entities and processes. This layer has the theoretical background of healthcare and medical sciences, followed by computational modeling theories and techniques.

The analytics layer is where intelligence functions are deployed to enable predictive analytics, pattern recognition, and decision support. This layer utilizes machine learning algorithms, statistical tools, and artificial intelligence procedures to extract meaning from data and make forecasts based on such data. This layer is theoretically grounded in machine learning theory, statistical inference, and decision theory.

The application layer places the outcomes of the digital twin and its insights in front of the end users, i.e., healthcare professionals, patients, and healthcare administrators. Such a layer must take into consideration human–computer interaction theories, user experience design principles, and clinical workflow theories as means to enable effective utilization of the digital twin's capabilities.

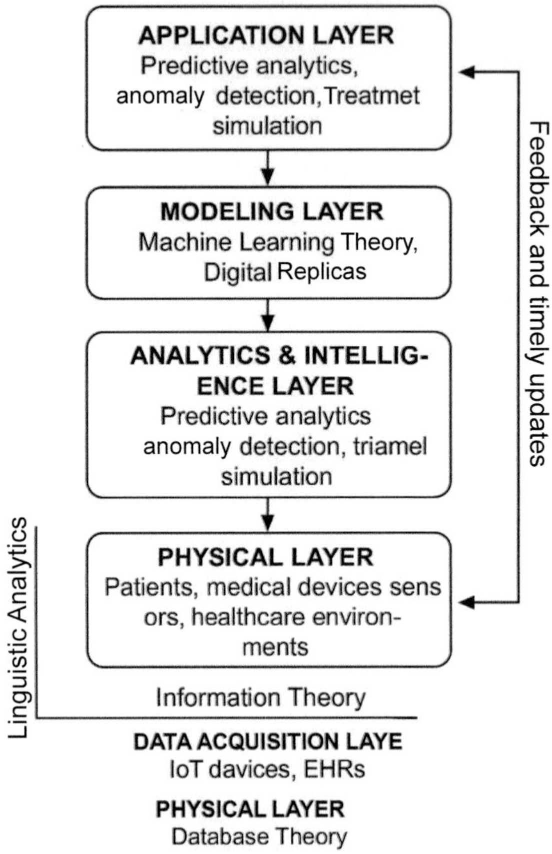

Figure 1: Conceptual Framework of Healthcare.

3. IoT Integration and Cyber-Physical Systems Theory

3.1 Theoretical Foundations of IoT in Healthcare

Cyber-physical systems (CPS) theory is based on a theoretical framework dedicated to the connection between computing systems and physical phenomena (Wang et al. 2022). The integration of Internet of Things technologies in healthcare digital twins is grounded in this theory, also referred to as cyberphysical systems theory (Wang et al. 2022). Within the healthcare setting, CPS theory helps guide the discussion of the potential usefulness of digital twins to incorporate IoT devices toward establishing a complete and real-time monitoring and control system.

There is a strong emphasis in cyber-physical systems theory on the strong coupling between the physical parts and the computing parts with feedback loops of the computational elements that monitor and control the physical processes. In healthcare digital twins, this involves temporal monitoring of patient physiological parameters, environmental conditions, and medical devices using IoT sensors, with the digital twin acting as the computational element that processes the information and feeds insights back to the physical system.

Distributed systems theory, which deals with the issues of managing and controlling multiple interconnected devices and systems, also informs the theoretical background of the approach to IoT integration into healthcare digital twins. The context of healthcare environments is characterized by a large number of IoT devices: wearable sensors, implantables, environmental monitors, and medical equipment, all of which depend on coordinated effort in providing a detailed overview of patient monitoring and care delivery (Liu et al. 2021). Table 3 presents a classification of IoT

Table 3: IoT Devices and Data Sources for Healthcare Digital Twins.

Device Type	Data Captured	Integration Purpose	Example Use Case
Wearable Sensors	Heart rate, blood pressure, temperature	Real-time health monitoring	Continuous glucose monitoring for diabetic patients
Implantable Devices	Neural signals, cardiac rhythm	Long-term internal monitoring	Pacemakers transmitting arrhythmia data
Environmental Sensors	Air quality, temperature, humidity	Contextual awareness of health environment	Asthma patient monitoring in polluted areas
Electronic Health Records (EHR)	Medical history, prescriptions, lab reports	Comprehensive view of clinical and historical data	Integration with digital twin for treatment personalization
Imaging Devices (MRI, CT)	Anatomical scans, physiological structures	Diagnostic model construction	Tumor growth modeling and treatment response prediction

devices and data sources, detailing the type of data they capture and their purpose within the digital twin ecosystem.

The network theory is another set of theoretical concepts that may be used to learn more about IoT device communication and data exchange within healthcare ecosystems. Some of the problems that the theory focuses on include network topology, network communication protocols, data routing, and network reliability, which are paramount in the successful implementation of IoT in healthcare digital twins. The theoretical framework emphasizes the significance of consistent, safe, and effective communication systems capable of supporting the high-frequency information flow required for real-time digital twin synchronization.

3.2 Data Fusion and Integration Theory

Data fusion in IoT-driven healthcare digital twins can be referred to the information fusion theory that discusses the complexity of integrating information generated by two or more sources to develop a complete picture of the system being governed. When applied in healthcare, this approach entails combining information collected by numerous IoT sensors, medical devices, electronic health records, and external data sources to develop a comprehensive picture of the patient's health status.

Data fusion theory covers various levels of integration, such as signal-level fusion, which involves integration of raw sensor data, through to decision-level fusion involving integration of high-level interpretations and recommendations. The theoretical framework indicates the need to consider the quality assessment of data, quantification of uncertainty, and resolution of conflicts in data merging from multiple sources. In healthcare digital twins, this means using complex algorithms capable of evaluating data sources by reliability and accuracy and resolving any conflicts that may occur when sensors report mutually inconsistent data.

The theory of sensor networking also offers more grounding on how IoT devices can be structured into networks and coordinated so that overall monitoring capabilities are enhanced. Among these are notions of sensor localization, data collection schemes, and energy-efficient communication patterns, especially for wearable and implantable medical devices with limited battery capacity.

3.3 Real-Time Systems Theory

The theory of real-time systems provides significant theoretical grounding in understanding how healthcare systems digitized through IoT can react to time-sensitive events and conditions. Timely response in healthcare applications is critical to patient safety, such as identifying cardiac arrhythmia, calculating vital signs during surgery, or alerting caregivers to possible adverse events.

The theory differentiates real-time systems regarded as hard, in which failure to meet deadlines can lead to disastrous outcomes, and soft, in which missed deadlines are tolerated with only degraded system performance. Within the context of healthcare digital twins, some applications may require hard real-time guarantees (e.g., closed-loop control systems regulating drug delivery or ventilator control systems), whereas others have soft real-time requirements (e.g., trend analysis and long-term health tracking).

Also considered under real-time systems theory are matters of task scheduling, resource allocation, and quality-of-service guarantees that must be met to ensure reliability in healthcare digital twins. In the theoretical framework, predictable system behavior, bounded response times, and fault-tolerance mechanisms that allow continued system operation even when components fail or the network is impaired are identified as important.

4. Machine Learning and AI Theoretical Foundations

4.1 Predictive Modeling Theory

Machine learning and artificial intelligence theories form the basis of the predictive functions of healthcare digital twins, as they allow the systems to be trained on data and predict emerging phenomena or possible outcomes (Hassanien et al. 2024). Predictive modeling theory provides the theoretical grounds for how digital twins are able to predict changes in patient health, outcomes of treatment programs, and adverse incidents using historical information and present data.

The keys to predictive modeling in healthcare digital twins include statistical learning theory, which answers basic questions on how training data is linked to predictive performance. The theory covers notions such as bias–variance tradeoffs, generalization bounds, and sample complexity, all essential in estimating reliable predictive models within healthcare settings. These theoretical bases help ensure that predictive models created to support healthcare digital twins generalize appropriately to new patients and conditions.

Supervised learning theory offers structured training schemes for building predictive models based on labeled historical data. In the healthcare context of digital twins, this may involve learning from past patient cases to predict future outcomes, such as the likelihood of disease progression, treatment response, or risk of complications. The theory covers matters such as feature selection, model selection and evaluation of their performance, which play a key role in developing competent predictive models.

Unsupervised learning theory can be applied to determine how digital twins identify patterns or structures within healthcare data when labels and targets are not explicitly provided. This is particularly suited for determining patient phenotypes, detecting abnormal conditions, and discovering previously uncharacterized associations among health measures. The theoretical background includes clustering, dimensionality reduction, anomaly detection, and related methods that support digital twins in analyzing complex healthcare data.

4.2 Deep Learning Foundations

The theory of deep learning offers advanced theoretical principles for formulating complex predictive models in healthcare digital twins. The theory addresses the ability of multilayered artificial neural networks to learn highly non-linear relationships in healthcare data, enabling more precise and accurate predictions and insights compared to conventional machine learning methods.

Theoretical justification: The universal approximation theorem suggests the application of deep neural networks to healthcare digital twins, showing that a neural network with a sufficiently wide neural network is able to approximate any continuous function to arbitrary accuracy. These are the theoretical backgrounds to develop deep learning models capable of detecting the intricate links among several physiological parameters, genetic and epigenetic factors as well as environmental factors that influence patient health.

Representation learning theory: Representation learning theory enables the automation of feature engineering and representation learning in deep learning models to improve raw healthcare data and make it more useful. It is especially valuable in healthcare digital twins, where the connections between various health measures may be complex and not clearly defined. The theoretical framework allows the creation of models capable of learning meaningful representations of patient health status directly from sensor data, medical images, and other raw health data.

4.3 Reinforcement Learning Theory

The theory of reinforcement learning provides the theoretical concepts for developing adaptive healthcare digital twins that can learn optimal treatment strategies by interacting with patients and healthcare environments. Unlike supervised learning, which is constrained by labeled training data, reinforcement learning allows digital twins to learn through trial and error and receive feedback in the form of rewards or penalties based on the outcomes of their actions.

Markov decision process theory offers a mathematical representation for modeling healthcare decision-making, in which the current condition of the patient dictates the action to take and the result of the action determines the subsequent state. This theoretical grounding enables the creation of digital twins that can discover optimal treatment procedures, medication administration methods, and care-flow planning schemes through experience.

The theory of multi-agent reinforcement learning extends these ideas to situations involving multiple interacting agents—for example, different healthcare providers, patients, or medical devices. This theoretical framework is especially applicable to healthcare digital twins that must interact with other systems and stakeholders to ensure comprehensive patient care.

5. Predictive Intelligence Frameworks

5.1 Theoretical Models for Health Prediction

Predictive intelligence in healthcare digital twins demands complex theoretical frameworks capable of handling the complexity and uncertainty of healthcare data and processes. Bayesian inference theory provides a foundation that helps digital twins make probabilistic forecasts about patient outcomes and treatment behaviors by incorporating uncertainty and prior knowledge into predictive models. Figure 2 shows the framework of predictive intelligence.

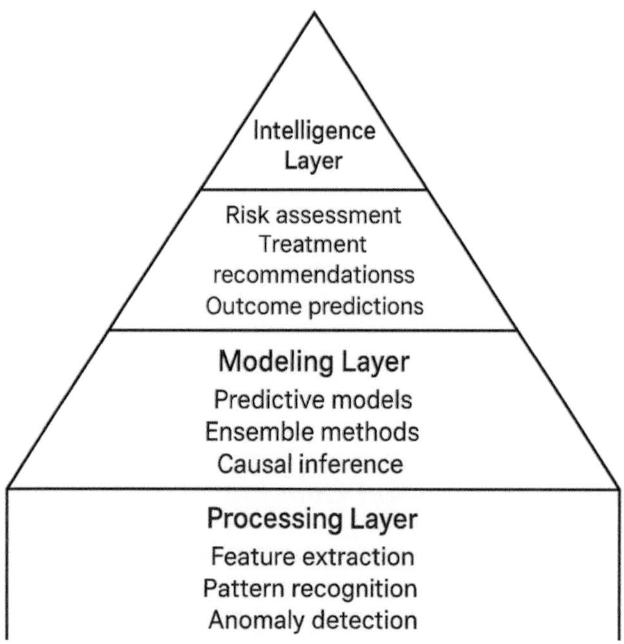

Figure 2: Predictive Intelligence Frameworks.

The bayesian network theory allows to produce a digital twin that has the ability to reason about the probability of various health outcomes with an observation of symptoms, test results, patient characteristics, etc. The theoretical aspects involve, among others, the conditional independence, belief propagation, and causal inference concepts which are relevant in ensuring the development of feasible predictive models in a healthcare setting.

The theory of time series analysis offers the foundation for understanding how healthcare digital twins can predict future health states based on historical trends and patterns. Temporal dependencies are common in healthcare data, as recent observations often influence near-future behavior; time series theory offers the mathematical approach to model these dependencies. Concepts such as autoregressive models, state-space models, and dynamic systems equip digital twins with the capability to capture time-based trends in patient health informatics.

5.2 Causal Inference Theory

The theory of causal inference offers invaluable theoretical support in the creation of healthcare digital twins that can not only predict whatever outcome, but are also able to comprehend the association of various interventions and their influence towards health outcomes. In healthcare, especially, this is part of the reason that knowing the cause and effect is necessary to guide sound treatment decisions and prevent dangerous courses of action.

The causal diagram theory offers a graphical way of expressing causal associations among various variables, enabling healthcare digital twins to reason about the consequences of different interventions on patient outcomes. The theoretical ground includes concepts such as confounding, mediation, and effect modification, which play a vital role in developing the correct causal model in healthcare settings.

The theory of counterfactual reasoning allows digital twins to operate under assumptions about what would happen if circumstances were different, such as asking what the outcome would have been if an alternative treatment had been adopted. This theoretical basis is essential for creating digital twins capable of guiding clinical decision-making by analyzing the potential outcomes of various treatment options.

5.3 Ensemble Methods Theory

The theory of ensemble methods provides a foundation for merging several predictive models to obtain superior performance compared to a single model. This is especially applicable in healthcare digital twins, where the complexity and variability of healthcare data may require different modeling approaches to obtain reliable predictions.

The ensemble method theory is represented by multiple approaches such as bagging, boosting, and stacking approaches with other theoretical grounds and assumptions. Bagging theory deals with combining models trained on various subsets of data to reduce the variance and deliver better generalization performance. Boosting theory describes how one can train models sequentially in order to fix the mistakes of earlier models to minimize bias and, in general, perform better.

Stacking theory provides a way to combine multiple models using a meta-learner, which learns how to best blend the predictions of the base models. These theories support the creation of complex ensemble systems that leverage the strengths of various modeling approaches to improve predictive performance in healthcare digital twins.

6. Challenges and Theoretical Considerations

As digital twin systems in healthcare span multiple disciplines, Figure 3 maps the interconnected theoretical foundations that support their development. This integration map offers a holistic view of how various theories—from real-time systems to interoperability—collectively contribute to the functionality, reliability, and scalability of healthcare digital twins.

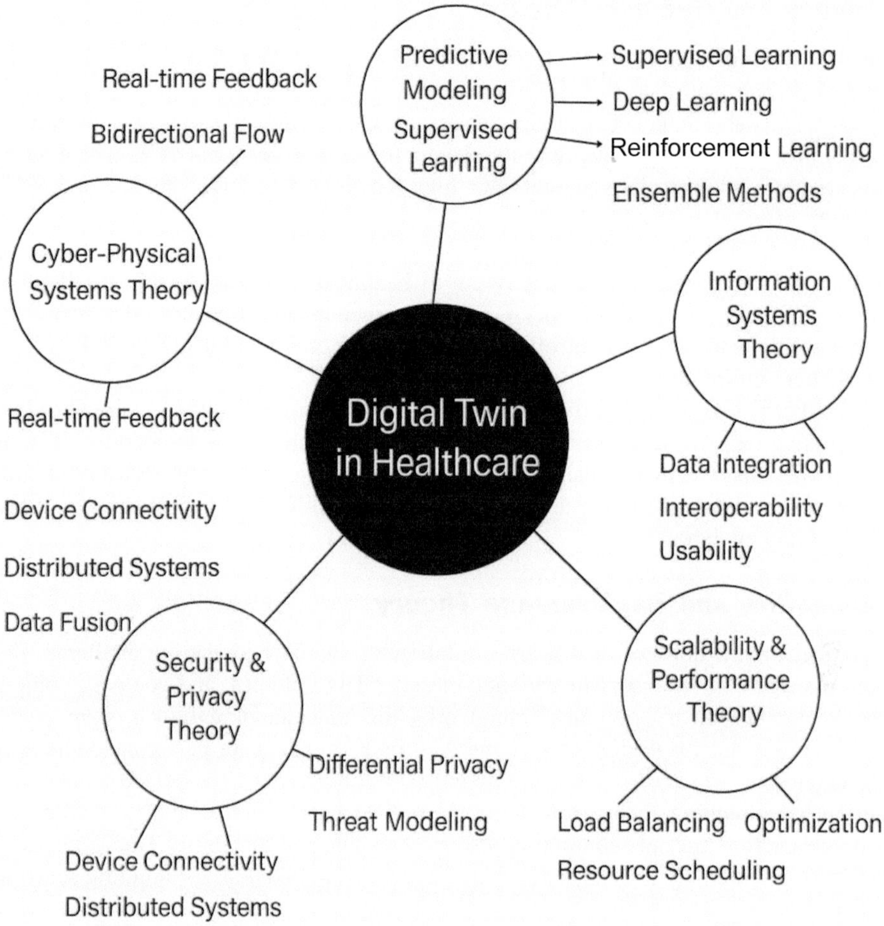

Figure 3: Theoretical Integration Map of Digital Twins in IoT-Driven Healthcare.

6.1 Data Quality and Uncertainty Theory

The theoretical underpinnings of healthcare digital twins must address major issues related to data quality and uncertainty in healthcare settings. Information quality theory offers guidelines for interpreting and managing data quality problems such as accuracy, completeness, consistency, and timeliness. These concerns are particularly complex in healthcare digital twins because of the variety of data sources, the depth of healthcare processes, and the significance of clinical decisions.

Uncertainty quantification theory provides a theoretical basis for understanding and managing uncertainty in healthcare digital twins. The theory covers several types of uncertainty, including aleatory uncertainty (due to the inherent randomness of healthcare processes) and epistemic uncertainty (due to incomplete knowledge or limited data). Recognizing these forms of uncertainty is important for developing digital twins that provide high-quality forecasts while acknowledging the limitations of available data and models.

The error propagation theory concerns the influence of the uncertainties of input values and the parameters of the model on the accuracy of predictions and suggestions provided by digital twins. This theory is necessary to develop healthcare digital twins that are able to offer appropriate confidence intervals and uncertainty estimates for their forecasts so that healthcare professionals can make accurate decisions based on the reliability of the data presented.

6.2 Privacy and Security Theoretical Frameworks

Theoretical reasoning behind digital twins in healthcare should acknowledge the considerable challenges of privacy and data security associated with the gathering, storage, and processing of confidential health data (Suhail et al. 2022). Privacy theory provides concepts for identifying various categories of privacy threats and developing relevant protection mechanisms. In a healthcare digital twin, this involves safeguarding patient identities, health information, and behavioral patterns against unauthorized access or disclosure.

Theoretical tools such as differential privacy are used to formulate the foundations for implementing privacy-preserving digital twins in healthcare, offering viable intelligence while protecting individual privacy. The theory provides mathematical guarantees of privacy protection and offers algorithms for adding controlled noise to data or query outputs to limit the disclosure of individual patient information.

Security theory concerns identifying the problems of securing healthcare digital twins against attacks, including data breaches, system compromise, and malicious manipulation of digital twin models or predictions. The theoretical background includes concepts such as threat modeling, attack surface analysis, and defense-in-depth, all of which are significant for creating secure healthcare systems supported by digital twins.

6.3 Scalability and Performance Theory

The theoretical implications of healthcare digital twins should address the challenge of scaling systems to manage large numbers of patients, massive data volumes, and complex computational demands. Scalability theory provides frameworks for understanding how system performance changes with scale and for designing systems that maintain adequate performance as operational demands increase.

Distributed computing theory offers the theoretical basis for distributing healthcare digital twin workloads across multiple computing resources to enhance scalability. Concepts such as load balancing, data partitioning, and consistency management are essential for building scalable digital twin systems.

Performance theory enables the structured analysis and optimization of healthcare digital twin performance using concepts like throughput, latency, and resource consumption. The theory incorporates queuing theory, which helps in understanding how system performance degrades under increased load, and optimization theory, which provides strategies for improving performance through more efficient scheduling and resource allocation.

6.4 Interoperability Theory

The theory of interoperability provides the theoretical knowledge required for healthcare digital twins to interact effectively with existing healthcare information systems and with other digital twins. The theory covers technical, semantic, and organizational interoperability, each with its own challenges in healthcare settings.

Technical interoperability theory deals with ensuring that systems with varying technology architectures, communication protocols, and data formats can work together. The theory also provides guidelines for developing standards-based integration strategies that enable seamless data sharing between healthcare digital twins and other systems.

The theory of semantic interoperability deals with the issues of how to guarantee that the data and information moved between systems keep their meaning and context. This is especially difficult in healthcare, where one clinical concept could have different representations in two different systems or contexts. The theory involves a subset of ontology engineering, semantic web technologies, and knowledge representation concepts necessary to create interoperable healthcare digital twins.

7. Future Directions and Emerging Theories

7.1 Quantum Computing Applications

With the development of quantum computing, new theoretical backgrounds have emerged that can transform healthcare digital twins in the future (Shafique et al. 2021). Theories of quantum computing offer conceptual frameworks for understanding how quantum mechanical effects can be harnessed to address computational problems that are intractable for classical computers. Quantum computing may make it possible to model more complicated biological systems within healthcare digital twins, accelerate treatment protocol optimization, and provide more accurate predictions of treatment outcomes.

The quantum machine learning theory investigates the potential of quantum algorithms to enhance the power of machine learning within healthcare digital twins. According to the theory, some forms of machine learning problems can be solved exponentially faster on quantum computers because quantum algorithms can solve them in super-polynomial time, including optimizing complex treatment plans and pattern recognition in high-dimensional healthcare data.

The theory of quantum cryptography offers novel strategies for securing digitally twinned healthcare systems that may be more resilient than conventional cryptographic techniques. The theory applies the properties of quantum mechanics to provide information-theoretic security guarantees, offering greater protection for sensitive health information associated with a digital twin system.

7.2 Neuromorphic Computing Frameworks

The neuromorphic computing theory offers new theoretical premises for building healthcare digital twins that more closely mimic the behavior and structure of biological neural networks. The theory deals with implementing artificial neural networks using brain-inspired computing systems that can be more efficient and optimized for processing noisy, complex, and time-varying data common in healthcare applications.

The theory of spike-based neural networks offers guidelines for building neuromorphic healthcare digital twins, in which information processing occurs through discrete events (spikes) rather than continuous signals. This approach can be particularly appropriate for processing data generated as discrete events, as in IoT sensors and medical devices.

Neuroscientific plasticity theory provides the foundation for developing adaptive healthcare digital twins that can learn and evolve based on changes in patient conditions and treatment outcomes. The theory addresses how artificial neural networks may undergo forms of plasticity—such as synaptic plasticity, structural plasticity, or homeostatic plasticity—to create more robust and flexible healthcare digital twin systems.

7.3 Edge Computing and Distributed Intelligence

The theory of edge computing provides theoretical guidelines for distributing healthcare digital twins across various edge devices, cloud-based servers, and hybrid systems (Chen et al. 2024). It deals with the tradeoffs between centralized and distributed computing, considering aspects such as latency, bandwidth, privacy, and reliability.

The theory of federated learning offers the possible development of healthcare digital twins without the requirement of gathering data at central points. The solution can be especially useful in the healthcare field, where the confidentiality and security of data can restrict the possibility of centralizing sensitive medical information (Suhail et al. 2022). The theory of distributed artificial intelligence offers more extensive concepts on how several learning agents can be used to achieve holistic capability in a healthcare digital twin. Such problems in the theory as coordination, communication, and conflict resolution are important in the conceptualization of effective healthcare complex multi-agent digital twin systems.

8. Conclusion

The theoretical background of digital twins in predictive intelligence developments in IoT-powered health ecosystems is an elaborate and still developing realm that is embodied by numerous fields of study, such as computer science, engineering, healthcare informatics, and medical sciences. The chapter has discussed the major theoretical frameworks supporting the preparation and deployment of healthcare digital twins, which can serve as a complete basis for understanding their powers and drawbacks.

The theoretical background of digital twins in healthcare includes proven theories of cyber-physical systems, complex adaptive systems, information systems, and those that relate to the field of healthcare (patient safety, clinical workflow, and compliance). IoT technologies further complicate theoretical properties, necessitating a distributed systems framework, network theory, and real-time systems frameworks in order to guarantee dependable and effective use.

The artificial intelligence performance of healthcare digital twins rests on machine learning and artificial intelligence theories that allow their systems to understand data and make proper projections concerning future health results. Such theoretical backgrounds incorporate supervised learning, unsupervised learning, and reinforcement learning, which have certain applications within healthcare settings. The predictive powers of any healthcare digital twin are improved by the development of ensemble methodologies and causal inference capacities.

The issues of healthcare digital twins—such as data quality, privacy, security, scalability, and interoperability—necessitate complex theoretical foundations that are able to meet the specific needs of healthcare applications. These issues are further complicated by the sensitivity of healthcare decision-making processes and the need to implement systems capable of delivering reliable, accurate, and timely information in support of clinical decision-making.

In the future perspective, new theoretical grounds emerging with quantum computing, neuromorphic computing, and edge computing bring new possibilities for improving healthcare digital twins. Such new theories have the possibility of facilitating more advanced modeling, enhanced efficiency, and better integration with established healthcare systems.

The theory behind the issues discussed in this chapter forms the foundation of real-life executions of healthcare digital twins that can revolutionize the provision of healthcare because of the predictive intelligence they offer. Nevertheless, translating these theoretical bases into practical systems requires critical attention to implementation issues, validation needs, and compliance with current healthcare operational processes. In future studies, it is recommended to address the gap between research possibilities and practical applications so that healthcare digital twins may effectively reach their potential in improving patient outcomes.

Development of further changes in theoretical grounds in this area will depend on progress in computing technologies, better comprehension of complex biological scenarios, and increasing access to high-quality healthcare data. With the growing maturity of these theoretical underpinnings, further advanced development of healthcare-based digital twins should be possible, capable of making more effective predictions, forming improved treatment proposals, and achieving better patient outcomes.

The interdisciplinarity of healthcare digital twins demands persistent cooperation among computer scientists and engineers, health professionals, and domain experts to transform theoretical progress into effective implementations solving real-life problems in medicine.

References

Abouella, A., Dhouib, D., Feki, N., Grissa, I., Kharroubi, A., and Elleuch, W. (2024). Digital twins' advancements and applications in healthcare, towards precision medicine. *Computers in Biology and Medicine*, 157: 108630. https://doi.org/10.1016/j.compbiomed.2024.108630.

Alshammari, N. (2024). Advancing health care with digital twins: Meta-review of applications and implementation challenges. *Journal of Medical Internet Research*, 27(1), e69544. https://doi.org/10.2196/69544.

Chen, J., Wang, X., Li, Y., Zhang, S., Liu, H., and Kumar, A. (2024). Generative AI-driven human digital twin in IoT-healthcare: A comprehensive survey. *Computer Networks*, 238, 110108. https://doi.org/10.1016/j.comnet.2024.110108

Hassanien, A. E., Rizk, R. Y., and Snášel, V. (2024). Machine learning and deep learning-based approach in smart healthcare: Recent advances, applications, challenges and opportunities. *Cluster Computing*, 27(3): 2791–2815. https://doi.org/10.1007/s10586-023-04134-2.

Kharroubi, A., Dhouib, D., Feki, N., Grissa, I., Abouella, A., and Elleuch, W. (2024). Digital twins for health: A scoping review. *npj Digital Medicine*, 7(1): 67. https://doi.org/10.1038/s41746-024-01073-0.

Liu, Y., Zhang, L., Yang, Y., Zhou, L., Ren, L., Wang, F. et al. (2021). A comprehensive survey on machine learning-based big data analytics for IoT-enabled smart healthcare system. *Mobile Networks and Applications*, 26(1): 234–252. https://doi.org/10.1007/s11036-020-01700-6.

Nadeem, M., Kostic, S., Dornhöfer, M., Weber, C., and Fathi, M. (2025). A comprehensive review of digital twin in healthcare in the scope of simulative health-monitoring. *SAGE Open Medicine*, 13. https://doi.org/10.1177/20552076241304078.

Pires, F., Cachada, A., Barbosa, J., Moreira, A. P., and Leitão, P. (2019). Digital twin in industry 4.0: Technologies, applications and challenges. pp. 721–726. *In: 2019 IEEE 17th International Conference on Industrial Informatics (INDIN)*. IEEE. https://doi.org/10.1109/INDIN41052.2019.8972134.

Qian, C., Liu, X., Ripley, C., Qian, M., Liang, F., and Yu, W. (2022). Digital twin for healthcare systems. *Frontiers in Digital Health*, 4: 1253050. https://doi.org/10.3389/fdgth.2023.1253050.

Rasheed, A., San, O., and Kvamsdal, T. (2020). Digital twin: Values, challenges and enablers from a modeling perspective. *IEEE Access*, 8: 21980-22012. https://doi.org/10.1109/ACCESS.2020.2970143.

Sahal, R., Breslin, J. G., and Ali, M. I. (2020). Big data and stream processing platforms for Industry 4.0 requirements mapping for a predictive maintenance use case. *Journal of Manufacturing Systems*, 54: 138–151. https://doi.org/10.1016/j.jmsy.2019.11.004.

Shafique, M., Theocharides, T., Bouganis, C. S., Hanif, M. A., Khalid, F., Hafız, R. et al. (2021). An overview of next-generation architectures for machine learning: Roadmap, opportunities and challenges in the IoT era. *Proceedings of the 2021 Design, Automation & Test in Europe Conference* (pp. 827–832). https://doi.org/10.23919/DATE51398.2021.9474087

Suhail, M., Hussain, R., Jurdak, R., and Hong, C. S. (2022). Trustworthy digital twins in the industrial internet of things with blockchain. *IEEE Internet Computing*, 26(6): 58–67. https://doi.org/10.1109/MIC.2021.3132449.

Viceconti, M., and Hunter, P. (2016). The virtual physiological human: Ten years after. *Annual Review of Biomedical Engineering*, 18: 103–123. https://doi.org/10.1146/annurev-bioeng-071114-040609.

Wang, J., Liang, F., Zhou, H., Yang, M., and Wang, Q. (2022). A survey on digital twin: Architecture, enabling technologies, security and privacy, and future prospects. *IEEE Internet of Things Journal*, 10(17): 14965–14987. https://doi.org/10.1109/JIOT.2023.3263909.

20

Strategic Management in the Smart Era

Industry 5.0 and Digital Finance

Shilpa A. C.,[1] *Kiran Kumar M.*,[2,*] *Soujanya L.*[3] and *Saravanapriya M.*[4]

1. Introduction

The introduction of Industry 4.0 represented a watershed moment in the story of global industry, heralding a digital transformation era of cyber-physical systems, automation, and real-time data exchange. This digital revolution synthesized a number of key technologies including the Internet of Things (IoT), Cloud Computing, Big Data Analytics, and Artificial Intelligence (AI) in an approach to improve industrial productivity and operational efficiency. However, as new technologies emerge, Industry 5.0 has brought new challenges and opportunities, moving the focus in a more human-centric, sustainable, and resilient manner away from pure automation and digitalization. Industry 5.0 places undeniable importance on the role of human intelligence, creativity, and ethics as fundamental to future production and service systems. Acknowledging the impact of humans partnering with machines in a collaborative way, co-creating personalized value, and contributing to sustainable innovation are foundational values and concepts that inform a more integrated and complete image of industrial progress. In parallel to these developments in industry, the global financial ecosystem is experiencing its own rapid transformation with the emergence of more digital finance (Zizic et al. 2022). Digital finance includes a wide array of innovations in the form of fintech platforms, blockchain transactions, digital currency, financial analytics using AI, and decentralized finance (DeFi). All of these developments are not only taking market share from traditional banking and financial services, but also driving financial inclusion, lowering transactional costs, and enhancing the speed and security of cross-border financial engagements. In many emerging economies, mobile-

[1] BGS Institute of Technology, Adichunchanagiri University, Karnataka, India.
Email: shilpaac@bgsit.ac.in, 0009-0000-1037-8478
[2] CMS Business School, Jain (Deemed-to-be University), Bangalore, India.
[3] S-Vyasa Deemed to be University , Bengaluru, India.
Email: soujanyanirav@gmail.com, 0009-0005-6765-6703
[4] Department of Management Studies, KGiSL Institute of Technology, Thudiyalur Road, Saravanampatti, Coimbatore – 641 035.
Email: saravanapriya.m@kgkite.ac.in, https://orcid.org/0000-0003-1632-8687
* Corresponding author: kiranm287@gmail.com, 0000-0003-2084-9418

DOI: 10.1201/9781003655190-24

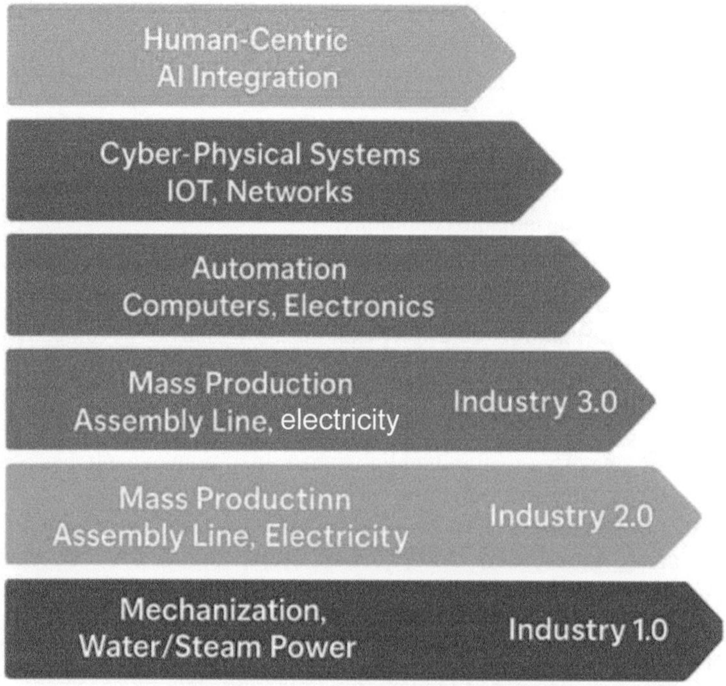

Figure 1: Evolution of Industry Paradigms.

based digital financial services have played a key role in addressing the divide between formal financial institutions and the underserved. As markets evolve, companies will increasingly leverage digital finance in their core operations to compete, enhance customer experience, and keep pace with market and regulatory shifts. Figure 1 illustrates the transition from Industry 1.0 to Industry 5.0, highlighting the technological and strategic shifts from mechanization and mass production to human-centric, AI-integrated smart systems.

In a rapidly changing environment, the complexity and importance of strategic management is increasingly evident. The ability of traditional strategic models, which based their strength on relatively stable environments and foresight, is not able to keep up with the pace of technological disruption and socio-economic volatility. Organizations need to be adapting to agile and data-driven strategic management—one that is capable of managing real-time dilemmas through new technologies with ethically based considerations. Aligning strategic objectives with the transformational capacity of Industry 5.0 and the power of digital finance is not an option; it is now a prerequisite for sustainable success. Just as strategic management in the smart era must make sense of the complex relationships of digital transformation, human capital, financial innovation, and ethical governance, interest in Industry 5.0 and digital finance in their own right is growing; however, there remains a gap in the literature exploring the consequences of both together for strategic management (Leng et al. 2022). Existing work is focused on either industrial transformation from a techno-strategic perspective or on financial innovations in isolation—the existing literature does not holistically position dual forces at the forefront of the strategic thinking and decision-making processes of organizations and the long-term resilience of organizations. Dealing with this deficiency is important, because organisations that do not systematically incorporate these changes will be at a disadvantage in advancing innovation, engagement, and sustainable value creation. The principal goal of this research chapter is to examine how strategic management could shift in response to the simultaneous development of Industry 5.0 and digital finance. It is possible to identify the strategic models, tools, and talents necessary to promote business in the intelligent era. The questions that underpin the study will include the following:

- What are the implications of the transition from Industry 4.0 to Industry 5.0 for strategic priorities in organisations?
- How does digital finance affect corporate strategy, risk, and value creation?
- What strategic management models can integrate technological innovation with human-inclusive and financially sustainable conditions?

The remainder of the chapter is designed around the following sections: Section 2 reviews theoretical context and literature around the distinctions and elements of Industry 5.0 and digital finance. In this introductory literature review, we examine how each has developed in isolation from each other and note that it is only now that the notions of digital finance and Industry 5.0 converge. Section 3 provides an examination of strategic management within an Industry 5.0 context, focusing primarily on human–machine interaction, ethical ethos for leaders, the notion of personalization, and the possible role of data. This section highlights the role of human capital as a changing driver of transformation and the strategic relationship between the human workforce and intelligent systems. Section 4 looks at digital finance's impact on strategic management and provides strategic implications for fintech, blockchain, DeFi, robo-advisors, and CBDCs considering financial inclusion, compliance issues, cybersecurity, and an organization's ability to respond and adapt with agility in relation to current or emerging trends. Section 5 provides examples of case studies demonstrating applied strategic transformation of both Industry 5.0 and digital finance, ideally from organizations with a distinguished reputation within their field like Siemens, Bosch, Fujitsu, Paytm, and Revolut, within the transformational capacity of CBDC implementations like eNaira and e-RUPI. Section 6 includes challenges and opportunities associated with embedding Industry 5.0 and digital finance into strategy—focusing on the realities of needing to cope with legacy systems, cultural inertia, ethical considerations, and real-time engagement. In Section 7 we propose a strategic management framework for the smart era, presented as five interrelated pillars to build the foundation of strategy: technology, people, finance, governance, and ethics—such a framework will enable us to enact sustainable and competitive advantage in an increasingly complex, digital environment. Finally, Section 8 concludes the research.

2. Theoretical Background and Literature Review

Digital finance is the application of digital technologies to financial services and functions. Digital finance revolutionizes how individuals and organizations access, manage, and use financial resources. It combines a wide range of other innovations, such as financial technology (fintech), blockchain-based assets and infrastructures, AI applications in finance, algorithmic trading, digital wallets, and CBDCs. These are not just reshaping financial markets, they are also reshaping how businesses and governments exercise their strategic capabilities. Fintech has made new business models possible, such as peer-to-peer loans, mobile banking, and buy now, pay later, thereby forcing traditional financial institutions to rapidly innovate and keep pace with faster and more customer-centric alternatives. Blockchain has enabled new ways to achieve transparency and maximum security around transactions through decentralized and immutable ledgers, digital smart contracts, tokenized assets, and DeFi platforms. CBDCs represent a new policy and infrastructure shift for central banks to offer a digital version of existing fiat currency with maximum control and oversight (Nahavandi 2019). One major difference between traditional finance and digital finance is their respective operational philosophy. Traditional finance is hierarchical, heavily regulated, and methodical about forging ahead with new innovations. Digital finance is not regulated as intensely as traditional finance, has rapidly evolving strategies through the use of digital networks, and is all about scalability.

It takes advantage of cloud computing services to handle vast amounts of transactions and data while supporting real-time financial decisions. The analysis of big data enhances understanding of customers, credit risk, and administration of investment opportunities. Internet of Things (IoT) devices collect financial data in real time throughout the supply chain and business operations, thus allowing for dynamic pricing, current and online insurance options, and adaptive lines of credit. Ultimately, AI and machine learning are capable of handling tasks from automated fraud detection and asset management to customizing financial services for customers. Cybersecurity is a fundamental enabler of digital finance because of the associated risks posed by data breaches, financial fraud, and cyberattacks that are becoming commonplace. Strategic planning also needs to incorporate digital identity management strategies, zero-trust architectures, and ongoing compliance with data protection legislation such as GDPR and the new DPDP Act in India. Thus, while digital finance appears to be purely operational, its impact is wider and intentional depending on where it is pursued; it touches on matters strategic (Özdemir and Hekim 2018). It is likely to shape key areas of the business, e.g., risk management, capital allocation, and places of value and engagement with stakeholders. Strategic leaders need to realize and prepare to drive change to remain competitive in a financial landscape increasingly defined by decentralization, digital assets, and intelligent systems. The increased financial activity in Industry 5.0 and digital finance creates broader implications for strategic management. Organizations will need to adopt a more holistic, sustainable, and anticipative strategic posture. The next sections of this chapter seek to explain how these changes are actively transforming the strategic landscape and what models and frameworks can assist organizations in transitioning successfully to the era of smart.

Personalization represents another important characteristic of Industry 5.0. In contrast to Industry 4.0's mass customization processes, 5.0's hyper-personalization is centered around AI and human creativity. In areas such as healthcare, retail, and manufacturing, customers are demanding products and services that are personal to their preferences and values. Supply chains must offer the flexibility associated with mass customization; full-scale predictive and analytics systems; and be able to respond quickly based on the alignment of technology and humans. A company's attention to sustainability and resilience is equally essential. The COVID-19 pandemic as well as climate crises and geopolitical tensions have exposed the fragility of global supply chains and the impact of non-resilient supply chain systems. There are also arguments that Industry 5.0 is keenly focused on uplifting non-resilient systems to resilient systems. This suggests that Industry 5.0 delves into themes linked to decentralization, transparency, ethically governed industrial practices, and practices cognizant of the circular economy, green manufacturing, and social innovation. They are now being developed into a strategy. A sustainability or purpose-driven mission is also an emergent reconciliation of the ESG (Environmental, Social, and Governance) and sustainable development (SDGs) remit of organizations encouraging longer-term business strategies.

3. Strategic Management in Context of Industry 5.0

One of the most important phenomena of Industry 5.0 is the reconfiguration of the human–machine relationship, which moves from seeing automation and artificial intelligence as substitutes for human labor to one of collaborative intelligence, in which machines support and extend people's capabilities. This shift represents a significant opportunity to rethink strategic leadership, workforce policy, and organizational culture more generally. Strategic leadership in an Industry 5.0 world requires a shift away from command-and-control type leadership and toward leadership that is more collaborative, humane, agile, or phronetic. Leaders should cultivate and enable conditions that allow human creativity to emerge and interact positively with technological capabilities. This will require organizations to foster a culture of continuous learning, experimentation, and innovation. Decision-making should also change. While the analytical cadence of data and algorithms facilitates more timely decisions, we should still recognize that human intuition, moral judgment, and emotional intelligence matter, particularly when decisions arise in situations of ambiguity or moral complexity

(Van Erp et al. 2024). This suggests that strategic leaders will still have to find ways to blend data-based insights about what is desired into human-valued outcomes when seeking to form long-term organizational objectives. From a perspective of workforce strategy, the emergence of smart technologies should signal significant investments in human capital.

Knowledge workers must be able to integrate new skill sets that will have them productively engaging with advanced tools like collaborative robots (cobots), machine learning platforms, and augmented/virtual reality systems. At this time, reskilling and upskilling must not be drive components of talent strategy but essential for organizational nimbleness and invention. Workers need to be trained with digital literacy, systems thinking, reading sources of data, and ethical reasoning to extract the maximum value from the opportunities that human–machine collaboration affords. Additionally, organizations need to design roles that balance automation with serious human input to add value and increase employee satisfaction and motivation. This reconfiguration of human–machine interaction also lends insight into how we will structure decision-making at various levels of the organization. Traditional models dictate that many individual decisions would exist in a hierarchy of decision-making and be sequential in nature. Nowadays, applications of artificial intelligence and intelligent systems can aggregate large amounts of data across multiple sources, at considerable pace, offering demonstrably valuable recommendations inexplicably faster. However, we must distinguish that strategic management must support an organizational goal of human fulfillment. Over-reliance on machines without a mechanism for human governance fails to consider the context for decision-making, implores respect for the decision-making environment, and incorporates ethics into the decision-taking process. Hence the job of strategic management is to shape the governance arrangement between managers and machines; for example, management can decide how many of the recommendations should be pursued, post-processing by the machine. Table 1 compares classical, digital-era, and smart-era strategic management models based on decision-making structure, planning horizon, value focus, tools employed, and workforce roles, highlighting the shift toward agility and human–AI collaboration in the Industry 5.0 era.

Data has now become ubiquitous to strategic thinking within Industry 5.0. Today's organizations operate in an era of continuous real-time data collected from multiple sources: IoT devices, internal enterprise systems, activities related to customer engagement, and active listening points within their market. This constant stream of data allows organizations to design strategic frameworks around predictive and prescriptive analysis of data, monetizing it while developing strategies. Today, strategic managers can use data analytics, artificial intelligence, and machine learning to begin making meaningful patterns in numbers based on the past. They can anticipate the future and strategically plan accordingly regarding resource allocation and sustainability. As organizations can now collect and analyze real-time data, they can strategically get decentralized and react quickly to changing conditions. In addition, they can now conduct follow-up on their operational performance and market conditions, and measure how customer sentiment develops to engage their stakeholders for proactive reasons at the right time in the right way (Rejeb et al. 2025). Predictive data analysis takes this a step further in forecasting the future based on present and historical data at the same time. In supply chain management, for example, several firms such as FedEx and DHL utilize the

Table 1: Comparison of Strategic Management Models.

Criteria	Classical Strategy	Digital Strategy	Smart Era Strategy
Decision-making	Top-down	Data-driven	Human-AI collaboration
Planning Horizon	Long-term	Agile cycles	Real-time adaptation
Value Focus	Profit	Efficiency & growth	Human-centric sustainability
Tools	SWOT, PESTLE	Big Data, Cloud	AI, Blockchain, IoT
Workforce Role	Operational	Digital contributors	Co-creators with intelligent systems

outcome of predictive analytic tools to reduce strain on their products, so that they can plan for and react to disruptions and know how to mitigate the issue in advance. In customer engagement, firms have successfully made use of machine learning algorithms to predict emerging consumer behavior in their preferences, which allows firms to adapt their offer or change their messages as a way to respond quickly and get ahead of their competitors. Data analytic technologies also provide avenues for personalization and customization. In a world where customers now expect everything to be tailored to their individual needs and every interaction focused on their experiences, organizations can achieve mass customization through data. AI-enabled recommendation engines, digital twins, real-time customer profiling tools, etc., enable firms to transition away from mass production and standardization to hyper-personalized value propositions. Hyper-personalization not only enables customization resulting in greater customer satisfaction but also results in strategic differentiation in saturated markets (Tyagi 2023a, Tyagi 2023b).

The implications for the development of strategy are significant. Strategic planning is no longer an annual or quarterly effort based on static forecasts but a dynamic, continual process that is repeating and feedback-driven. Organizations need to develop internal capabilities regarding data, analytics, and collaboration across functions. The strategic manager will also need to develop fluency in their own interpretation and use of data in an ethical context to ensure that the insights are responsibly guided in usage. Our increasing reliance on data also highlights important questions around ownership, privacy, and bias. Strategic decisions, therefore, will have to grapple with the ethical and legal implications of data collation and use and will need to reflect increasing regulatory propositions and societal expectations. Table 2 outlines the core technological enablers of digital finance such as blockchain, AI/ML, IoT, and cloud computing and maps them to their strategic roles and example applications in contemporary financial systems.

Sustainability and ethics are no longer afterthoughts when it comes to strategic management; they are primary motivators of value creation and competitive advantage in the Industry 5.0 world. Stakeholders are demanding greater accountability and purpose due to the increased awareness of climate change, social inequality, and governance failures. As a result, ESG considerations must be part of the strategic management decision-making framework. To incorporate ESG into strategy, decision-making must be about more than compliance or "doing good" for corporate social responsibility purposes (Yadav et al. 2024). It is about aligning organizational goals with multi-faceted societal and environmental well-being over the long term. From an environmental standpoint, this could mean rethinking areas like product design, material and labor sourcing, energy consumption, and waste as they relate back to purpose-defined ecometrics objectives. From a social standpoint, it could involve prioritizing diversifying, equity, and inclusion in organizational workforce structure, while ensuring fair and equitable labor practices and policies across the supply chain. From a governance standpoint, this includes transparent structures of decision-making, ethical leadership, and mechanisms to deter corruption, bribery, and/or malfeasance.

Integrating ESG into strategy leads to undeniable advantages. Research shows that ESG-based companies perform above average in the long run not only for financial returns, but also innovation, employee engagement, and brand loyalty. There are also capital access constraints due to ESG, as institutional investors are committed to companies who demonstrate sustainability leadership. In turn, this creates both an opportunity and an obligation for strategic managers to adopt a new

Table 2: Key Enablers of Digital Finance and Their Strategic Functions.

Enabler	Strategic Function	Example Application
Blockchain	Transparency, trust	Smart contracts, DeFi
AI/ML	Predictive decision-making	Robo-advisory, fraud detection
IoT	Real-time data flow	Insurance risk monitoring
Cloud Computing	Scalability and flexibility	Digital banking platforms
Cybersecurity	Risk mitigation	Threat monitoring systems

identity of value creation greater than short-term profit maximization. That said, sustainability and ethical strategy are continuously challenged in a continuously automated world augmented by AI. As machines assume decision-making and replace certain labor functions, organizations are faced with moral or ethical judgment calls. For instance, automating portions of the workforce involves an efficiency dimension and a social dimension relating to worker dislocation, productivity, and unrest. Reliance on algorithmic decisions highlights lack of transparency, accountability, and bias. Strategic managers will need to ensure that AI systems are developed and used in a fair manner, and do not discriminate or violate users' autonomy.

A further ethical consideration also arises from the collection and usage of individual data. Although data drives personalization and efficiency, it can also jeopardize an individual's privacy and autonomy. In addition, regulations associated with personal user data, like the GDPR in Europe and the Digital Personal Data Protection (DPDP) Act in India, are establishing how organizations can, legally and ethically, work with user data. This creates further complexity for strategic managers to navigate, balancing being innovative whilst considering ethical issues and limits relative to the vast technological possibilities. Consequently, there is a more complex, multidirectional approach to strategic management and organizational behavior focused on accommodating human-centered interfaces reflecting new technological innovations and the capabilities for real-time data, along with ethical and sustainable governance (Abdel-Basset et al. 2024). Strategic management needs to reinforce multidimensional approaches that demonstrate the value of strategic management in the context of Industry 5.0 that incorporates cooperation between humans and technology, realizes the innovation pipeline of ideas aligned to bigger social and environmental stakeholder needs. Strategic managers collaborating with technological assets must advance strategic agility, realize collaboration between humans and robots, and must place their strategic objectives in social and environmental surroundings—about the society and the planet—if we are to describe our impact/competitive positioning. Table 3 presents a side-by-side view of the key strategic challenges and corresponding opportunities organizations face in the transition to Industry 5.0 and digital finance integration, emphasizing areas like technology, workforce, finance, governance, and market dynamics.

Table 3: Challenges vs. Opportunities in Smart Era Strategic Management.

Category	Strategic Challenge	Strategic Opportunity
Technology	Legacy system integration	Real-time strategy execution
Workforce	Reskilling needs	Human-AI synergy
Finance	Regulatory uncertainty	Inclusive financial services
Governance	Ethical dilemmas in AI	Purpose-driven strategy
Market	Resistance to change	Hyper-personalization

4. Integration of Digital Finance in Strategic Management

Financial technologies are ceasing to be methods of operational execution and are becoming methods of strategic enablers. These technologies are enabling firms to rethink their business models, reach new customers, and differentiate their offerings in increasingly competitive spaces. Blockchain is perhaps the most significant of these technologies in that it provides a foundation of trust, security, and transparency on which models can be constructed and evaluated. Blockchain offers a trustless distributed ledger mechanism that is not altered/changed once it is created. Since it has permanent transactional history and records, blockchain makes it easier to eliminate the need for intermediaries, which diminish accountability in transactions. This can prove operationally strategic since the opportunities for optimizing supply chains, conducting quasi-autonomous contract enforcement with smart contracts, and using inscribed data trails for audit reporting in the name of regulatory compliance for ESG requirements are immense. DeFi, which—depending on the authoritative view of its boundaries—is built largely on blockchain platforms and the provision of financial services, stands to increase the strategic opportunity even further. DeFi is a fundamental technology that removes the use of an intermediary in a transaction, where individuals can use these services for peer-to-peer lending and borrowing, for both lenders and insured parties for underwriting insurance claims, and open protocols utilized for trading because most likely those assets are blockchain fungible tokens. For firms, this means new sources of funding, liquidity management, tokenized assets, and new ways products and services can be monetized. This also means that startups and SMEs have access to financial services that historically were gated by high barriers to entry that are now democratizing participation and innovation. The strategic implications from organizations adopting DeFi are being nimble, agile, and community-focused. These changes are especially impactful in the context of emerging digital-native markets (Crnobrnja et al. 2023).

Robo-advisors are yet another disruptive force for strategic financial planning. These platforms, which are AI-driven, allow for investment advisory and portfolio management through automated planning based on user characteristics, goals, and risk/return profiles. Robo-advisory services provide streamlined services that can be delivered at a significantly lower cost, and they allow wealth management and financial services to reach a larger percentage of the population. For firms, robo-advisors not only amplify customer service quality but also allow for significant scale, in addition to data-based customer personalization. Strategic management can utilize important information derived from the analytics of robo-advisors to optimize customer segmentation, cross-sell financial products, and structure better financial literacy campaigns. Analogous is CBDCs, which are a more centralized form of financial disruption. CBDCs are distributed and regulated centrally by national central banks, which grants them the benefits of digital transactions alongside the steadiness of fiat currency. CBDCs can reduce friction costs in transactions, diminish settlement costs, and extend government control over monetary policy. From a strategic management perspective, all businesses should mentally prepare for the impact CBDCs will have on ordinary transactions: implications for their digital payment costs; modifications to required regulatory transaction reports; middlemen adjusting; and operational costs for international movement of funds. The roll-out of CBDCs has already started to disrupt the financial services industry in jurisdictions like China, Nigeria, and India, and firms in these markets should consider these regulatory developments in their long-range strategic plans. Altogether, these financial technologies are transforming how competition occurs in banking, investment, and risk management. Traditional banks are being forced to adopt more flexible, customer-focused business models, while investment firms are using AI and blockchain to enhance asset management. Strategic managers need to brace for transformation and realign the firms' internal capacities and capabilities with the external opportunity technologies while constantly scanning the horizon for more transformative financial technologies. Figure 2 depicts the synergetic relationship between human workers, AI systems, and robotics in Industry 5.0. It emphasizes an ecosystem that works together to improve productivity, innovation, and human-centric value creation.

Digital finance presents opportunities for transformation, along with significant strategic risks. Regulatory ambiguity, cybersecurity threats, and ethical dilemmas create impediments that can lead to loss of trust, disruption of services, and exposure to legal liability. Therefore, operationalizing digital finance within the strategic management aspect of firms requires an integrated process involving a thorough examination of risk as well as regulations. The regulatory environment is evolving rapidly to keep step with innovations in digital finance. For example, in the European Union, the General Data Protection Regulation (GDPR) imposes stringent rules around data privacy, consent, and cross-border data flows. These are considerations that must be implemented in any digital finance strategy. Similarly, in Europe, the Revised Payment Services Directive (PSD2) requires open banking and allows banks to provide customer data to licensed third parties in a secure and transparent manner. This opens opportunities for partnerships with fintech firms but also raises issues of compliance and data protection. In India, the DPDP Bill proposes comprehensive data protection regimes that will require firms to ensure adequate processing is conducted legally, minimize personal data, and ensure remediation for the user (Chrifi-Alaoui et al. 2025).

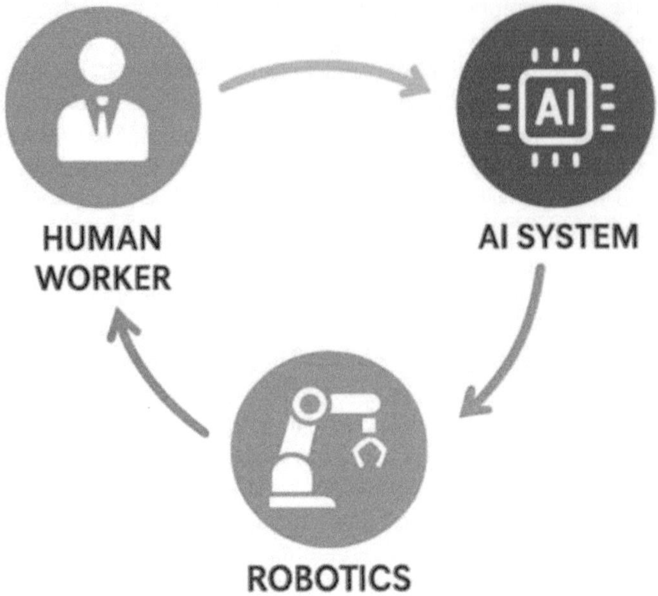

Figure 2: Human-Machine Collaboration Model in Industry 5.0.

Strategic managers must consider this complicated regulatory environment with vision and flexibility. Compliance with regulations is more than a "legal" obligation: it is a strategic differentiating factor. Companies that implement strong data governance, ethical artificial intelligence models, and customer policies transparently can differentiate themselves as trustworthy, pro-social actors in this ethical marketplace. And now, with new regulatory reforms and compliance burdens growing across the financial sector, it is highly advisable that firms consider their interaction with regulation as a strategic action—by pursuing industry coalitions, firms can find themselves aligned with regulatory visions long before regulations arise. The same is true for cybersecurity perspectives. The industry, which is moving away from traditional paper and financial transactions that promote face-to-face contact, creates serious liability issues from hacking, ransomware, and identity theft born from digital engagement and manipulation. Hence, firms must take a strategic view on planning for cybersecurity. Not only are the financial costs of cyberattacks egregious, the reputational impact of cyberattacks can create strategic distance post-duress as firms scramble to regain trust and compliance with regulators. Subsequent consequences cannot be overlooked; failure to strategize on cybersecurity can bring organizations strategic obstacles or, respectively, strategic collapse.

The risk of fraud is also increased with the digital finance environment. With real-time transactions, anonymous wallets, and algorithmic trading systems, these new technologies leave the door open to exploitation for financial crime. Strategic managers must employ tools that enhance fraud detection and engage in partnerships with cybersecurity firms and regulators that can detect suspicious activity. Risk management should be integrated into the strategic planning cycle and should not be treated as a reactive or compliance function. A proactive stance on digital risk can help firms better protect themselves and could also serve to differentiate firms in an environment that increasingly values trust and security from customers and investors. One of the most exciting dimensions of digital finance is its ability to enhance financial inclusion and support strategies that can be scaled globally. In developing economies, where access to traditional banking services is stymied by infrastructure, geography, or socio-economic factors, digital finance can serve as a pathway to financial empowerment. Management strategies should not consider financial inclusion as an afterthought, philanthropic, or only part of a three-year plan, but rather foundational, longer-term, stable, persistent, and an important aspect of growth, innovation, and increasing market space (Tyagi 2023a). Digital finance can create inclusive pathways to growth by providing interactions that afford access to financial services that are accessible and affordable and work through mobile platforms, digital wallets, and, in particular, banking "agents." In markets such as Kenya with M-Pesa, India with Paytm, and Bangladesh with bKash, mobile money services have completely transformed access to payment, savings, and microloan products. They also complete these transactions with much lower transaction costs, improved engagement and convenience, and, lastly, a built system that allows consumers access to formal markets that had yet to be available for underserved populations and markets. So strategically, it would only make these firms that are integrating with or building from these ecosystems gain access to great new customer segments that might not meet their customer "strategy," actionable data opportunities, and the ability to build brand loyalty in rapidly growing markets.

Digital finance enables an array of strategic opportunities involving cross-border payment systems. Cross-border payments have historically lagged on many fronts—the transactions are slow, opaque, and expensive, resulting from the various intermediaries and regulatory requirements. Disruption in payments with blockchain-enabled remittance models and stablecoins demonstrates a faster, cheaper, and more transparent system of value creation and transfer, which has proven instrumental for diaspora communities remitting to their home countries or for businesses engaging in global e-payments. Strategic managers will need to determine how these systems impact elements such as pricing, collateral management, and compliance while assessing strategic partnership options that serve to bolster their cross-border offering. Additionally, for developing economies, mobile banking has become a viable option for the market to integrate multiple service options—loans, payments, insurance, investment—into a single environment enabled by behavioral and real-time data and analytics. Strategic managers have an opportunity in the bundling of financial products, ecosystem partnerships, and educating customers about financial literacy that could move their products into the daily financial lives of a customer and reduce friction. Seamless delivery systems and the establishment of network effects create deeper connections with customers and promote growth.

Integrating financial inclusion into global strategy dovetails with sustainability and ESG objectives. More inclusive financial practices foster gender equality, lower poverty rates, and develop entrepreneurship. From a longevity perspective, the strategic adoption of financial inclusion may lead to improved ESG ratings, attracting more impact investors and promoting brand reputation. It also reduces long-term systemic risks by developing more stable, diverse, and resilient markets. Adding financial inclusion into digital finance as part of strategic management is a significant transition (Sriman and Kumar 2022). Technologies such as blockchain and robo-advisors are changing how financial strategies are formulated and executed. Thus, while complexities in regulations and cyber risks require a more rigorous and anticipatory type of strategic planning, digital finance creates new opportunities for inclusive growth and scalability globally, granting organizations the ability to reach unrealized markets and create sustainable social impact. Strategic managers who address this transition systematically—balancing innovation with governance, agility with security, and profit with purpose—will be the most competitively powerful organizations in the smart era.

5. Case Studies and Study Examples

Siemens, Bosch, and Fujitsu are at the forefront of Industry 5.0 implementation using emerging technologies and human-centric principles. A globally renowned innovator in industrial automation and digitalization, Siemens is a recognized leader in "digital twin" technology, allowing the creation of virtual replicas of physical assets, processes, or systems. Although part of the Industry 4.0 movement, Siemens has moved beyond that to include real-time human feedback and AI to further improve systems within the digital twin environment and now offers digital twin systems that allow for real-time indicators such as predictive maintenance with databases of prior filters and other operational efficiencies, while allowing for customization that considers individual customer needs (Durigan Junior and Laurindo 2023). By cultivating technology and human functions in unison,

Siemens has created a mechanism of sustainable, intelligent manufacturing that accounts for both peak production and customized demands. Bosch has also advanced itself in the Industry 5.0 area with cobots as part of its smart factories, which perform repetitive or precise tasks alongside human operators and allow employees to focus on higher-order thinking and innovation.

Bosch's focus on human-machine collaboration contributes to the principles of safety, ergonomics, and worker welfare, and fits well with the broader objective of Industry 5.0, which is human empowerment. Bosch is also integrating AI and IoT into its logistics and supply chain systems to enable quick and timely decision-making while being able to respond to market demands. It is the clever application of advanced automation with human-in-the-loop design that has helped Bosch sustain its competitive advantage in the face of a rapidly evolving industry context (Verhoef et al. 2019). Fujitsu is another excellent example of Industry 5.0, especially through its principle of human-centered innovation. Fujitsu's Human Centric Experience Design (HXD) framework beautifully integrates technology, design thinking, and co-creation practices to confront complex societal problems. Fujitsu's AI-driven systems in areas like healthcare and public transportation have a specific focus on the human experience. For example, Fujitsu is utilizing AI and IoT technologies to monitor health in elderly care while allowing caregivers to make emotionally dynamic decisions. The mix of empathy in care practices and digital intelligence is indicative of a broader strategic approach toward currency, fully inclusive resilience, and personalization that embodies the core of the Industry 5.0 vision.

On the digital finance side, firms like Paytm, Revolut, and Square, and initiatives like eNaira and e-RUPI, are helping redefine the financial landscape and reconceptualizing the strategic thought process in both mature and emerging markets. Paytm, a leading Indian fintech platform, has moved beyond a mobile wallet provider to a fully fledged digital ecosystem encompassing payments, banking, insurance, lending, and wealth management. Paytm's success has been attributed to its expansion of access to financial services in areas of rural and semi-urban development, which have been underserved. Paytm has already established itself as India's financial super-app through its aggressive market positioning, user interface, and integration with the Unified Payments Interface (UPI). Paytm's real-time data analytics further enables hyper-personalization, which improves customer retention and enables further engagement opportunities. Paytm has also maintained relevance in public-private digital financial ecosystems, specifically in India, with alignment to government-led initiatives like Digital India and e-RUPI vouchers (Tyagi and Abraham 2021).

Revolut, a neobank based in the United Kingdom, has shaken up traditional banking with a mobile-first approach to financial services and borderless offerings. Its highly flexible set of offerings—like multi-currency accounts, cryptocurrency trades, AI-driven budgeting tools, real-time analytics, just to name a few—lends itself to a younger, digitally engaged consumer base that propels its strategic model of continuous product development and regulatory agility in scaling. By combining regulated and unregulated supply features enabled by decentralized technologies and financial inclusion with affordable services, Revolut shows how fintech can evolve with digital finance in large-scale competitive ways designed around users' needs. Square (now Block Inc.) is yet another disruptor. Square began as a point-of-sale solution for small businesses and has transformed into a broader financial services platform through payment processing services, payroll, invoicing, lending to small businesses, and now digital wallets through its Cash App platform (Sunio et al. 2023). Square's strategic shift has integrated early on the technology and currency features of blockchain. It now allows users to buy and transact in cryptocurrency on its Cash App, bringing digital currency to the average consumer. Its unique dual strategy of complementing small business efforts with assisting individual financial independence illustrates an innovative plan for joining technology with financing prospects of inclusion. Figure 3 illustrates the core components of the digital finance ecosystem—Blockchain, DeFi, CBDCs, and Mobile Payments—and highlights their strategic roles in enabling distributed ledgers, financial inclusion, and real-time digital transactions within the smart economy.

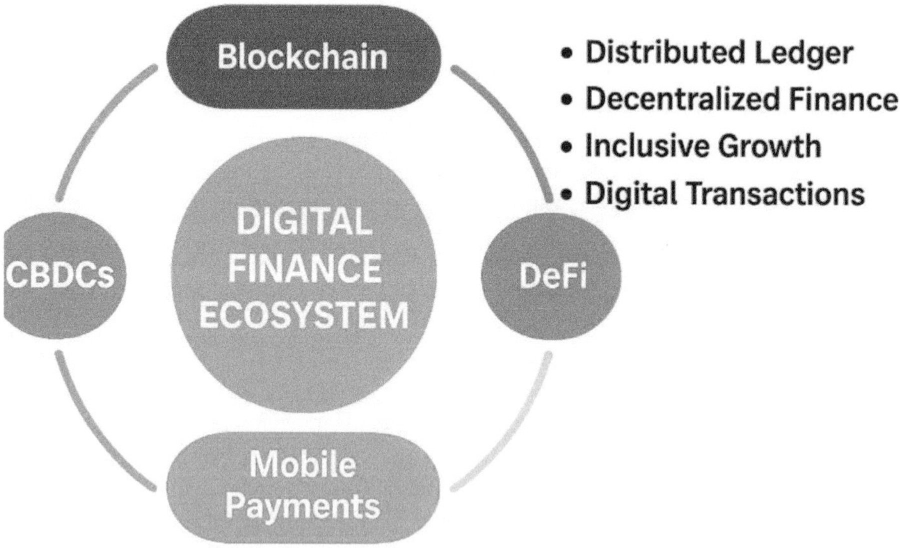

Figure 3: Digital Finance Ecosystem and Its Strategic Touchpoints.

CBDCs provide noteworthy instances of digital finance in operation. Nigeria's eNaira, launched by the Central Bank of Nigeria, is one of the first CBDCs in Africa. eNaira aims to enhance financial inclusion and payment efficiency by providing a safe and low-cost means for Nigerians to store and transmit value digitally. eNaira is integrated with banks and local mobile money solutions, which advances major national strategies—whether reducing dependence on cash, enhancing tax collection, or improving the effectiveness of monetary policy (Wamba and Queiroz 2020). In India, the e-RUPI is not a full CBDC but a government-led pilot project that provides digital vouchers to target subsidies. e-RUPI is transmitted by SMS or QR code to ensure that welfare benefits are delivered to citizens without any leakage or misuse. For strategic planners in public and/or private agencies, these types of initiatives demonstrate how digital finance introduces operational efficiencies, improves governance and accountability, and allows citizens to interact more efficiently with government agencies while reducing transaction costs (Bresciani et al. 2021).

6. Challenges and Opportunities

While organizations incorporate Industry 5.0 and digital finance into their strategic framework, they are faced with a changing landscape of challenges and opportunities. The ability for organizations to take advantage of new technologies paired with human-centered principles has tremendous potential for value creation, as long as organizations are able to navigate strategic challenges. These challenges often consist of technology disruptions, legacy systems, culture, and ethics. However, organizations and leaders that are willing to innovate and adapt can create new sources of personalized products, real-time strategic responsiveness, and purpose-oriented growth (Teece 2022). With rapid technology disruption a key strategic challenge in all organizations, Industry 5.0 represents serious change; human-machine collaboration and digital finance may require firms to adopt technologies that are inherently different from what may exist at a firm (Tiemann et al. 2018). Unless an organization has the ability to exploit these advancements, software legacy systems, and IT infrastructure, it will be difficult to grasp the impact of new solutions on existing organizations. In general, the demands of contemporary digital platforms upon legacy software may offer incompatibility in terms of interoperability, increased risk of downtime, data silos, and inefficiency. From a strategic planning framework perspective, this requires firms to undertake a significant overhaul regarding IT modernization and architecture, which can require large financial and human investments of time to establish the basis for formulating a modernization strategy (Liboni et al. 2019). Compounding those change problems is inertia. Although people may broadly appreciate the promise of Industry 5.0 and digital finance, firms also have issues with inertia. For example, an employee who has experience working within traditionally structured activities may seek to resist change based on fear related to losing their jobs or relevance. They can be in concert with the traditional forms of behavior-related change as they do not want to upset the status quo and sow instability. Equally, middle managers responsible for overseeing related traditional operations may not want to lose their operational routine and control to digitalization or automated features; they will be reluctant to hold staff accountable to new performance metrics—such heuristics represent challenges for adopting change. This cultural resistance can interfere with the adoption of strategic innovations, slow or stall decision processes, and demotivate employees. If organizations do not make conscious attempts to create a culture of experimentation, learning, and inclusivity, strategic changes may stall, or worse, change will miss its mark (Akter et al. 2020).

Security and ethical challenges complicate the strategic environment. As organizations embrace interdependence between businesses metaphorically, they each face interconnectedness by adopting interconnected systems, cloud-based platforms, and real-time data analytics that open them to a broader range of cybercriminal attacks, data breaches, and fraud risks. Digital finance adds multiple new risk vectors, including identity theft, fraudulent electronic or unauthorized transactions, smart contract hacking and vulnerabilities, and systemic risks rare and unique to decentralized finance environments. The strategic management of cybersecurity must then go beyond technical protections and become foundational to corporate governance and involve board governance mechanisms, alongside ethical approaches, including transparency with stakeholders. In addition, the ethical issues related to automated decision-making and AI, including algorithmic bias and erosion of human agency in digitally automated settings, present fundamentally difficult dilemmas for organizations. Organizations typically need to manage innovation with accountability, which requires strategic clarity and stakeholder relationships, in addition to, and probably most importantly, the formal institutionalization of ethical review mechanisms. Beyond the more specific forms of perplexity, many organizations struggle with the challenges associated with regulatory compliance when integrating new technologies (Whitehead et al. 2025). The rapidly evolving regulatory landscape, including protection of digital data regulations like GDPR, financial directives that include payment networks, applications to strengthen consumer protections from innovative technologies like autonomous services, and eventually a national regulatory framework such as the Government of India's DPDP Bill, requires organizations to anticipate change. If organizations do not pay close attention to regulatory shifts, they can face compliance and reputational risk implications, as well as misaligned strategies. Organizations must develop nimble governance systems that can traverse the regulatory maze of the legalities of digital transformation without eroding the trust of their relevant stakeholders.

Though the challenges are considerable, the implementation of Industry 5.0 and digital finance into strategic management also creates major opportunities for competitive advantage and sustainable growth. One of the most interesting opportunities is the possibility of providing hyper-personalized products, services, and experiences. Hyper-personalization becomes possible through the use of real-time data, machine learning, and user-centered design, allowing organizations to move from mass production to individually customized value propositions (Tyagi and Sreenath 2021). For example, in manufacturing, an organization may produce completely individualized products based on what the customer personally prefers. Whereas in finance, they may offer customized investment portfolios or customized credit solutions. Hyper-personalization is not only a means of improving customer delight; it can also provide organizations access to niche markets and differentiation in cluttered industry offerings. A second major opportunity is in providing real-time adaptations to strategy. Planning traditionally relied upon annual strategic reviews or long-term forecasting, neither of which is effective if the business environment is increasingly volatile and uncertain.

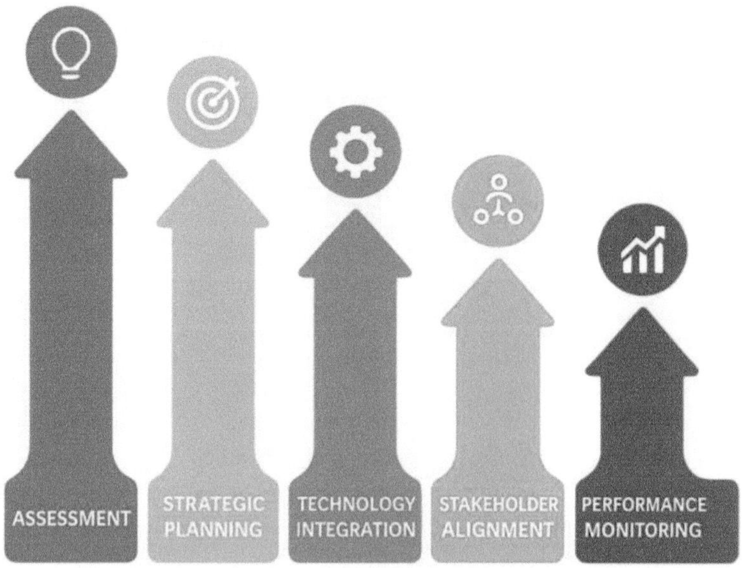

Figure 4: Strategic Transformation Roadmap for the Smart Era.

Digital capabilities associated with Industry 5.0 and digital finance—predictive analytics, digital dashboards, and AI simulations—provide the capacity for organizations to track performance, predict shocks, and adjust strategies in real time (refer Figure 4). This evolution from static strategic projects to dynamic strategic responsiveness provides organizations with the flexibility to advance innovation through experimentation, iteration, and scaling of ideas. Real-time or responsive strategic capabilities help organizations manage risks quickly, meaning they can anticipate and either react quickly to emerging risks or adapt, adjust, and shape opportunities (Madsen and Berg 2021). In addition, they may have to pivot quickly to mitigate losses. The potential that could be most valuable to organizations in some aspects is the unique opportunity in the smart era to produce human-centric value. Industry 5.0 emphasizes human creativity, human well-being, and human collaboration. These ecosystems seek a better balance of the human experience and human impact rather than emphasizing efficiency or profitability with minimal impact on people or the planet. The principles of inclusivity, sustainability, and ethical intent behind the human-centric lens in decision-making—if part of the strategic decision-making process—can generate better brand reputation, create a pipeline for talent, and build stakeholder-focused long-term value. The push for financial inclusion in the digital finance space is a prime example of how technology can be utilized to overcome systemic barriers. Organizations can extend affordable and available financial services to allow the unbanked or underbanked to gain a foothold through their business initiatives in inclusive economic development. Figure 5 outlines a five-phase strategic transformation roadmap including Assessment, Strategic Planning, Technology Integration, Stakeholder Alignment, and Performance Monitoring, which is designed to guide organizations through the systematic adoption of Industry 5.0 and digital finance technologies.

Furthermore, the proposed emergence of digital finance and Industry 5.0 enables new ecological strategies, wherein firms can work as consortia with fintech startups, research organizations, policymakers, and community organizations to co-design novel systems solutions. As part of this collaborative approach, they can share capital and services, shorten time to market, and learn collectively, thereby moving from competition to advantage. Strategic ecological systems are especially valuable in tackling multifaceted global issues such as climate change, digital inequality, and public health, where no one actor represents a holistic response. To conclude, the intersection of Industry 5.0 and digital finance in strategic management leaves us with a double-edged sword of

Figure 5: Integrated Strategic Management Framework for the Smart Era.

disruption and opportunity (Ciucu-Durnoi et al. 2024). While organizations have major obstacles to overcome in relation to legacy systems and culture, cybersecurity and ethical complexities, they also have an unprecedented opportunity to see personalization, agility, and a focus on value to individuals and groups. The firms that create a competitive financial edge will be those organizations that lead transformational change with intention, not just investing in technology, but in people, governance, and purpose. As strategic horizons expand toward increasingly smart and connected eras, the ability to compete through complexity with clarity will separate the leaders from the laggards.

7. Framework for Strategic Opportunities in the Smart Era

In order to manage the ambiguities of the smart era, organizations must find a new strategic management system that connects the significant elements of Industry 5.0 with the vitalizing aspects of digital finance. The strategy system cannot be the typical one that is seen today—very rigid, linear, silo-based, etc.—to manage the rapidly paced, interrelated realities present in today's digital economy. Due to this new environment, rethinking strategy management to be dynamic and human-centric must be embraced—one that brings together technology, people, finance, governance, and ethics into one strategic architecture for endurance management that can maintain and sustain competitive advantage with purpose in the 21st century. The principle of integrating dynamic interaction between more complex technologies and human capabilities lies at the center of this interconnected approach. Strategy management must therefore embrace a more intelligent integration of digital technologies, like artificial intelligence and machine learning, blockchain, IoT, and robotics, such that it is framed in a way that enhances human creativity, decision-making, and well-being. In this landscape, technology is not only a means of reducing cost or automating tasks but also a means of innovating, customizing, and differentiating. For example, AI and machine learning can assist managers with evidence and data for their decision-making, real-time analytics can help with executive decision-making, AI can enable customized and personalized product development, and blockchain can ensure secure and transparent value chains. Personalization is important because it will be organized into a framework to ensure organizations remain competitive. Consequently, technologies must not only be selected but implemented and scaled based on an appreciation of what the organization can do now and in the future, and the needs of the market.

The center of gravity of the framework is exactly about people - employees, customers, the community, and other stakeholders - as co-creators of value. In Industry 5.0, the workforce is positioned as a strategic asset, not merely as passive doers of technologies, therefore making effective use of the creative and intuitive abilities of people who empathize with their context as a route to sustainable growth (Espina-Romero et al. 2023). This requires strategic management to invest in continuous learning, reskilling, and upskilling programs to prepare workers to work alongside intelligent systems in an increasingly collaborative fashion. Leadership models need to evolve to become more inclusive, emotionally intelligent, participatory, and capable of decentralized decision-making with a culture of innovation and psychological safety. Strategies from a customer perspective should emphasize hyper-personalization, accessibility, and responsiveness, as a result of recognizing the multiple needs and preferences of digital consumers.

Finance is a critical component of the framework. The emergence of fintech, digital wallets, CBDCs, robo-advisors, and DeFi platforms has transformed finance into an architecture that allows business to operate. Strategic leaders should now see finance not only as a means of allocating resources but also as a strategic enabler of flexibility, scalability, and inclusion. Access to alternative financing models, real-time liquidity management, and smart contract-enabled operations allows organizations to innovate faster. The digital finance toolbox also provides access to new revenue streams, reduces transaction costs, and expands access to underserved markets (Rame et al. 2024). The governance pillar for organizational accountability and resilience is the final essential element of the proposed framework. Without effective governance systems, organizations can only apply actions to some challenges but are likely to miss useful solutions captured by the governance system. The smart era has complexities in business environments because of the speed of innovation, rapid cycles of change, data-driven ecosystems, and demanding stakeholders that require agile and transparent governance. The board and executive team must be digitally savvy to read trends and evaluate their strategic consequences, and scenario planning, stakeholder mapping, and dynamic risk review need to be embedded within strategic agency. The governance system will also need to build relationships across ecosystems, including technology partners, regulators, and civil society organizations. Such relationships and partnerships can help organizations cope with regulatory change, synchronize with national public policies, and develop solutions to systemic challenges such as climate change and digital inequality (Gladysz et al. 2023). Figure 5 depicts a complete strategic management framework for the Smart Era: Technology, People, Finance, Governance, and Ethics. The framework distinguishes and emphasizes relationships and balance among Technology, People, Finance, Governance, and Ethics and has affinity with strategically sustainable, human-centric, and innovation-led strategic decisions based on Industry 5.0 and digital finance.

In conclusion, ethics must always be the moral compass of strategic management in the era of smartness. When powerful technologies are implemented, lavish ethical issues arise and these can represent anything from algorithmic bias, exploitation of personal and private information, employee replacement, to workforce exclusion based on digital skills, to name a few. Consequently, any strategic framework must include ethical foresight as a standard part of decision-making. Ethical foresight can take many forms, including building ethical clarity where the ethical parameters for automation by AI are clearly established, agreeing to data consent for data protection, promising to be inclusive in innovation, and agreeing to advance environmental and social causes within the corporate purpose. Furthermore, ethical integrity is an increasingly critical source of competitive advantage, impacting consumer trust, employee loyalty, investor trust, and regulatory favor. The scope of this integrated framework can apply to multiple sectors, with each sector facing particularly context-related challenges and opportunities. In manufacturing, we see Industry 5.0 technologies enabling mass customization, predictive maintenance, circular economy practices; and we see digital finance tools enabling models of procurement, payments to suppliers, payments to service providers, and performance-based contracts. For the financial services sector, the application of converging technologies including Artificial Intelligence, Blockchain, and Open Banking is changing how we engage with customers, how we detect fraud, and how we manage portfolios. Each technology needs a new strategy depending on how quickly they develop and how fast we need to innovate in the new opportunities that accompany digital provenance. In healthcare, intelligent systems are changing diagnostic accuracy and can also personalize diagnosis and treatment and optimize the use of human and, in many cases, scarcer resources. Digital payments platforms, portable payment platforms, and new health financing facilitate greater access wherever people are and therefore improve the patient experience.

In each sector, technology, human capital, financial innovation, governance discipline, and ethics are five pillars that influence consortium outcomes effectively. Organizations adopting and using the five pillars can develop a sustainable competitive advantage, create resilient ecosystems, and have an impact on a better society. In conclusion, the recommended strategic management framework in the smart era provides a holistic response toward Industry 5.0 and digital finance with the effective coordination of technology, people, finance, governance, and ethics. With this robust framework, organizations will be more effective in navigating complexity, driving innovation, and creating sustainable, innovative, and inclusive value generation. This holistic perspective will help strategic leaders define what a smart, sustainable, and human-centered future can look like as their industries continue to advance.

8. Conclusion

In transitioning to the smart era, the integration of Industry 5.0 and digital finance requires a radical reinvention of strategic management. Rather than relying solely on automation, systems will need to rely on human-machine collaboration that balances creativity, ethics, personalization, and resilience. At the same time, the emergence of digital finance, which includes blockchain, AI-powered analytics, decentralized platforms, and central bank digital currencies, forces firms to consider technology agility when making strategic decisions. More importantly, this chapter illustrated how traditional strategy models will be completely inadequate to incorporate the rapid technological and financial disruptions we will continue to face, and instead championed a multidimensional strategy model built on five interconnected pillars: technology, people, finance, governance, and ethics. Firms will need to invest in organizational human capital, inclusive leadership, agile and transparent governance, and prioritize ethical foresight as a primary strategic asset. Digital finance must be considered more than just a transactional layer of finance, but as a transformative vehicle for the following three goals: financial inclusion, risk agility, and strategic differentiation. Taken collectively, firms that embrace it as part of their model as design principles can fuse the values of Industry 5.0 with the tools and guidance enabled by digital finance to enable real-time responsiveness, create sustainable value, and enhance stakeholder trust over the long term. Strategic sustainability is no longer a question of selecting between innovation or responsibility, but harmonizing responsibility with innovation. The firms of the future will be those who lead by selecting the harmonious design in their strategy which is tech-infused, ethics-infused, and human-centric.

References

Abdel-Basset, M., Mohamed, R., and Chang, V. (2024). A multi-criteria decision-making framework to evaluate the impact of Industry 5.0 technologies: Case study, lessons learned, challenges and future directions. *Information Systems Frontiers*. https://doi.org/10.1007/s10796-024-10472-3.

Akter, S., Michael, K., Uddin, M. R., McCarthy, G., and Rahman, M. (2020). Transforming business using digital innovations: The application of AI, blockchain, cloud and data analytics. *Annals of Operations Research,* 308(1–2): 7–39. https://doi.org/10.1007/s10479-020-03620-w.

Bresciani, S., Huarng, K.-H., Malhotra, A., and Ferraris, A. (2021). Digital transformation as a springboard for product, process and business model innovation. *Journal of Business Research,* 128: 204–210. https://doi.org/10.1016/j.jbusres.2021.02.003.

Chrifi-Alaoui, C., Bouhaddou, I., Benabdellah, A. C., and Zekhnini, K. (2025). Industry 5.0 for sustainable supply chains: A fuzzy AHP approach for evaluating the adoption barriers. *Procedia Computer Science,* 253: 2645–2654. https://doi.org/10.1016/j.procs.2025.01.324.

Ciucu-Durnoi, A. N., Delcea, C., Stănescu, A., Teodorescu, C. A., and Vargas, V. M. (2024). Beyond Industry 4.0: Tracing the path to Industry 5.0 through bibliometric analysis. *Sustainability,* 16(12): 5251. https://doi.org/10.3390/su16125251.

Crnobrnja, J., Stefanovic, D., Romero, D., Softic, S., and Marjanovic, U. (2023). Digital transformation towards Industry 5.0: A systematic literature review. pp. 269–281. *In: IFIP Advances in Information and Communication Technology*. https://doi.org/10.1007/978-3-031-43662-8_20.

Durigan Junior, C. A., and Laurindo, F. J. B. (2023). IT strategic alignment in the decentralized finance (DeFi): CBDC and digital currencies. https://doi.org/10.5748/19contecsi/pse/itm/6915.

Espina-Romero, L., Guerrero-Alcedo, J., Avila, N. G., Sánchez, J. G. N., Hurtado, H. G., and Li, A. Q. (2023). Industry 5.0: Tracking scientific activity on the most influential industries, associated topics, and future research agenda. *Sustainability,* 15(6): 5554. https://doi.org/10.3390/su15065554.

Gladysz, B., Tran, T.-A., Romero, D., Van Erp, T., Abonyi, J., and Ruppert, T. (2023). Current development on the Operator 4.0 and transition towards the Operator 5.0: A systematic literature review in light of Industry 5.0. *Journal of Manufacturing Systems,* 70: 160–185. https://doi.org/10.1016/j.jmsy.2023.07.008.

Leng, J., Jiang, P., Xu, K., Liu, Q., Wang, Q., and Zhao, J. (2022). Industry 5.0: Prospect and retrospect. *Journal of Manufacturing Systems,* 65: 279–295. https://doi.org/10.1016/j.jmsy.2022.09.017.

Liboni, L. B., Cezarino, L. O., Jabbour, C. J. C., Oliveira, B. G., and Stefanelli, N. O. (2019). Smart industry and the pathways to HRM 4.0: Implications for SCM. *Supply Chain Management: An International Journal,* 24(1): 124–146. https://doi.org/10.1108/SCM-03-2018-0150.

Madsen, D. Ø., and Berg, T. (2021). An exploratory bibliometric analysis of the birth and emergence of Industry 5.0. *Applied System Innovation,* 4(4): 87. https://doi.org/10.3390/asi4040087.

Nahavandi, S. (2019). Industry 5.0 A human-centric solution. *Sustainability,* 11(16): 4371. https://doi.org/10.3390/su11164371.

Özdemir, V., and Hekim, N. (2018). Birth of Industry 5.0: Making sense of big data with artificial intelligence, the Internet of Things and next-generation technology policy. *OMICS: A Journal of Integrative Biology,* 22(1): 65–76. https://doi.org/10.1089/omi.2017.0194.

Rame, R., Purwanto, P., and Sudarno, S. (2024). Industry 5.0 and sustainability: An overview of emerging trends and challenges for a green future. *Innovation and Green Development,* 3(4): 100173. https://doi.org/10.1016/j.igd.2024.100173.

Rejeb, A., Rejeb, K., Zrelli, I., and Süle, E. (2025). Industry 5.0 as seen through its academic literature: An investigation using co-word analysis. *Discover Sustainability, 6*(1). https://doi.org/10.1007/s43621-025-01166-0.

Sriman, B., and Kumar, S. G. (2022). Decentralized finance (DeFi): The future of finance and DeFi application for Ethereum blockchain-based finance market. pp. 1–9. *In: 2022 International Conference on Advances in Computing, Communication and Applied Informatics (ACCAI)*. https://doi.org/10.1109/accai53970.2022.9752657.

Sunio, V., Mendejar, J., Nery, J. R., and Carlos, J. P. (2023). Generalization of digital innovation for financial inclusion by means of market creation through regulation and governance. *Global Transitions,* 5: 29–39. https://doi.org/10.1016/j.glt.2023.04.001.

Teece, D. J. (2022). The evolution of the dynamic capabilities framework. pp. 113–129. *In: FGF Studies in Small Business and Entrepreneurship* (). https://doi.org/10.1007/978-3-031-11371-0_6.

Tiemann, I., Breuer, H., Fichter, K., and Freund, F. L. (2018). Sustainability-oriented business model development: Principles, criteria and tools. *International Journal of Entrepreneurial Venturing,* 10(2): 256. https://doi.org/10.1504/IJEV.2018.10013801.

Tyagi, A. K. (2023a). Decentralized everything: Practical use of blockchain technology in future applications. pp. 19–38. *In:* Elsevier eBooks. https://doi.org/10.1016/b978-0-323-96146-2.00010-3.

Tyagi, A. K. (2023b). Dew Computing. In Advances in systems analysis, software engineering, and high performance computing book series (pp. 332–345). https://doi.org/10.4018/978-1-6684-8531-6.ch017.

Tyagi, A. K., and Abraham, A. (2021). Recent trends in blockchain for information systems security and privacy. In CRC Press eBooks. https://doi.org/10.1201/9781003139737.

Tyagi, A. K., and Sreenath, N. (2021). Cyber Physical Systems: Analyses, challenges and possible solutions. Internet of Things and Cyber-Physical Systems, 1: 22–33. https://doi.org/10.1016/j.iotcps.2021.12.002.

Van Erp, T., Carvalho, N. G. P., Gerolamo, M. C., Gonçalves, R., Rytter, N. G. M., and Gladysz, B. (2024). Industry 5.0: A new strategy framework for sustainability management and beyond. *Journal of Cleaner Production,* 461: 142271. https://doi.org/10.1016/j.jclepro.2024.142271.

Verhoef, P. C., Broekhuizen, T., Bart, Y., Bhattacharya, A., Qi Dong, J., Fabian, N. et al. (2019). Digital transformation: A multidisciplinary reflection and research agenda. *Journal of Business Research,* 122: 889–901. https://doi.org/10.1016/j.jbusres.2019.09.022.

Wamba, S. F., and Queiroz, M. M. (2020). Blockchain in operations and supply chain management: Benefits, challenges and future research opportunities. *International Journal of Information Management,* 52: 102064. https://doi.org/10.1016/j.ijinfomgt.2019.102064.

Whitehead, J., Hashim, M. A. M., Tlemsani, I., and Gilani, S. A. M. (2025). Strategic leadership 5.0: Reality or illusion? *Journal of the Knowledge Economy.* https://doi.org/10.1007/s13132-025-02785-8,

Yadav, N. N., Maurya, N. A., and Mehta, N. A. (2024). When industry meets trustworthy AI: A systematic review of AI for Industry 5.0. *International Journal of Advanced Research in Science Communication and Technology*, 616–622. https://doi.org/10.48175/ijarsct-17683.

Zizic, M. C., Mladineo, M., Gjeldum, N., and Celent, L. (2022). From Industry 4.0 towards Industry 5.0: A review and analysis of paradigm shift for the people, organization and technology. *Energies,* 15(14): 5221. https://doi.org/10.3390/en15145221.

21

AI Enabled Smart Banking Systems

An Analysis from Society's Perspective

Aswatha Kumar,[1] *Chidambaram N.*,[2] *Smrity Prasad*[3] and *Amit Kumar Tyagi*[4,*]

1. Introduction to AI and Smart Banking Systems

In the past few years, the banking industry has moved towards a huge transformation due to advancements in technology, particularly the integration of AI into financial services. AI uses a range of technologies, including machine learning, natural language processing (NLP), and data analytics, which enable systems to learn from experience, analyze vast amounts of data, and make informed decisions (Svoboda 2023). As banks aim to enhance operational efficiency, reduce costs, and improve customer experiences, AI has emerged as a cornerstone of innovation in the sector. Smart banking systems, characterized by their ability to provide personalized services and streamline processes, have gained traction among consumers and financial institutions alike. These systems use AI to automate routine tasks, enhance fraud detection, optimize lending processes, and provide efficient financial advice. For example, chatbots powered by NLP facilitate customer interactions, while predictive analytics inform risk assessments and investment strategies. Hence, despite the advantages that AI-enabled banking systems provide, their implementation raises essential questions regarding ethical issues, data privacy, and societal impact. The intersection of technology and finance necessitates an important examination of how these advancements affect consumer trust, accessibility, and the overall financial landscape. As AI continues to evolve, understanding its implications from a societal perspective becomes important in ensuring that the benefits are equitably distributed and aligned with the values of the communities they serve. This chapter aims

[1] School of Management Studies, Reva University, Bengaluru, India.
Email: ashwath2meet@gmail.com

[2] Department of Management Studies, KGiSL Institute of Technology, Thudiyalur Road, Saravanampatti Coimbatore – 641 035.
Email: chidambaram.n@kgkite.ac.in, https://orcid.org/0009-0007-1767-385X

[3] Department of Statistics and Data Science, CHRIST UNIVERSITY, Bengaluru-560029.
Email: smritykashvi@gmail.com

[4] National Forensic Sciences University, Gandhinagar, Gujarat, India.

* Corresponding author: amitkrtyagi025@gmail.com

DOI: 10.1201/9781003655190-25

to discuss the integration of AI in smart banking systems, analyzing its impact on various consumers and addressing the challenges and opportunities that arise from this technological evolution.

1.1 Importance of AI in Banking

The importance of AI in banking is multifaceted, transforming traditional banking practices and paving the way for innovative financial services. Below are several key areas highlighting the importance of AI in the banking sector:

Enhanced Customer Experience: AI enables banks to deliver personalized services tailored to individual customer needs. Through data analysis, banks can provide customized product recommendations, targeted marketing campaigns, and personalized communication, which fosters stronger customer relationships and increases satisfaction (refer Figure 1).

Operational Efficiency: AI automates routine tasks such as transaction processing, data entry, and customer inquiries, reducing the time and cost associated with these processes. This operational efficiency allows banks to allocate resources more effectively and focus on strategic initiatives.

Fraud Detection and Risk Management: AI algorithms can analyze transaction patterns in real time to identify anomalies and potentially fraudulent activities. By using machine learning techniques, banks can enhance their fraud detection systems, minimizing losses and protecting customer assets. Also, AI assists in risk assessment by evaluating creditworthiness and predicting loan defaults, enabling banks to make informed lending decisions.

Data-Driven Decision Making: AI enables banks to use big data for better decision-making. Advanced analytics provide insights into market trends, customer behavior, and operational performance, allowing banks to adapt their strategies quickly and effectively in a competitive landscape.

Cost Reduction: By automating processes and improving efficiency, AI can lead to substantial cost savings for banks. This reduction in operational costs can be passed on to consumers in the form of lower fees and better interest rates, ultimately enhancing the overall customer experience.

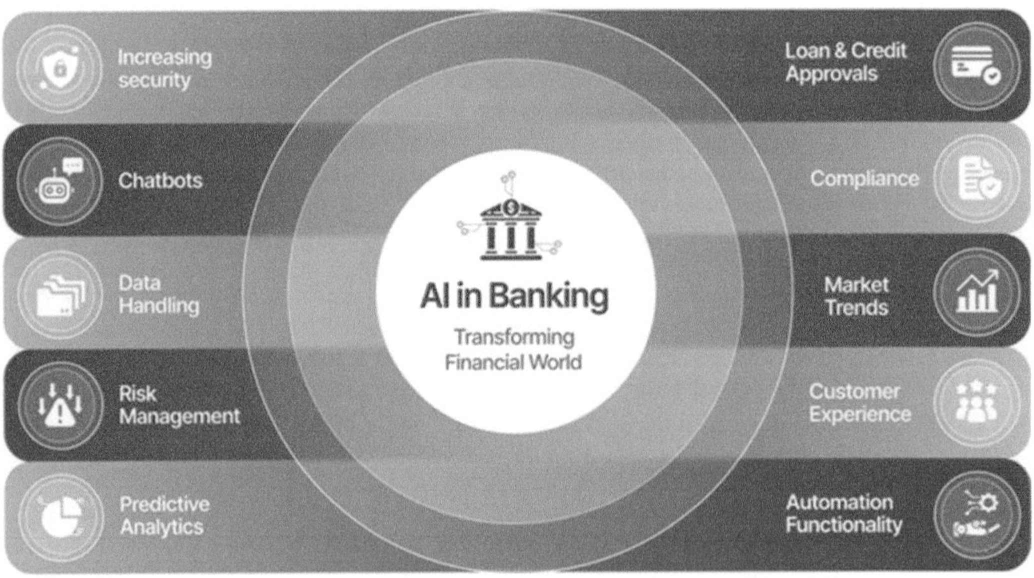

Figure 1: AI Role in Banking.

Innovation and Competitive Advantage: AI promotes innovation in product and service offerings, allowing banks to stay ahead of the curve. By using AI technologies, banks can introduce new features such as robo-advisors, AI-driven investment platforms, and smart payment solutions, thereby gaining a competitive edge in the rapidly evolving financial landscape.

Scalability: AI systems can easily scale to accommodate growing customer bases and increased transaction volumes. This scalability ensures that banks can maintain service quality and performance even during peak times or when expanding their operations (Arner et al. 2016).

In summary, the integration of AI in banking is not merely a trend but an important component of the industry's evolution. By improving customer experiences, optimizing operations, and driving innovation, AI positions banks to navigate the challenges of a digital economy and meet the evolving needs of their customers.

1.2 Organization of the Work

This work has been organized in 9 sections.

2. Understanding AI-Enabled Smart Banking Systems: Definition, Features, Types

AI-enabled smart banking systems refer to the integration of AI technologies into banking services and operations to enhance efficiency, improve customer experiences, and automate processes. These systems use AI algorithms, machine learning, and data analytics to deliver personalized services, streamline operations, and optimize decision-making in real time. AI-enabled smart banking systems possess several key features that set them apart from traditional banking systems:

- **Personalization**: Utilizing data analytics, these systems can analyze customer behavior and preferences, allowing banks to tailor their offerings to individual needs, from personalized product recommendations to customized financial advice (Ghandour 2021).
- **Automation**: Many routine banking processes, such as transaction processing, customer inquiries, and compliance checks, can be automated through AI, leading to faster service and reduced operational costs.
- **Predictive Analytics**: AI systems can forecast trends and behaviors based on historical data, aiding banks in making proactive decisions related to marketing, risk assessment, and investment strategies (Gyau et al. 2024).
- **NLP**: NLP capabilities enable AI chatbots and virtual assistants to interact with customers in real time, answering queries, providing information, and resolving issues effectively.
- **Enhanced Security**: AI-powered systems can monitor transactions for unusual patterns or activities, improving fraud detection and enhancing overall security measures.
- **Scalability**: These systems can adapt to increasing volumes of transactions and data without compromising performance, making them suitable for both small and large financial institutions.

Further, types of AI-enabled smart banking systems are:

- AI-enabled smart banking systems can be categorized into several types based on their applications and functionalities:
- **Customer Service Chatbots**: AI-driven chatbots are employed by banks to provide 24/7 customer support. They handle inquiries, assist with transactions, and provide product recommendations, thereby enhancing customer engagement and satisfaction.
- **Robo-Advisors**: These AI systems provide automated investment advice based on individual financial goals and risk tolerance. By analyzing market trends and customer data, robo-advisors

provide personalized portfolio management, making investing more accessible to a broader audience.

- **Fraud Detection Systems**: AI algorithms monitor transactions in real time to detect anomalies and suspicious activities, reducing the risk of fraud. These systems adapt to new threats by learning from historical data patterns.
- **Credit Scoring and Risk Assessment Tools**: AI technologies analyze a wide array of data, including non-traditional data sources, to evaluate creditworthiness and assess risks associated with lending. This results in more accurate and fair lending decisions.
- **Personal Finance Management Apps**: AI-enabled apps help customers track spending, manage budgets, and set financial goals. By analyzing spending patterns, these tools provide actionable insights to improve financial health.
- **Algorithmic Trading Systems**: AI systems are employed to execute trades at optimal times based on market conditions and predictive analytics. These systems can process vast amounts of market data in real time, allowing for swift and informed trading decisions.

In summary, AI-enabled smart banking systems are revolutionizing the banking landscape by enhancing service delivery, optimizing operations, and providing personalized financial solutions. Such applications and features enable banks to meet the evolving needs of customers while navigating the complexities of a digital economy.

2.1 The Role of Data Analytics in Smart Banking in Today's Era

In the contemporary banking landscape, data analytics has emerged as an important component of smart banking systems, transforming how financial institutions operate and engage with customers. With the exponential growth of data generated from various sources, such as transactions, customer interactions, social media, and market trends, banks are using data analytics to derive actionable insights, improve decision-making, and enhance overall service delivery (Lazo and Ebardo 2023, Sheth et al. 2022). Some of the key roles that data analytics plays in smart banking today are:

2.1.1 Customer Insights and Personalization

Data analytics enables banks to gain deep insights into customer behavior and preferences by analyzing transaction histories, demographics, and engagement patterns. This information allows banks to provide personalized services and products, such as tailored financial advice, customized marketing strategies, and individualized loan offerings. By understanding customers' needs, banks can foster stronger relationships and enhance customer satisfaction (Gomber et al. 2018).

2.2.2 Risk Management and Fraud Detection

Advanced data analytics tools play a vital role in identifying potential risks and detecting fraudulent activities. By using machine learning algorithms, banks can analyze large datasets to identify unusual transaction patterns and flag potential fraud in real time. Additionally, predictive analytics helps assess credit risk by evaluating borrower profiles and market conditions, allowing banks to make informed lending decisions and minimize defaults.

2.2.3 Operational Efficiency

Data analytics streamlines banking operations by optimizing various processes. Through data-driven insights, banks can identify inefficiencies, manage resources better, and improve workflows. For example, analyzing customer service data can highlight areas needing improvement, such as reducing response times or enhancing service delivery channels. As a result, banks can increase productivity and lower operational costs.

2.2.4 Enhanced Decision-Making

Data analytics enables banks to make informed strategic decisions by providing insights into market trends, customer behavior, and competitive positioning. By using analytical tools, financial institutions can evaluate the performance of products and services, assess the effectiveness of marketing campaigns, and forecast future market conditions. This data-driven decision-making provides agility and adaptability in a rapidly changing financial landscape.

2.2.5 Predictive Modeling and Forecasting

Predictive analytics enables banks to anticipate future trends and customer behaviors by analyzing historical data and identifying patterns (Prakash et al. 2024). This capability is essential for effective planning and resource allocation. For example, banks can forecast loan demand, predict economic downturns, or identify emerging market opportunities, allowing them to strategize effectively and mitigate risks.

Hence, in today's era, data analytics is integral to the success of smart banking systems. By harnessing the power of data, financial institutions can enhance customer experiences, improve operational efficiency, and navigate complex market dynamics. As technology continues to evolve, the role of data analytics in banking will only become more pronounced, driving innovation and shaping the future of financial services.

3. Societal Perspectives on AI in Banking in Today's Era

As AI increasingly integrates into banking systems, its societal implications become an important area of discussion. Understanding how various users perceive AI in banking is essential for using a balanced and inclusive approach to technological advancement. Below are key societal perspectives on the use of AI in banking today:

3.1 Trust and Transparency

One of the primary issues among consumers regarding AI in banking is trust. Many individuals worry about the transparency of AI algorithms, particularly in decision-making processes related to credit scoring, lending, and fraud detection. There is a demand for clarity about how AI systems function, the data they utilize, and how decisions are made. Building trust requires financial institutions to be open about their AI practices and to implement measures that promote accountability and ethical use of data.

3.2 Data Privacy and Security

With the increasing reliance on data analytics, concerns about data privacy and security have escalated. Customers are apprehensive about how their personal information is collected, stored, and used by AI systems (Gomber et al. 2018). High-profile data breaches have amplified these concerns, leading to a call for stricter regulations and robust security measures. Society demands that banks prioritize data protection to ensure that customer information remains confidential and secure.

3.3 Bias and Fairness

AI systems can inadvertently perpetuate biases present in the data they are trained on, leading to unfair treatment of certain groups of people. This issue is particularly important in areas like credit scoring and lending, where algorithmic biases can result in discriminatory practices. Societal perspectives emphasize the need for fairness and equity in AI systems, advocating for continuous evaluation and adjustment of algorithms to eliminate biases and promote inclusivity.

3.4 Job Displacement vs. Job Creation

The automation of banking processes through AI raises concerns about job displacement. While AI can enhance efficiency and reduce costs, it may also lead to the reduction of certain job roles, particularly those involving routine tasks. Conversely, AI also has the potential to create new jobs in areas such as AI maintenance, data analysis, and cybersecurity. Society grapples with the need to strike a balance between embracing technological advancement and addressing potential workforce disruptions.

3.5 Consumer Empowerment

AI has the potential to empower consumers by providing access to personalized financial services and tools that enhance financial literacy. For example, AI-driven applications can provide tailored budgeting advice, investment strategies, and real-time insights into spending habits. Society views these developments as opportunities for increased financial awareness and engagement, promoting more informed decision-making among consumers.

3.6 Regulatory and Ethical Issues

As AI continues to shape the banking landscape, there is a growing call for regulatory frameworks that govern its use. Society advocates for regulations that ensure ethical AI practices, protect consumer rights, and maintain market integrity. Policymakers are urged to engage with various users, including consumers, financial institutions, and technologists, to develop comprehensive guidelines that encourage innovation while safeguarding societal interests.

3.7 Impact on Financial Inclusion

AI has the potential to enhance financial inclusion by providing access to banking services for underserved populations. By using AI-driven technologies, banks can develop products tailored to meet the needs of individuals without traditional banking relationships. However, society remains cautious about ensuring that these technologies do not create new barriers to access, emphasizing the need for equitable distribution of AI benefits across diverse demographic groups.

Note that in today's era, societal perspectives on AI in banking reflect a complex interplay of trust, privacy, fairness, and empowerment. As financial institutions adopt AI technologies, they must address the issues and aspirations of various consumers to build a banking ecosystem that is not only efficient and innovative but also ethical and inclusive. Engaging in transparent dialogue and adopting responsible practices will be important in navigating the societal implications of AI in banking, ultimately ensuring that the benefits of technological advancement are shared widely and equitably.

4. The Impact of AI on Banking Services Today: Pros and Cons from Industry and Society Perspectives

The integration of AI into banking services has reshaped the financial landscape, providing several benefits while also presenting major challenges.

Analyzing the pros and cons from both industry and societal perspectives provides a comprehensive understanding of AI's impact on banking today. Some of the benefits of AI in banking are (also refer Figure 2):

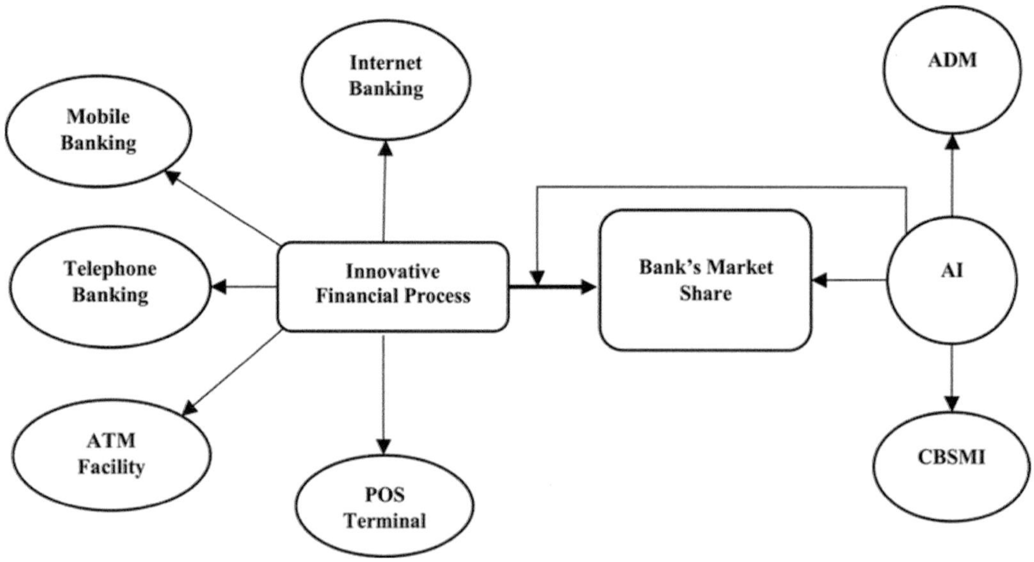

Figure 2: Role of AI in Smart Banking Solutions.

4.1 Enhanced Efficiency and Cost Reduction

From an industry perspective, AI automates routine tasks, such as customer service inquiries, data entry, and compliance checks, improving operational efficiency. This automation reduces labor costs and allows banks to allocate resources to more strategic initiatives. Whereas, from a societal perspective, customers benefit from faster service delivery and improved response times, leading to enhanced satisfaction and loyalty. Lower operational costs can also translate to reduced fees and better interest rates for consumers.

4.2 Personalized Customer Experiences

From an industry perspective, AI enables banks to analyze customer data to provide tailored services and recommendations. This personalization enhances customer engagement and drives revenue growth. Whereas, from a societal perspective, consumers enjoy personalized financial products that meet their unique needs, increasing accessibility to banking services and improving financial literacy through targeted advice.

4.3 Improved Risk Management and Fraud Detection

From an industry perspective, AI systems enhance the ability to detect fraud and assess credit risk by analyzing vast amounts of data in real time. This proactive approach minimizes financial losses and strengthens overall security. Whereas, from a societal perspective, consumers feel safer knowing that advanced AI technologies are in place to protect their accounts from fraudulent activities, increasing trust in the banking system.

4.4 Data-Driven Decision Making

From an industry perspective, AI allows banks to use data analytics for informed decision-making regarding lending, investments, and product development. This data-driven approach enhances competitiveness and responsiveness to market changes. Whereas, from a societal perspective, better-informed banks can develop products that better serve the public interest, leading to more equitable access to financial services and improved economic stability.

Some of the disadvantages of AI in banking are:

i. **Data Privacy Issues**: From an industry perspective, the reliance on vast amounts of personal data raises concerns about data security and the potential for breaches. Banks must invest in robust cybersecurity measures to protect sensitive information (Prakash et al. 2024). Whereas, from a societal perspective, consumers are increasingly worried about how their data is collected, used, and stored. High-profile data breaches have heightened fears regarding privacy, leading to distrust in banking institutions.

ii. **Algorithmic Bias**: From an industry perspective, AI systems may unintentionally perpetuate biases present in historical data, leading to discriminatory practices in lending and customer service. Banks face reputational risks if these biases are not addressed. Whereas, from a societal perspective, discriminatory outcomes can exacerbate inequalities, preventing marginalized groups from accessing essential financial services. This issue raises ethical concerns about fairness and inclusivity in the banking sector.

iii. **Job Displacement**: From an industry perspective, while AI creates new roles in data science and technology, it also poses a threat to jobs involving routine tasks. Financial institutions must navigate workforce changes and retrain employees to adapt to new technologies. Whereas, from a societal perspective, employees may face job loss or the need for retraining, leading to economic insecurity for individuals and communities. Society must address these transitions to ensure that workers are supported during technological changes.

iv. **Overreliance on Technology**: From an industry perspective, heavy reliance on AI can lead to vulnerabilities, especially if systems fail or are compromised. Banks must maintain human oversight to mitigate risks associated with technology malfunctions. Whereas, from a societal perspective, consumers may become overly dependent on technology for financial management, potentially losing essential skills in personal finance. Additionally, technological failures can disrupt services and erode consumer trust.

Hence, the impact of AI on banking services is profound, presenting both opportunities and challenges from industry and societal perspectives. While AI enhances efficiency, personalization, and risk management, it also raises concerns about data privacy, bias, and job displacement. To maximize the benefits of AI while minimizing its drawbacks, banks must adopt responsible practices, prioritize transparency, and engage with users/consumers to create a banking environment that serves the needs of society as a whole.

5. Popular Issues in AI-Enabled Banking

As AI continues to revolutionize the banking sector, several important issues have emerged that warrant attention from financial institutions, regulators, and consumers alike. These issues raise ethical, operational, and technological challenges that impact the implementation and effectiveness of AI in banking services. Below are some of the most pressing issues:

A. **Data Privacy and Security**: The use of AI in banking relies heavily on vast amounts of personal data, raising major concerns about data privacy and security. With increasing cyber threats and high-profile data breaches, consumers are wary of how their information is collected, stored, and used. Banks must prioritize robust cybersecurity measures and comply with data protection regulations to maintain customer trust and safeguard sensitive information (Prathap et al. 2024a, 2024b, Arumugam et al. 2024, Vittala et al. 2024a, 2024b, 2024c, Hossain and Khatun 2020).

B. **Algorithmic Bias**: AI systems can inadvertently perpetuate biases present in historical data, leading to discriminatory outcomes in areas like lending and credit scoring. If the data used to train AI models reflects societal biases, these biases can manifest in the decision-making processes of AI systems, potentially resulting in unfair treatment of certain groups. Addressing algorithmic bias is important for promoting fairness and inclusivity in banking services.

C. **Regulatory Compliance**: The rapid pace of AI innovation often outstrips the ability of regulatory frameworks to adapt. Banks must navigate a complex regulatory landscape, ensuring compliance with existing laws while also anticipating future regulations that may emerge as AI technologies evolve. There is a pressing need for clear guidelines that govern the ethical use of AI in banking and protect consumer rights.

D. **Consumer Trust and Transparency**: Building trust in AI systems is essential for widespread adoption in banking. Consumers often lack understanding of how AI algorithms function and make decisions. There is a growing demand for transparency in AI processes, where banks must clearly communicate how AI is used, what data is collected, and how decisions are made. Establishing transparency can help alleviate consumer concerns and build trust in AI-enabled services.

E. **Financial Inclusion**: While AI has the potential to enhance financial inclusion by providing access to banking services for underserved populations, there is a risk that it may also create new barriers. For example, reliance on technology may exclude individuals without access to digital devices or the internet. Ensuring that AI-driven banking solutions are accessible to all segments of society is an important challenge for financial institutions.

F. **Integration Challenges**: Implementing AI solutions in existing banking infrastructure can be complex and resource-intensive. Banks may face challenges related to system integration, data quality, and the alignment of AI technologies with organizational goals. Effective change management strategies and investment in training and development are necessary to overcome these integration issues.

Hence, the popular issues surrounding AI-enabled banking highlight the need for careful consideration and proactive measures as financial institutions navigate this transformative landscape. By focusing on challenges related to data privacy, bias, compliance, and ethical issues, banks can harness the potential of AI to enhance services while safeguarding consumer interests and promoting trust in the financial system.

5.1 Explainability Challenges in Deep Learning Models

Among the most urgent problems in the implementation of AI-based banking systems is the absence of explainability, which is an inherent limitation of most advanced AI models, especially deep learning algorithms. These kinds of models are commonly known as black boxes since, unlike human decision-making—where a clear thinking process can be traced—these models are often unable to offer clear explanations of their results in a manner human beings can understand. This creates ethical, operational, and regulatory challenges in critical financial processes including credit scoring, loan provision, and fraud identification, where outcomes have a significant effect on the financial well-being of customers. Customers, regulatory authorities, and other stakeholders are increasingly wanting to know how an AI-driven decision is made, especially in situations where the results may reinforce or act as a source of unfair treatment against individuals, resulting in a demand for transparency regarding how a certain decision is reached.

This matter has been identified by global regulatory bodies, and there is active advocacy toward the research and implementation of explainable AI (XAI) frameworks. Organizations like the Federal Reserve and the European Banking Authority (EBA) have noted that financial institutions need to ensure that AI-based systems are not only powerful but also reasonable and transparent in their functioning. Such organizations have advocated the need to ensure that algorithmic decision processes can be explained to end users, compliance officers, and regulators. As a response, banks and fintech firms are investing in XAI methodologies that can make some components of the AI black box visible, providing a more transparent picture of the factors influencing algorithmic decision-making processes.

Local Interpretable Model-agnostic Explanations (LIME) and SHapley Additive exPlanations (SHAP) are techniques that have become central in that regard. These approaches offer an idea of which characteristics or parameters played substantial roles in the final output of a model, which in turn enables stakeholders to query and appreciate the rationale behind certain decisions. An example would be a credit scoring system in which SHAP may be used to identify which factor—e.g. income, credit history, or employment status—contributed the greatest weight in determining a customer's creditworthiness. Such transparency is important not only to meet regulatory expectations, but it also plays a significant role in boosting customer confidence, especially among individuals who might perceive algorithmic decisions as arbitrary or biased. Also, explainability in AI systems contributes to internal accountability within banking institutions. Compliance departments and risk management teams will be in a better position to audit and validate AI models to ensure that they operate within established ethical and legal frameworks. Along with the increasing awareness of algorithmic biases, the expectation to prove that AI systems do not further contribute to existing social inequalities also increases. Transparency tools enable banks to locate and counteract inadvertent biases in data sets and model training processes to ensure fairness in service provision.

The principle of explainability is becoming a key component rather than an optional attribute in the ethical and responsible application of AI in the banking industry. Model performance and interpretability will be key issues to balance as AI transforms the future of the financial services industry. New regulations may result in regulators rejecting AI systems, consumers distrusting them, and courts challenging them. Consequently, transparency and human-centered design must be emphasized when developing AI innovations in the banking industry to guarantee trust, ensure compliance, and enhance fairness and accountability in the digital financial world.

6. Important Challenges Facing AI-Enabled Smart Banking Systems

As the banking industry increasingly adopts AI-enabled smart systems, it faces a variety of important challenges that can impede the successful implementation and operation of these technologies. Addressing these challenges is essential for banks to use AI's potential fully while ensuring compliance, security, and customer satisfaction. Below are some of the major challenges:

A. **Data Privacy and Security**: The reliance on vast amounts of sensitive customer data raises major concerns about privacy and security. Banks must ensure compliance with regulations such as GDPR and CCPA, which mandate strict data protection standards (Kumar et al. 2024, Vittala et al. 2024c, Arumugam et al. 2024). Cybersecurity threats, including data breaches and identity theft, pose serious risks to customer trust and institutional integrity. Financial institutions must implement robust security measures to protect data and maintain consumer confidence.

B. **Integration with Legacy Systems**: Many banks operate on legacy systems that may not be compatible with modern AI technologies. Integrating AI solutions with existing infrastructure can be technically challenging and resource-intensive, requiring substantial investment in upgrades and training. Successful integration is essential for maximizing the benefits of AI while minimizing disruptions to operations.

C. **Consumer Trust and Transparency**: Building consumer trust in AI-enabled systems is important for widespread adoption. Many customers remain skeptical about the use of AI in banking due to concerns about transparency, data handling, and decision-making processes. Financial institutions must prioritize clear communication about how AI technologies operate, how customer data is utilized, and the measures in place to protect consumer interests.

D. **High Implementation Costs**: Implementing AI technologies can be costly, particularly for smaller financial institutions. The expenses associated with technology acquisition, integration, and ongoing maintenance can create major barriers to entry. Banks must evaluate the return on investment (ROI) for AI initiatives to justify the costs and ensure alignment with strategic goals.

E. **Changing Customer Expectations**: As AI technologies evolve, customer expectations regarding banking services continue to rise. Consumers increasingly demand personalized experiences, instant responses, and seamless interactions across channels. Banks must continually innovate and adapt their offerings to meet these expectations, which can be challenging in a rapidly changing technological landscape.

Hence, the important challenges facing AI-enabled smart banking systems require careful consideration and strategic planning by the research.

7. Future Trends in AI and Smart Banking

As the banking industry continues to evolve, the integration of AI in smart banking systems is expected to shape the future landscape rapidly (Arumugam et al. 2024, Tyagi et al. 2024, Vittala et al. 2024b, Ambekar et al. 2024). The following trends are likely to define the future of AI in banking, enhancing operational efficiency, customer experience, and financial inclusivity:

- **Enhanced Personalization**: AI will drive hyper-personalization in banking services, allowing institutions to tailor offerings based on individual customer behavior, preferences, and financial histories. Advanced algorithms will analyze customer data to provide customized product recommendations, personalized financial advice, and targeted marketing campaigns, enhancing customer engagement and satisfaction.
- **Voice-Activated Banking**: Voice-activated technologies will become increasingly prevalent in smart banking. Customers will be able to perform banking transactions, inquire about account balances, and receive assistance through voice commands via smart speakers and mobile devices. This trend will enhance accessibility and convenience, particularly for tech-savvy consumers.
- **AI-Driven Fraud Detection and Prevention**: With the growing sophistication of cyber threats, AI will play an important role in fraud detection and prevention. Machine learning algorithms will analyze transaction patterns in real time to identify unusual behavior and potential fraud attempts. Continuous learning capabilities will enable these systems to adapt and improve over time, reducing false positives and protecting customer assets more effectively.
- **Robo-Advisors for Investment Management**: The use of AI-powered robo-advisors will become more mainstream, providing automated investment management services to a broader audience. These platforms will use algorithms to assess risk tolerance, investment goals, and market conditions to provide personalized investment strategies, making wealth management more accessible to individuals with varying levels of financial knowledge.
- **Improved Regulatory Compliance through AI**: AI technologies will assist banks in navigating complex regulatory environments by automating compliance processes. NLP and machine learning will enable institutions to monitor transactions for compliance violations, analyze regulatory updates, and generate necessary reports, reducing the risk of non-compliance and enhancing operational efficiency.
- **Integration of Blockchain and AI**: The integration of AI with blockchain technology will create secure, transparent, and efficient banking processes. AI can enhance smart contracts, automate transaction verification, and improve identity verification processes. This convergence will likely lead to faster and more secure financial transactions, particularly in cross-border payments.

- **AI in Credit Scoring and Lending Decisions**: Predictive analytics and machine learning will revolutionize credit scoring and lending processes. By analyzing a broader range of data points, including alternative data (e.g. social media activity, payment history on utilities), AI can provide a more accurate assessment of creditworthiness. This trend will facilitate greater access to credit for underserved populations and small businesses.
- **Chatbots and Virtual Assistants**: The use of AI chatbots and virtual assistants will continue to expand in customer service. These tools will become increasingly sophisticated, capable of handling complex inquiries and providing support in multiple languages. Enhanced sentiment analysis will enable chatbots to better understand customer emotions, leading to more empathetic and effective interactions.
- **Sustainability and Ethical AI**: As sustainability becomes a focal point for consumers and regulators, banks will increasingly adopt AI solutions that support sustainable practices. AI can analyze environmental, social, and governance (ESG) data to guide investment decisions and assess the sustainability of financial products. Ethical issues in AI development, including fairness and transparency, will also gain prominence, prompting banks to prioritize responsible AI practices.
- **Augmented Analytics for Decision-Making**: Augmented analytics, powered by AI, will transform how banks analyze data and make decisions. Advanced analytics tools will enable business users to derive insights without needing extensive data science expertise. This democratization of data access will lead to more informed decision-making across the organization, enhancing agility and responsiveness.

Hence, the future of AI in smart banking is poised for effective transformation, driven by advancements in technology and changing consumer expectations. By focusing on these trends, banks can enhance their operational efficiency, improve customer experiences, and position themselves for success in an increasingly digital landscape. As AI continues to evolve, financial institutions must remain agile and proactive in adapting to emerging trends to meet the needs of their customers and the market.

7.1 Integration of Blockchain and AI for Evolving Customer Expectations/Fulfill Requirements

As customer expectations continue to evolve in the digital age, the integration of blockchain technology and AI in banking provides a compelling solution to meet these demands (Shah et al. 2024, Arumugam et al. 2024, Prathap et al. 2024a). This integration can enhance transparency, security, efficiency, and personalization in financial services, addressing the needs of modern consumers who prioritize trust, speed, and tailored experiences.

- **Enhanced Security and Trust**: Blockchain technology provides a decentralized and immutable ledger, which enhances the security of transactions and data. By integrating AI, banks can analyze transaction patterns and detect anomalies, reducing the risk of fraud. This combination enhances trust among customers, who are increasingly concerned about the safety of their financial information. With AI algorithms monitoring blockchain transactions in real time, any suspicious activity can be quickly identified and addressed.
- **Transparent Transactions**: Customers today demand transparency in financial dealings. The integration of AI with blockchain can provide consumers with clear visibility into the transaction process. AI can generate real-time insights and predictive analytics regarding transaction statuses, fees, and delivery times. By providing customers detailed information about their transactions, banks can enhance customer satisfaction and loyalty.

- **Faster and More Efficient Processes**: Combining blockchain with AI can streamline various banking processes. For example, smart contracts powered by AI can automate the execution of agreements when predefined conditions are met, eliminating the need for intermediaries and reducing transaction times. This efficiency addresses customer expectations for quick service, particularly in areas such as loan processing, cross-border payments, and asset transfers.
- **Personalized Financial Services**: AI excels at analyzing large datasets to derive insights about customer behavior and preferences. When integrated with blockchain, AI can access a wide range of verified customer data, enabling banks to provide personalized services tailored to individual needs. For example, banks can use AI to assess a customer's financial health through blockchain-verified transactions and credit history, allowing for customized loan products, investment advice, and financial planning.
- **Improved Customer Experience**: The synergy of blockchain and AI can lead to innovative customer experiences. For example, AI-driven virtual assistants can guide customers through blockchain transactions, providing real-time support and answering queries related to payments, account status, and product offerings. Enhanced customer service can improve overall satisfaction and foster stronger relationships between banks and their clients.
- **Financial Inclusion**: The integration of blockchain and AI has the potential to enhance financial inclusion by providing underserved populations with access to banking services. Blockchain can provide a decentralized, secure method for storing and transferring value, while AI can assess creditworthiness using alternative data sources, such as mobile payment history and social media activity. This dual approach allows banks to provide financial products to individuals without traditional credit histories, helping to bridge the gap for those previously excluded from the financial system.

Note that the integration of blockchain and AI in banking is poised to reshape how financial institutions meet evolving customer expectations. By enhancing security, transparency, efficiency, and personalization, this combination provides a pathway for banks to adapt to the demands of modern consumers. As technological advancements continue to progress, embracing this integration will be essential for banks aiming to remain competitive and responsive to their customers' needs in an increasingly digital landscape.

7.2 The Role of Augmented Reality and Virtual Reality with AI in Smart Banking

As the banking industry evolves, the integration of Augmented Reality (AR) and Virtual Reality (VR) with AI presents exciting opportunities for enhancing customer engagement, service delivery, and overall banking experiences (Arumugam et al. 2024, Tyagi et al. 2024, Prathap et al. 2024, 2024b). By combining these advanced technologies, banks can create immersive environments that cater to the evolving needs and expectations of tech-savvy consumers.

- **Enhanced Customer Experience**: AR and VR can transform the way customers interact with banking services. Through immersive experiences, banks can engage customers in innovative ways.
- **Virtual Branches**: VR can create virtual banking environments where customers can interact with digital avatars of banking staff. This immersive experience allows customers to discuss products, conduct transactions, and receive personalized financial advice in a simulated bank setting, mimicking the physical branch experience without the need to visit in person.
- **Interactive Product Demos**: AR can provide customers with interactive demonstrations of financial products. For example, customers can visualize mortgage options by overlaying

potential property values and financing scenarios in their physical environment, helping them make more informed decisions.

- **Personalized Financial Planning**: Integrating AI with AR and VR can enable banks to provide personalized financial planning tools. Using VR simulations, customers can visualize their financial futures based on different scenarios, such as retirement planning or investment strategies. AI algorithms can analyze individual financial data and goals, generating customized simulations that guide customers through various financial decisions in an engaging way.
- **Training and Development for Employees**: AR and VR can serve as effective training tools for bank employees by creating realistic training environments that prepare staff for various customer interactions and scenarios.
- **Customer Interaction Simulations**: VR can simulate customer interactions, allowing employees to practice handling inquiries, resolving issues, and providing personalized service. This training can enhance staff confidence and effectiveness in real-life situations.
- **Product Knowledge Training**: AR can overlay information about banking products and services in real time during training sessions, helping employees quickly learn about offerings and improve their ability to assist customers.
- **Enhanced Security and Authentication**: AR can enhance security measures in banking by providing customers with interactive ways to authenticate transactions. For example, customers could use AR to visualize and verify transactions before approving them, reducing the risk of fraud. AI can analyze behavioral patterns to identify potential anomalies, ensuring that any suspicious activities are flagged for further investigation.
- **Remote Banking Services:** The combination of AR, VR, and AI allows banks to provide remote services that bring the banking experience to customers' homes. For example, customers can conduct virtual meetings with financial advisors through VR platforms, where AI can provide real-time data analysis and recommendations. This approach enhances convenience and accessibility, catering to the growing demand for digital banking solutions.
- **Financial Education and Literacy:** AR and VR technologies can play an important role in improving financial literacy among consumers. Banks can create immersive educational experiences that explain complex financial concepts, such as investing, savings, and budgeting. AI can tailor these experiences to individual learning styles and needs, making financial education more engaging and effective.

Note that the integration of Augmented Reality and Virtual Reality with AI in smart banking has the potential to revolutionize customer engagement and service delivery. By providing immersive experiences, personalized financial planning tools, and enhanced training capabilities, banks can meet the evolving expectations of customers in a digital-first world. As these technologies continue to advance, their adoption will be essential for financial institutions seeking to enhance their competitive edge and build deeper relationships with their customers.

7.3 AI in Sustainable Banking Practices

As the global focus on sustainability intensifies, the banking sector is increasingly recognizing the importance of integrating sustainable practices into its operations. AI is playing an important role in driving these sustainable banking initiatives by enabling better decision-making, enhancing resource efficiency, and supporting environmentally friendly financial products (Arumugam et al. 2024, Vittala et al. 2024a). Below are key areas where AI is contributing to sustainable banking practices:

- **Environmental Risk Assessment**: AI algorithms can analyze vast amounts of data to assess environmental risks associated with lending and investment decisions. By incorporating environmental, social, and governance (ESG) factors into their risk models, banks can evaluate the potential impacts of climate change and environmental degradation on their portfolios. This data-driven approach enables banks to make more informed lending decisions, aligning their operations with sustainability goals.

- **Sustainable Investment Strategies**: AI facilitates the development of sustainable investment strategies by analyzing trends and performance metrics related to ESG criteria. Machine learning models can identify investment opportunities in companies that prioritize sustainability, such as those focused on renewable energy, waste reduction, and social responsibility. This enables banks to provide green investment products that appeal to environmentally conscious investors.
- **Energy Efficiency in Operations**: AI technologies can optimize energy usage within bank branches and data centers, reducing their carbon footprints. By implementing AI-driven energy management systems, banks can monitor and analyze energy consumption patterns in real time, identifying inefficiencies and recommending strategies for reduction. This not only lowers operational costs but also contributes to overall sustainability efforts.
- **Enhanced Customer Engagement for Sustainability Initiatives**: Banks can use AI to promote sustainable financial products and services effectively. By analyzing customer behavior and preferences, AI can provide personalized recommendations for green loans, sustainable investment options, and eco-friendly banking practices. Engaging customers through targeted campaigns builds a culture of sustainability and encourages responsible financial behaviors.
- **Supply Chain Transparency**: AI can enhance transparency in banking supply chains by tracking and analyzing the sustainability practices of vendors and partners. This helps banks ensure that their supply chains align with sustainability goals and adhere to ethical practices. AI-driven analytics can identify areas of risk or non-compliance, allowing banks to take corrective actions.
- **Carbon Footprint Monitoring:** Banks can utilize AI to track and measure their carbon footprints, enabling them to set and achieve sustainability targets. AI can analyze data related to energy consumption, travel emissions, and waste production, providing insights that help banks implement strategies for reducing their environmental impacts. By transparently reporting on their carbon footprints, banks can enhance their credibility and commitment to sustainability.
- **Sustainable Product Development**: AI can facilitate the development of innovative financial products that promote sustainability. For example, banks can create green bonds or eco-friendly loan products designed specifically for financing renewable energy projects or energy-efficient home renovations. By analyzing market demand and customer preferences, AI can guide product development to align with sustainability trends.

Hence, the integration of AI in sustainable banking practices provides important opportunities for financial institutions to contribute to environmental and social well-being while enhancing their operational efficiency and competitiveness. By using AI technologies, banks can better assess risks, develop sustainable products, and engage customers in responsible financial practices. As sustainability continues to be a priority for consumers and regulators alike, the role of AI in driving these initiatives will become increasingly vital in shaping the future of banking.

7.4 AI and Quantum Computing in Financial Forecasting

AI, coupled with QC, will be a game changer in the capabilities of data analysis, risk modeling, and forecasting that the banking sector is able to achieve. In its early stages of development, quantum computing (on qubits, not conventional bits) can handle exponentially greater quantities of data at once, which is optimal for complex simulations such as in finance. Coupled with AI algorithms, quantum systems provide upgrades in quickly and accurately performing quantitative tasks such as Monte Carlo simulations, value-at-risk (VaR) modeling, and portfolio optimization. Quantum computing can also vastly improve AI-based fraud detection, where millions of transactions occur daily; quantum computing capabilities could allow analysis of several probability vectors simultaneously and detect anomalies. Such synergy not only enhances security but also results in a large decrease in false positives. Quantum-enhanced AI is able to parallel-process enormous and diverse data variables—e.g., interest rates, economic indicators, geopolitical events, and consumer behavior—on an extremely large scale, generating highly precise predictions on market dynamics and lending risk activities. With such innovations, it is possible to make better investment decisions, credit scoring assessments, and strategic financial planning. Nevertheless, quantum computing has some setbacks despite its potential, namely scalability, error rates, and high infrastructure costs. However, large financial institutions and fintech firms are increasingly experimenting with quantum-as-a-service (QaaS) systems and running pilot projects, and it is expected that the coming marriage of artificial intelligence and quantum computing will quickly become a foundational component of next-generation banking and financial services, providing orders of magnitude improvements in computing performance, modeling, forecasting, and risk management.

8. Developing Ethical Frameworks for AI Use in Banking

As the banking sector increasingly adopts AI technologies, the need for robust ethical frameworks becomes paramount. These frameworks are essential for ensuring that AI systems operate transparently, fairly, and responsibly while protecting customer rights and maintaining public trust. In the near future, banks will need to focus on several key areas when developing ethical frameworks for AI use (Somisetti et al. 2020):

A. Transparency and Explainability: Banks must prioritize transparency in their AI systems, ensuring that customers/consumers understand how decisions are made. This involves:

- **Clear Communication**: Financial institutions should communicate the workings of AI algorithms, especially those used in credit scoring, lending decisions, and risk assessments. Providing insights into the data and methodologies used can help demystify AI processes for consumers.
- **Explainable AI (XAI)**: Implementing XAI techniques allows banks to develop models that not only produce accurate results but also provide clear rationales for their outputs. This is particularly important in regulated environments where customers may need to understand the basis for financial decisions affecting them (Tripathi et al. 2021, Vishnuram et al. 2022).

B. Fairness and Bias Mitigation: AI systems can inadvertently perpetuate existing biases if not carefully managed. To develop ethical frameworks that promote fairness:

- **Bias Audits**: Regular audits of AI systems should be conducted to identify and mitigate biases in training data and algorithmic outputs. Banks should ensure that diverse data sets are used to train AI models, reflecting a broad range of customer demographics.
- **Fair Lending Practices**: Establishing guidelines that promote fair lending practices is important. Banks should use AI to enhance, rather than hinder, access to credit for marginalized communities, ensuring that algorithmic decisions do not discriminate based on race, gender, or socioeconomic status (Jangir et al. 2024).

C. Data Privacy and Security: Protecting customer data is a fundamental ethical obligation for banks using AI. Some of the key issues include:

- **Data Governance**: Developing clear policies regarding data collection, usage, and sharing is essential. Banks should implement strong data governance frameworks that outline how customer data is handled and safeguarded.

- **Informed Consent**: Customers should have a clear understanding of how their data will be used in AI processes. Banks must obtain informed consent, ensuring that customers are aware of the implications of sharing their data (Sharma et al. 2024).

D. Accountability and Responsibility: Establishing accountability for AI decisions is important to maintaining trust. Banks should:

- **Designate AI Ethics Committees**: Banks can create internal committees responsible for overseeing AI initiatives, ensuring that ethical issues are incorporated into the development and deployment of AI systems.
- **Establish Clear Accountability Protocols**: Defining accountability structures helps clarify who is responsible for decisions made by AI systems. This includes outlining processes for addressing grievances related to automated decisions and providing avenues for customers to appeal decisions made by AI (Shaikh et al. 2024).

E. Collaboration and Industry Engagement: Developing ethical frameworks for AI in banking should involve collaboration with various consumers, including:

- **Industry Partnerships**: Banks can collaborate with industry peers, technology providers, and regulatory bodies to share best practices and develop standardized ethical guidelines for AI use.
- **Public Consultation**: Engaging with customers and advocacy groups in the framework development process can provide valuable insights into community expectations and concerns regarding AI use in banking.

Finally, the development of ethical frameworks for AI use in banking is essential to fostering trust, ensuring fairness, and protecting customer rights. Note that by prioritizing transparency, fairness, data privacy (Khushboo et al. 2021, Somisetti et al. 2020, Subasree et al. 2022, Mapanga et al. 2017, Agarwal et al. 2022, Vishnuram et al. 2022, Jangir et al. 2024, Sharma et al. 2023, Agarwal et al. 2023, Prasad et al. 2021), accountability, collaboration, and continuous monitoring, banks can create ethical AI systems that align with societal values and expectations. In the near future, a proactive approach to ethics in AI will not only safeguard the interests of customers but also position banks as responsible and trustworthy financial institutions in an increasingly automated world.

9. Role of International Regulatory Bodies

Artificial intelligence (AI), and the ongoing spread of AI in the global banking industry, raises new opportunities and regulatory challenges to unprecedented levels. Digital banking involves cross-border transactions, and the fast-growing world of AI technologies has surpassed existing regulatory systems, making international coordination and integration imperative. International regulatory institutions have assumed a new height in developing ethical guidelines, setting technical standards, and providing safe environments to implement AI into the banking ecosystem across different countries. The Organisation for Economic Co-operation and Development (OECD) is one of the most outstanding bodies leading the creation of high-level AI principles. These principles reinforce the importance of transparency, human values, robustness, accountability, and fairness in AI systems. The OECD has proposed "trustworthy AI," which implies reliable and safe use of AI, as well as protecting human rights and democratic principles. These guidelines aim not only to serve as aspirations but also as practical foundations for national governments to shape regulatory policies and for financial institutions to guide their AI development strategies. Equally, the International Organization for Standardization (ISO) has taken a technical approach to AI governance by developing an emerging standard, ISO/IEC 42001, focused on AI management systems. This measure will be adopted as a global standard and implement a holistic framework for risk mitigation, explainability limitations, and ethical use of AI in sectors such as finance. The OECD and ISO initiatives combined will eventually form the foundation of a common global perspective on responsible AI, providing much-needed consistency in approaches banks can apply across borders.

Another body that has acknowledged the strategic value of aligning regulations on AI is the Group of Twenty (G20), and this has become increasingly relevant as the banking sector becomes more internationalized. The G20 is working on creating frameworks to facilitate ethical AI use in digital finance through its finance policy body, the Financial Stability Board (FSB). They are striving to find the right balance between promoting innovation and ensuring compliance with regulations related to data sovereignty, data security, and consumer protection. Considering the patchwork regulatory environment that exists today, the FSB is promoting interoperability among national regulatory systems to eliminate regulatory arbitrage, where companies exploit softer regulations in certain areas. By articulating shared values and fostering multilateral cooperation with regional regulators, the FSB aims to ensure that AI systems are not only innovative and efficient but also aligned with national legislation and consumer rights. Another significant actor in this sphere is the Bank for International Settlements (BIS), which supports central banks and supervisory authorities through its Innovation Hub. The BIS explores how AI can be incorporated into regulatory technology (RegTech) and supervisory technology (SupTech) to support oversight and compliance throughout the financial system. RegTech technologies have the potential to streamline compliance requirements for banks (e.g., transaction surveillance for anti–money laundering (AML)), while SupTech solutions enable regulators to analyze large amounts of data and detect potential warning indicators of systemic risk. Through the BIS, central banks can share knowledge, build capacities, and stay aware of the new risks and opportunities that AI technologies bring. Their efforts emphasize not only advancing technology but also ensuring responsible innovation through international collaboration.

Besides these institutions, there are other multilateral organizations that are engaging the global society in the discourse on AI ethics and governance, including United Nations (UN), World Economic Forum (WEF), regional regulatory agencies such as European Union (EU) among others. EU Artificial Intelligence Act is an example, which classifies applications of AI into risk-based tiers with strict requirements for high-risk AI applications including its use in financial services. It is probable that such regulatory blueprints will change the world standards and force banks all over to meet new levels of transparency, equity, and accountability. International regulation entities are essential towards spearheading moral development of AI in the banking sector. These organizations encourage cooperation to advance standard setting, and responsible innovation to make the AI systems safe and transparent in cross-border practice. With the further development and implementation of AI technologies into financial processes, international harmonization will become essential to prevent regulatory gaps, reduce systemic risks, and foster an inclusive, fair, and trustworthy global banking environment. Without well-coordinated efforts, the advantages of AI could be overshadowed by risks related to privacy violations, algorithmic bias, and unequal access to financial opportunities. Hence, regulatory coherence to support global finance is not just a compliance requirement but a strategic necessity for sustaining digital finance.

10. Case Studies in AI in Global Smart Banking

With the maturation of AI technologies, financial institutions worldwide continue to build on them in an attempt to redesign customer experiences, enhance operational efficiency, and strengthen risk management. In this section, real-world examples of successful implementation of AI-enabled smart banking systems in leading banks are displayed.

A. **JPMorgan Chase - United States:** JPMorgan Chase has heavily invested in the use of AI in several banking processes, including its COiN (Contract Intelligence) platform. COiN uses machine learning to analyze legal documents and derive important data points within seconds, a task that previously required 360,000 labor hours per year. The bank has also integrated AI-driven fraud detection solutions that detect and prevent fraud on a large scale by monitoring transaction behavior in real time. Moreover, JPMorgan Chase applies natural language processing (NLP) to create chatbots and virtual assistants to improve customer care.

B. **ICICI Bank -India:** In Indian retail banking, ICICI Bank has been an early adopter of AI. It has released iPal, a chatbot based on NLP that can process more than a million queries each month with 90 percent accuracy. In addition, ICICI applies AI in credit risk modeling, predictive modeling, and loan disbursement automation. It has also implemented robotic process automation (RPA) in 750 business processes, which has decreased manual efforts and errors. Using AI to develop mobile banking applications has significantly enhanced the user interface and improved customer satisfaction ratings.

C. **DBS Bank- Singapore:** DBS Bank has been recognized as one of the most digitalized banks in the world. It applies Artificial Intelligence and big data in predictive analytics, chatbots, and customized financial planning tools. The company has an AI platform called Customer Connect, which uses behavior analysis to send hyper-personalized messages to users. Talent management is another area where DBS uses AI-guided systems to enhance organizational workflow. Such efforts have led DBS to be recognized as the world's best digital bank by both Euromoney and The Banker.

D. **BBVA-Spain:** Banco Bilbao Vizcaya Argentaria (BBVA) has deployed AI to automate financial advice, anti-fraud measures, and credit decisions. It created an internal AI tool named Blue that provides proactive banking recommendations and spending insights. The AI tools used by BBVA also improve cybersecurity by enabling real-time behavioral anomaly detection, providing customers with protection against phishing and malware attacks.

E. **China Construction Bank (CCB) -China:** CCB has developed a fully AI-driven bank branch in Shanghai that features service robots equipped with facial recognition technology to greet customers. Services such as opening an account, issuing a card, and providing mortgage consultations are performed in an automated manner. CCB also applies AI in risk management, robo-advisors, and face-recognition-based authentication via mobile apps.

11. Conclusion

The growth of AI-enabled smart banking systems represents a major milestone in the evolution of financial services, providing unparalleled efficiencies and personalized user experiences. However, this research underscores the complex societal implications of such advancements. While consumers recognize the benefits of AI in terms of convenience and enhanced services, their trust in these systems is often undermined by concerns over data privacy, security, and potential biases inherent in AI algorithms. The findings indicate that fostering public trust requires financial institutions to prioritize transparency and ethical practices in AI deployment. Regulatory bodies must also play an important role in establishing frameworks that safeguard consumer rights and ensure fair access to AI technologies. Furthermore, ongoing public engagement and education are essential to demystify AI applications in banking and address societal apprehensions. In summary, the successful integration of AI in banking is not solely a technological challenge but a societal one, necessitating collaboration among stakeholders to build systems that are not only innovative but also aligned with the values and needs of the society they serve. As we move forward, it is imperative to advocate for responsible innovation that balances technological advancement with ethical considerations, ultimately leading to a more inclusive and trustworthy banking ecosystem.

References

Agarwal, D., and Tripathi, K. (2022). A framework for structural damage detection system in automobiles for flexible insurance claim using IoT and machine learning. pp. 5–8. *In:* 2022 International Mobile and Embedded Technology Conference (MECON). *IEEE*. https://doi.org/10.1109/MECON53876.2022.9751889.

Agarwal, D., Tripathi, K., and Kumar, K. (2024). Concepts of artificial intelligence and its application in modern healthcare systems. CRC Press. https://doi.org/10.1201/9781003333081.

Ahmadi, S. (2024). A Comprehensive Study on Integration of Big Data and AI in Financial Industry and its Effect on Present and Future Opportunities. *International Journal of Current Science Research and Review*, 07(01). https://doi.org/10.47191/ijcsrr/v7-i1-07.

Ambekar, A., Anusha, K. M., Kumar, M. K., and Tyagi, A. K. (2024). Smart banking solutions for modern society: Opportunities and challenges. In Convergence of technology and operations management in modern businesses. *IGI Global*. https://doi.org/10.4018/979-8-3693-6200-6.ch011.

Arner, D. W., Barberis, J., and Buckley, R. P. (2016). Fintech: Evolution and regulation. *Journal of Financial Transformation*, 44: 5–12. https://www.capco.com/insights/articles/fintech-evolution-and-regulation.

Arumugam, S. K., Seranmadevi, R., Susendiran, S., and Tyagi, A. K. (2024). Blockchain-IoT-based solutions for smart banking and finance systems: Architecture, issues, challenges, and way ahead for the next generation. In Convergence of technology and operations management in modern businesses. *IGI Global*. https://doi.org/10.4018/979-8-3693-6200-6.ch008.

Arumugam, S. K., Susendiran, S., Seranmadevi, R., and Tyagi, A. K. (2024). Artificial intelligence-enabled innovative and advanced solutions for banking and finance sector: Potential, benefits, opportunities, and challenges. In Convergence of technology and operations management in modern businesses. IGI Global. https://doi.org/10.4018/979-8-3693-6200-6.ch007.

Arumugam, S. K., Tyagi, A. K., Tiwari, S., and Kumari, S. (2024). The role of futuristic technologies in building sustainable environment. In Creating AI synergy through business technology transformation. *IGI Global*. https://doi.org/10.4018/979-8-3693-4187-2.ch012.

Ghandour, A. (2021). Opportunities and Challenges of Artificial Intelligence in Banking: Systematic Literature review. *TEM Journal*, 1581–1587. https://doi.org/10.18421/tem104-12.

Gomber, P., Kauffman, R. J., Parker, C., and Weber, B. W. (2018). On the Fintech Revolution: Interpreting the Forces of Innovation, Disruption, and Transformation in Financial Services. *Journal of Management Information Systems*, 35(1): 220–265. https://doi.org/10.1080/07421222.2018.1440766.

Gyau, E. B., Appiah, M., Gyamfi, B. A., Achie, T., and Naeem, M. A. (2024). Transforming banking: Examining the role of AI technology innovation in boosting banks financial performance. *International Review of Financial Analysis*, 96: 103700. https://doi.org/10.1016/j.irfa.2024.103700.

Jain, N. R. (2023). Role of artificial intelligence in banking and finance. *Journal of Management and Science*, 13(3): 1–4. https://doi.org/10.26524/jms.13.27.

Jangir, J., Tripathi, K., Punetha, N., and Srivastava, A. (2024). An efficient meta-heuristic dimensionality LSTM model for dimensionality reduction and clustering for attack classification. pp. 1–5. *In:* 2024 International Conference on Intelligent Systems for Cybersecurity (ISCS). *IEEE*. https://doi.org/10.1109/ISCS61804.2024.1058135.

Kumar, M. K., Srinidhi, V., Shamanth, B. S., and Tyagi, A. K. (2024). Modern smart Banking with Cutting-Edge Technologies. pp. 73–94. *In: Advances in Business Information Systems and Analytics Book Series*. https://doi.org/10.4018/979-8-3693-4187-2.ch004.

Kumar, N. C. M., Addula, S. R., Seranmadevi, R., and Tyagi, A. K. (2024). Advanced banking solutions for Industry 5.0: From industry's perspective. In Creating AI synergy through business technology transformation (Chapter 1). IGI Global. https://doi.org/10.4018/979-8-3693-4187-2.ch001.

Lazo, M., and Ebardo, R. (2023). Artificial intelligence adoption in the banking industry: Current state and future prospect. *Journal of Innovation Management*, 11(3): 54–74. https://doi.org/10.24840/2183-0606_011.003_0003.

M, V., and K, M. S. (2024). *Smart Banking: AI and IoT Transformations of Financial Services* (pp. 1–6). https://doi.org/10.1109/ictbig64922.2024.10911355.

Mapanga, I., Kumar, V., Makondo, W., Kushboo, T., Kadebu, P., and Chanda, W. (2017). Design and implementation of an intrusion detection system using MLP-NN for MANET. In 2017 IST-Africa Week Conference (IST-Africa) (pp. 1–12). IEEE. https://doi.org/10.23919/ISTAFRICA.2017.8102374.

Midha, S., and Tripathi, K. (2019). Extended TLS security and defensive algorithm in OpenFlow SDN. pp. 141–146. *In:* 2019 9th International Conference on Cloud Computing, Data Science & Engineering (Confluence). IEEE. https://doi.org/10.1109/CONFLUENCE.2019.8776607.

Prakash, M., Tyagi, A. K., Arumugam, S. K., and Rawat, A. (2024). Internet of Things for building a smart and sustainable environment: A survey. In Achieving sustainability with AI technologies (Chapter 2). IGI Global. https://doi.org/10.4018/979-8-3693-5247-2.ch002.

Prasad, P., and Tripathi, K. (2021). Natural scene text localization and removal: Deep learning and Navier-Stokes inpainting approach. pp. 1–5. *In:* 2021 IEEE International Conference on Electronics, Computing and Communication Technologies (CONECCT). IEEE. https://doi.org/10.1109/CONECCT52877.2021.9622609.

Prathap, B. N., Devaraju, N., Chidambaram, N., and Tyagi, A. K. (2024). Modern banking solutions in Industry 5.0: Evolution, impacts, and potential applications for the future. In Convergence of technology and operations management in modern businesses. *IGI Global*. https://doi.org/10.4018/979-8-3693-6200-6.ch013.

Prathap, B. N., Kulkarni, P. G., Harish, N., and Tyagi, A. K. (2024). Modern banking and finance systems using artificial intelligence: A view from citizen's perspective. In Convergence of technology and operations management in modern businesses. *IGI Global*. https://doi.org/10.4018/979-8-3693-6200-6.ch004.

Shah, M. D., Seranmadevi, R., Jose, S., and Tyagi, A. K. (2024). Blockchain for banking and finance: A survey. In Convergence of technology and operations management in modern businesses. *IGI Global*. https://doi.org/10.4018/979-8-3693-6200-6.ch009.

Shaikh, A. A., Kumar, A., Mishra, A., and Elahi, Y. A. (2024). A study of customer satisfaction in using banking services through Artificial Intelligence (AI) in India. *Public Administration and Policy*, 27(2): 167–181. https://doi.org/10.1108/pap-05-2023-0060.

Sharma, S., Verma, P., Singh, R., and Tripathi, K. (2024). Advancements in facial expression recognition: A comprehensive analysis of techniques. In Chauhan, N., Yadav, D., Verma, G. K., Soni, B., and Lara, J. M. (Eds.). Machine learning, image processing, network security and data sciences (MIND 2023, Communications in Computer and Information Science, Vol. 2128). Springer. https://doi.org/10.1007/978-3-031-62217-5_18.

Sheth, J. N., Jain, V., Roy, G., and Chakraborty, A. (2022). AI-driven banking services: the next frontier for a personalised experience in the emerging market. *International Journal of Bank Marketing*, 40(6): 1248–1271. https://doi.org/10.1108/ijbm-09-2021-0449.

Somisetti, K., Tripathi, S., and Verma, J. K. (2020). Design, implementation, and controlling of a humanoid robot. pp. 831–836. *In:* 2020 International Conference on Computational Performance Evaluation (ComPE) (). IEEE. https://doi.org/10.1109/ComPE49325.2020.9200020.

Subasree, S., Sakthivel, N. K., Tripathi, K., Agarwal, D., and Tyagi, A. K. (2022). Combining the advantages of radiomic features-based feature extraction and hyperparameters-tuned RERNN using LOA for breast cancer classification. Biomedical Signal Processing and Control, 72(Part A), 103354. https://doi.org/10.1016/j.bspc.2021.103354.

Svoboda, A. (2023). The impact of artificial intelligence on the banking industry. *Journal of Banking and Finance Management*, 4(1): 7–13. https://doi.org/10.22259/2642-9144.0401002.

Tripathi, K., Bhateja, N., and Dhillon, A. (2021). Changing the conventional banking system through blockchain. In The smart cyber ecosystem for sustainable development (Chapter 20). *Wiley Scrivener*. https://doi.org/10.1002/9781119761655.ch20.

Tyagi, A. K. (2019, February 22). Building a smart and sustainable environment using Internet of Things. In Proceedings of International Conference on Sustainable Computing in Science, Technology and Management (SUSCOM), Amity University Rajasthan, Jaipur, India, February 26–28, 2019.

Tyagi, A. K., Kumari, S., Tiwari, S., and Subbiah, P. (2024). Industry 5.0 at the core of Society 5.0: A revolution in the modern era of the health care sector. In Sustainable innovation for Industry 6.0. IGI Global. https://doi.org/10.4018/979-8-3693-3140-8.ch017.

Vishnuram, G., Tripathi, K., and Tyagi, A. K. (2022). Ethical hacking: Importance, controversies and scope in the future. In 2022 International Conference on Computer Communication and Informatics (ICCCI) (pp. 1–6). *IEEE*. https://doi.org/10.1109/ICCCI54379.2022.9740860.

Vittala, K. R. P., Ahmad, S. S., Seranmadevi, R., and Tyagi, A. K. (2024). Emerging technology adoption and applications for modern society towards providing smart banking solutions. In Enhancing medical imaging with emerging technologies. IGI Global. https://doi.org/10.4018/979-8-3693-5261-8.ch018.

Vittala, K. R. P., Arumugam, S. K., Satish Kumar, N., and Tyagi, A. K. (2024). Artificial intelligence: Blockchain integration for modern business. In Advancing software engineering through AI, federated learning, and large language models. IGI Global. https://doi.org/10.4018/979-8-3693-3502-4.ch006.

Vittala, K. R. P., Hemamalini, V., Seranmadevi, R., and Tyagi, A. K. (2024). The road to smart banking: Opportunities and challenges for the next generation. In Convergence of technology and operations management in modern businesses. IGI Global. https://doi.org/10.4018/979-8-3693-6200-6.ch002.

Vittala, K. R. P., Kumar, K. M., Seranmadevi, R., and Tyagi, A. K. (2024). Artificial intelligence–Internet of Things integration for smart marketing: Challenges and opportunities. In Advancing software engineering through AI, federated learning, and large language models. IGI Global. https://doi.org/10.4018/979-8-3693-3502-4.ch019.

Vittala, K. R. P., Nagarathnamma, J., Chidambaram, N., and Tyagi, A. K. (2024). Smart financial services through smart banking in the era of Industry 5.0: Opportunities and challenges. In Creating AI synergy through business technology transformation IGI Global. https://doi.org/10.4018/979-8-3693-4187-2.ch010.

Index

L

M

P

R

S

T

For Product Safety Concerns and Information please contact our EU representative GPSR@taylorandfrancis.com Taylor & Francis Verlag GmbH, Kaufingerstraße 24, 80331 München, Germany